Proteomics Sample Preparation

Edited by
Jörg von Hagen

Related Titles

J. E. Van Eyk, M. J. Dunn (Eds.)

Clinical Proteomics

From Diagnosis to Therapy

2007

ISBN: 978-3-527-31637-3

M. Hamacher, K. Marcus, K. Stühler, A. van Hall, B. Warscheid, H. E. Meyer (Eds.)

Proteomics in Drug Research

2006

ISBN: 978-3-527-31226-9

F. Azuaje, J. Dopazo (Eds.)

Data Analysis and Visualization in Genomics and Proteomics

2005

ISBN: 978-0-470-09439-6

D. Figeys (Ed.)

Industrial Proteomics

Applications for Biotechnology and Pharmaceuticals

2005

ISBN: 978-0-471-45714-5

M. H. Hamdan, P. G. Righetti, D. M. Desiderio, N. M. Nibbering

Proteomics Today

Protein Assessment and Biomarkers Using Mass Spectrometry, 2D Electrophoresis, and Microarray Technology

2005

ISBN: 978-0-471-64817-8

J.-C. Sanchez, G. L. Corthals, D. F. Hochstrasser (Eds.)

Biomedical Applications of Proteomics

2004

ISBN: 978-3-527-30807-1

S. Mitra (Ed.)

Sample Preparation Techniques in Analytical Chemistry

2003

ISBN: 978-0-471-32845-2

R. Westermeier, T. Naven

Proteomics in Practice

A Laboratory Manual of Proteome Analysis

2002

ISBN: 978-3-527-30354-0

Proteomics Sample Preparation

Edited by
Jörg von Hagen

WILEY-
VCH

WILEY-VCH Verlag GmbH & Co. KGaA

The Editor

Dr. Jörg von Hagen
Merck KGaA
Performance & Life Science Chemicals
Chromatography and Bioscience
Frankfurter Strasse 250
64271 Darmstadt
Germany

All books published by Wiley-VCH are carefully produced. Nevertheless, authors, editors, and publisher do not warrant the information contained in these books, including this book, to be free of errors. Readers are advised to keep in mind that statements, data, illustrations, procedural details or other items may inadvertently be inaccurate.

Library of Congress Card No.: applied for

British Library Cataloguing-in-Publication Data
A catalogue record for this book is available from the British Library.

Bibliographic information published by the Deutsche Nationalbibliothek
The Deutsche Nationalbibliothek lists this publication in the Deutsche Nationalbibliografie; detailed bibliographic data are available in the Internet at http://dnb.d-nb.de.

© 2008 WILEY-VCH Verlag GmbH & Co. KGaA, Weinheim

All rights reserved (including those of translation into other languages). No part of this book may be reproduced in any form – by photoprinting, microfilm, or any other means – nor transmitted or translated into a machine language without written permission from the publishers. Registered names, trademarks, etc. used in this book, even when not specifically marked as such, are not to be considered unprotected by law.

Cover   WMXDesign GmbH, Heidelberg
Typesetting   Thomson Digital, Noida, India
Printing   betz-druck GmbH, Darmstadt
Binding   Litges & Dopf GmbH, Heppenheim

Printed in the Federal Republic of Germany
Printed on acid-free paper

ISBN: 978-3-527-31796-7

To my family
Vanessa, Annika, and Roswitha

Contents

Preface *XXI*

List of Contributors *XXIII*

List of Abbreviations *XXXI*

Part I **Perspectives in Proteomics Sample Preparation** *1*

1 **Introduction** *3*
   N. Leigh Anderson

2 **General Aspects of Sample Preparation for Comprehensive Proteome Analysis** *5*
   Sven Andrecht and Jörg von Hagen
2.1 The Need for Standards in Proteomics Sample Preparation *5*
2.2 Introduction: The Challenge of Crude Proteome Sample Analysis *6*
2.3 General Aspects: Parameters which Influence the Sample Preparation Procedure *8*
2.3.1 Technical Dependent Aspects for Sample Preparation in Proteomics *9*
2.3.2 Sample-Dependent Aspects for Sample Preparation in Proteomics *11*
2.3.2.1 Enrichment or Depletion Strategy *11*
2.3.2.2 Sample Recovery and Standardization *12*
2.3.2.3 Quantification, Internal Standards and Spiking *13*
2.3.2.4 Calculating the Amount of Sample for Proteomic Approaches *13*
2.3.2.5 Developing Procedures for Different Model Systems: From Bench to Bedside *15*
2.3.2.6 Sample Matrix *15*
2.3.2.7 Localization of Target Protein *15*
2.3.3 An Example of Subcellular Protein Extraction *17*
2.3.3.1 Subcellular Extraction and Monitoring the Redistribution of Regulatory Proteins *17*

Proteomics Sample Preparation. Edited by Jörg von Hagen
Copyright © 2008 WILEY-VCH Verlag GmbH & Co. KGaA, Weinheim
ISBN: 978-3-527-31796-7

2.4	Summary and Perspectives 17	
	References 19	

3 **Proteomics: A Philosophical Perspective** *21*
Erich Hamberger

3.1	Introduction: "In the Beginning was the Word" 21	
3.2	The Experiment as a Scientific Method and a Tool of Cognition 23	
3.2.1	The Experiment Historically Viewed 23	
3.2.2	The Experiment Theoretically Viewed 23	
3.2.3	The Entanglement Between Theory and Experiment 25	
3.3	The Experiment as a Method (Tool) of Cognition Within the Scope of Biology: The So-Called "Life Sciences" 26	
3.4	Proteomics as a Cognition-Theoretical Challenge 30	
3.4.1	Cognition-Theoretic Support from Physics 31	
3.4.2	The "Pietschmann Axioms" of the Experiment in Biological View 32	
3.5	Conclusion 36	
	References 37	

Part II **Methods** *41*

4 **Mass Spectrometry** *43*

4.1	A Practical Guideline to Electrospray Ionization Mass Spectrometry for Proteomics Application 43

Jon Barbour, Sebastian Wiese, Helmut E. Meyer, and Bettina Warscheid

4.1.1	Introduction 43
4.1.1.1	Electrospray Ionization 43
4.1.1.2	Nano-Electrospray Ionization 45
4.1.1.3	ESI-MS Instrumentation 45
4.1.1.4	Protein Identification Strategies 46
4.1.2	Sample Preparation 47
4.1.2.1	Purification 48
4.1.2.2	Protein Digestion 51
4.1.3	ESI-MS Analysis 52
4.1.3.1	Protein Analysis by ESI-MS 52
4.1.3.2	Peptide Analysis by Nano-HPLC/ESI-MS 53
4.1.4	Application Example of ESI-MS in Proteomics 60
4.1.5	Concluding Remarks 64
4.1.6	Recipes and Methods 64
4.1.6.1	MeOH/Chloroform Protein Precipitation to Remove Salts and Detergents 64
4.1.6.2	Preparation and Washing of a Crude Membrane Pellet 65
4.1.6.3	Proteolytic Digestion and Peptide Extraction 65

4.1.6.4	Off-Line Analysis of Intact Proteins 67	
4.1.6.5	ESI Sample Preparation Checklist 67	
	References 68	

4.2	Sample Preparation for the Application of MALDI Mass Spectrometry in Proteome Analysis 73	
	Andreas Tholey, Matthias Glückmann, Kerstin Seemann, and Michael Karas	
4.2.1	Introduction 73	
4.2.2	Sample Preparation for MALDI-Based Protein Identification 75	
4.2.2.1	Selection of the MALDI Matrix 75	
4.2.3	Sample Preparation 78	
4.2.4	LC-MALDI 84	
4.2.5	Application Example 85	
4.2.5.1	Gel-Based Workflow 85	
4.2.5.2	Application Example: LC-MALDI Workflow 87	
4.2.6	Summary 88	
4.2.7	Perspectives 89	
4.2.8	Recipes for Beginners 90	
4.2.8.1	Sample Spotting Techniques 90	
4.2.8.2	Sample Cleaning Procedures 90	
	References 91	

4.3	Sample Preparation for Label-Free Proteomic Analyses of Body Fluids by Fourier Transform Ion Cyclotron Mass Spectrometry 95	
	Cloud P. Paweletz, Nathan A. Yates, and Ronald C. Hendrickson	
4.3.1	Introduction 95	
4.3.2	Perspective 99	
4.3.3	Recipe for Beginners 100	
4.3.3.1	Step-By-Step Instructions 102	
	References 103	

4.4	Sample Preparation for Differential Proteome Analysis: Labeling Technologies for Mass Spectrometry 105	
	Josef Kellermann	
4.4.1	Introduction 105	
4.4.2	Isotopic Labeling of Peptides and/or Proteins 107	
4.4.2.1	Stable Isotope Labeling of Proteins in Cell Culture 107	
4.4.2.2	Chemical Isotopic Labeling of Peptides or Proteins 108	
4.4.2.3	Spiking of Labeled Peptides 111	
4.4.3	Summary 111	
4.4.4	Perspectives 112	
4.4.5	Recipe for Beginners 112	
	References 114	

4.5	Determining Membrane Protein Localization Within Subcellular Compartments Using Stable Isotope Tagging *118*
	Kathryn S. Lilley, Tom Dunkley, and Pawel Sadowski
4.5.1	Introduction *118*
4.5.2	Preparation and Treatment of Samples in the Early-Stage LOPIT Protocol *120*
4.5.2.1	Preparation and Fractionation of Organelles *120*
4.5.2.2	Carbonate Washing of Fractions to Lyse Organelles, and Removal of Soluble and Peripheral Proteins *123*
4.5.2.3	iTRAQ Labeling *124*
4.5.3	Application of LOPIT to Map the Organelle Proteome of *Arabidopsis* *125*
4.5.4	Summary *126*
4.5.5	Recipe for Beginners *127*
	References *127*
5	**Electrophoresis** *129*
5.1	Sample Preparation for Two-Dimensional Gel Electrophoresis *129*
	Walter Weiss and Angelika Görg
5.1.1	Introduction *129*
5.1.2	General Aspects of Sample Preparation for 2-DGE *130*
5.1.2.1	Cell Disruption *130*
5.1.2.2	Sample Clean-Up *132*
5.1.2.3	Protein Solubilization *135*
5.1.3	Application Samples *137*
5.1.3.1	Mammalian Tissues *138*
5.1.3.2	Microbial Cell Cultures *138*
5.1.3.3	Plant Cells *138*
5.1.4	Summary *139*
5.1.5	Perspective *140*
5.1.6	Recipes for Beginners *141*
	References *142*
5.2	Sample Preparation for Native Electrophoresis *144*
	Ilka Wittig and Hermann Schägger
5.2.1	Introduction *144*
5.2.2	Sample Preparation: General Considerations *146*
5.2.2.1	Choice of Detergent and Detergent/Protein Ratio *146*
5.2.2.2	Choice of Ionic Strength and pH for Sample Solubilization *147*
5.2.2.3	Storage of Biological Membranes *148*
5.2.2.4	Effects of Adding Coomassie Dye to Sample and/or Cathode Buffer for BNE *149*
5.2.3	Applications *150*
5.2.3.1	Solubilization of Bacterial Membranes, Yeast and Mammalian Mitochondria *150*

5.2.3.2	Homogenization and Solubilization of Mammalian Cells and Tissues *150*	
5.2.3.3	Recipe for Beginners: Mass Calibration Ladder for BNE *151*	
5.2.4	Summary and Perspectives *151*	
	References *152*	
5.3	Sample Preparation for LC-MS/MS Using Free-Flow Electrophoresis *155*	
	Mikkel Nissum, Afsaneh Abdolzade-Bavil, Sabine Kuhfuss, Robert Wildgruber, Gerhard Weber, and Christoph Eckerskorn	
5.3.1	Introduction *155*	
5.3.2	The Problems of Sample Preparation: The Pros and Cons *157*	
5.3.2.1	Separation *157*	
5.3.2.2	Extraction *158*	
5.3.2.3	Media Composition *158*	
5.3.3	Application Example *159*	
5.3.3.1	Reagents *159*	
5.3.3.2	Sample Preparation for FFE *160*	
5.3.3.3	FFE Separation of Peptides *160*	
5.3.3.4	RPLC-MS/MS Analysis *161*	
5.3.3.5	Data Processing *161*	
5.3.4	Summary *161*	
5.3.5	Perspective *165*	
5.3.6	Recipe for Beginners *166*	
5.3.6.1	FFE Set-Up Procedure *166*	
5.3.6.2	Pre-Experimental Quality Control (QC) *167*	
5.3.6.3	Experiment *168*	
	References *168*	
5.4	Sample Preparation for Capillary Electrophoresis *171*	
	Ross Burn and David Perrett	
5.4.1	Introduction *171*	
5.4.2	Sample Preparation *173*	
5.4.2.1	Sample Collection and Storage *174*	
5.4.2.2	Sample preparation for CE *174*	
5.4.2.3	Sample Concentration *175*	
5.4.2.4	Off-Line Preconcentration *175*	
5.4.2.5	On-Line Preconcentration *175*	
5.4.2.6	Desalting *176*	
5.4.2.7	Analyte Modification *176*	
5.4.3	Background Electrolyte *177*	
5.4.4	Capillary Preparation *178*	
5.4.4.1	Capillary Dimensions *178*	
5.4.4.2	Capillary Conditioning *178*	
5.4.4.3	Capillary Coating *179*	
5.4.5	Summary *179*	

5.4.6	Perspective	179
5.4.7	Recipe for Beginners	180
5.4.7.1	Method 1: Analysis of Human Serum/Plasma by CZE	180
5.4.7.2	Method 2: Analysis of Tryptic Digests by CZE	182
5.4.7.3	Method 3: Analysis of Proteomes by CIEF	183
	References	185
6	**Optical Methods**	**187**
6.1	High-Throughput Proteomics: Spinning Disc Interferometry (SDI)	187

Patricio Espinoza Vallejos, Greg Lawrence, David Nolte, Fred Regnier, and Joerg Schreiber

6.1.1	Proteomics as a Tool for Health Assessment	187
6.1.2	Translational Proteomics	188
6.1.3	The Principles of Spinning Disc Interferometry	189
6.1.3.1	The Spinning Disc	189
6.1.3.2	Why Spin?	191
6.1.3.3	In-line Quadrature	192
6.1.3.4	Scaling Mass Sensitivity	194
6.1.4	The Spinning Disc as a High-Throughput Immunological Assay Platform	196
6.1.4.1	Immunological Assays Using a Disc Array Format	197
6.1.4.2	Assay Formats	198
6.1.4.3	Assay Protocols	199
6.1.5	Types of Assay that Fit the Spinning Disc	200
6.1.5.1	Assay Structure	200
6.1.5.2	Assay Development Kit (ADK)	201
6.1.6	Assay and Sample Processing	201
6.1.6.1	High-Throughput System	201
6.1.6.2	The ADK	203
6.1.7	Conclusions	205
	References	206
6.2	Optical Proteomics on Cell Arrays	208

Andreas Girod and Philippe Bastiaens

6.2.1	Introduction	208
6.2.2	A Description of the Problem with Regards to Sample Preparation	211
6.2.2.1	General Remarks	211
6.2.2.2	Cell Line Selection	211
6.2.2.3	Sample Preparation	212
6.2.2.4	Choice of Transfection Reagent	212
6.2.2.5	Nucleic Acid Preparation	213
6.2.2.6	Sample Scale: How Many Duplicates are Required?	213
6.2.2.7	Choice of Microarrayer/Microspotting System (Spotter)	213
6.2.2.8	Layout Design (Spotting Pattern)	214

6.2.3	Summary *215*	
6.2.4	Perspectives *215*	
6.2.5	Sample Preparation: Short Protocol *215*	
6.2.5.1	Recommended Equipment and Consumables *215*	
6.2.5.2	Preparation of the Source Plate *216*	
6.2.5.3	Spotting *217*	
6.2.5.4	Cell Culture and Experiment *217*	
	References *218*	
6.3	Sample Preparation by Laser Microdissection and Catapulting for Proteome Analysis *219*	
	Karin Schütze, Andrea Buchstaller, Yilmaz Niyaz, Christian Melle, Günther Ernst, Kerstin David, Thorsten Schlomm, and Ferdinand von Eggeling	
6.3.1	Introduction: Laser Microdissection and Functional Proteomic Research *219*	
6.3.2	The Relevance of Pure Starting Material for Proteomics *219*	
6.3.3	Examples of Combined LMPC and Proteomic Analyses *220*	
6.3.3.1	LMPC and Preeclampsia *220*	
6.3.3.2	LMPC and Renal Cell Carcinoma *221*	
6.3.3.3	LMPC and Hepatocellular Carcinoma *221*	
6.3.3.4	LMPC and Brain Disorders *221*	
6.3.3.5	LMPC and Plant Biology *221*	
6.3.4	LMPC Adapted for Proteomic Applications *222*	
6.3.5	LMPC Combined with SELDI-TOF MS: A Promising Approach for Patient-Specific Analyses *224*	
6.3.6	Correlation of Gene and Protein Expression: The Best Data Capture for Comprehensive Diagnosis *228*	
6.3.7	Recipe for Beginners *228*	
6.3.7.1	Patients and Specimens *228*	
6.3.7.2	Laser Microdissection of Tissue Sections *230*	
6.3.7.3	ProteinChip Array Preparation and Analysis *230*	
6.3.8	Summary and Outlook *231*	
	References *232*	
6.4	Sample Preparation for Flow Cytometry *234*	
	Derek C. Davies	
6.4.1	Introduction *234*	
6.4.2	Sample Preparation for Flow Cytometry *236*	
6.4.2.1	Preparation from Cells in Suspension *236*	
6.4.2.2	Preparation from Adherent Cells *237*	
6.4.2.3	Preparation from Solid Tissue *237*	
6.4.2.4	General Considerations *237*	
6.4.3	Identification of Relevant Cells *238*	
6.4.4	Cell Sorting *238*	
6.4.4.1	Cells and Samples *238*	
6.4.4.2	Cytometer Considerations *239*	

6.4.5	Application Example	240
6.4.6	Summary	242
6.4.7	Perspectives	242
6.4.8	Recipes for Beginners	242
6.4.8.1	Cultured Suspension Cells	242
6.4.8.2	Adherent Cells	242
6.4.8.3	Solid Tissue	243
	References	243

7 Chromatography 245

7.1 Sample Preparation for HPLC-Based Proteome Analysis 245
Egidijus Machtejevas and Klaus K. Unger
7.1.1 Introduction 245
7.1.2 Problems Related to Direct Sample Injection in HPLC 246
7.1.3 Trial and Error Selection of the Sample Preparation Method 247
7.1.4 Classical Approaches 249
7.1.5 Specific Approaches Applied to Sample Clean-Up in Proteomics 250
7.1.5.1 Miniaturized Extraction Techniques 250
7.1.5.2 Most Abundant Component Depletion 250
7.1.5.3 Affinity-Enrichment Approaches 251
7.1.6 On-Line Sample Clean-Up Approaches 252
7.1.7 Restricted Access Technology 254
7.1.8 Application Example: The Case Study 259
7.1.9 Conclusion and Perspectives 260
References 262

7.2 Sample Preparation for Two-Dimensional Phosphopeptide Mapping and Phosphoamino Acid Analysis 265
Anamarija Kruljac-Letunic and Andree Blaukat
7.2.1 Introduction 265
7.2.2 Important Aspects in Sample Preparation Procedures 265
7.2.3 Application Example 267
7.2.4 Summary 267
7.2.5 Perspective 269
7.2.6 Recipe for Beginners 269
7.2.6.1 2-D Phosphopeptide Mapping Procedure 269
7.2.6.2 Phosphoamino Acid Analysis Procedure 271
References 271

8 Structural Proteomics 273

8.1 Exploring Protein–Ligand Interactions by Solution NMR 273
Rudolf Hartmann, Thomas Stangler, Bernd W. König, and Dieter Willbold
8.1.1 Introduction 273
8.1.2 Localization of Interaction Sites by Chemical Shift Perturbation (CSP) Mapping 274

8.1.3	Saturation Transfer Difference Spectroscopy	*276*
8.1.4	Ligand Screening by NMR	*278*
	References	*279*
8.2	Sample Preparation for Crystallography	*281*
	Djordje Musil	
8.2.1	Introduction	*281*
8.2.2	Use of Recombinant Proteins in Crystallization	*282*
8.2.3	Protein Solubility and Crystallization	*284*
8.2.4	Protein Crystallization	*286*
8.2.5	Practical Examples	*291*
	References	*292*

9 Interaction Analysis *295*

9.1	Sample Preparation for Protein Complex Analysis by the Tandem Affinity Purification (TAP) Method	*295*
	Bertrand Séraphin and Andrzej Dziembowski	
9.1.1	Introduction	*295*
9.1.2	The Problem with Regards to Sample Preparation: The Pros and the Cons	*296*
9.1.3	Application Example	*300*
9.1.4	Summary	*301*
9.1.5	Perspective	*301*
9.1.6	Recipe for Beginners	*301*
	References	*302*
9.2	Exploring Membrane Proteomes	*303*
	Filippa Stenberg and Daniel O. Daley	
9.2.1	Introduction	*303*
9.2.2	Defining Membrane Proteomes	*303*
9.2.3	Separation of Membrane Proteomes	*304*
9.2.4	Experimental Identification of Membrane Proteins	*307*
9.2.5	Mapping Membrane Interactomes	*307*
9.2.6	Structural Analysis of Membrane Proteomes	*308*
9.2.7	Summary and Perspective	*309*
9.2.8	Recipe for Beginners	*311*
9.2.8.1	Sample Preparation	*311*
9.2.8.2	BN–PAGE	*311*
9.2.8.3	SDS–PAGE	*312*
	References	*312*

10 Post-Translational Modifications *317*

10.1	Sample Preparation for Phosphoproteome Analysis	*317*
	René P. Zahedi and Albert Sickmann	
10.1.1	Introduction	*317*
10.1.2	General Sample Preparation	*317*

10.1.3	Reduction of Sample Complexity	318
10.1.3.1	Gel Electrophoresis	318
10.1.3.2	Isoelectric Focusing	318
10.1.4	Methods for Phosphopeptide/Protein Enrichment	319
10.1.4.1	Immunoprecipitation	319
10.1.4.2	Immobilized Metal Ion Affinity Chromatography (IMAC)	319
10.1.4.3	Metal Oxides	320
10.1.4.4	Cation-Exchange Chromatography	321
10.1.4.5	Derivatization Approaches	322
10.1.5	Summary	323
10.1.6	Perspective	324
10.1.7	Recipe for Beginners: IMAC	324
	References	325

10.2 Sample Preparation for Analysis of Post-Translational Modifications: Glycosylation 328
David S. Selby, Martin R. Larsen, Miren J. Omaetxebarria, and Peter Roepstorff

10.2.1	Introduction	328
10.2.2	Advantages and Disadvantages of Different Sample Preparation Methods	331
10.2.3	Example Applications of Enrichment Methods	334
10.2.3.1	ZIC-HILIC Microcolumns for Preparation of N-Linked Glycan-Containing Samples	334
10.2.3.2	Titanium Dioxide Microcolumns for Enrichment of Sialic Acid-Containing Glycopeptides and Glycosylphosphatidylinositol Lipid-Anchored Peptides	336
10.2.4	Summary	337
10.2.5	Perspective	338
10.2.6	Recipe for Beginners: Enrichment of Glycopeptides with a HILIC Microcolumn	338
10.2.6.1	Materials	338
10.2.6.2	Procedure: Purification of Glycopeptides	339
10.2.6.3	Procedure: Deglycosylation of N-Linked Glycopeptides	339
	References	340

11 Species-Dependent Proteomics 343

11.1 Sample Preparation and Data Processing in Plant Proteomics 343
Katja Baerenfaller, Wilhelm Gruissem, and Sacha Baginsky

11.1.1	Introduction	343
11.1.2	Plant-Specific Considerations in Proteomics	344
11.1.2.1	Cell Walls	344
11.1.2.2	Plastids	344
11.1.2.3	Protein Extraction from Plant Tissue	345
11.1.2.4	Extraction from Recalcitrant and Resistant Tissue	345

11.1.2.5	Dynamic Range Limitations 346
11.1.2.6	Proteomics in As-Yet Unsequenced Organisms 346
11.1.3	Sample Preparation Protocols 347
11.1.3.1	Cell Wall Protein Extraction 348
11.1.3.2	Plastid Isolation 349
11.1.3.3	Protein Extraction with TCA/Acetone 350
11.1.3.4	Phenol Extraction 351
11.1.3.5	Serial Extraction 351
11.1.3.6	Extraction from Recalcitrant and Resistant Tissue 352
11.1.3.7	Extraction and Fractionation with Polyethylene Glycol (PEG) 353
11.1.3.8	Stages Following Protein Extraction 353
11.1.4	MS/MS Data Processing for Unsequenced Organisms 354
11.1.5	Concluding Remarks 355
	References 356
11.2	Sample Preparation for MudPIT with Bacterial Protein Samples 358
	Ansgar Poetsch and Dirk Wolters
11.2.1	Introduction 358
11.2.2	The MudPIT Technology 359
11.2.3	Membrane Proteins and MudPIT 361
11.2.4	Quantitative MudPIT 363
11.2.5	Limitations of MudPIT 364
11.2.6	Pitfalls of MudPIT 365
11.2.7	Summary 365
11.2.8	Perspective 365
11.2.9	Recipe for Beginners: MudPIT: Soluble and Membrane Proteins 366
	References 368
11.3	Sample Preparation for the Cell-Wall Proteome Analysis of Yeast and Fungi 371
	Kai Sohn, Ekkehard Hiller, and Steffen Rupp
11.3.1	Introduction 371
11.3.2	Description of the Problem with Regards to Sample Preparation 372
11.3.3	Application Example 373
11.3.4	Summary 375
11.3.5	Perspective 376
11.3.6	Recipe for Beginners 376
11.3.6.1	Cultures 376
11.3.6.2	Preparation of Soluble Cell-Surface Proteins 376
11.3.6.3	Preparation of Peptides from Covalently Linked Cell-Wall Proteins 377
	References 378
12	**The Human Proteosome** 379
12.1	Clinical Proteomics: Sample Preparation and Standardization 379
	Gerd Schmitz and Carsten Gnewuch
12.1.1	Introduction 379

12.1.2	The Preanalytical Phase: Sample Preparation, Standardization, and Quality Management	*380*
12.1.2.1	Standardization of the (Pre)-Analytical Process	*381*
12.1.3	Proteomics in Body Fluids	*382*
12.1.3.1	Techniques for Proteomic Analysis	*382*
12.1.3.2	Applications	*383*
12.1.3.3	Preparation of Clinical Samples for Fluidic Proteomics	*386*
12.1.4	Cellular Proteomics (Cytomics)	*389*
12.1.4.1	Sample Preparation and Standardization for Clinical Cytomics	*389*
12.1.4.2	Tissue Arrays	*390*
12.1.4.3	Bead-Based Immunoassays for Protein Analysis	*391*
12.1.4.4	Preparative Methods	*391*
12.1.4.5	Clinical Applications in Cytomics	*399*
12.1.5	Conclusion	*404*
	References	*405*
12.2	Stem Cell Proteomics	*412*
	Regina Ebert, Gabriele Möller, Jerzy Adamski, and Franz Jakob	
12.2.1	Introduction	*412*
12.2.2	Stem Cell Niches	*413*
12.2.3	Why Study Proteomes in Stem Cells?	*413*
12.2.4	Technical Challenges and Problems	*414*
12.2.4.1	Stem Cell Preparation	*414*
12.2.4.2	Cultivation	*415*
12.2.4.3	Treatment	*415*
12.2.4.4	Whole-Cell Proteome	*415*
12.2.4.5	Secretory Proteome	*415*
12.2.5	Recipes for Beginners	*417*
12.2.5.1	Whole-Cell Lysate	*417*
12.2.5.2	Secretory Proteome Procedure	*418*
12.2.5.3	Labeling with ^{35}S	*418*
12.2.5.4	Ethanol Precipitation	*419*
12.2.5.5	TCA Precipitation	*419*
	References	*419*
13	**Bioinformatics**	*423*
13.1	Bioinformatics Support for Mass Spectrometric Quality Control	*423*
	Knut Reinert, Tim Conrad, and Oliver Kohlbacher	
13.1.1	Introduction	*423*
13.1.2	Problem description	*423*
13.1.2.1	Signal Processing Pitfalls	*424*
13.1.2.2	Map Quality Control	*425*
13.1.2.3	Statistical Validation Results	*425*
13.1.3	Quality Assessment for One-Dimensional (1-D) MS Data	*426*
13.1.3.1	Filter	*427*

13.1.4	Application Example: Absolute Quantification of an Unknown Peptide Content *428*	
13.1.5	Summary *429*	
13.1.6	Perspective *430*	
13.1.7	Recipe for Beginners *430*	
13.1.7.1	Acquiring the Raw Data *430*	
13.1.7.2	Preprocessing the Data *431*	
13.1.7.3	Analyzing the Preprocessed Data *431*	
	References *431*	
13.2	Use of Physico-Chemical Properties in Peptide and Protein Identification *433*	
	Anastasia K. Yocum, Peter J. Ulintz, and Philip C. Andrews	
13.2.1	Introduction *433*	
13.2.2	Isoelectric Point *434*	
13.2.3	Ion-Exchange Chromatography *436*	
13.2.4	Reversed-Phase Chromatography *438*	
13.2.5	Mass Accuracy *442*	
13.2.6	Summary *443*	
	References *444*	

Index *449*

Preface

Why is there a need to consider *Sample preparation* in proteomics? Following the successes of the genome era, researchers have switched their efforts to analyzing complex protein mixtures, hopefully to obtain deeper insights into the molecular development of diseases by comparing whole proteomes from healthy versus disease tissues, body fluid samples, or other sources. Proteomics was born on the waves of these advances and, as a consequence, enormous investments were made in many attempts to unravel the proteome for biomarker identification. The first wave of proteomics resulted in a re-arming of the laboratories which, by this time, no longer required vastly expensive equipment such as mass spectrometers. Inevitably, this surge of interest led to a vast number of reports in which biomarkers had, supposedly, been identified. The second wave of proteomics has been characterized more by the establishment of diverse methods and their combination, as so-called "standard proteomic workflows". Today, this subset of methodologies, databases and workflows appears largely to have been optimized, and the numbers of applications for the funding of studies and grants which include the catchword "proteomics" are rapidly increasing as the research teams continue their quests for meaningful data. Yet, the best way to obtain high-quality data and ensure consistency is not only to perform analyses in replicate but also – and more importantly – to standardize the methods of sample preparation.

What is meant by the term "proteomics"? Whilst this is to some extent a philosophical question, the answer depends heavily on an individual's point of view. Some researchers describe proteomics as a unique scientific area for the analysis of whole proteomes, as notably do clinical proteome scientists. Others define proteomics as a subset of methodologies that are valuable in the analysis of proteins, as proteins represent the most common drug targets today and are the molecules closest to the point of invention in living cells. Despite these differences of opinion, common sense among the scientific community decrees that sample preparation procedures must be kept as simple as possible. In this way, such procedures will go hand in hand with high accuracy and standardization. Clearly, proteomics – in contrast to genomics, which embraces sensitivity, abundance and a combination of different methods – depends on the state of the biological sample

Proteomics Sample Preparation. Edited by Jörg von Hagen
Copyright © 2008 WILEY-VCH Verlag GmbH & Co. KGaA, Weinheim
ISBN: 978-3-527-31796-7

itself. The main question, therefore, is how to create an optimal workflow for each particular experimental set-up.

This book will provide those scientists on the third wave of proteomics – whether researchers or simply users of protein biochemical methodologies – with a comprehensive overview of the different requirements for sample preparation when using today's technologies. Hopefully, it will also provide any "beginners" in proteomics with some very brief "recipes" designed by well-known experts in each particular field.

I believe that this book will "sensitize" the need for sample preparation in proteomics, and will illustrate – with many useful practical applications – the problems which stem from the complexity of whole proteome samples. In this way it will provide solutions for those scientists who are new to this intriguing field of proteomics.

Jörg von Hagen

List of Contributors

Afsaneh Abdolzade-Bavil
BD Diagnostics – Preanalytical
Systems
Am Klopferspitz 19a
82152 Planegg/Martinsried
Germany

Jerzy Adamski
German Research Center for
Environmental Health
Institute of Experimental Genetics
Ingolstädter Landstrasse 1
85764 Neuherberg
Germany

N. Leigh Anderson
CEO, Plasma Proteome Institute
P.O. Box 53450
Washington, DC 20009-3450
USA

Sven Andrecht
Merck KGaA
Performance & Life Science
Chemicals
Chromatography and Bioscience
Frankfurter Strasse 250
64271 Darmstadt
Germany

Philip C. Andrews
University of Michigan Medical School
Department of Biological Chemistry
National Resource for Proteomics
and Pathways
300 North Ingalls Street
Ann Arbor, MI 48109-0606
USA

Sacha Baginsky
ETH Zürich
LFW E51.1
Universitätsstrasse 2
8092 Zürich
Switzerland

Jon Barbour
Ruhr-University Bochum
Medical Proteome-Center
Universitätsstrasse 150
44801 Bochum
Germany

Katja Bärenfaller
ETH Zürich
Institute of Plant Sciences
Universitätsstrasse 2
8092 Zürich
Switzerland

Rudolf Hartmann
Heinrich-Heine-Universität Düsseldorf
Institut für Physikalische Biologie
40225 Düsseldorf
Germany

Ronal C. Hendrickson
Merck & CO.
Merck Research Laboratories
Rahway, NJ
USA

Ekkehard Hiller
Fraunhofer Institute for Interfacial
Engineering and Biotechnology (IGB)
Department of Molecular
Biotechnology
Nobelstrasse 12
70569 Stuttgart
Germany

Franz Jakob
University of Würzburg
Orthopedic Center for
Musculoskeletal Research
Orthopedic Department
Brettreichstrasse 11
97074 Würzburg
Germany

Michael Karas
Johann Wolfgang Goethe University
Institute of Pharmaceutical Chemistry
Max-von-Laue-Strasse 91
60438 Frankfurt
Germany

Josef Kellermann
MPI für Biochemie
Proteinanalysis
Am Klopferspitz 19 a
82152 Martinsried/Planegg

Oliver Kohlbacher
Universität Tübingen
Wilhelm-Schickard-Institut
für Informatik
Sand 14
72076 Tübingen
Germany

Bernd W. König
Heinrich-Heine-Universität Düsseldorf
Institut für Physikalische Biologie
40225 Düsseldorf
Germany
and
Forschungszentrum Jülich
Institut für Neurowissenschaften
und Biophysik, Biomolekulare NMR
52425 Jülich
Germany

Anamarija Kruljac-Letunic
EMBL Heidelberg
Cell Biology and Biophysics
Meyerhofstrasse 1
69117 Heidelberg
Germany

Sabine Kuhfuss
BD Diagnostics – Preanalytical Systems
Am Klopferspitz 19a
82152 Martinsried/Planegg
Germany

Martin R. Larsen
University of Southern Denmark
Department of Biochemistry
and Molecular Biology
Campusvej 55
5230 Odense M
Denmark

Greg Lawrence
Quadraspec Inc.
3000 Kent Avenue
West Lafayette, IN 47906
USA

Kathryn S. Lilley
University of Cambridge
Department of Biochemistry
Downing Site
Cambridge CB2 1QW
United Kingdom

Egidijus Machtejevas
Johannes Gutenberg University
Institute of Inorganic and
Analytical Chemistry
Duesbergweg 10–14
55099 Mainz
Germany

Christian Melle
Medical Faculty at the
Friedrich Schiller University
Core Unit Chip Application (CUCA)
Institute of Human Genetics and
Anthropology
07740 Jena
Germany

Helmut E. Meyer
Ruhr-University Bochum
Medical Proteome-Center
Universitätsstrasse 150
44801 Bochum
Germany

Gabriele Möller
German Research Center
for Environmental Health
Institute of Experimental Genetics
Ingolstädter Landstrasse 1
85764 Neuherberg
Germany

Djordje Musil
Merck KGaA
Merck Serono Research
NCE-Tech / LDT / MIB
Frankfurter Strasse 250
64293 Darmstadt
Germany

Mikkel Nissum
BD Diagnostics – Preanalytical Systems
Am Klopferspitz 19a
82152 Martinsried/Planegg
Germany

Yilmaz Niyaz
P.A.L.M. Microlaser Technologies
GmcB
Am Neuland 9
82147 Bernried
Germany

David Nolte
Purdue University
Department of Physics
525 Northwestern Avenue
West Lafayette, IN 47907-2036
USA

Miren J. Omaetxebarria
University of The Basque Country
Department of Biochemistry and
Molecular Biology
Faculty of Science and Technology
Sarriena z/g
48940 Leioa
Spain

Cloud P. Paweletz
Merck & CO.
Merck Research Laboratories
Rahway NJ
USA

David Perrett
Queen Mary University London
St. Bartholomew's Hospital Medial College
West Smithfield
London EC1A 7BE
United Kingdom

Ansgar Poetsch
Ruhr-University Bochum
Medical Proteome-Center
Universitätsstrasse 150
44801 Bochum
Germany

Fred Regnier
Purdue University
Department of Chemistry
560 Oval Drive
West Lafayette, IN 47907-2084
USA

Knut Reinert
Freie Universität Berlin
Institut für Informatik
Takustrasse 9
14195 Berlin
Germany

Peter Roepstorff
University of Southern Denmark
Department of Biochemistry
and Molecular Biology
Campusvej 55
5230 Odense M
Denmark

Steffen Rupp
Fraunhofer Institute for Interfacial Engineering and Biotechnology (IGB)
Department of Molecular Biotechnology
Nobelstrasse 12
70569 Stuttgart
Germany

Pawel Sadowski
University of Cambridge
Cambridge Centre for Proteomics
Department of Biochemistry
Downing Site
Cambridge CB2 1QW
United Kingdom

Hermann Schägger
Universitätsklinikum
Zentrum der Biologischen Chemie
Molekulare Biochemie
Theodor-Stern-Kai 7, Haus 26
60590 Frankfurt
Germany

Thorsten Schlomm
University Clinic Hamburg-Eppendorf
Department of Urology
20246 Hamburg
Germany

Gerd Schmitz
University Hospital Regensburg
Institute for Clinical Chemistry and Laboratory Medicine
Franz-Josef-Strauss-Allee 11
93053 Regensburg
Germany

Joerg Schreiber
Quadraspec Inc.
3000 Kent Avenue
West Lafayette, IN 47906
USA

Karin Schütze
Carl Zeiss Microlmaging GmbH
Am Neuland 9
82347 Bernried
Germany

Kerstin Seemann
Merck KGaA
Analytical Development and
Bioanalytics
64293 Darmstadt
Germany

David S. Selby
Harrison Goddard Foote
Belgrave Hall, Belgrave Street
Leeds LS2 8DD
United Kingdom

Bertrand Séraphin
CGM–CNRS UPR2167
Equipe Labellisée La Ligue
Avenue de la Terrasse
91198 Gif sur Yvette Cedex
France

Albert Sickmann
Rudolf-Virchow-Center for
Experimental Biomedicine
Protein Mass Spectrometry and
Functional Proteomics
Versbacher Strasse 9
97078 Würzburg
Germany

Kai Sohn
Fraunhofer Institute for Interfacial
Engineering and Biotechnology (IGB)
Department of Molecular
Biotechnology
Nobelstrasse 12
70569 Stuttgart
Germany

Thomas Stangler
Heinrich-Heine-Universität Düsseldorf
Institut für Physikalische Biologie
40225 Düsseldorf
Germany
and
Forschungszentrum Jülich
IBI-2
Institut für Naturwissenschaften
52425 Jülich
Germany

Filippa Stenberg
Stockholm University
Department of Biochemistry and
Biophysics
Svante Arrhenius väg 12
10691 Stockholm
Sweden

Andreas Tholey
Universität des Saarlandes
Technische Biochemie
Functional Proteomics Group
Campus A 1–5
66123 Saarbrücken
Germany

Peter J. Ulintz
University of Michigan Medical School
Bioinformatics Graduate Program
National Resource for Proteomics
and Pathways
300 North Ingalls Street
Ann Arbor, MI 48109-0606
USA

Klaus K. Unger
Johannes Gutenberg University
Institute of Inorganic and Analytical
Chemistry
Duesbergweg 10–14
55099 Mainz
Germany

Bettina Warscheid
Ruhr-University Bochum
Medical Proteome-Center
Universitätsstrasse 150
44780 Bochum
Germany

Gerhard Weber
BD Diagnostics
Am Klopferspitz 19a
82152 Martinsried/Planegg
Germany

Walter Weiss
Technical University Munich
Proteomics Department
Am Forum 2
85350 Freising-Weihenstephan
Germany

Sebastian Wiese
Ruhr-University Bochum
Medical Proteome-Center
Universitätsstrasse 150
44801 Bochum
Germany

Robert Wildgruber
BD Diagnostics
Am Klopferspitz 19a
82152 Martinsried/Planegg
Germany

Dieter Willbold
Heinrich-Heine-Universität Düsseldorf
Institut für Physikalische Biologie
40225 Düsseldorf
Germany
and
Forschungszentrum Jülich
Institut für Naturwissenschaften
IBI-2 / Molekulare Biophysik II
52425 Jülich
Germany

Ilka Wittig
Universitätsklinikum Frankfurt
Zentrum der Biologischen Chemie
Molekulare Bioenergetik
Theodor-Stern-Kai 7
60590 Frankfurt
Germany

Dirk Wolters
Ruhr-Universität Bochum
Analytische Chemie NC 4/72
Biomolekulare Massenspektrometrie
Universitätsstrasse 150
44801 Bochum
Germany

Nathan A. Yates
Merck & CO.
Merck Research Laboratories
Rahway, NJ
USA

Anastasia K. Yocum
University of Michigan Medical School
Michigan Center for Translational
Pathology
Department of Pathology
1150 West Medical Center Drive
Ann Arbor, MI 48109-0606
USA

René P. Zahedi
Rudolf-Virchow-Center for
Experimental Biomedicine
Protein Mass Spectrometry and
Functional Proteomics
Versbacher Strasse 9
97078 Würzburg
Germany

List of Abbreviations

2-DE	two-dimensional electrophoresis
BGE	background electrolyte
CE	capillary electrophoresis
CGE	capillary gel electrophoresis
CHAPS	3-[(3-cholamidopropyl)dimethylammonio]-1-propanesulfonate hydrate
cIEF	capillary isoelectric focusing
cITP	capillary isotachophoresis
CLOD	concentration Limit of Detection
CZE	capillary zone electrophoresis
DC	direct current
DMSO	dimethyl sulfoxide
EDTA	ethylenediaminetetraacetic acid
EOF	electroosmotic flow
FIA	field-amplified jnjection
HPLC	high-performance liquid chromatography
HUPO	Human Proteome Organisation
ITP	isotachophoresis
LE	leading electrolyte
LOD	limit of detection
MEKC	micellar electrokinetic chromatography
MIP	molecular imprinted polymer
MS	mass spectrometry
OPA/NAC	phthaldialdehyde/N-acety-L-cysteine
PC	personal computer
pI	isoelectric point
PVA	polyvinyl alcohol
RPLC	reversed-phase liquid chromatography
SDS	sodium dodecyl sulfate
SPE	solid-phase extraction
SPME	solid-phase micro-extraction

TE	terminating electrolyte
TEMED	N,N,N',N'-tetramethylethylenediamine
Tris	Tris(hydroxymethyl)aminomethane
UF	ultrafiltration
UV	ultraviolet
Vis	visible

Part I
Perspectives in Proteomics Sample Preparation

1
Introduction

N. Leigh Anderson

A lot can happen to a protein in the time between its removal from an intact biological system and its introduction into an analytical instrument. Given the increasing sophistication of methods for characterizing many classes of post-translational modification, an increasing variety of protein-modifying processes need to be kept under control if we are to understand what is biology, and what is noise. Hence, the growing importance of sample preparation in proteomics. One might justifiably say that the generation of good samples is half the battle in this field.

Fortunately, proteomics provides us with good methods for studying sample preparation issues. Two-dimensional electrophoresis of plasma, for example, provides a visual protein fingerprint that allows the immediate recognition of sample handling issues such as clotting, platelet breakage, and extended storage at $-20\,°C$ (instead of $-80\,°C$). A deeper exploration of plasma using mass spectrometry-based methods provides a more comprehensive picture, though perhaps more difficult to understand.

Unfortunately, despite the power of these methods, we do not know as much about sample quality and sample processing as we need to. The general attitude to these issues in proteomics has been to focus on the standardization of a few obvious variables and hope that the power of the analytical methods allows the sought-for differences between sample groups to shine through. This short-cut approach is likely to be problematic. Not only do the unrecognized effects of sample preparation differences add noise to the background against which the biological signal must be detected, but the sample preparation effects themselves are occasionally confused with biology. Well-informed skeptics correctly suspect that variables as basic as how blood is drawn or stored can generate spurious biomarker signals if the case and control samples are not acquired in exactly the same way. At this point we do not have adequate definitions of what "in exactly the same way" actually means for any given analytical platform.

These problems point to a need to take sample preparation (including initial acquisition through all the steps leading up to analysis) as a mission-critical issue, worthy of time and effort with our best analytical systems. Published data on

Proteomics Sample Preparation. Edited by Jörg von Hagen
Copyright © 2008 WILEY-VCH Verlag GmbH & Co. KGaA, Weinheim
ISBN: 978-3-527-31796-7

differences between serum and plasma protein composition, the effect of blood clotting, is interesting but very far from definitive – and in fact specialists in blood coagulation can offer a host of reasons why this process is not easily controllable (and hence not especially reproducible) in a clinical environment. Even a process as widely relied on as tryptic digestion is not really understood in terms of the time course of peptide release or the frequency with which "non-tryptic" peptides are generated – aspects which are critical for quantitative analysis. These and a host of similar issues can be attacked systematically using the tools of proteomics, with the aim of understanding how best to control and standardize sample preparation. In doing so, we will learn much about the tools themselves, and perhaps resolve the paradox surrounding the peptide profiling (originally SELDI) approach: that is, why it seems to be so successful in finding sample differences, but so unsuccessful in finding differences that are reproducible. Perhaps peptide profiling is the most sensitive method for detecting sample preparation artifacts: if it is, then it may be the best tool to support removal of these artifacts and ultimately the best way to classify and select samples for analysis by more robust methods.

Obviously, it is time to take a close look at sample preparation in proteomics. The reader is encouraged to weigh what is known against what is unknown in the following pages, and contemplate what might be done to improve our control over the complex processes entailed in generating the samples that we use.

2
General Aspects of Sample Preparation for Comprehensive Proteome Analysis

Sven Andrecht and Jörg von Hagen

2.1
The Need for Standards in Proteomics Sample Preparation

Sample preparation is not the only step – but it is one of the most critical steps – in proteome research. The quality of protein samples is critical to generating accurate and informative data. As proteomic technologies move in the direction of higher throughput, upstream sample preparation becomes a potential bottleneck. Sample capture, transportation, storage, and handling are as critical as extraction and purification procedures. Obtaining homogeneous samples or isolating individual cells from clinical material is imperative, and for this standards are essential. Advances in microfluidic and microarray technologies have further amplified the need for higher-throughput, miniaturized, and automated sample preparation processes.

The need for consistency and standardization in proteomics has limited the success of solutions for proteomics sample preparation. Without effective standards, researchers use divergent methods to investigate their proteins, making it unrealistic to compare their data sets. Until standards emerge, the continual generation of randomized data sets is likely to contribute to the increasing complexity of proteomics research as well as sample preparation.

Proteomics aims to study dynamically changing proteins expressed by a whole organism, specific tissue or cellular compartment under certain conditions. Consequently, two main goals of proteomics research are to: (1) identify proteins derived from complex mixtures extracted from cells; and (2) quantify expression levels of those identified proteins. In recent years mass spectrometry (MS) has become one of the main tools to accomplish these goals by identifying proteins through information derived from tandem mass spectrometry (MS/MS) and measuring protein expression by quantitative MS methods. Recently, these approaches have been successfully applied in many studies, and can be used to identify 500 to 1000 proteins per experiment. Moreover, they can reliably detect and estimate the relative expression (proteins differentially expressed in different conditions) of high- and medium-abundance proteins, and can measure absolute protein expression (quantitation) of

Proteomics Sample Preparation. Edited by Jörg von Hagen
Copyright © 2008 WILEY-VCH Verlag GmbH & Co. KGaA, Weinheim
ISBN: 978-3-527-31796-7

As a point of reference, most eukaryotic cells contain approximately 20 000 proteins that have an average molecular weight of 50 kDa. Enzymatic digestion yields approximately 30 peptides per protein, or roughly 6 000 000 unique peptides. Certainly, these numbers present at the moment technical challenges in terms of analytical sample throughput, detection and data analysis. The sample preparation techniques of greatest interest in expression proteomics focus on prefractionating and enriching proteins before their separation.

Subcellular fractionation can be conducted as a means to enrich specific organelles and fractions, and thus enable the visualization of significantly more proteins than can be detected in a whole homogenate. However, an increase in workload will result for each additional fraction to be analyzed. Also, sample processing and solubilization will vary for each fraction, and additional troubleshooting is generally needed for optimal reproducibility. Some research groups may be concerned that subcellular fractions prepared from frozen tissue may not be pure due to a loss of membrane integrity during the freeze–thaw process. However, a substantial gain can be realized through the enrichment of subcellular organelles that extends the reach of the researcher in seeking differentially expressed proteins. Nevertheless, researchers should be acutely aware of the need for follow-up studies to validate their assignment of proteins to particular subcellular locations or fractions.

This book provides researchers with a practical report of how advanced proteomics studies can be used in common laboratory settings by giving a step-by-step description for the sample preparation of the most popular proteomics approaches, and by focusing on reproducible methods and data analysis. We discuss the difficulties and potential sources of variability associated with each phase of proteomics studies, and how these can be addressed. With regard to automated methods, these difficulties include: a high rate of false identifications; the limitations of identifying low-abundance proteins in complex biological samples; the development of analysis methods and statistical models using simple samples; and a lack of validation and a lack of sensitivity and specificity measurements for methods applied to complex samples. Therefore, to maximize protein identification and quantitation and to accelerate biological discovery, it is vital to optimize the methods of sample preparation with regard to the relevant parameters described in the following subsection, as well as separation, data processing and analysis. Finally, we illustrate the utility and value of these methods by describing several diverse proteomics-based studies of diverse species, organisms and up-coming technologies in the field of proteome research.

2.3
General Aspects: Parameters which Influence the Sample Preparation Procedure

Before starting upon your proteomic efforts, its important to spend a few minutes examining the following list, which provides an impression of the parameters which have direct influence on the experimental set-up. Moreover, you might consider additionally any special requisites for your special application, which are not mentioned and described in the following chapters.

- Consider the read-out of the experiment (the aim of the experiment)
- Technically dependent aspects for sample preparation in proteomics:
  - The number of samples (single measurements/high-throughput screening)
  - Schedule for experimental setting
  - Costs per analysis
  - Technical equipment in the laboratory (hardware available/sensitivity)
  - Use of chemicals/detergents and their effect on down-stream applications and alternatively mechanical disruption methods
  - Bioinformatical support (data flow handling)
- Sample-dependent aspects for sample preparation in proteomics:
  - Enrichment or depletion strategy: Evaluate the abundance and dynamic range of proteins of interest
  - Sample recovery/standardization/storage conditions
  - Internal standards (spiking) for quantitative proteome analysis or comparison of diverse samples
  - Depending on target protein calculating the amount of sample per analysis (analyte concentration level(s), sample size)
  - Developing a procedure which works for different species (Bench to Bedside)
  - Sample matrix (origin of sample and isolation of the sample to analyze)
  - Localization of target protein (organelle, membrane, cytosolic) has impact on the extraction strategy

Some of these aspects mentioned above, which have a direct impact on the experimental set-up of complex proteome analysis, will be discussed in more detail in the following sections. However, we can begin by drawing some conclusions to the technical aspects of whole-proteome analysis experiments.

2.3.1
Technical Dependent Aspects for Sample Preparation in Proteomics

Before starting the analysis it is important to describe the aim of the experiment(s) and thus the number of experiments required for reliable results, including the number of analytes measured in triplicate or, at minimum, duplicate. If the aim is to study effects on signal transduction events (biodiscovery), the *number of samples* is about 10 to 100 protein samples in sum to discover the proteins involved in a particular signal transduction pathway, or to analyze changes of protein post-translational modifications. If the aim of the test series is to identify a disease-related biomarker, then a significant larger number of samples must be analyzed (up to several thousands) and the identified candidates must be validated under clinical conditions. In both cases, the proteins *a priori* are not known and thus the approach is labor-intensive, because the requirements for the experimental set-up is to separate in parallel, with a very high resolution, complex protein samples (e.g., a liver lysate with over 3×10^4 different proteins at a given time in a single cell). A further challenge is to characterize the identified protein by sequencing or MS-analysis. The trend in proteome analysis is to quantitate proteins in crude samples; therefore, a set of different labeling strategies is commercially available for MS-based approaches as well as for gel-based methods. These techniques allow the researcher to compare several samples in parallel. Although this often implies a higher accuracy and better reproducibility, and also allows the data to be validated in several *independent* experiments, it often causes a bottleneck such that the current problem is not efficiently solved. So, based on the considerations described above, the *schedule* for a basic experiment is from days up to several months or years in biomarker discovery.

If a larger number of protein samples is planned in order to analyze a technique, then a platform selection is needed. Therefore, in addition to the *costs per analysis* (budget) the number of samples per month and the *technical equipment* available in the laboratory must be evaluated. Based on the objective facts with regard to gel separation equipment or MS amenities, the number of expected samples and the sensitivity, the optimized workflow for the proteome analysis must be assigned according to the recommendations described in Figure 2.2. If the appropriate workflow is assigned, the next decision to make is which type of single steps should be combined to answer the question worded in the aim of the experiment. One of the major pitfalls with regard

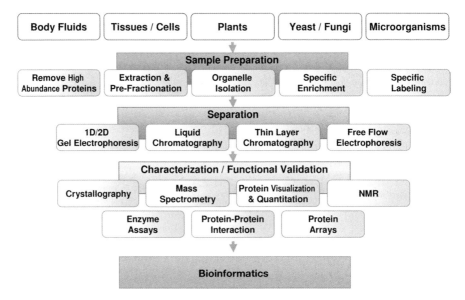

Figure 2.2 The proteomics workflow. When passing from the specimen (first stage), single or multiple steps within the different stages are feasible for deep proteomics in comparative studies or signal transduction analysis. Following sample preparation, the next level is separation before characterization or functional validation and bioinformatics. The separation techniques are not ensnared in all proteomics workflows.

to the selected steps is that, often, a simple combination of several single steps is not easily implementable due to the fact that often-applied buffer components such as *chemicals/detergents* or high salt concentrations are not directly compatible with the down-stream workflow [19]. Therefore, it is very important to read very carefully all of the information provided within the chapters for each particular technique, including the recipe for beginners. Last – and by no means least – the researchers should also consider how to analyze the raw data obtained from the experiment performed. With regards to MS, many different databases (*bioinformatical support*) are used, although the quality of these bioinformatical tools remain the subject of controversy, as the data sets used are often *in-silico*-based, which means that the peptides were bioinformatically digested outgoing from the genomic sequences. One subsequent problem is that post-translational modifications are only taken into account to a limited extent, because until now they have not been directly linked with the amino acid sequences of all known proteins. Another problem is that today MS is capable of examining crude protein mixtures; however, as no guidelines have yet been established among the scientific community with regards to acceptable probability scores and special algorithms and decoy databases to obtain credible results, these must then be validated using a variety of techniques. Within this field, the suppliers of MS systems are currently extremely active developing bioinformatical tools for crude extract analysis. Thus, the research team which uses MS most prominently must also be aware that the bioinformatical section at the end of the proteome analysis is probably the most time-consuming procedure, as it may take weeks to perform and account for 50 to 80% of the entire experimental procedural costs.

After having discussed the technical points worthy of mention in proteome analysis, the following scheme should provide an impression of the currently used techniques and the workflow involved.

Both, standard methods and novel techniques are explained in currently available, dedicated protocols [14,21], although these focus more on the technique than on the sample preparation procedure. The first step of sample preparation may be the key to a successful experiment, as small variations in the sample preparation will later be amplified in each successive step, ultimately to have a major impact on technical variance. Particularly in biomarker discovery it is essential to maintain a very narrow technical variance; otherwise, the biological variance – which is of utmost significance for identifying marginal changes in the protein patterns of healthy versus diseased samples – cannot be measured or described.

After introducing the technical aspects of proteome analysis and a graphical overview of the most prominent techniques currently used in proteomics, attention is now focused on the analytical sample itself.

2.3.2
Sample-Dependent Aspects for Sample Preparation in Proteomics

2.3.2.1 Enrichment or Depletion Strategy

With regard to the experimental set-up, proteome analysis is not a simple combination of routine methods. Rather, proteome analysis necessitates the precise planning of experiments and the desired read-out. Depending on the optimal workflow it is essential to consider the ideal sample preparation procedure for the planned sequential methodological scheme. With respect to the abundance of the target protein, several diverse schemes may be appropriate. For low-abundance proteins such as prostate-specific antigen (PSA) or other tumor markers it is useful to enrich specific protein fractions or to deplete high-abundant proteins such as immunoglobulins or the most prominent protein HSA using affinity-based capture molecules such as dyes or antibodies [3]. Depletion strategies are also a field of interest when interaction partners of drug metabolic relevant proteins are analyzed. Therefore, the depleted protein fraction is analyzed using conventional techniques including two-dimensional gel electrophoresis (2DGE) or MS analysis to identify proteins, which bind for example, to HSA (the *albuminome*).

The enrichment strategy implies that the proteins of interest are previously known. Thus, for the identification of novel disease-related biomarkers this strategy is not directly of great use for many approaches. If there is a suggestion that the desired protein belongs to a defined group of molecules, for example kinase substrates, then the enrichment of phosphoproteins or peptides represents an attractive strategy to proceed straight to the use of diverse capture materials which allow for an efficient enrichment of the sub-proteome. Most of these approaches are used to identify drug targets, which undergo post-translational modifications (e.g., phosphorylation, glycosylation, ubiquitination). Besides antibody-based enrichments, some state-of-the-art techniques are available in proteomics (predominantly immobilized metal-affinity chromatography (IMAC)-based enrichment procedures [15]) for capturing either phosphopeptides (e.g., using zirconium, iron, titanium dioxide) or sugar-binding proteins (e.g., lectins) for glyco analysis. However, this situation will continue to change as attention is focused increasingly on the function of post-translational modifications in signal transduction, and this hopefully will provide further insight into the development of diseases.

2.3.2.2 Sample Recovery and Standardization

The quality of the data obtained from any study will be limited by the quality of the specimens used; hence, the integrity of samples is an important consideration. The ability to delineate and identify the changes that may occur to specimens as a result of differences in storage conditions is pivotal, especially when a larger sample cohort from different inquisitional (technical or personnel) is compared. As noted in many publications and presentations, one major factor of reliable proteome analysis itself is the step before the intrinsic experimental analysis, namely the sample recovery. Proteome analysis with samples obtained from the same origin result in different subsets of identified proteins if, for example, the method of choice for blood recovery was performed using EDTA-preserved blood rather than citrated blood. If larger cohorts of samples are to be analyzed it is absolutely necessary to maintain the experimental deviations at an absolute minimum. Thus, it is not only the nature of the blood preservatives that is important; additional points which might affect the quality and reproducibility of proteome analysis include the anticoagulant used, the presence or absence of blood clotting, the time and temperature of sample storage, and the number of freeze–thawing cycles of the samples, as well as the recovery procedure itself [9,10,22].

2.3.2.3 Quantification, Internal Standards and Spiking

The strong desire for quantitative information in proteomics has nurtured the development of MS-based analytical methods capable of determining protein abundances. Quantification is achieved by comparing the MS signal intensities of the reference with an endogenous peptide that is generated upon proteolytic cleavage of the target protein. In an analogous manner, the level of post-translational modification at a distinct residue within a target protein can be determined. Among the strengths of absolute quantification are low detection limits reaching subfemtomole levels, a high dynamic range spanning approximately up to ten orders of magnitude, only a few steps for sample clean up, and a fast and straightforward method development. Recent studies have demonstrated the compatibility of absolute quantification with various MS readout techniques (e.g., ICAT, iTRAQ, ICPL and others) and sample purification steps such as one-dimensional gel electrophoresis, size-exclusion chromatography, isoelectric peptide focusing, strong cation exchange and reversed phase or affinity chromatography. Under ideal conditions, quantification errors and coefficients of variation below 5% have been reported. However, suitable reference peptides must be chosen, and any protein digestion must be near-complete in order to avoid severe quantification errors. Within the ensemble of MS-based quantification methods, absolute quantification is the method of choice in cases where absolute numbers, many repetitive experiments or precise levels of post-translational modifications are required for a few, preselected species of interest. Consequently, prominent application areas include biomarker quantification, the study of post-translational modifications such as phosphorylation or ubiquitination, and the comparison of concentrations of interacting proteins. For gel-based quantification, the differential in-gel electrophoresis (DIGE) approach is becoming increasingly popular (for further details, see Chapter 5.1). In addition to the labeling technologies available for MS-analysis, many different techniques have been established, including stable isotope labeling, both *in vitro* [11,24,25] and *in vivo* [7], and 2DGE with post staining [8,17] or prelabeling [12,23]. Details of these methods, together with their strengths and weaknesses, are provided in the appropriate chapters of this book.

2.3.2.4 Calculating the Amount of Sample for Proteomic Approaches

Before conducting proteomic experiments, the researcher must be aware of the experimental set-up and the questions to be answered by the analysis or the particular experiment. Based on this strategy, it is important to consider the target protein or protein family and the concentration of the target protein within the sample. If the plan is to analyze body fluid samples then, depending on the workflow, it is recommended that the high-abundance proteins be depleted (see Section 2.3.2.1). The depletion of high-abundance proteins is a controversial subject, as high concentration of proteins may mask lower-abundance proteins, which makes them harder to detect. Moreover, the maximum loading capability is a limiting factor for several methods because, when almost 20 high-abundance proteins cover 90% and more of the protein freight, the number of low-abundance molecules will be orders of magnitude below the current detection limits for all techniques used in proteome research. If the concentration is to be expected within nanomolar or femtomolar ranges (e.g., as for interleukins, tumor markers such as PSA or transcription factors), then the target proteins should be enriched by any reliable method. Hence, depletion is often used as the preliminary step in the sample preparation of body fluids. Besides these initial considerations of abundance, it is also important to calculate the amount of protein needed for the experiment. In current detection methods, such as immunological detection or MS, the lower limit of detection (LLOD) is located in the femtomolar range for a target protein (see Figure 2.1). Therefore, in order to calculate the required cell number or sample volume of body fluid, the following brief calculation indicates the likely minimum sample required for proteomic approaches.

The current LLOD in MS is about 10 femtomol. If the plan is to use 10^5 cells for a single experiment, then the copy number per cell of the target protein must be at least 10^4 in order to achieve the 10^{-14} molar detection limits for the sensitive MS analysis. The high copy number of 10 000 molecules per cell is more likely for a high-abundance protein than for a transcription factor or another rare protein type. Based on this calculation, the strong limitations in reducing cell numbers for whole proteome research become apparent. With currently used techniques, the analyses are generally restricted to seeking changes above a threshold which is determined by the system's experimental noise [13]. Thus, the method of analysis will limit the sensitivity of the system, as biologically relevant changes smaller than the threshold cannot be detected. A future challenge here is, on one hand, to lower the limit of detection, and on the other hand to develop reliable methods to separate the protein content that is not useful for this certain analysis from the information containing protein equivalent sample fraction. Before collecting the sample, great care must be taken regarding the number of replicates, and the loss of protein resulting from each particular method as a sum of all steps.

The number of experiments should also be taken into account when planning proteome studies. Especially, the loss of protein must be evaluated seriously; although in many cases experiments are started with a given amount of sample, when the different steps are combined the sample loss may not be considered and the LLOD may fall short of the analytical detection system. Often, when clinical samples are used there is no possibility of scaling up the sample size. When using plasma or serum this is not problematic, but biopsies are often limited in their sample size. In this case, the researcher must consider the experiment very carefully because, as described above (based on the LLOD of different techniques) it is not possible to obtain an insight into the whole proteome of low-abundance proteins. If this is not taken into account when planning the experiment, the database search will only identify high-abundance proteins, which often do not deliver the expected information regarding changes in protein expression. Depending on the question to be answered within an analysis, it is sometimes important to take care of the "low-hanging fruits". In other words, whilst changes in the expression profile of high-abundance proteins are valuable for further validation, it is impossible to obtain a deep insight into a low-abundance proteome within a given sample.

2.3.2.5 Developing Procedures for Different Model Systems: From Bench to Bedside

Before starting to establish a dedicated workflow for a particular problem in proteome research, there is one additional aspect for concern after the identification of a putative biomarker: whether – and how – the procedure could be transferred easily from laboratory (bench) to the automated screening of larger cohorts in clinical trials (bedside)?

Such a development would depend on parameters such as the model system used (species specificity), sample accessibility, sample amount, cost-per-analysis, and the availability of databases for data interpretation.

For example, if a depletion or affinity-enrichment sample preparation procedure is developed using laboratory animals, for subsequent transfer to human samples, then only those techniques should be used that are easily adaptable from the laboratory animal to the human specimen.

An additional aspect here is that, in animal experiments expensive affinity purification products may be used up to several hundred times. However, such repetition would not be possible for clinical samples due to the risk of cross-contamination, and the use of high-cost, disposable products would be essential.

2.3.2.6 Sample Matrix

Only quality samples can lead to quality or, as Denis Hochstrasser of the University of Geneva claimed at the 2006 HUPO Meeting, "Garbage in" leads to "Garbage out".

Protein solubilization and cell lysis of a given sample are key to an effective analysis. The complete recovery and enrichment of selective fractions permit global and targeted proteomics. The origin of the analyte is important, because the extraction procedure for whole-proteome approaches is extremely closely related to the structure or nature of the sample. In the case of body fluid samples, the researcher must consider whether to use blood, serum or plasma for the analysis, and the advantages here are discussed in terms of the desired technique in the following chapters. Today, proteomics analysis schemes require upstream and downstream reagent and instrument compatibility for integrated 2DGE, multi-dimensional chromatography and MS platforms. In the analysis of tissue samples, in addition to the dissociation of complex tissue structures with collagenases, followed by subcellular extraction procedures, laser capture microdissection (LCM) is becoming increasingly important for enriching cell fractions of interest. Indeed, without such an enrichment strategy it would be impossible to obtain a deep insight with a small sample size, even with sensitive techniques such as MS. Other species such as fungi, yeast, plants or bacteria each require dedicated procedures to destroy the outer cell barrier and to isolate proteins from the membrane into the target organelle. These specialized topics are discussed in detail in the following sections.

2.3.2.7 Localization of Target Protein

Subcellular fractionation is an essential step among enrichment techniques in proteomics research, and is of special importance for the analysis of intracellular organelles and multiprotein complexes. Subcellular fractionation is a flexible and adjustable approach that results in reduced sample complexity [18] and can be combined efficiently with high-resolution 2DGE and MS analysis, as well as with gel-independent techniques.

Fractionation techniques to isolate distinct subcellular compartments have been among the standard strategies established in biochemistry-oriented laboratories for decades. The efficiency of the subcellular fractionation was assessed based on the determination of marker enzyme activities, and a major analytical goal was the identification of single new proteins specifically localized to the subcellular structure. However, due to the limited power of protein identification techniques in traditional protein chemistry, the systematic characterization of the protein subsets specific to subcellular compartments was time-consuming, of limited sensitivity, or even impossible. With the increasing degree of complexity, organisms acquire a broader repertoire of options to meet environmental challenges. This increased complexity of organisms is realized at two levels:

- Not all cells of the organism serve the same purpose; the organism contains several different subsets of cells with distinct properties, for example neurons, germ cells, or epithelial cells.
- Within a given cell, functions such as the storage of genetic material, degradation of proteins, or the provision of energy-rich metabolites to fuel cellular reactions are compartmentalized. Different subcellular compartments contain different and compartment-specific subsets of gene products in order to provide suitable biochemical environments, in which they exert their particular function.

The identification of subsets of proteins at the subcellular level is therefore an initial step towards understanding cellular function. Some subsets of proteins have been shown to be associated with subcellular structures only in certain physiological states, but are localized elsewhere in the cell in other states. Among the possible mechanisms that underlie such conditional association, are included protein translocation between different compartments, the cycling of proteins between the cell surface and intracellular pools, or shuttling between nucleoplasm and cytoplasm. In many cases, the initial states of developing diseases are likely to be characterized by translocation events that precede alterations in gene expression. For comparative studies, in order to elucidate the molecular basis of biological processes, the analysis of dynamic changes of the subcellular distribution of gene products is necessary. In order to be able to monitor these changes, the classic proteome analysis approach must be modified. Performing proteomics at a subcellular level is an appropriate strategy for this kind of analysis, as it is suited to the way in which cells are organized [1,5,6].

Subcellular Proteomics at the Tissue Level Many current proteome analysis projects are aimed at the comparative analysis of tissue samples. Yet, tissue samples are more complex than samples from cultured cells, as any tissue contains not only many different cell types but also structural material such as connective tissue that may not be the target of the analysis.

Different subcellular compartments contain not only different but also compartment-specific subsets of gene products in order to provide suitable biochemical environments, in which they exert their particular function. The identification of subsets of proteins at the subcellular level therefore represents an initial step towards the understanding of cellular function. Indeed, there are subsets of proteins that are associated with subcellular structures only in certain physiological states, but localized elsewhere in the cell in other states.

2.3.3
An Example of Subcellular Protein Extraction

2.3.3.1 Subcellular Extraction and Monitoring the Redistribution of Regulatory Proteins

The developed subcellular proteome extraction procedure [1] should be suitable for investigating changes in the subcellular localization of regulatory proteins impacted by experimental or disease parameters. In order to demonstrate this application, the well-described translocation of NFκB from the cytosol to the nucleus upon stimulation of cells with TNFα [4,16] was chosen. NFκB translocation, when studied in A-431 cells stimulated for different time periods with TNFα, was detected and quantified using densitometric analysis of immunoblots. A time-course analysis demonstrated measurable translocation of NFκB from the cytoplasm to the nucleus in as little as 5 min, with a stronger response observed at 15 min (Figure 2.3), whilst the control protein calpain did not translocate between fractions upon TNFα-stimulation of cells. Thus, the subcellular extraction method allows the assessment of spatial rearrangements of proteins, as shown for NFκB (see Figure 2.3).

2.4
Summary and Perspectives

Proteomics, which currently is in its exponential growth phase, provides an excellent tool to study variations in protein expression between different states and conditions in signal transduction analysis, as well in clinical research. The fact remains that whilst a gene sequence and protein function cannot be correlated, proteomics is the next natural step after genomics. In recent years, extensive progress has been made into proteome analysis, including interaction studies using experimental technologies such as MS, NMR and computational biology. Post-translational modifications generate tremendous diversity, complexity and heterogeneity of gene products, and their determination and identification of its diverse functions in the development of diseases represents one of the main challenges in proteomics research within the next years. Combinations of affinity-based enrichment and extraction methods, multi-dimensional separation technologies and MS are particularly attractive for the systematic investigation of post-translationally modified proteins.

It is apparent that the current 2-D technology has its limitations for proteome analysis. Most whole-protein separation of cell lysates was originally accomplished

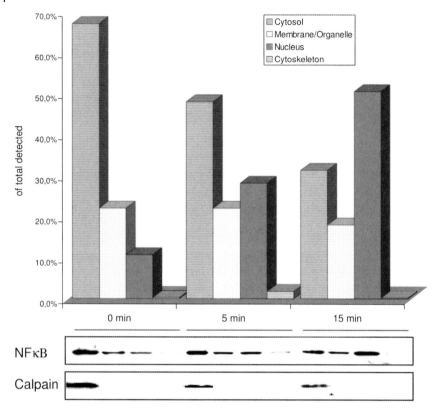

Figure 2.3 Feasibility demonstration of exploiting the subcellular extraction procedure to detect and measure redistribution of proteins affected by disease or experimental factors. NFκB redistribution was recorded during a time course analysis in A-431 cells that were stimulated with TNFα by quantification of immunoblots generated with the subcellular fractions and anti-NFκB antibodies. The time course analysis demonstrates a measurable translocation of NFκB from the cytoplasm to the nucleus within 5 min, with a stronger response observed at 15 min upon cell stimulation. In contrast, the control protein calpain does not change its topology.

using 1DGE or 2DGE, both of which separate proteins based on molecular weight and isoelectric point. However, these methods are biased against proteins with pI extremes, high molecular weights and hydrophobic properties. Recent approaches have involved the use of liquid chromatography separation by size exclusion, strong cation/anion exchange or hydrophobic interaction (reverse-phase chromatography). Individual proteins can also be isolated from complex mixtures using immunoprecipitation, although a technology which is superior to 2DGE for global profiling is yet to emerge. Taking into account factors such as cost, availability and ease of use, it is believed that in present times, 2DGE represents one of the most apposite approaches towards the methodical characterization of proteomes. Moreover, cell proteomes are complex and would require both 2-D-based and non-2-D-based technologies to help decipher protein function. This will remain a vast – and very exciting – challenge in proteomics research for years to come. Finally, interpretation of the data generated would provide meaningful conclusions with the successful integration of computer routines. Structural proteomics, along with the high-throughput chemistry and screening, forms an integrated platform to investigate the mechanisms that underpin the modern drug discovery process. The eventual goal of proteomics is to typify the information tide through protein networks. Although, this information may be seen as a cause – or even as a corollary – of disease processes, taken together it would provide a more complete picture to improve our present understanding of health and disease.

Abbreviations

HTP	high-throughput
ICAT	isotope-coded affinity tags
ICPL	isotope-coded protein label
IMAC	immobilized metal-affinity chromatography
iTRAQ	isobaric tags for relative and absolute quantification
kDa	kilodalton
LCM	laser capture microdissection
LLOD	lower limit of detection
MS	mass spectrometry
MS/MS	tandem mass spectrometry
PTM	post-translational modification

References

1 Abdolzade-Bavil, A. *et al.* (2004) Convenient and versatile subcellular extraction procedure, that facilitates classical protein expression profiling and functional protein analysis. *Proteomics*, **4** (5), 1397–1405.

2 Anderson, N.L. and Anderson, L.G. (2003) The human plasma proteome: history character, and diagnostic prospects. *Mol. Cell. Proteomics*, **2** (1), 50–67.

3 Björhall, K. *et al.* (2005) Comparison of different depletion strategies for improved resolution in proteomic analysis of human serum samples. *Proteomics*, **5** (1), 307–317.

4 Butcher, B.A., Kim, L., Johnson, P.F. and Denkers, E.Y. (2001) *Toxoplasma gondii* tachyzoites inhibit proinflammatory cytokine induction in infected macrophages by preventing nuclear translocation of the transcription factor NF-kappa B. *J. Immunol.*, **167**, 2193–2201.

5 Coligan, J.E., Dunn, B.M. and Ploegh, H. L. (eds.) (2002) *Current Protocols in Protein Science*, John Wiley & Sons, New York.

6 Corthals, G.L. *et al.* (2000) The dynamic range of protein expression: a challenge for proteomic research. *Electrophoresis*, **21** (6), 1104–1115.

7 Everley, P.A. *et al.* (2004) Quantitative cancer proteomics: stable isotope labeling with amino acids in cell culture (SILAC) as a tool for prostate cancer research. *Mol. Cell. Proteomics*, **3** (7), 729–735.

8 Fievet, J. et al. (2004) Assessing factors for reliable quantitative proteomics based on two-dimensional gel electrophoresis. *Proteomics*, **4** (7), 1939–1949.

9 Franzen, B. et al. (1993) Nonenzymatic extraction of cells from clinical tumor material for analysis of gene expression by two-dimensional polyacrylamide gel electrophoresis. *Electrophoresis*, **14** (10), 1045–1053.

10 Franzen, B. et al. (1995) Sample preparation of human tumors prior to two-dimensional electrophoresis of proteins. *Electrophoresis*, **16** (7), 1087–1089.

11 Gygi, S.P. et al. (1999) Quantitative analysis of complex protein mixtures using isotope-coded affinity tags. *Nat. Biotechnol.*, **17** (10), 994–999.

12 Hu, Y. et al. (2003) Proteome analysis of *Saccharomyces cerevisiae* under metal stress by two-dimensional differential gel electrophoresis. *Electrophoresis*, **24** (9), 1458–1470.

13 Karp, N.A. et al. (2005) Impact of replicate types on proteomic expression analysis. *J Proteome Res.*, **4** (5), 1867–1871.

14 Kellner, R., Lottspeich, F. and Meyer, H.E. (1999) *Microcharacterization of Proteins*, 2nd edn, John Wiley & Sons, New York.

15 Kweon, H.K. and Hakansson, K. (2006) Selective zirconium dioxide-based enrichment of phosphorylated peptides for mass spectrometric analysis. *Anal. Chem.*, **78** (6), 1743–1749.

16 Mejdoubi, N., Henriques, C., Bui, E. and Porquet, D. (1999) NF-kappaB is involved in the induction of the rat hepatic alpha1-acid glycoprotein gene by phenobarbital. *Biochem. Biophys. Res. Commun.*, **254**, 93–99.

17 Smejkal, G.B. et al. (2004) Comparison of fluorescent stains: relative photostability and differential staining of proteins in two-dimensional gels. *Electrophoresis*, **25** (15), 2511–2519.

18 Snape, S. et al. (1990) Subcellular localization of recently-absorbed iron in mouse duodenal enterocytes: identification of a basolateral membrane iron-binding site. *Cell Biochem. Funct.*, **8** (2), 107–115.

19 Wang, H. (2005) Development and evaluation of a micro- and nanoscale proteomic sample preparation method. *J. Proteome Res.*, **4** (6), 2397–2403.

20 Wasinger, V.C. et al. (1995) Progress with gene-product mapping of the Mollicutes: Mycoplasma genitalium. *Electrophoresis*, **16** (7), 1090–1094.

21 Westermeier, R. and Naven, T. (2002) *Proteomics in Practice: A Laboratory Manual of Proteome Analysis*, John Wiley & Sons, New York.

22 West-Nielsen, M. et al. (2005) Sample handling for mass spectrometric proteomic investigations of human sera. *Anal. Chem.*, **77** (16), 5114–5123.

23 Yan, J.X. et al. (2002) Fluorescence two-dimensional difference gel electrophoresis and mass spectrometry based proteomic analysis of *Escherichia coli*. *Proteomics*, **2** (12), 1682–1698.

24 Yao, X. et al. (2001) Proteolytic ^{18}O labeling for comparative proteomics: model studies with two serotypes of adenovirus. *Anal. Chem.*, **73** (13), 2836–2842.

25 Zhou, H. et al. (2002) Quantitative proteome analysis by solid-phase isotope tagging and mass spectrometry. *Nat. Biotechnol.*, **20** (5), 512–515.

3
Proteomics: A Philosophical Perspective
Erich Hamberger

3.1
Introduction: "In the Beginning was the Word"

In the beginning was the word: *Proteom*. And it was used frequently. Indeed, soon it was used so extensively that actual biological research could not be thought of without this term. Although it is only slightly more than a decade since Marc Wilkins, in 1994, offered this new key-term to the biological scientific community for common use, it was taken with unexpected willingness. Today, in 2007, the research field of proteomics is expanding rapidly and new approaches are constantly being developed in different ways. This in turn has led to the introduction of new journals dedicated to these concepts and technologies, and to the organization of high-level international symposia. (At the time of writing, the International Conference on Proteomics: Bridging the Gap Between Gene Expression and Biological Function was held in October 2006, in Luxembourg. This symposium highlighted the current and future status of proteomic research fields and techniques; see: http://proteomlux2006.lippmann.lu/).

How was such development possible? Moreover, the term "proteom" does not describe anything new – just as the terms double helix or Plank's affect quota. It simply gave the name to an obvious circumstance: the context of the quantitative totality of the proteins in a cell, a tissue or an organism, which means the whole of all expressed proteins and its respective concentration under certain conditions. Concerning this issue, Rehm remarked that, "... reality, not the word, has changed the conditions. The reason is that the thrill of science has changed creeping its direction but nobody noticed that until the word proteom was exposed all of a sudden." [1]. This enforces the question of how far the thrill of science has changed its direction, so that the proposal to name the protein context of a cell or a tissue proteom let develop a new area of research, which seemed quickly "ready for the mainstream".

Proteomics Sample Preparation. Edited by Jörg von Hagen
Copyright © 2008 WILEY-VCH Verlag GmbH & Co. KGaA, Weinheim
ISBN: 978-3-527-31796-7

Rhem also noted that:

> "Einer der Felsen, die den Strom ablenkten, besteht aus den neuen MALDI (matrix-assisted laser-desorption ionization) Massenspektrometer. Sie bestimmen die MG (molecular weights) von Proteinen und Peptiden schneller und mit hoher Genauigkeit und geben der Proteinforschung einen Ruch von Hai Täk. Weniger spektakulär, aber von ähnlicher Reichweite war die Einführung der immobilisierten pH-Gradienten in die IEF (isoelectronic focusing). Endlich aber war der Molekularbiologie mit dem Ende des Human Genom Projects die große Vision abhanden gekommen." H. Rehm [1]

In my view, only parts of Rehm's arguments are valid, as some have to be completed. Surely, mass spectrometers and other technical progresses provided the possibility of a more effective protein research. But improved technical possibilities alone are maybe a rock, which turns aside the flux, but are not yet able to fix the direction of the turning away. Analogous is valid – in my opinion – concerning the argument of Rehm, which says that with the end of the Human Genome Project, molecular biology has lost its "big vision". This may explain the tendency to avert from the molecular area of research, yet not automatically turn towards the quantitative (contextual) whole of proteins in a cell, respectively a tissue, as a dynamic living whole.

This scientific theoretical turning is – in my view – a result of the circumstance that just the specific detailed molecular genetic research showed the importance of the understanding of the cell or tissue as a whole, specifically for the understanding of the individual gene activity. Or in other words: An endeavor/exertion to understand the gene activity became slowly more and more the necessity for the understanding of the (individual) gene-activation. In this way, Zimmer writes, in his work *Jenseits der Gene. Proteine. Verständnis zum Schlüssel des Lebens* (2005):

> "It's all in the genes – 'Die Gene bestimmen alles.' Diese einmal sehr populäre Aussage würde heute wahrscheinlich nur noch wenige Genforscher so pauschal gelten lassen. Vielmehr richtet sich der Blick immer stärker auf die Wechselwirkungen von Genen, Nukleinsäuren und Proteinen. ... Zur Genetik tritt nun allmählich die sogenannte Epigenetik hinzu, die eben dieses Wechselspiel untersucht." D. Zimmer [2]

One of the theoretical problems in this context is that the genome is seen as extensive static, and that the proteom has a dynamic character. In the first case the focus of cognition is fragmentary and "literal"-oriented, whereas in the other case the focus is holistic, and context-oriented. How can these necessarily different experimental approaches be brought in relation?

Against the background of the effective interactions between genes, nucleic acids and proteins! Before we examine this central question, it seems advisable to me, especially with the background of a philosophic (and historic) perspective, to examine the scientific method in general, which is called "experiment".

3.2
The Experiment as a Scientific Method and a Tool of Cognition

3.2.1
The Experiment Historically Viewed

Derived from the Latin word *experimentum* (proof, attempt, experience; based on the verb try, prove), the word experiment has established as one of the central terms of modern science since the 17th century. Although experiments were made also in the past, in modern times experimentation developed into an excellent column of scientific gaining of cognition respectively progress. One – if not *the* – authoritative role is seen in Galileo Galilei (1564–1642). As generally known, Galilei was not of the opinion of that time, because in his view the Earth circled around the sun, and not the other way round. The representatives of the church exhorted Galilei often, that he should teach this point of view not as "truth", but only as a "hypothesis"; whereas in this context it is important to know, that – according to the common view of that time – each (scientific) hypothesis was not seen as truth, because it was only seen as a construct of the human mind. Galilei had not the purpose "... to fight against the solemn truth, but he did not want that a hypothesis is devalued in principle." [3]. For Galilei, the experiment was an available fund to prove his hypothesis concerning the soundness. Against this background, he postulated "safe acknowledge" through the prediction of results of experiments. Many historians share the opinion with Pietschmann, that "... nature science is born in a modern sense when Galilei changed the difference 'truth-hypothesis' into the difference 'truth-knowledge'" ([3], p. 75).

3.2.2
The Experiment Theoretically Viewed

After the 17th century the experiment was seen as both a central criterion for scientific hypothesis and theories, and as a source of new cognition, that could help to gain further hypothesis and theories. What can be understood from a good scientific experiment, should be shown with the help of Pietschmann [3], who proposed three axioms of an experiment as unalterable characteristics: reproducibility; quantification; and analysis.

Reproducibility means, in comparison to simple observations, the circumstance, that a result in science can only hope for recognition if repetition of the experiment, when made by the same or other scientists, leads to the same results.[1]

---

[1] In that context, Pietschmann provides an example of the expendable evidence of the W- and Z-Bosons (with all their postulated attributes), from the recent history of physics: "1984 erhielten Carlo Rubbia und Simon Van der Meer den Nobelpreis für Physik, für their decisive contributions to the large project, which led to the discovery of the field particles W and Z, communicators of weak interaction.' Voraussetzung dafür war jedoch, daß das Experiment makellos, also reproduzierbar war! Daher arbeitete am CERN nicht nur die Gruppe um Carlo Rubbia mit mehreren hundert Mitarbeitern aus fast 40 Laboratorien, genannt 'UA1' (for 'underground area 1'); eine zweite Gruppe (UA2) mußte die Ergebnisse von UA1 bestätigen, erst dann waren sie anerkannt und reif für den Nobelpreis." [3].

that was open for description. However, not until better technical instruments were available did the experimental research of the living begin to spread [3]. As a result, the exploration of the living was for a long time marked through description.

During the early days of modern scientific dealing with the living, morphology was the central cognition aspect of biological and zoological research – with the first aim to understand a type as a living-configuration as a whole.[12] The experience of the biologist – which is always a reflected and interpreted one [11] – was in the case of morphologic cognition interest more focused on the "Zusammenschau der [anschaulichen] gestaltlichen Fülle als . . . [auf die] . . . Einordnung der Details in einen Begriff oder eine[r] Theorie."[13] ([12]; see also [13,14]).

Kummer refers to this central theoretical aspect in concise shortness, when he says: "Soweit Biologie Morphologie ist, ist Erfahrung im Sinne von Anschauung alles. [12]". Beside this main cognitive interest to describe morphological figures, biology as a modern natural science was always also a science of figure-dissection, of anatomy. More generally expressed, an analytic science especially in the research area of medicine.[14] With regards to the modern beginning of the anatomical exploration of the living, and in accordance with criterions of modern science, it is mostly Andreas Vesalius (1514–1564) who is named, and especially his work *De humani corporis fabrica*, which appeared for the first time in 1543. This book is a successful interplay between image and word, in which Vesalius did his utmost to make the pictures as realistic as possible. The purpose of the book was to discover and show, through illustrations of an anatomic dissection of the organism "human being", the level of the naturally given organs, and so to achieve a contribution to the scientific description of the living creature "man".

Approximately 100 years later, Rene Descartes (1596–1650), one of the key-persons of modern scientific culture, recommended in his *Discours sur la methode* (1637), the analysis, means the dissection, the division of the whole, which one wants to understand as the best way of cognition. With that, the research became always smaller parts socially acceptable, connected with the thrilling question where the dissected principle comes to an end, means when there is no further division possible. Descartes is of the opinion – call it u-topic or visionary – that the method of dissection finally leads to the point where the divided is only subjected to

---

[12] Against this background it is better to understand why, in the 17th century, especially the empirists under the physicians and the natural scientists were highly skeptical concerning the invention of the microscope. Catherine Wilson reveals this context in her book *The Invisible World. Early Modern Philosophy and the Invention of the Microscope* [40].

[13] Kummer notes in this context: "Die Entwicklung eines 'systematischen Feingefühls' bzw. der visuellen Fähigkeit zur Fallunterscheidung vor aller begrifflichen Festlegung ... wurde insbesondere von botanischen Morphologen immer als die unverzichtbare Grundlage ihrer Wissenschaft angesehen." [12]. Analogously, [28]: "Bis vor kurzem herrschte unter den Naturwissenschaftlern uneingeschränkt die Überzeugung, die in dem Satz Poincarés ausgedrückt ist: "Wenn ein Forscher über unendlich viel Zeit verfügte, genügte es, ihm zu sagen: Schaue, aber schaue gut." [28].

[14] Compare to this the third-shown axiom of the experiment.

laws of physics. In his opinion, reaching this point would mean that science is at the happy end.

But not all contemporaries were content with the ideas of Descartes, and from that time onwards, both more scientific analysis took place and paralleled the technique which goes with it (different analysis devices) was developed. The showing of details of the developing in the science, and describing how the analytic vision of the Modernity led to the exploration of the cell and the chromosomes, and so on until the decoding of the genetic code in connection with the human genome project, would be beyond the scope of this chapter. A general aspect which is connected with that should not be omitted from mention because it seems – especially cognition theoretical – very important. With the raising of the wish to discover more and more profound areas of the living through analysis lead while the modern age – one is inclined to say consequently – simultaneous to a turning away from the idea that life/health is a relational structure or a balance of body-humors.

Fischer refers to this important circumstance, when he tells: "Die Erklärung von Krankheiten durch ein mangelhaft balanciertes Zusammenspiel von vier Körpersäften hielt sich über zweitausend Jahre lang, und sie wurde erst im 18. Jahrhundert abgelöst, als die von Vesalius offengelegte Ebene der Organe allgemein von den Wissenschaftern er- und anerkannt wurde." [15]. In the middle of the 19th century (1858), Rudolf Virchow (1821–1902) finally formulated, after conducting 10 years of analytic research behind the microscope, some form of "analytic illness-theory": the concept of "cell-pathology", or the imagination that the cause of illnesses is placed in the cells. In the 20th century not only the scientific area "medicine" but also the classical "life sciences" zoology and biology were focused primarily on the different functional areas of the living, with the main purpose being able to understand better/decode the mechanism "life" by understanding the single parts of it. Portmann notes in this context, in 1974: "Wer heute die Fülle der biologischen Bücher und die Arbeitsgebiete in der Praxis überblickt, wer etwa das Vorlesungs-verzeichnis einer Hochschule im biologischen Sektor überfliegt, der wird feststellen, wie klar die Akzente auf Physiologie, auch auf Entwicklungsphysiologie liegen, auf Biochemie und Biophysik, auf Molekuarbiologie sowie Mikrobiologie und Genetik. Wie weit sind wir weg von der Welt der Gestalten, dem ursprünglichen Objekt der zoologischen wie der botanischen Morphologie." [16].

This system of cognition changed to such an extent, that the actual objects or "epistemic things" (Rheinberger) of scientific research in biology became less and less graphic. One could say: it developed from an interpreted experience to an unclear theoretical model. With that change, concerning the central subject of biological cognition interest, the set of biological research changed necessarily too. As the experiment – respectively the interplay between theory and experiment – became more and more important, an increasing movement from the "soft" describing experience-science to the "hard" experimental-science took place. The summit – and parallel – turning point in this regard shows the Human Genome Project, because it was not possible to dissect the living any further.

In this sense, the view of Fischer [15] seems plausible, as it describes the modern analytic/static exploration of the living as the era from Vesalius (1543) up to the realization/finalization of the Human Genome Project in 2001. Or to make it short: from "Vesalius to HUGO". Fischer verbally writes, that with Vesalius and his book *De humani corporis fabrica*, "... die Expedition in das Innere unseres Lebens [beginnt,] – wobei an dieser Stelle angemerkt werden kann, dass unsere Gegenwart gerade das Ende des forschenden Feldzugs erlebt, und zwar in Form des humanen Genomprojekts, mit dessen Hilfe die Reihenfolge der chemischen Bausteine offen gelegt wird, die das Erbgut einer menschlichen Zelle ausmachen. Den Bau des menschlichen Körpers [des Lebendigen] noch genauer zu erkunden, macht allein deshalb kaum Sinn, weil die dazu gehörenden Moleküle längst aus dem Blickwinkel der Chemie ins Auge gefasst worden und somit bekannt sind. Die dazugehörige Wissenschaft der Anatomie, deren Name sich vom griechischen Wort für Zerschneiden (anatome) ableitet, hat ihren Rahmen damit weitgehend abgesteckt und kann mit dem weiteren Zerschneiden aufhören." [15].

Against this background it becomes clear, which challenge represents the scientific concentration on the phenomenon "proteome". Not only regarding the difficulty, to image big (static) quantities all at once, but especially on the basis of the circumstance, that the proteom represents the moment-reception of a dynamic process, and on top of that: of a relational dynamic process. With that we arrived at this point again, from which we took our exit: Proteomics as a science-/cognition-theoretical challenge.

3.4
Proteomics as a Cognition-Theoretical Challenge

What is now the concrete scientific challenge of proteomic research? It consists not only of instrumental-technique difficulties which should be solved [1] but also (respectively above all) of fundamentally new theoretical questions. How can something living be experimentally researched, while it is inaccessible to direct viewing, as is the proteome? As mentioned above, the targets of scientific research in molecular biology and genetics became during the 20th century increasingly unillustrative. But, they were at least static targets – in other words, they were basically constant and could be researched analogously to abiotic/inorganic cognition subjects in physics and chemistry. This fact leads to the tendency to understand biotic/organic processes in a mechanistic way. In other words, these research areas, which are especially focused on living processes, are in danger to understand living entities out of non-living entities with the result, that they lose the ability to see its specific lively subject of research. This background was verbalized by Fischer: "Alles Wissen beginnt mit unveränderlich wirkenden Gegebenheiten, auch wenn sie tatsächlich – eigentlich – anders sind und sich wandeln." [15].

3.4 Proteomics as a Cognition-Theoretical Challenge

With these remarks it should be shown, which fundamental epistemic challenge the "proteomic turn", the turning to the dynamic proteom-context represents.

3.4.1
Cognition-Theoretic Support from Physics

I see scientific-theoretic support especially from physics. Why is that? Because physics was touched by a similar main problem concerning the comprehension of the atomic/subatomic basic structure, which was solved by the conception of the quantum mechanics in the first quarter of the 20th century (Bohr/Heisenberg, 1927). In other words: Physics has perhaps a cognitional-theoretic help-potential, because of the circumstance, that physics have finished with the "atomic-dissected" exploration of the (inorganic) material almost a century earlier than did biology. I want to complete this thought with the observation of the quantum physicist Dürr in the context of his scientific comparison between biology and physics in the 20th century which is: "Prinzipielle Grenzen der fragmentierenden, reduktionistischen Betrachtungsweise waren dort deutlich geworden. Zerlegbare Substanz offenbarte auf seltsame Weise holistische Züge." ([17], see also [18]).

Against this background, proteomic research seems to require – in its essence – a "post-analytic" way of cognition, inasmuch as it focuses especially on interactions/(hierarchic) relational structures of the cognition-interest. Because of that, this new research-area should not forget to look at the great experiences which the quantum-physics (-mechanics) has collected almost a century before.[15] (See in this context [19], 1932 and [20–24]).

But to keep it in order: At first it can generally be told, that "das erste Problem, dem wir begegnen, wenn wir den eigentlichen Bereich der Physik verlassen [that means: the area of inorganic/a-biotic], die Frage [ist], welchen Platz die lebenden Organismen in der Beschreibung der Naturwissenschaften einnehmen." [25].

So, Bohr asks the crunch question: Can biology, the science of the living (life), be reduced to physics (and chemistry) respectively the cognition-methods which come with it, or not? He himself has twice answered the question in the negative. In his opinion is, as Röhrle remarks, "der Begriff des Lebens in der Biologie ebenso elementar wie derjenige des Wirkungsquantums in der Physik." [25,26].

What does that mean? What does Bohr want to express? He refers to the fact, that an adequate cognition of the living is only possible in the way that one does without wanting to discern this living with primary physical/chemical methods, means wanting to discern with "a-biotic" methods. To express it in other words:

---

[15] With the following quotation Dürr makes clear, that with this not since long-ago accomplished methodical entry of the 'classical' physics in biology is meant: "Im Gegensatz zu Sheldrake, der sich an der Unangemessenheit der heute bevorzugten physikalischen Beschreibung biologischer Phänomene reibt, betone ich eher die Bedeutung der Physik für die Biologie. Die Physik, die ich dabei meine, ist allerdings eine andere als die heute von den Biologen unterlegte, die sie von den Chemikern übernommen haben." [17].

Life cannot be explained purely in physico-chemical terms. Meyer-Abich remarks, citing Bohr: "Danach scheinen der physikalischen Beschreibung biologischer Phänomene ähnliche Schwierigkeiten entgegenzustehen wie der Raumzeitbeschreibung atomarer Phänomene. Genau so wie es nur auf Grund der prinzipiellen Komplementarität zwischen der Anwendbarkeit des Zustandsbegriffs und der raumzeitlichen Verfolgung der Atomteilchen möglich ist, in sinnvoller Weise von der charakteristischen Stabilität der Atomeigenschaften Rechenschaft abzulegen, so dürfte die Eigenart der Lebenserscheinungen und insbesondere der Selbststabilisierung der Oranismen untrennbar mit der prinzipiellen Unmöglichkeit einer eingehenden Analyse der physikalischen Bedingungen, unter denen das Leben sich abspielt, verknüpft sein. ... Die Komplemantarität zwischen Raumzeitbeschreibung und Anwendbarkeit des Zustandsbegriffs ... läßt sich auch als Unmöglichkeit einer eingehenden Analyse der raumzeitlichen Verhältnisse, unter denen die Zustandsentwicklung sich abspielt, zum Ausdruck bringen. Genau so wie zwischen Raumzeitbeschreibung und Anwendbarkeit des Zustandsbegriffs wird dann von einer Komplementarität zwischen physikalischer Beschreibung und Anwendbarkeit des Begriffs 'Leben' die Rede sein können." [27].

3.4.2
The "Pietschmann Axioms" of the Experiment in Biological View

Before we concentrate on the specific context of the experiment concerning the exploration of the living, let us remember the general axioms of the experiment as Pietschmann proposed them:

- (subject-variable) reproducibility,
- (qualitative) quantification,
- (constructive) abstraction/analysis.

In my view, it can be generally postulated that: The more alive (complex) the cognition-subject, the more difficult seems to be the reproducibility, the more important is the subjective sensitivity of the researcher. Fischer writes in this context: "Enzyme haben etwas von Primadonnen an sich. Wenn sie auftreten, sind sie grandios, aber wer sie vernachlässigt und beim Umgang mit ihnen den kleinsten Fehler begeht, verliert sofort ihre Zuneigung. Sie verstummen, was heißt, sie zeigen keine biochemische Aktivität mehr und bleiben verschwunden." ([15], see also [28,29]).

Arthur Kornberg, the Nobel prize winner of 1959, says in his autobiography *For the love of enzymes: The odyssey of a Biochemist* (1989), that only fantastic enzymes are existing. Everyone who does not love them cannot get along with them. Schneider [30] pleads for the legitimacy of interactive (subjective) experiences[16] in the natural sciences. He speaks of the overcoming of a pure "technomorph look" also in natural science, because for him this look is an unnecessarily

---

[16] In demarcation to (neutral) adventures.

narrowing. He thinks that a biomorph look, which does justice to the circumstance that the living is being explored, especially in the context of life sciences is appropriate.

With regard to the axiom of (qualitative) quantification it can be postulated: The more alive (complex) a cognition-subject, the more crossed over seem to be quantity and quality. Corresponding to the shown distinction of Heitler (see section 3.3) concerning the characteristics of differentiation between the living and the non-living, it must be noted that cognition of the living – however quantified – necessarily performs measuring sizes which, in proportion to cognition of the non-living material, in a higher sense never are only quantitative measuring sizes, but show in an over-crossed sense qualitative aspects. In short, quantification means in the context of the living in an extensive sense quantification of qualitative figure (see [26,31]).

Now, we should concentrate more on the third of the proposed experimental axioms of (constructive) abstraction/analysis concerning the context of the living. For this stands the thesis, that the more alive (complex) a cognition-subject, the more constructive the abstraction/analysis has to be. We remember: Constructive abstraction means – in the classic understanding – the dissection/reduction of the subject to a conclusive model. In this sense, Lord Kelvin (William Thomson) defines the understanding of a study-object, what is here called "constructive abstraction", as follows: "I am never content until I have constructed a mechanical model of the subject I am studying. If I succeed in making one, I understood; otherwise I am not." (Thomson, in [32], V).

In quantum mechanics, however, such a simply exemplary understanding of a cognition-object is no longer possible. Feynman shows this circumstance when in his *Lectures on Physics* concerning the (constructive-abstract) understanding of quantum mechanics he considers: "Even the experts do not understand it the way they would like to, and it is perfectly reasonable that they should not, because all off direct, human experience and of human intuition applies to large objects."
(Feynman, 1965, 1-1; cited in [32], V)

If by (constructive) abstraction/analysis gaining an unopposed, figurative imagination of a subject is meant – in the sense of Lord Kelvin – "than", so Pietschmann literally, "quantum mechanics cannot be understood" ([32], VI). Against this background, Pietschmann defines "understanding" in a wider context, of which I think that it can be possibly pioneering for the exemplary comprehension in the context of proteomic research. He notes that:

"Ich will aber den Begriff 'verstehen' weiter fassen; wenn wir aus einem Gegenstand – z. B. der Quantenmechanik[17] – alle Widersprüche, die wir eliminieren können, entfernt haben, aber bei denjenigen Widersprüchen, die dann noch übrig bleiben, erkannt haben, warum sie nicht zu eleminieren sind, und wir sie überdies handhaben können, dann haben wir diesen Gegenstand in einem weiteren Sinne auch 'verstanden'. Auf Anschaulichkeit im klassischen Sinne müssen wir dann freilich verzichten." ([32], VI)

---

[17] I wish to add here: Or "epistemic things" of proteomic research.

vollständigen und umfassenden Begriff der Art formulieren würde. Das liegt wohl letztlich darin begründet, daß das Statische und Dynamische (Sein und Werden) komplementär einander gegenüberstehen." [34].[21]

3.5
Conclusion

With these transdisciplinary remarks, I wish to give a signal to prove, in which way (s) quantum theoretical insights may be able to help in establishing new experimental "description-rooms" [36] in proteomic research. As shown in Section 3.2.4 with help of Fleck [6] and Rheinberger [36], the daily praxis of biological research does not focus on a single experiment, but rather on an experimental system. As shown in Section 3.4.2 by Overhage [34], the cognition of living entities need – in analogy to the description context of quantum phenomena – two (parallel valid) systemic views at the cognition object, which are not integrable in one cognition system. Against this background, I suggest enlarging the term of experimental-system – according to [37] – to the term experimental-structure for all these cognition contexts, where two parallel valid epistemic pictures are unalterable, to describe a cognition object or "epistemic thing" in a relevant way; so that, for example, methodical excluding consideration kinds, just like the substance isolation on the one hand and the process isolation on the other hand, can be seen together "complementary" described, without being pressed into one cognition system.

Finally, I wish to consent to the question concerning the principal cooperation between science and humanities (philosophy). Here, we are standing in front of a mutual completion condition, which is – nowadays – for the first time difficult to show. On the one hand (nature), science is rightly value-free, as it just explores non-living and living phenomena given from nature with the help of observation respectively experiment. On the other hand – through this progress of research – an increasing number of possibilities for the technical transformation of the world are developing; which opens new dimensions especially for the modern biology. Questions concerning the value and sense of these deeds arise. In other words: More

---

[21] Overhage to this context literal: "Sobald man eine Art, auch eine polytypische, in ihren Merkmalen und ihrer genetisch-geographischen Isolation beschreibt, wird sie zu etwas Statischem, das ihr Werden unberücksichtigt läßt. Legt man dagegen den Akzent auf die Phylogenese der Art, also auf ihren in der Zeit ablaufenden Werdegang, dann beginnen die morphologischen Merkmale und die ausbalancierten genetischen Zustände der Populationen an Schärfe zu verlieren oder ganz zu verschwinden. Die zur Herausarbeitung der Eigenart und Merkmalskombination einer Art verwendeten Begriffe eignen sich nicht für das Erfassen der in der Zeit ablaufenden dynamischen Vorgänge. Es liegt also an, dem eminent historischen Wesen der Art', wie Schwarz (1960) sagt,, daß eine Definition, die gleichzeitig sozusagen ein Rezept zur Unterscheidung der Arten liefert, nicht möglich ist'. Oder anders ausgedrückt: Eine Art der Population kann nur solange scharf abgegrenzt werden, als sie nicht, als Chronotyp und überhaupt nicht als historisches Wesen' betrachtet wird. Die Zeit muß als stillstehend [a-temporal], das heißt, das genetische Erbe der Art als unveränderlich, aber auch die Einflüsse der Umwelt als konstant gedacht werden." ([34], see also [41])

and more value-centered basic questions respectively problems, which cannot be answered by biology or (nature) science themselves, are developing especially through value-free (nature) science research. The more necessary it seems to me that the science undergoes a constant critical reflection: for example through making an transdisciplinary exchange with (intercultural) philosophy, humanities, also concerning theological aspects. Thinking about that nature sciences are also based on culture-specific assumptions, which they cannot look at methodically. Or with the words of Pietschmann: "Einerseits gibt es die Versuchung, alles [naturwissenschaftlich] Ausgeschlossene als 'nur subjektiv' und daher allgemein als weniger bedeutsam zu erklären, andererseits gibt es den Versuch, das mit der naturwissenschaftlichen Methode Nicht-Erfaßbare trotzdem erfassen zu wollen. Ich kann nicht genug betonen, dass dieses Dilemma aus den unerwartet reichen und vielfältigen Möglichkeiten der Naturwissenschaft stammt und nicht etwa aus ihrer Schwäche! Es wäre Aufgabe einer kritischen Wissenschaftstheorie, die sich selbst als Teil der Philosphie bestimmt, das von der Naturwissenschaft Ausgeschlossene (und diesen Ausschluß selbst) zu reflektieren. Eine echte Partnerschaft von Philosophie und Naturwissenschaft, ohne daß sich eine der anderen unterwirft, könnte obiges Dilemma (vielleicht) lösen." ([3]; see also: [38,39]).

I think especially the proteomic research is able – in close "cognition touch" with the results of one hundred years of quantum theoretical research – to establish new guiding experimental conceptions for a better understanding of life in all its varieties.

References

1 Rehm, H. (2002) *Der Experimentator: Proteinchemie, Proteomics*, 4th edn., Spektrum Akademischer Verlag, Heidelberg, Berlin.
2 Zimmer, D. (2005) *Jenseits der Gene. Proteine. Verständnis zum Schlüssel des Lebens*, Klett-Cotta, Stuttgart.
3 Pietschmann, H. (1996) *Phänomenologie der Naturwissenschaft. Wissenschaftstheoretische und philosophische Probleme der Physik*, Ibera, Berlin, Heidelberg, New York.
4 Rheinberger, H.-J. (1992) *Experiment – Differenz – Schrift*, Basilisken-Presse, Marburg/Lahn.
5 Popper, K. (1984) *Logik der Forschung*, 8th edn., Akademie-Verlag, Tübingen.
6 Fleck, L. (1980) *Entstehung und Entwicklung einer wissenschaftlichen Tatsache. Einführung in die Lehre vom Denkstil und Denkkollektiv (1935)*, Suhrkamp, Frankfurt/Main.
7 Bachelard, G. (1988) *Der neue wissenschaftliche Geist (1934)*, Suhrkamp, Frankfurt/Main.
8 Lee, T.D. (1987) History of the weak interactions. *CERN Courier*, Jan./Feb., 27, 12.
9 Heitler, W. (1970) *Naturphilosophische Streifzüge. Vorträge und Aufsätze*, Vieweg Friedrich & Sohn Verlag, Braunschweig.
10 Bertalanffy, L. von (1990) *Das biologische Weltbild. Die Stellung des Lebens in Natur und Wissenschaft (1949)*, Francke, Wien, Köln.
11 Audretsch, J. (2002) Erfahrung und Wirklichkeit. Überlegungen eines Physikers, in *Was ist Erfahrung? Theologie und Naturwissenschaft im Gespräch*, Karlsruhe (eds J. Audretsch and K. Nagorni), Evangelische Akademie Baden, pp. 9–34.
12 Kummer, C. (2004) Wie viel Erfahrung braucht die Biologie? in *Spiel mit der*

Wirklichkeit. Zum Erfahrungsbegriff in den Naturwissenschaften (eds R. Esterbauer, E. Pernkopf and M. Schönhart), Königshausen & Neumann, Würzburg, pp. 67–80.

13 Troll, W. (1928) Organisation und Gestalt im Bereich der Blüte, Springer, Berlin.

14 Lorenz, K. (1965) Gestaltwahrnehmung als Quelle wissenschaftlicher Erkenntnis. in ders.: Über tierisches und menschliches Verhalten. 2nd ed., Wissenschaftliche Buchgesellschaft, München, pp. 255–300.

15 Fischer, E.P. (2006) Die Bildung des Menschen. Was die Naturwissenschaften über uns wissen, Ullstein, Berlin.

16 Portmann, A. (1974) An den Grenzen des Wissens. Vom Beitrag der Biologie zu einem neuen Weltbild, Wien, Fischer-Taschenbuch-Verlag, Düsseldorf.

17 Dürr, H.-P. (1997) Sheldrakes Vorstellungen aus dem Blickwinkel der modernen Physik, in Rupert Sheldrake in der Diskussion. Das Wagnis einer neuen Wissenschaft des Lebens, (eds H-.P. Dürr and F-.T. Gottwald), Bern, Scherz, München, Wien, pp. 224–249.

18 Held, C. (2002) Die Bohr-Einstein-Debatte und das Grundproblem der Quantenmechanik, in Verschränkte Welt. Faszination der Quanten (ed. J. Audretsch), Wiley-VCH, Weinheim, pp. 55–75.

19 Bohr, N. (1957) Light and Life Lecture 1932, in Atomic Physics and Human Knowledge, Ox Bow Press, New York.

20 Delbrück, M. (1935) Die Natur der genetischen Mutationen und die Struktur der Gene, in Nachrichten aus der Biologie der Gesellschaft der Wissenschaften, I.

21 Fröhlich, H. (1968) Longe-Range Coherence and Energy Storage in Biological System. International Journal of Quantum Chemistry, 2 (5), 641–649.

22 Fröhlich, H. (1969) Quantum Mechanical Concepts in Biology, in Contributions to Physics and Biology (ed. M. Marois), Amsterdam.

23 Jordan, P. (1932) Quantenmechanik und die Grundprobleme der Biologie und Psychologie. Naturwissenschaften, 20.

24 Jordan, P. (1941) Die Physik und das Geheimnis des organischen Lebens, Kornberg, A. (1989), For the Love of Enzymes. The Odyssey of a Biochemist. Cambridge, Harvard University Press, London.

25 Bohr, N. (1985) Atomphysik und menschliche Erkenntnis. Aufsätze und Vorträge aus den Jahren 1930 bis 1961, Braunschweig, Vieweg Friedrich & Sohn Verlag.

26 Röhrle, E. (2001) Komplementarität und Erkenntnis. Von der Physik zur Philosophie, Literatur-Verlag, Münster.

27 Meyer-Abich, K.-M. (1965) Korrespondenz, Individualität und Komplementarität. Eine Studie zur Geistesgeschichte der Quantentheorie in den Beiträgen Niels Bohrs, Steiner Franz Verlag, Wiesbaden.

28 Fleck, L. (1983) Erfahrung und Tatsache. Gesammelte Aufsätze, Suhrkamp, Frankfurt/Main.

29 Raus, S. (2006) Erkenntnisgewinnung in der Kommunikationswissenschaft und die Rolle des Forschers als Subjekt. Dipl. Arb., Salzburg.

30 Schneider, H.J. (2004) Erfahrung und Erlebnis. Ein Plädoyer für die Legitimität interaktiver Erfahrungen in den Naturwissenschaften, in Spiel mit der Wirklichkeit. Zum Erfahrungsbegriff in den Naturwissenschaften (eds R. Esterbauer, E. Pernkopf and M. Schönhart), Königshausen & Neumann, Würzburg, pp. 231–248.

31 Witzany, G. (2006) Plant communication from biosemiotic perspective. Differences in abiotic and biotic signal perception determine content arrangement of response behavior. Context determines meaning of meta-, inter- and intra-organismic plant signaling. Plant Signaling and Behavior, 1 (4), 169–178.

32 Pietschmann, H. (2003) Quantenmechanik verstehen. Eine Einführung in den Welle-Teilchen-Dualismus für Lehrer und Studierende, Berlin, Heidelberg, Springer, New York.

33 Dürr, H.-P. (2005) *Auch die Wissenschaft spricht nur in Gleichnissen. Die neue Beziehung zwischen Religion und Naturwissenschaften*, 2nd edn., Herder, Freiburg.

34 Overhage, P. (1966) Das Problem der Art. *Stimmen der Zeit*, **177**, 369–378.

35 Overhage, P. (1965) *Die Evolution des Lebendigen. Die Kausalität*, Freiburg, Basel, Herder, Wien.

36 Rheinberger, H.-J. (2001) *Experimentalsysteme und epistemische Dinge. Eine Geschichte der Proteinsynthese im Reagenzglas*, Suhrkamp, Göttingen.

37 Rombach, H. (1965/66) *Substanz, System, Struktur. Die Ontologie des Funktionalismus und der philosophische Hintergrund der modernen Wissenschaft*, 2 Vols, Alber.

38 Daston, L. (2000) Die Kultur der wissenschaftlichen Objektivität? in *Naturwissenschaft, Geisteswissenschaft, Kulturwissenschaft: Einheit–Gegensatz–Komplementarität*, 2nd edn., Göttingen, (ed. O.G. Oexle), pp. 9–40.

39 Hamberger, E. (2004) Transdisciplinarity. A Scientific Essential, in *Signal Transduction and Communication in Cancer Cells* (ed. L.H. Bradlow et. al.), Annals of the New York Academy of Sciences, Bd. 1028, New York, pp. 487–496.

40 Wilson, C. (1995) *The Invisible World. Early Modern Philosophy and the Invention of the Microscope*, Princeton University Press, Princeton.

41 Wehrt, H. (1991) Komplementarität und Geschichtlichkeit, in *Ökologie und Humanökologie* (eds H. Wehrt and R. Heege), Frankfurt, pp. 145–228.

Part II
Methods

4
Mass Spectrometry

4.1
A Practical Guideline to Electrospray Ionization Mass Spectrometry for Proteomics Application

Jon Barbour, Sebastian Wiese, Helmut E. Meyer, and Bettina Warscheid

4.1.1
Introduction

Although traditional ionization methods, such as electron impact ionization, facilitate the mass spectrometric analysis of relatively volatile molecules with low molecular weight (MW), they are generally not applicable to the analysis of large biomolecules. Consequently, it was with the discovery of "soft" ionization techniques – such as matrix-assisted laser desorption/ionization (MALDI) and electrospray (ES) – that this significant limitation was overcome, thereby revolutionizing the field of biological mass spectrometry (MS). Ionization by ES was first described during the 1960s [1], but it was not until some 20 years later that its application to the analysis of biomolecules was implemented. John Fenn is accredited with such realization, and was later awarded the 2002 Nobel Prize in Chemistry for his efforts [2,3]. In his initial investigations, Fenn and his coworkers showed that multi-protonated species of intact proteins are produced when spraying a diluted solution of protein in a high-voltage electrostatic field gradient [2,4]. Since then, ES has been proven as a most effective method by which to generate stable, multiply charged gas-phase ions from a solution of polar and thermally labile biomolecules of low volatility. Moreover, sequence specific information can be readily obtained from ES-derived peptide ions through the application of collisional activation techniques. In providing these capabilities, ES has greatly extended the application of biological MS, and thereby has advanced MS-based proteomics to one of the most powerful and promising tools in the biosciences.

4.1.1.1 Electrospray Ionization

In electrospray, an analyte is diluted in a suitable solvent (e.g., CH_3OH, CH_3CN, or CH_2Cl_2) that is sprayed from a fine capillary needle in the presence of a strong electric

Proteomics Sample Preparation. Edited by Jörg von Hagen
Copyright © 2008 WILEY-VCH Verlag GmbH & Co. KGaA, Weinheim
ISBN: 978-3-527-31796-7

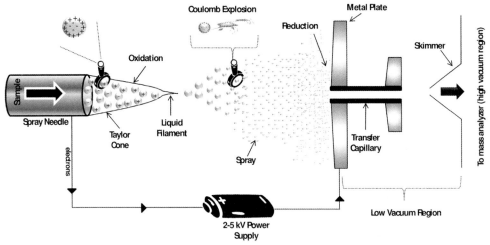

Figure 4.1 Diagrammatic representation of the electrospray ionization (ESI) process. For details, please refer to the text.

field (3–5 kV cm^{-1}). This produces a mist of ionized droplets (∼10 μm in diameter), the surface charge of which depends on whether the needle carries a positive or negative voltage potential. In this process, the interface plate of the MS instrument acts as counter electrode to the spray needle, having a voltage potential of 2 to 6 kV. A schematic representation of the electrospray ionization (ESI) process is illustrated in Figure 4.1. In the positive ion mode, positive ions are enriched at the surface of a droplet, whereas negative ions are driven towards the center. These positive charges repel each another whilst being concomitantly "pulled" by the electric field and, in doing so, exceed the surface tension of the liquid, leading to formation of the characteristic "Taylor cone" [5]. The tip of the Taylor cone extends into a micron-size filament until it reaches the Rayleigh limit, which is determined by the Coulomb forces of the accumulated positive charges and the surface tension of the solution. Evaporation of the volatile solvent results in increased Columbic repulsion between the positive charges which, in due time, exceed the liquid surface tension (i.e., the Raleigh limit), causing the droplet to "explode" into nanometer-sized daughter droplets (Coulomb explosion) [6]. Further solvent evaporation propagates this cascade until any ions still present in the liquid phase become completely desolvated. The nebulization process occurs at atmospheric pressure, and can be supported by a sheath gas. In addition, a nitrogen curtain gas streams from the interface into the region where ionization occurs and hinders neutral molecules from entering the high-vacuum region of the mass spectrometer. Notably, during the entire process, very little residual energy is retained by the analyte, which remains stable in the gas phase. Electrospray is therefore considered an even softer ionization technique as compared to matrix-assisted laser desorption/ionization mass spectrometry (MALDI), causing virtually no fragmentation of analyte ions, and is routinely applied to the analysis of proteins, peptides, lipids, oligonucleotides, and carbohydrates.

One of the main features that distinguishes ES from other ionization techniques is that it is favorable to producing multiply charged species, the extent of which increases generally with molecular weight. For peptides and proteins which become readily protonated, this multi-charging phenomenon is mainly dependent on the availability of basic amino acid sites [2]. A major benefit of multiple charging is effectively to extend the mass range of analysis in proportion to the extent of the multiplicity of such charging. This enables the use of mass analyzers with only a modest m/z-range, such as quadrupole or ion trap instruments. In case the mass analyzer provides adequate mass resolution, the charge state of a compound can be directly derived from the difference of m/z-values of its isotopes in the mass spectrum.

Despite the continuous advancements in the design of the ESI source, its general configuration and basic working principle have remained unchanged. ES-derived ions are efficiently conveyed to a high-vacuum region of the mass spectrometer via a low-pressure ion desolvation and transport system. Within the low-pressure region there exists typically a transfer capillary that facilitates the separation of analyte ions from any remaining solvent – a process termed "declustering". In conventional ESI, a relatively high flow rate, in the range of microliters per minute, is used. Such a high rate of sample consumption proved unpopular in proteomics and prompted the development of ESI at low nanoliter per minute flow rates, namely nano-ESI [7,8].

4.1.1.2 Nano-Electrospray Ionization

In nano-ESI, a small volume (1–4 µL) of the sample dissolved in a suitable volatile solvent (concentration of 1–10 pmol µL^{-1}) is sprayed from a metal-coated glass capillary with an inner diameter of about 1 µm. The flow rate of solute and solvent using this procedure is very low (30–1000 nL min^{-1}). A sample volume of 1 µL dissolved in MeOH/H$_2$O (1 : 1, v/v) may be analyzed for up to 20 min. Nano-ESI generates much smaller charged droplets (nanometer range), leading to increased ionization efficiency and greater sensitivity [9,10] as compared to conventional ES. Sensitivity can be further enhanced through direct coupling of the nano-ESI source with a high-performance liquid chromatography (HPLC) system operated at a flow rate of approximately 200 nL min^{-1} [11].

4.1.1.3 ESI-MS Instrumentation

Due to the continuous process of ion generation using ES, the ESI source is typically coupled to quadrupole or ion trapping instruments [12,13]. Although these mass analyzers exhibit a limited mass range, the soft ionization along with the multi-charging characteristics of ES enables ready measurement of the m/z-ratios of intact biomolecules. However, in order to obtain information on the primary structure of peptides and proteins, more sophisticated experiments within the mass spectrometric device must be performed. Sequence-specific data on these compounds can be generated with mass spectrometers that allow for ion isolation, gas-phase fragmentation, and mass analysis of the generated fragment ions.

The fragmentation of electrosprayed peptides is typically performed at low-energy collision-induced dissociation (CID) conditions, thereby inducing effective cleavage of the amide bonds between the amino acid residues (for a detailed review on peptide sequencing, see Refs. [14–16]). The overall process is generally referred to as tandem MS (MS/MS) analysis. In tandem-in-space experiments, the fragmentation of peptide ions by CID, for example, requires the combined use of two mass analyzers with the same or two different ion-separation principles. For example, when using a quadrupole time-of-flight (QTOF) instrument, ions of a particular m/z-ratio are selected in the first quadrupole (Q1), and then fragmented by collisional activation with nitrogen or argon atoms in the second quadrupole (Q2), operated in the "RF-only" mode (Q2 is referred to as the collision cell). The produced fragment ions are then eventually analyzed by the reflector time-of-flight (TOF) mass analyzer. By using this hybrid instrument, peptides at the low femtomole level can be measured with a mass resolution of about 10 000 FWHM (full width at half maximum), and with mass accuracies of about 10 ppm in both the MS and MS/MS modes.

In contrast, ion trap instruments allow for tandem-in-time experiments; ion isolation, fragmentation, and mass analysis take place in a single-stage device, but at different times [17]. Since the ion trap is a benchtop-sized, rugged, and low-cost mass analyzer which allows for highly sensitive MS and most efficient MS/MS analyses of electrosprayed ions at high scan speed, it is one of the most common instrumentations used in proteomics. Moreover, recent advancements in ion trap technology have resulted in the capability to: (i) perform mass analysis with extended m/z-range, improved mass resolution, higher scan speed, and increased sensitivity; (ii) acquire low-mass ions in MS/MS scans; and (iii) alternately conduct CID and electron transfer dissociation (ETD) experiments for the most informative sequencing of electrospray-derived peptide ions (for a recent review on ion trap MS, see Ref. [18]).

Moreover, new trapping instruments such as linear ion traps – either as stand-alone devices [19] or in combination with Fourier-transform ion cyclotron resonance (FTICR) MS [20], as well as the new Orbitrap mass analyzer [21–24] – have already taken a leading role in current proteomics, For more detailed information on these advancements in ESI-MS instrumentation, see Refs. [25–27].

4.1.1.4 Protein Identification Strategies

In proteomics, a plethora of strategies can be followed for the identification of proteins in biological samples by ESI-MS. In order to reduce the complexity of protein samples prior to MS analysis, methods such as subcellular fractionation by ultracentrifugation, protein affinity purification, phase partitioning, polyacrylamide gel electrophoresis (PAGE), as well as liquid chromatography (LC), are widely used. Following fractionation and/or separation, the proteins are usually converted to a set of peptides by proteolytic digestion, and the generated peptide mixtures are then analyzed using nano-HPLC/ESI-MS/MS. An alternative approach, known as the "shotgun" method, is initially to digest a protein mixture (e.g., with trypsin) and subsequently to subject the highly complex proteolytic peptide mixture to two-dimensional (2-D) LC combining strong cation-exchange (SCX) and reversed-phase (RP) chromatography coupled online to ESI-MS/MS [28,29]. Although in all these approaches the sample complexity is significantly increased by protein digestion, the main reasons for subjecting peptides rather than intact proteins to MS analysis are based on the following facts: (i) MS sensitivity and mass accuracy are best in the nominal mass range below 2500 Da; (ii) informative sequence information is most readily obtained from peptides of up to 20 amino acid residues using low-energy CID; and (iii) peptides are more soluble than proteins and therefore easier to handle, to separate chromatographically, and to electrospray.

Following peptide analyses, tailored algorithms such as SEQUEST™ [30] are used for large-scale MS/MS data interpretation by searching sequence databases. Through the application of adequate quality criteria (e.g., scores, thresholds, use of decoy methods as detailed in Section 4.1.3.2), the use of such search strategies enables both correct peptide identification and reliable protein assembly. However, in order to fully characterize an entire protein and/or to detect post-translational modifications (PTMs) in proteins, more sophisticated MS-based strategies must be followed, though these are beyond the scope of this chapter (for further information, see Chapters 10.1 and 10.2).

4.1.2
Sample Preparation

Since samples are sprayed from solution at atmospheric pressure, ESI represents a suitable technique by which LC methods can be directly coupled with MS. Online LC/ESI-MS offers a robust and automated technique to "clean", to separate effectively, and to analyze protein/peptide samples within a single experiment. Nevertheless, the strategies that are classically important in the preparation of "crude" biological samples remain fundamentally important when analyzing proteins or peptides by (LC/)ESI-MS. Sample preparation workflows used in MS-based proteomics typically include multiple steps such as sample desalting, concentration, subfractionation, and further separation by gel electrophoresis and/or chromatography. The workflow of choice is dependent on both the sample characteristics (e.g., cell culture, tissue, available amount, pH and temperature stability, hydrophobicity, lipid content) and the biological question to be addressed (e.g., analysis of membrane and/or soluble proteins, protein interactions, post-translationally modified proteins). In order to maintain the solubility and integrity of the protein samples as far as possible, an array of buffers, reductants, detergents, and inhibitors is typically used from the early steps of preparation. As such compounds are invariably incompatible with (LC/)ESI-MS, the initial sample must be adequately purified before analysis. An overview of the most frequently used purification steps is provided in a step-like fashion in the following paragraphs.

4.1.2.1 Purification

Sample purification may be separated into two steps:

- Clarification to remove constituents such as particulates, lipids, salts, and polymers which usually interfere with downstream processing and analysis.
- Enrichment, fractionation, and/or separation to reduce the complexity of samples and/or effectively to increase the amount of target proteins.

Although the latter step is optional, it is often indispensable in MS-based proteomics.

Clarification The starting material from biological samples is typically a crude cell lysate which, for further protein analysis, needs to be "cleaned". Details of the most common methods for the clarification and/or concentration of proteins from cell lysates, which are often used in combination, are listed in Table 4.1.

Enrichment, Fractionation, and Separation Following the clarification of cell lysates, it is recommended that additional enrichment and/or fractionation steps be performed. These additional steps are often essential to be able adequately to address any biological questions by MS-based proteomic methodology. One major challenge, for instance, is the ability to identify – as well as to quantify – low-abundant proteins such as membrane receptors which are promising drug targets [31]. To further exemplify, the dynamic range in human plasma extends over 10 orders of magnitude [32], and consequently the effective depletion of highly abundant constituents such as albumin and immunoglobulin is necessary to monitor proteins of considerably lower abundance by MS methodology [33]. Furthermore, for the most sensitive ESI-MS analysis of proteins and peptides, salts and electrolytes should be carefully removed from the samples as these cause ion suppression, even at very low concentrations [34].

Sample complexity may be reduced either at the level of the protein, or at the level of the peptide, though a combination of different methods is often used. For example, protein samples are initially fractionated by ion-exchange chromatography

Table 4.1 Example procedures for clarifying and concentrating proteins.

Method	Principle of method	Example
Precipitation	An agent is added to the lysate which alters the solvation potential of proteins and lowers the solubility of the protein/solute.	Addition of salts (e.g., ammonium sulfate), organic solvents (e.g., EtOH, MeOH, chloroform, acetone) or non-ionic hydrophilic polymers (e.g., dextran, polyethylene glycol).
Centrifugation	Removal of particulates and debris according to mass.	Low-spin centrifugation, typically at $1000 \times g$ for 15 min.
Filtration	Proteins are separated according to size and shape.	Dialysis, ultrafiltration (e.g., spinfilters), gel filtration.

(or by one-dimensional (1-D) PAGE) and, after proteolytic digestion, the resulting peptide samples are further separated and eventually analyzed using nano-HPLC/ESI-MS/MS.

Complexity Reduction at the Level of the Protein Depending on the individual technique applied, fractionation at the level of the protein allows for maintaining information concerning post-translational modifications, polymorphisms, functional groups as well as the cellular location and interactions of proteins. An overview of the common techniques used to fractionate, to enrich and/or to separate samples at the protein level for subsequent MS analysis is provided in Table 4.2. For further details on sample preparation strategies at the protein level, see Ref. [35].

Subcellular fractionation by ultracentrifugation, combined with MS-based analysis, enables one specifically to enrich or diminish the proteins associated with distinct cellular compartments (e.g., nucleus, mitochondria, peroxisomes, Golgi vesicles) and, at the same time, to obtain valuable information concerning the cellular location of proteins. (For a detailed overview of this approach, see

Table 4.2 Example procedures for enrichment and fractionation of proteins before ESI-MS analysis.

Method	Principle of method	Example
Ultracentrifugation	Separation of proteins, protein complexes, and cellular structures according to mass and/or density.	Continuous/discontinuous density gradient ultracentrifugation using sucrose, glycerol, or Percoll™.
Tagging	Proteins of interest are equipped with an affinity tag ("bait") used to, e.g., identify interacting proteins.	Tandem Affinity Purification (TAP), His-tags, GST-tags.
Phase-partitioning	Proteins are separated according to their charge and hydrophobicity.	Aqueous biphasic system (e.g., PEG-dextran or Triton-PEG/Dextran), triphasic system comprised of organic solvent and salt (e.g., ammonium sulfate-butanol). May be applied with additional affinity step, e.g., lectin-coupled dextran to further enrich glycosylated proteins.
Chromatography	Proteins are resolved according to physico-chemical properties (pI, hydrophobicity, size, charge) or via specific binding sites (carbohydrate binding, metal binding, antibody epitopes).	Ion-exchange, reversed-phase, size-exclusion, affinity chromatography (e.g., immunoaffinity, IMAC-immobilized metal affinity chromatography, and lectin-affinity).
1-D/2-D PAGE	Proteins are resolved according to isoelectric point and/or molecular weight.	2-D difference gel electrophoresis (DIGE), Blue native gel electrophoresis, 16-BAC/SDS–PAGE, SDS–PAGE.

Chapter 11.) In case of a need to identify interacting proteins of a target protein by ESI-MS, a variety of tags [e.g., glutathione (GST), His, Protein A] can be used to "fish" protein complexes from the cell lysates. Notably, phase partitioning has recently regained popularity as an effective tool for enriching integral plasma membrane proteins before LC/ESI-MS analysis [36]. Whilst affinity and size-exclusion chromatography methods are widely used to purify target proteins or intact protein complexes, RP and ion-exchange chromatography are usually employed effectively to reduce the complexity of cell lysates by fractionation. In a typical proteomics workflow, the collected fractions are then subjected to in-solution tryptic digestion, and the resulting peptides are eventually analyzed using nano-RPLC/ESI-MS/MS. Alternatively, chromatography can be replaced by gel electrophoresis (e.g., SDS–PAGE) for protein separation prior to HPLC/ESI-MS. In order to separate membrane proteins by gel electrophoresis, it is preferable to use methods such as SDS–PAGE or 16-benzyldimethyl-*n*-hexadecyl-ammonium chloride (BAC)/SDS-PAGE [37,38]. In contrast, 2-D gel electrophoresis, using isoelectric focusing in the first dimension and SDS–PAGE in the second dimension, represents a powerful method for the separation of soluble proteins. Due to the high resolving power of 2-D PAGE, the respective samples can be readily analyzed by MALDI-MS. However, for the reliable identification of proteins of very low concentration in gel spots, it may be advantageous to subject the proteolytic digests to nano-HPLC/ESI-MS/MS.

Complexity Reduction at the Level of the Peptide In this approach, protein samples or whole-cell lysates are digested *in-solution*; the resultant peptide mixture is then fractionated by capillary chromatography and analyzed either on-line or off-line by LC/MS. In addition, sample complexity may be further reduced through the addition of an orthogonal dimension. SCX chromatography of peptides is often performed before RPLC in order to resolve protonated peptides according to charge in the first dimension. Peptides displaced from the SCX phase by a gradient containing an increasing salt concentration (e.g., ammonium acetate) are subsequently separated by RPLC in the second dimension. A variety of SCX/RPLC separation systems have been developed; for further information, the reader is referred to a comprehensive review by Fournier *et al.* [39]. Strong anion-exchange (SAX) chromatography may also be coupled to RPLC for the separation of complex peptide mixtures [40], and recently an anion and cation mixed-bed ion exchange approach was reported [41]. The latter approach, which is based on MudPIT, is reported to significantly improve peptide/protein identification rates [41]. RPLC is also commonly coupled to an off-line affinity purification step. This facilitates the enrichment or depletion of a "population of peptides" which share comparable physico-chemical characteristics, such as specific post-translational modifications. Phosphopeptide enrichment may be performed using metal ion-affinity chromatography (IMAC) [42,43] or titanium dioxide chromatography [44] before RPLC/ESI-MS/MS. In addition, glycopeptide enrichment may be achieved using lectin affinity chromatography for N-glycosylated [45] or O-glycosylated [46] peptides before separation by RPLC.

4.1.2.2 Protein Digestion

A wide variety of endoproteases can be used for protein digestion, although most commonly trypsin is employed as it generates peptides which are highly amenable to MS(/MS) analysis. Trypsin preferentially cleaves peptide bonds that are C-terminal to the basic amino acid residues arginine and lysine, and thus generates peptides with an average size of 800 to 2000 Da. Peptides of such size can be analyzed with high sensitivity, as well as being readily dissociated by collisional activation in order to obtain sequence informative data in ESI-MS/MS experiments. Moreover, due to the high basicity of lysine and, in particular, of arginine, tryptic peptides are efficiently protonated when electrosprayed. Depending on the preceding workflow, the enzymatic digestion of proteins is performed either in-gel or in-solution, as briefly discussed below.

In-Gel Digestion Following the separation of samples by 1-D or 2-D gel electrophoresis, proteins are fixed (50% MeOH/2% phosphoric acid) and visualized using an MS-compatible stain, usually Coomassie Blue [47], or silver employing a glutaraldehyde-free protocol [48,49]. Visualized proteins are excised from the gel and the respective gel bands or spots are washed and dehydrated before being trypsinized. In order to ensure efficient hydrolysis of those proteins embedded in the gel matrix, a relatively high enzyme concentration ($\sim$12.5 ng μL^{-1}) is generally used. The generated proteolytic peptides can subsequently be released from the gel matrix, by for example 50% acetonitrile (CAN)/5% formic acid (FA) being used as an extraction buffer, in combination with sonication. The efficiency of proteolytic in-gel digestion, combined with peptide extraction, depends on a variety of factors, including: (i) the physico-chemical properties of the proteins and resulting peptides (e.g., degree of hydrophobicity, size, amino acid sequence); (ii) the composition, size, and thickness of the gel pieces; (iii) the composition of the extraction buffer (e.g., acetonitrile (ACN) concentration); (iv) the type of enzyme and its specific activity; (v) the general reaction conditions (e.g., temperature, time, ratio of enzyme to substrate); (vi) the type of protein stain (e.g., Coomassie or silver); and (vii) the degree of care taken with respect to sample handling (e.g., adsorptive losses due to the inappropriate use of plastic ware).

One significant advantage of in-gel protein digestion is that any contaminants (e.g., detergents, salts) are already removed during electrophoresis, and that the generated peptide samples can be readily subjected to (LC/)ESI-MS analysis. However, the effectiveness of this procedure can be limited due to poor accessibility of the protease to proteins and/or inefficient release of peptides from the gel matrix, as well as inadequate storage of gels. As an excellent alternative to gel electrophoresis combined with in-gel digestion, proteins can be directly digested in-solution, which is usually followed by 1-D or 2-D LC to effectively separate the resulting peptide mixture before ESI-MS.

In-Solution Digestion For in-solution digestion, a compromise between protein solubilization and enzyme activity must be reached. Detergents should be generally avoided but, depending on the physico-chemical characteristics of the protein lysate, organic solvents and/or chaotropes may be used. For the analysis of proteins which are difficult to solubilize or denature (e.g., membrane proteins), 8 M urea may be used in conjunction with the resilient protease Lys-C; subsequently, trypsinization is performed at a reduced concentration of chaotrope (2 M urea) [28]. Following acidification using FA it may be possible directly to perform LC/ESI-MS, although it is generally recommended that an additional desalting step be performed, for which commercially available solid-phase extraction tips are routinely used [50,51]. As a promising alternative to double proteolytic digestion in urea, organic solvents can be used to solubilize and effectively digest hydrophobic proteins in-solution. Blonder et al. employed 60% (v/v) buffered MeOH (60% CH_3OH/40% 50 mM NH_4HCO_3, pH 7.9) in combination with trypsin (enzyme/protein ratio 1:20) to effectively digest membrane proteins from mammalian cells for subsequent 2-D LC/ESI-MS/MS analysis [52]. In order to enhance proteolytic digestion, enzymes that are chemically immobilized or physically adsorbed to a stationary phase can be used. When the endoproteases are immobilized on HPLC cartridges, the proteins can be digested on-line and at a faster, automated rate [53,54].

4.1.3
ESI-MS Analysis

4.1.3.1 Protein Analysis by ESI-MS

The analysis of intact proteins by ESI-MS represents an important method of characterizing proteins. Indeed, the method is employed in many forms in modern laboratories, from controlling the success of protein purification, to the study of post-translational modifications, and the investigation of non-covalent protein complexes. Although the coverage of all aspects of protein analysis by ESI-MS is beyond the scope of this chapter, attention will be focused here on the basic observations that can be made when analyzing electrosprayed proteins. Information relating to the more advanced investigations on intact proteins using LC/ESI-MS combined with "top-down" sequencing as well as ESI-MS analysis of intact protein complexes, is available elsewhere [55–60].

When electrosprayed, proteins become multiply protonated [2], whereupon they are generally not observed as single peaks with a certain charge state but rather appear as a series of peaks of different charge states in the mass spectrum, with each peak representing a species of ions. The multiple charges are statistically distributed upon the available basic amino acid sites of the protein, the number of which – as well as the spray conditions applied – mainly determine the maximum charge state of the electrosprayed protein. Proteins are typically sprayed from an acidified aqueous solution with a high concentration of organic solvent (e.g., ACN, MeOH). The high organic solvent content (e.g., 50% MeOH) and the addition of acid (e.g., 1–2% acetic acid, FA, or propionic acid) generally facilitates the transfer of proteins into the gas phase and protonation, resulting in higher ion currents. In any case, sample solutions should be kept free of buffers and detergents as these compounds compete for protons and may completely suppress the ionization of the proteins. It is also important to note that the presence of salts (e.g., sodium chloride) in the sample leads to extensive adduct formation, which further increases the complexity and thus exacerbates the interpretation of mass spectra derived from electrosprayed proteins.

Calculation of the charge states of a protein – and thus the protein's intact mass – is usually based on the distance between its different multiply charged peaks. This approach is chosen since mass differences between isotopic peaks at high charge states are usually too small to be well resolved by MS (exceptions are high-resolution mass analyzers, such as FTICR instruments). As an application example, the ESI-QTOF-MS spectrum of the GTP-binding nuclear protein Ran1B is shown in Figure 4.2a. In order to calculate the intact mass of Ran1B, two differently charged peaks, $m_1 = 1105.20$ amu and $m_2 = 935.27$ amu, are chosen. The peak m_2 exhibits four additional charges compared with m_1, as indicated by corresponding peak series in the spectrum. Accordingly, two equations can be set up as follows: $m_1 = \frac{M+z}{z}$ and $m_2 = \frac{M+(z+4)}{z+4}$, where M is the intact mass of the protein and z the charge state of the peak, labeled m_1. Next, both equations are solved for M and set equal. Solving the resulting formula for z results in the equation $z = \frac{4^*(m_2-1)}{m_1-m_2}$ by which the charge state z of the peak m_1 is calculated. The intact mass M of the protein is eventually determined by $M = m_1{}^*z - z$. In the example provided here, a charge state of + 22 was calculated for m_1 and, accordingly, an intact mass of 24292.4 Da was measured for Rna1B.

Nowadays, most vendors of ESI-MS instruments provide the user with suitable algorithms that allow for automated charge deconvolution, which results in enhanced ion statistics by combining the peak intensities for all detected charge states. The deconvoluted spectra usually exhibit both higher mass resolution and better signal-to-noise (S/N) ratios, thereby facilitating the determination of intact protein masses with greater accuracy. Moreover, such algorithms are basically essential when evaluating ESI-MS spectra of intact protein mixtures and/or when peak distributions are further complicated by salt adducts. The deconvoluted ESI-MS spectrum of Rna1B analyzed on a QTOF instrument, with an average mass of 24292.2 Da ($z = 0$) is shown in Figure 4.2b. In this example, the intact mass of Rna1B was determined with an error of 0.3 Da. Additional, low-abundant peaks observed at higher m/z-ratios in the deconvoluted spectrum derive from the formation of Na^+-adducts (+22 amu), as well as minor impurities. At this point, it should be noted that the use of high-resolution ESI-MS instruments (e.g., FTICR) enables ready measurement of the monoisotopic masses of proteins, thereby proving a valuable means for the detection of, for example, protein modifications.

4.1.3.2 Peptide Analysis by Nano-HPLC/ESI-MS

Peptide-based proteomics is currently the most mature and widely used strategic track for large-scale protein identification and characterization by ESI-MS/MS. The major benefits of the analysis of peptides compared with proteins are the

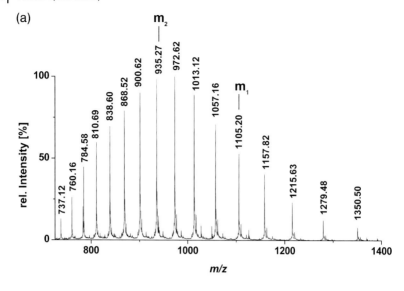
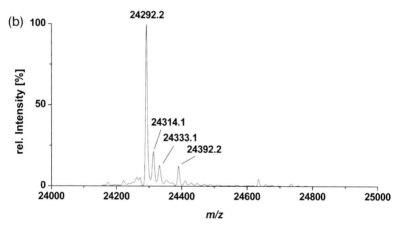

Figure 4.2 (a) Nano-ESI-MS spectrum of the protein Rna1B acquired on a quadrupole time-of-flight instrument. Peaks labeled m_1 and m_2 were used for the manual calculation of the protein's intact mass. (b) The corresponding deconvoluted ESI-MS spectrum of Rna1B, showing an average mass of 24292.2 Da ($z=0$).

ability to: (1) perform mass analysis with higher sensitivity and mass accuracy; (2) readily obtain sequence-specific information in MS/MS experiments; and (3) easily couple LC-based separation techniques and ESI-MS. For the analysis of complex peptide mixtures, nano-HPLC/ESI-MS/MS is arguably the most common, effective, and elegant method in current proteomics. Here, those aspects of nano-HPLC are briefly discussed which are of particular relevance to MS-based

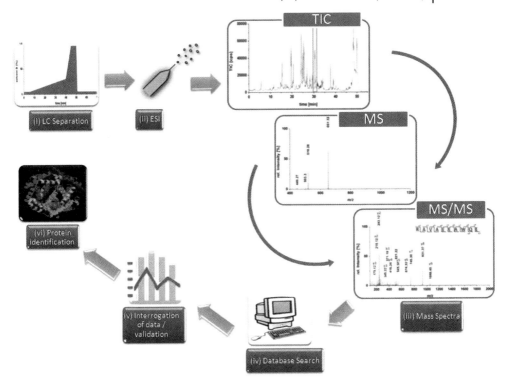

Figure 4.3 Protein identification by nano-HPLC/ESI-MS/MS combined with bioinformatics. Peptides separated by nano-HPLC (i) are directly ionized by nano-ESI (ii) and visualized as a total ion chromatogram (TIC), in which peptide ion intensity is plotted as a function of peptide retention time (iii). Precursor ions (MS) are automatically selected for MS/MS analysis in a data-dependent manner (iii). Acquired MS/MS datasets are used to query sequence databases using tailored search algorithms, e.g., SEQUEST™ or Mascot (iv). Database search results are statistically evaluated in order to estimate the rate of false-positive hits (v), leading to a qualified list of identified proteins (vi).

peptide analysis (for further details on HPLC, see Chapter 7.1). The entire workflow, from peptide separation by HPLC to the acquisition of spectra in a data-dependent manner, culminating in protein identification, is illustrated diagrammatically in Figure 4.3.

Nano-HPLC Many factors influence the performance of nano-HPLC systems, including practicable decisions concerning the specification of the pump, injector, filters, tubing, and detector. Here, those aspects which are of particular relevance to nano-HPLC/ESI-MS are briefly discussed.

- The Stationary Phase and Column
  The application of RP chromatography to peptide separation was first reported in 1976 [61]. In RPLC, peptides bind to a hydrophobic stationary phase in the presence of a polar, hydrophilic, mobile phase. Thus, the analyte is separated

according to its hydrophobicity, with more hydrophilic species eluting earlier in the LC run. The most commonly used hydrophobic adsorbents with different lengths of n-alkylated ligands are: n-octadecyl (C18), n-octyl (C8), and n-butyl (C4). C18 columns are typically used to resolve peptides and hydrophilic samples, whereas C8 and C4 materials are preferentially employed for the analysis of rather hydrophobic samples and/or proteins. Spray capillaries can be directly packed with RP material, and the eluting peptides are directly ionized into the mass spectrometer. It is more common, however, to use commercially available columns, of which there are a plethora to choose from. The use of an additional "trap" or pre-column, which precedes the separation column, is also recommended [62]. Trap columns are used in column-switching mode and allow fast loading, concentration, and desalting of the sample. This is due to the shorter length and larger inner diameter (i.d.) of the trap column (e.g., 300 μm i.d. × 5 mm) relative to the separation column. Such pre-columns assist the reproducible analysis of complex peptide mixtures, although it is important to note that they should demonstrate high loading capacities and low void volumes [62,63].

In general, a reduction of the stationary phase particle size (i.d. <2 μm) enhances column resolution [64], although this emergent technology (ultra-performance chromatography) is not yet widely adopted by proteomic laboratories [65]. Currently, most investigators use 75 to 100 μm i.d. RP columns for the separation of peptides for subsequent nano-ESI-MS/MS analysis. In recent years, the use of multidimensional-nano-LC/MS technologies has emerged, of which MudPIT is the most renowned [28]. In classical MudPIT, a fused capillary column is packed "back-to-back" with SCX and RP materials. MudPIT and other multidimensional strategies, when combined with nano-ESI-MS/MS, enable the separation and identification of complex peptide mixtures. For further details, the reader is referred to Ref. [39].

- The Mobile Phase: Solvents
  The composition of the mobile phase is of fundamental importance to LC/ESI-MS. It must facilitate adequate separation of analytes on the stationary phase whilst satisfying the conditions required for efficient ionization. In order to achieve a stable ES, the applied voltage, conductivity, flow rate, and liquid surface tension must be properly balanced. In the positive ion mode, this may be achieved using conductive solutions comprising a moderately polar organic solvent (e.g., MeOH or ACN) in water. Peptide protonation is promoted by adding a weak acid to HPLC buffers. In regard to this, FA has evolved as the most commonly used acid and is typically added at concentrations in the range of 0.01% to 0.1%. Strong ion-pairing agents such as trifluoroacetic acid (TFA) should be avoided at all costs, as these induce detrimental ion suppression during ESI [66]. However, a concentration of 0.1% TFA may be used as a mobile phase ion-pairing agent during loading of the pre-column [62], as this assists efficient "trapping" of peptides before "flushing" onto the analytical column. For example, peptide samples may be loaded onto the pre-column using 0.1% TFA followed by RPLC separation

applying a gradient from 5 to 30% ACN in 0.1% FA at a flow rate of ~200 nL min^{-1} for 60 min.

- Maintaining a Good S/N Ratio: The Danger of Contaminants
  In spite of the fact that nano-ESI tolerates impurities to a greater degree than high-flow ES [9], due care must nevertheless be taken to avoid even trace amounts of contaminants. Detergents form adducts with peptides and proteins, and generally lead to considerable signal suppression. In most cases, the detergent concentration should be less than 0.01%, although a higher tolerance to zwitterionic detergents such as CHAPS has been reported [67]. Even though low concentrations of salt can facilitate ionization through adduct formation, higher concentrations cause ion suppression, masking the analyte signal. Sodium ions, for example, often arise from biological samples, buffers, and even from glassware and analytical-grade solvents. If phosphate or sulfate salts are present in the buffers, they can likely precipitate within the ion transfer system of the MS instrument, and this results in a significant loss in ion transmission efficacy and, thus, sensitivity. Furthermore, "HPLC-grade" solvents may not always be compatible with nano-HPLC/ESI-MS and must be checked for purity. In addition, certain plastic ware and even some stationary phase materials may release synthetic polymers [68]. In order to overcome such problems, the predominant use of clean glassware is recommended; however, if the use of HPLC-compatible plastic ware (e.g., siliconized or low-binding protein tubes) is unavoidable, these also offer excellent sample recovery.

Data-Dependent Acquisition LC/ESI-MS/MS represents a powerful tool for the effective separation and analysis of proteolytic digests, using an automated approach. When the peptides are eluted from the chromatographic column, they are transferred directly into the nano-ES ion source via a fused silica capillary to become ionized, and are then analyzed by MS/MS. Low-energy CID is commonly used for peptide fragmentation, although ETD-capable instruments are now available commercially, providing a most promising means for the identification of phosphopeptides [69]. In order to most efficiently "sample" the eluting peptides, scripts for automated precursor ion selection and subsequent fragmentation can be created by using the control software of the MS instrument. In such experiments, a survey MS scan from m/z 400 to 1200, for example, is acquired, followed by the sequential isolation and fragmentation of typically the three most intense peptide peaks observed in the respective MS spectrum (see Figure 4.3). Due to multiple charging of peptides derived from ES, MS/MS scans are usually acquired in the range from m/z 100 to 2000. This series of experiments is then performed in an iterative process over the whole course of the gradient.

One important characteristic of LC/MS experiments is the "duty cycle", which refers to the time over which the survey scan followed by the respective MS/MS scans takes place. QTOF instruments typically provide cycle times of about 1 s for each MS/MS scan, whereas linear ion trap instruments can allow for the acquisition of MS/MS spectra with a scan cycle time of 0.3 s. Consequently, when acquiring the same number of MS/MS spectra, the duty cycle of linear ion traps is significantly faster compared with QTOF instruments, and this results in a significantly higher peptide sampling rate over the entire course of an LC gradient. However, ion traps usually exhibit unit mass resolution (~2000), and thus provide limited mass accuracy, which can only be increased by reducing the scan speed. In LC/ESI-MS/MS runs, adequate mass resolution is advantageous as it enables one directly to identify the charge state of the peptides. This facilitates the automated selection of only multiply charged peptides that provide most sequence-informative MS/MS spectra using low-energy CID. Moreover, it generally provides the ability to adjust collision energies according to the m/z-ratio of peptides (referred to as the "rolling collision energy"). It is important to note that, as peptides exhibit peak widths of about 0.5 min during nano-HPLC, there is a high probability that they are subjected to MS/MS on multiple occasions. Hence, in order to reduce redundancy in ESI-MS/MS data while simultaneously increasing the probability of dissociating peptides of lower abundance, exclusion limits of about 0.5 to 1.2 min should be applied to previous precursor ions.

When analyzing peptide samples of high complexity, it is generally beneficial to stretch the length of the LC gradient, for example, from 1 to 3 h. This increases the effective dynamic range of MS/MS analysis. In addition, the application of a fast MS/MS scan cycle time is recommended to maximize the peptide sampling rate. The use of fast scanning instruments, such as linear ion traps, also provides the possibility of increasing the number of MS/MS scans following each survey scan; the acquisition of 10 MS/MS spectra was recently reported [70]. However, less-complex samples composed of low-abundant constituents (as is often the case for purified protein complexes) requires the performance of LC/ESI-MS/MS of the highest sensitivity. In such cases, scan cycle times are generally prolonged to improve ion statistics and thus sensitivity in MS(/MS) analyses. In order to circumvent the overfilling of ion trap devices, the respective survey and fragmentation spectra are typically the sum of consecutively acquired scans (e.g., seven and four individual microscans for MS and MS/MS spectra, respectively). This results in spectra exhibiting improved S/N ratios; however, "peptide undersampling" may occur due to the slower duty cycle applied in such LC/ESI-MS/MS experiments.

Besides LC/MS being used for protein identification, the acquired peptide MS/MS datasets can also be exploited to provide semi-quantitative information on proteins. At this point, it should be noted that most accurate relative or absolute quantification of proteins by MS is only achievable through the use of adequate internal standards. (For details on MS-based protein quantification, see section 4.4.) Nevertheless, a simple approach which is referred to as "peptide counting" may be used in order to obtain semi-quantitative information on proteins in LC/ESI-MS/MS analyses [71]. This method is based on the suggested correlation between the number of peptides (i.e., MS/MS spectra) sampled for a distinct protein and its relative abundance in the respective sample; the proteolytic digestion of highly abundant proteins results in highly abundant peptide species which are most likely to be selected for MS/MS. Consequently, the counting of MS/MS spectra allocated to a protein can be used as a quantitative measure of the relative protein abundance. In an alternative approach, termed "exponentially modified protein abundance index" (emPAI) [72,73], the number of peptides sampled in LC/ESI-MS/MS runs is normalized to the number of all possibly observable peptides of a protein. The emPAI method is implemented into the Mascot search algorithm (http://www.matrixscience.co.uk).

Interpretation of MS/MS Datasets Proteomic informatics is a key tool for the advanced analysis of large datasets acquired in LC/ESI-MS/MS experiments (for details, see Chapter 13). The principal approach to the interpretation of MS/MS datasets is to perform automated database searches using tailored algorithms such as SEQUEST™ [30], Phenyx [74], or Mascot [75]. The basic concepts of such algorithms are reviewed in Ref. [76]. During the course of database searches, measured and predicted peptide fragmentation spectra are matched, resulting in a list of predicted peptide sequences. In order to assess the quality of the retrieved peptide identifications, scores are calculated based on the likelihood that a match is a random hit. The use of adequate search parameter settings is a crucial aspect to successfully and reliably identify peptides. If possible, information should be included regarding: (i) enzyme (ii) specificity and number of missed cleavages; (iii) modifications (e.g., oxidation of methionine); (iv) mass tolerances for MS and MS/MS data according to the instrument used; and (v) which fragment ion series to match (i.e., y-, b-, and a-ion series for low-energy CID spectra). The most well-annotated – and thus preferential to use – databases are the International Protein Index (IPI, www.ebi.ac.uk), the SwissProt (www.expasy.org/sprot/), and the National Center for Biotechnology Information (NCBI, www.ncbi.nlm.nih.gov/) database. It is important to note that, depending on the search algorithm used, different minimum scores for peptide identifications apply. When employing Mascot, a cut-off score of about 15 to 23 is commonly used to reliably identify peptides based on database searches.

Subsequent to peptide identification, the retrieved data are filtered and assembled to proteins which are ranked according to their overall identification score. However, the discrimination between true constituents of the sample and false-positive hits remains a challenge. This is particular true for "shot-gun" proteomics, which often provide data for several hundreds of protein identifications in a single experiment. Applying rather stringent criteria, only those protein hits are reported to which at least two distinct peptides were reliably allocated. In addition, the presence of a protein isoform must be confirmed by the identification of at least one unique peptide.

During recent years, significant effort has been made to allow for most accurate and "user-independent" evaluation of large MS/MS datasets for protein identification. As a result, the so-called decoy-database concept was established [77–79], which facilitates the calculation of false-discovery rates (FDRs). To this end, uninterpreted MS/MS spectra are searched against a combined version of the original database and a duplicate of the target database composed of randomized (reversed or shuffled) protein sequences. By using this approach, the percentage of false-positive identifications can be estimated based on the number of matches derived from the randomized part of the combined target/decoy-database. Although different equations are currently used for calculating FDRs [79–81], there is a consensus of only reporting protein identifications with a confidence level of at least 95%.

4.1.4
Application Example of ESI-MS in Proteomics

Currently, there is no MS-based proteomics strategy that delivers total proteome coverage, whether it is at the multicellular, cellular, or even subcellular level. In order to obtain high proteome coverage, complementary approaches should be followed comprising, for example, density gradient centrifugation for subcellular fractionation, followed by (i) SDS–PAGE, in-gel tryptic digestion and nano-HPLC/ESI-MS/MS; as well as (ii) in-solution tryptic digestion and 1-D(/2-D)-LC/ESI-MS/MS.

Proteomics Analysis of Mammalian Peroxisomes
Peroxisomes are ubiquitous, single, membrane-bound organelles that are present in virtually all eukaryotic cells. They harbor approximately 50 different enzymes which enable them to execute a vast array of metabolic functions [82]. Failure in the biogenesis of peroxisomes, or deficiencies in the function of single peroxisomal proteins, ultimately leads to serious, often lethal diseases such as Zellweger syndrome or X-linked adenoleukodystrophy in humans [82,84]. Although, to date more than 15 inherited peroxisomal diseases are known – underscoring the metabolic importance of this small organelle – the present picture of mammalian peroxisomes remains limited. In order to gain deeper insight into the biochemistry and function of this organelle, investigations were undertaken to characterize this vital organelle by using proteomics methodology. To this end, peroxisomes from the kidneys of mice were purified by differential and Nycodenz equilibrium gradient centrifugation, as described previously [84,85]. For subsequent analyses, proteins from the peroxisomal peak fraction of the density gradient were precipitated using acetone. For in-solution tryptic digestion followed by nano-HPLC/ESI-MS/MS analyses, the proteins were resuspended in sample buffer containing 30 mM Tris/HCl, 2 M thiourea, and 7 M urea (pH 8.5) to a final protein concentration of about $1\,\mu g\,\mu L^{-1}$. Two different strategies were then followed for the proteomics analysis of peroxisomal samples, as detailed below.

Gel-Based Approach The peroxisomal proteins were separated by SDS–PAGE and, after colloidal Coomassie staining, the gel was uniformly cut into 2-mm slices. The resulting 144 gel slices were destained and proteins in-gel-digested using trypsin. Following peptide extraction and the removal of ACN *in vacuo*, samples were acidified by the addition of 5% FA and then analyzed by nano-HPLC/ESI-MS/MS on an high-capacity ion trap instrument (HCTplus; Bruker Daltonics, Bremen, Germany).

On-line nano-HPLC separation was performed as described previously [62]. Peptide MS/MS analyses were performed in a data-dependent fashion using low-energy CID. Only multiply charged peptide ions were selected for MS/MS, and the exclusion limits were automatically placed on previously selected m/z-ratios for 1.2 min. As the peptide samples were of limited complexity, survey spectra were a sum of seven individual scans ranging from m/z 300 to 1500, with a scanning speed of 8100 $(m/z)\,\mathrm{s}^{-1}$, while the MS/MS spectra were a sum of four scans ranging from m/z 100 to 2200 at a scan rate of 26 000 $(m/z)\,\mathrm{s}^{-1}$. As a result of LC/MS analyses of the 144 samples from SDS–PAGE, a total of 42 328 MS/MS spectra was extracted and merged to a combined peak list, which was then correlated with a target/decoy database of the mouse International Protein Index (Mouse IPI V3.15.1; www.ebi.ac.uk) database using Mascot. The ion trap MS/MS spectra were generally accepted with a Mascot cut-off score of 22.5 as well as a mass tolerances of 1.2 Da and 0.4 Da for MS and MS/MS experiments, respectively. Protein identifications were made on the basis of at least two peptides, with a confidence level above 95%.

Advanced MS-Based Approach Protein samples were dissolved in 50 mM NH_4HCO_3 to a final concentration of 0.1 µg µL^{-1}, after which in-solution tryptic digestion was carried out. For LC/MS analyses, the resulting peptide mixtures were diluted in 5% FA to 0.07 µg µL^{-1} and separated as described [62]. In order to increase the effective dynamic range of ESI-MS analysis, a gas-phase fractionation (GPF) was performed in the m/z-dimension for precursor ion selection in MS/MS scans [86]. To this end, each sample was analyzed three times, with an m/z-range of 300 to 1500 in the MS scan applying different overlapping narrow m/z-ranges (i.e., m/z 400–650, 600–850, and 800–1200) in each analysis for the selection of precursor ions in MS/MS scans. ESI-MS/MS analyses on the high-capacity ion trap instrument were performed as outlined above, with the exception that MS(/MS) spectra were a sum of two scans. In addition, an ESI-MS hybrid instrument was used which was composed of a linear ion trap and an FTICR analyzer (7-Tesla Finnigan LTQ-FT; Thermo Electron, Bremen, Germany) which was operated in the data-dependent mode as described previously [87]. Survey MS scans from m/z 300 to 1500 were acquired in the FTICR cell with $r = 25\,000$ at m/z 400, with a target accumulation value of 50 000 000. The three most intense ions were sequentially isolated for accurate mass measurements by an FTICR "SIM scan" (mass window ± 5 Da, resolution of 50 000, and target accumulation value of 100 000). Peptide fragmentation was carried out in the linear ion trap by low-energy CID and former precursor ions were dynamically excluded for 45 s. The total cycle time was approximately 3 s. As a result, a total of 12 106 MS/MS spectra was acquired on the ion trap, and 8802 MS/MS spectra on the LTQ-FT instrument. LTQ-FT mass spectra were searched with a mass tolerance of 2 ppm for precursor ions and 0.4 Da for fragment ions. Data processing for protein identification was performed as outlined above. When using LTQ-FT-MS/MS datasets, no false-positive hits were detected, provided that the proteins were identified on the basis of at least two peptides with a minimum Mascot score of 15.

Summary of Results The analysis of peroxisomal samples from mouse kidney using SDS–PAGE followed by nano-HPLC/ESI-MS/MS on an ion trap instrument resulted in the identification of 135 proteins, of which 48 (36%) are known to reside in mammalian peroxisomes (Figure 4.4a). The remaining proteins were mainly derived from the cytoplasm (19%) and from mitochondria (15%), due to the fact that peroxisomes cannot be purified to homogeneity. In addition, 25 proteins (19%), the functions and/or cellular localizations of which have yet to be determined, were identified using this approach. Among the latter, potentially new peroxisomal candidates may be included. Through semi-quantitative analysis of MS/MS data using

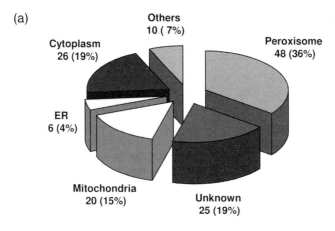

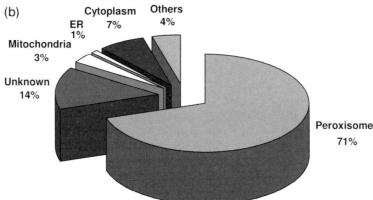

Figure 4.4 (a) Subcellular distribution of proteins identified in peroxisomal samples from mouse kidney using SDS–PAGE combined with in-gel tryptic digestion and nano-HPLC/ESI-MS/MS. (b) Percentage abundance of identified proteins from different cellular compartments as estimated on the basis of "peptide counting". For details, please refer to the text. ER, endoplasmic reticulum.

the method of "spectral counting" (see Section 4.1.3.2.1), it was further estimated that ~70% of all MS/MS spectra were derived from tryptic peptides of peroxisomal proteins. Thus, the analyzed sample was highly enriched for peroxisomes, indicating that the contamination by co-purifying subcellular structures/proteins was rather low (Figure 4.4b). Interestingly, the percentage abundance of "unknown" proteins was of the same magnitude (14%) as the combined percentage abundance of all other known contaminants (15%). Hence, it is tempting to speculate that new peroxisomal candidates were among those "unknown" proteins.

By using the "shot-gun" approach combined with GPF in the m/z-dimension for the selection of precursor ions in MS/MS experiments on the ion trap or hybrid LTQ-FT instrument, it was possible to identify 91 and 104 proteins, respectively. Whereas, 50 proteins could be identified in both datasets, the remaining proteins were unique to each dataset (Figure 4.5). When including the data from the SDS–PAGE approach, a total of 40 proteins was identified in all three methodical approaches, and 65 proteins were unique to the dataset obtained by the gel-enhanced method. When taken together, the ESI-MS/MS analyses resulted in the identification of 210 non-redundant proteins, 64 (30%) of which have been reported as being localized in mammalian peroxisomes. This accounts for approximately 85% of all genuine components of mouse/rat kidney peroxisomes currently reported. Furthermore, through the application of MS-based "protein correlation profiling" it was possible to identify 15 new candidates of peroxisomes from mouse kidney [86].

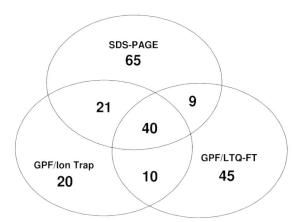

Figure 4.5 Venn diagram of protein identifications from mouse kidney peroxisomes obtained by SDS–PAGE combined with in-gel tryptic digestion and nano-HPLC/ESI-MS/MS and in-solution tryptic digestion, followed by nano-HPLC/ESI-MS/MS combined with gas-phase fractionation (GPF) in the m/z-dimension on an ion trap or a LTQ-FT instrument. For details, please refer to the text.

4.1.5
Concluding Remarks

The advent of nano-ESI has had a profound impact on the field of biological mass spectrometry and its application to proteomics. Today, by using sophisticated sample preparation procedures and compatible orthogonal fractionation strategies, it is possible to identify hundreds of proteins in a single experiment using ESI-MS methodologies. Indeed, state-of-the-art ESI-MS instrumentation, in combination with advanced MS/MS techniques, have helped to advance the concepts of "top-down" and "peptide" proteomics. For example, the recent introduction of the hybrid linear ion trap-Orbitrap mass spectrometer offers unrivaled mass accuracy ($\sim$2 ppm), a high resolving power (up to $\sim$150 000), and a high dynamic range ($>10^3$), without the need for a superconducting magnet and its associated maintenance requirements. Moreover, ion trap technology can now be used in conjunction with both CID and ETD, which facilitates the acquisition of highly comprehensive sequence-specific information on electrosprayed peptides. The application of ESI-MS/MS using ETD has already demonstrated the potential of this approach in areas of phospho-, glyco- and "top-down" proteomics. Moreover, substantial progress has also been made in proteomic bioinformatics, in terms of both real-time (e.g., generation of inclusion/exclusion lists, data-dependent acquisition methods) and post-hoc (e.g., database searching, data organization and evaluation) mass spectral analyses. Of particular note is here the use of target/decoy search strategies that ensure the reliable identification of peptides and proteins in MS-based proteomic studies, as well as the development of advanced software modules to facilitate label-free quantitative proteomics using LC/ESI-MS (/MS). In future, all of these advancements will help to better address the various salient challenges in proteomics, such as the identification of low-abundant proteins, which often have "challenging" physico-chemical properties, including hydrophobic G protein-coupled receptors or post-translationally modified proteins.

4.1.6
Recipes and Methods

4.1.6.1 MeOH/Chloroform Protein Precipitation to Remove Salts and Detergents

- This MeOH/chloroform precipitation protocol is based on the protocol published by Wessel and Flügge [88]:
  - Add 3 parts of MeOH and 1 part of chloroform to the sample. For example, to a sample volume of 100 µL, add 300 µL of MeOH and 100 µL of chloroform. Vortex thoroughly.
  - Add three parts (300 µL) of water, and vortex.
  - Centrifuge for 5 min at 6000–10 000 × g.
  - Remove the upper, aqueous layer, taking care not to disturb the interface where the protein is located.
  - Add an additional 3 parts of MeOH (300 µL), and vortex.
  - Centrifuge for 5 min at 6000–10 000 × g.

- Remove as much supernatant as possible without disturbing the pellet, and allow the pellet to air-dry. Alternatively, the pellet may be dried using a vacuum concentrator, for example, and stored for future use.
- Resuspend the pellet in an appropriate buffer (e.g., Laemmli buffer) for subsequent SDS–PAGE analysis [89].

4.1.6.2 Preparation and Washing of a Crude Membrane Pellet

- All steps are performed at 4 °C:
  - Add lysis buffer (e.g., 140 mM NaCl, 5 mM KCl, 2 mM $CaCl_2$, 2 mM $MgCl_2$, 10 mM HEPES, 10 mM glucose containing protease inhibitors) to the sample material.
  - Disrupt the starting material using a Dounce homogenizer; if necessary, sonicate for 10 s at 40 W (Ultrasonic stick).
  - Clarify the lysate by centrifugation at $1000 \times g$ for 10 min to pellet the cellular debris and nuclei.
  - Ultracentrifuge the supernatant for 2 h at $100\,000 \times g$.
  - Resuspend the crude membrane pellet in 100 mM sodium carbonate (pH 11.5); gently disrupt the pellet using a Dounce homogenizer.
  - Ultracentrifuge for 30 min at $100\,000 \times g$.
  - Remove the supernatant and neutralize the pellet by gently overlaying it with 50 mM ammonium bicarbonate (Ambic; pH ≈ 8.0).
  - Remove the Ambic and resuspend the pellet in an appropriate solubilization buffer.

4.1.6.3 Proteolytic Digestion and Peptide Extraction

- Tryptic in-gel digestion of proteins from silver-stained 2-D gels:
  - Excise the spots using a clean spot picker, and transfer to clean glass tubes.
  - Destain the spots in ca. 20 µL of 30 mM potassium ferrocyanide/100 mM sodium thiosulfate. Incubate for 1–2 min and remove the solution.
  - Incubate the spots with ca. 20 µL of 10 mM Ambic (pH ≈ 8) for 10 min; remove the solution.
  - Apply a dehydration solution (10 mM Ambic/ACN 1 : 1, v/v) for 10 min. Remove the solution.
  - Repeat the two previous steps.
  - Dry the gel spots *in vacuo*. At this point, the dried spots may be stored at $-80\,°C$.
  - Apply a sufficient volume (the spots should be covered) of trypsin dissolved in 10 mM Ambic [modified sequencing-grade trypsin; product number V5111 (Promega, Madison, WI, USA); final concentration 12.5 ng µL^{-1}] and incubate overnight at 37 °C.

- Tryptic in-gel digestion of proteins from coomassie-stained 1-D gels:
  - Excise the gel band(s), taking care to cut as close to the band as possible to minimize excess gel material; transfer to clean glass tubes. It is recommended that large gel bands are diced into small cubes.

- Add 100 μL double-distilled water (ddH$_2$O) to the band, vortex, and incubate for 10 min.
- Remove ddH$_2$O, add 40 μL of ACN/ddH$_2$O (1 : 1, v/v), and incubate for 15 min.
- Remove ACN/ddH$_2$O and add 40 μL of 100 mM Ambic (pH ≈ 8). Incubate for 10 min.
- Add 40 μL of ACN to obtain a final solution of 50 mM Ambic in 50% ACN. Incubate for 15 min.
- In the event that the gel pieces are not completely destained, repeat the previous two steps.
- Dry the gel pieces *in vacuo* (it is essential that they are completely dry).
- Add 40 μL of trypsin solution (see Section 4.1.6.3.1) and incubate for 45 min on ice. Add more of the trypsin solution if the pieces absorb all of the liquid.
- Remove any excess solution; add ca. 10 μL of 10 mM Ambic (enough to cover the gel pieces) and incubate overnight at 37°C.

Peptide Extraction

- Peptide extraction is performed as follows:
  - Following tryptic digestion, extract the peptides by adding 20 μL of extraction solution (5% FA/ACN; 1 : 1, v/v) and sonicating (10 min in a cooled sonication bath). Perform this step twice.
  - Pool the peptide extracts and remove ACN *in vacuo*; take care not to dry the sample as this may lead to sample loss.
  - Perform the nano-HPLC/ESI-MS/MS.

In-Solution Tryptic Digestion of Membrane Proteins

- This method is based on a protocol devised by Blonder *et al.* [52]:
  - Suspend the membrane pellet in 500 μL of 50 mM Ambic (pH ≈ 8).
  - Reduce the proteins by adding dithiothreitol (DTT) to a final concentration of 20 mM, and incubate for 15 min at 37°C.
  - Sonicate for 15 s using a sonication bath; cool on ice for 1 min.
  - Alkylate the proteins by adding iodoacetamide to a final concentration of 25 mM; incubate in the dark for 30 min at room temperature.
  - Perform a protein determination using the BCA assay.
  - Add MeOH to a final concentration of 60% (v/v) and sonicate for 1 min, followed by cooling on ice for 1 min. Repeat the sonication procedure three times.
  - Add modified sequencing-grade trypsin dissolved in 25 mM Ambic (enzyme to protein ratio of 1:100). Incubate the mixture overnight at 37°C with agitation (600 r.p.m.). Alternatively, incubation may be performed at 60°C for 5 h.
  - Quench the digestion by adding FA to a final concentration of 0.01%. Centrifuge at 13 000 × g for 5 min.
  - Remove the solvent and concentrate the peptide mixture *in vacuo*. Take care not to completely dry the sample.
  - Perform the nano-HPLC/ESI-MS/MS.

4.1.6.4 Off-Line Analysis of Intact Proteins

- The suggested procedure is as follows:
  - Purify the protein samples using either C_4-Ziptip™ (Millipore, Schwalbach, Germany), an RPLC system, or a protein precipitation technique (e.g., MeOH/chloroform precipitation as described in Section 4.1.6.1.1).
  - Dissolve the protein in 50% (v/v) MeOH and 2% (v/v) acetic acid to reach a final concentration of ca. 1 pmol μL^{-1}.
  - Load the solution into the off-line ESI needle, ensuring that no air is left at the tip of the needle.
  - Using a syringe, it may be necessary to apply a little air pressure to the end of the needle to promote a stable spray.
  - Apply a voltage of $+1.0$ to $1.5\,kV$ to the ESI needle.

4.1.6.5 ESI Sample Preparation Checklist

- General sample handling guidelines:
  - *Keratin contamination* is a common problem; take care to ensure that keratin contamination is minimized, for example by wearing gloves and long-sleeved laboratory coats, and by using clean tubes, tips, etc.
  - It is good laboratory practice to rinse polypropylene tubes and vials with acetonitrile, MeOH, and/or nano-pure water before preparing solutions, as well as before collecting or storing samples.
  - Be aware that concentrating large sample volumes ($>500\,\mu L$) may simultaneously enrich potential non-volatile contaminants. If no pre-column is used, samples may be concentrated and concomitantly "washed" using off-line reversed-phase purification methods such as C4- or C18-Ziptips (Millipore). Note that Ziptip-purification is not effective in removing detergents, polymers, and other large molecules which exhibit an affinity for reversed-phase material and, thus, interact with the hydrophobic stationary phase.
  - Samples must not contain particulates. For removal, centrifuge the sample (e.g., 13 000 r.p.m. for 5 min) or use size-exclusion membranes (e.g., Millipore Microcon centrifugal filter units).
- Sample properties:
  - Samples must be free of non-volatile buffers such as MES, MOPS, or HEPES. Recommended buffers: 50 mM Tris, 50 mM Ambic, or other volatile ammonium salts.
  - Samples must not contain salts (e.g., Na^+, K^+, Ca^{2+}), detergents (e.g., SDS, CHAPS), or stabilizers (e.g., polyethylene glycol, glycerol). Note that even trace amounts of detergent can have a deleterious effect on proteins/peptide ionization.
  - Unless a trap-column is used, acidify the samples using only a weak organic acid (e.g., FA or acetic acid). A typical sample comprises $\sim$50% water, $\sim$50% organic solvent (e.g., ACN), and $\sim$0.1% organic acid (e.g., FA). TFA, HCl, and H_2SO_4 should be avoided.

- Reagents and solvent system:
  - Reagents should be HPLC grade, and water should be of Milli-Q purity (18 MΩ de-ionized). All solutions should be filtered (e.g., 0.45 µm membrane filter disks; Millipore) to remove particulates.
  - All glassware and must be thoroughly cleaned using nano-pure water. HPLC-compatible plastic ware should be thoroughly washed and "pre-leached" in solvent (e.g., 50% MeOH) and then rinsed with Milli-Q water.
  - Acceptable solvents for ESI are: water, ACN, MeOH, isopropanol, and chloroform; however, MeOH and ACN are most recommended. Non-polar or non-volatile solvents must be avoided (e.g., dimethylformamide, dimethylsulfoxide).

Acknowledgments

These studies were supported by the FP6 European Union Project "Peroxisome" (LSHG-CT-2004–512018) and the Deutsche Forschungsgemeinschaft within the SFB 642. The authors thank Dr. Carsten Kötting and Sven Brucker for providing Ran1B protein, and Dr. Silke Oeljeklaus for critically reading the manuscript.

References

1 Dole, M., Mack, L.L., Hines, R.L., Mobley, R.C., Ferguson, L.D. and Alice, M.B. (1968) Molecular beams of macroions. *J. Chem. Physics*, **49**, 2240.

2 Fenn, J.B., Mann, M., Meng, C.K., Wong, S.F. and Whitehouse, C.M. (1989) Electrospray ionization for mass spectrometry of large biomolecules. *Science*, **246**, 64–71.

3 Fenn, J.B. (2002) Electrospray ionization mass spectrometry: How it all began. *J. Biomol. Tech.*, **13**, 101–118.

4 Meng, C.K., Mann, M. and Fenn, J.B. (1988) Of protons or proteins. *Z Phys D Atoms Molecules and Clusters*, **10**, 361–368.

5 Taylor, G. (1964) Disintegration of water drops in an electric field. *Proc. R. Soc. Lond. A, Math. Phys. Sci.*, **280**, 383–397.

6 Fenn, J.B., Mann, M., Meng, C.K., Wong, F.S. and Whitehouse, C.M. (1990) Electrospray ionization-principles and practice. *Mass Spectrom. Rev.*, **9**, 37–70.

7 Wilm, M., Shevchenko, A., Houthaeve, T., Breit, S., Schweigerer, L., Fotsis, T. and Mann, M. (1996) Femtomole sequencing of proteins from polyacrylamide gels by nano-electrospray mass spectrometry. *Nature*, **379**, 466–469.

8 Wilm, M. and Mann, M. (1996) Analytical properties of the nanoelectrospray ion source. *Anal. Chem.*, **68**, 1–8.

9 Schmidt, A., Karas, M. and Dulcks, T. (2003) Effect of different solution flow rates on analyte ion signals in nano-ESI MS or: when does ESI turn into nano-ESI? *J. Am. Soc. Mass Spectrom.*, **14**, 492–500.

10 Wilm, M. and Mann, M. (1994) Electrospray and Taylor-Cone theory Dole's beam of macromolecules at last? *Int. J. Mass Spectrom. Ion Process.*, **136**, 167–180.

11 Banks, J.F.J. (1996) High-sensitivity peptide mapping using packed-capillary liquid chromatography and electrospray ionization mass spectrometry. *J. Chromatogr. A*, **743**, 99–104.

12 Jonsson, A.P. (2001) Mass spectrometry for protein and peptide characterisation. *Cell. Mol. Life Sci.*, **58**, 868–884.

13 Mann, M., Hendrickson, R.C. and Pandey, A. (2001) Analysis of proteins and proteomes by mass spectrometry. *Annu. Rev. Biochem.*, **70**, 437–473.

14 Steen, H. and Mann, M. (2004) The ABCs (and XYZs) of peptide sequencing. *Nat. Rev. Mol. Cell. Biol.*, **5**, 699–711.

15 Medzihradszky, K.F. (2005) Peptide sequence analysis. *Methods Enzymol.*, **402**, 209–244.

16 Wysocki, V.H., Tsaprailis, G., Smith, L.L. and Breci, L.A. (2000) Mobile and localized protons: a framework for understanding peptide dissociation. *J. Mass Spectrom.*, **35**, 1399–1406.

17 Jonscher, K.R. and Yates, J.R. III (1997) The quadrupole ion trap mass spectrometer – a small solution to a big challenge. *Anal. Biochem.*, **244**, 1–15.

18 Brancia, F.L. (2006) Recent developments in ion-trap mass spectrometry and related technologies. *Expert Rev. Proteomics*, **3**, 143–151.

19 Schwartz, J.C., Senko, M.W. and Syka, J.E. (2002) A two-dimensional quadrupole ion trap mass spectrometer. *J. Am. Soc. Mass Spectrom.*, **13**, 659–669.

20 Wilcox, B.E., Hendrickson, C.L. and Marshall, A.G. (2002) Improved ion extraction from a linear octopole ion trap: SIMION analysis and experimental demonstration. *J. Am. Soc. Mass Spectrom.*, **13**, 1304–1312.

21 Makarov, A., Denisov, E., Kholomeev, A., Balschun, W., Lange, O., Strupat, K. and Horning, S. (2006) Performance evaluation of a hybrid linear ion trap/orbitrap mass spectrometer. *Anal. Chem.*, **78**, 2113–2120.

22 Hu, Q., Noll, R.J., Li, H., Makarov, A., Hardman, M. and Graham Cooks, R. (2005) The Orbitrap: a new mass spectrometer. *J. Mass Spectrom.*, **40**, 430–443.

23 Hardman, M. and Makarov, A.A. (2003) Interfacing the orbitrap mass analyzer to an electrospray ion source. *Anal. Chem.*, **75**, 1699–1705.

24 Makarov, A. (2000) Electrostatic axially harmonic orbital trapping: a high-performance technique of mass analysis. *Anal. Chem.*, **72**, 1156–1162.

25 Domon, B. and Aebersold, R. (2006) Mass spectrometry and protein analysis. *Science*, **312**, 212–217.

26 Smith, R.D. (2002) Trends in mass spectrometry instrumentation for proteomics. *Trends Biotechnol.* 20, S3-, S7.

27 Yates, J.R. III, (2004) Mass spectral analysis in proteomics. *Annu. Rev. Biophys. Biomol. Struct.*, **33**, 297–316.

28 Washburn, M.P., Wolters, D. and Yates, J.R. III (2001) Large-scale analysis of the yeast proteome by multidimensional protein identification technology. *Nat. Biotechnol.*, **19**, 242–247.

29 Wolters, D.A., Washburn, M.P. and Yates, J.R. III (2001) An automated multidimensional protein identification technology for shotgun proteomics. *Anal. Chem.*, **73**, 5683–5690.

30 Yates, J.R., III, Eng, J.K., McCormack, A.L. and Schieltz, D. (1995) Method to correlate tandem mass spectra of modified peptides to amino acid sequences in the protein database. *Anal. Chem.*, **67**, 1426–1436.

31 Weinglass, A.B., Whitelegge, J.P. and Kaback, H.R. (2004) Integrating mass spectrometry into membrane protein drug discovery. *Curr. Opin. Drug Discov. Devel.*, **7**, 589–599.

32 Anderson, N.L. and Anderson, N.G. (2002) The human plasma proteome: history character, and diagnostic prospects. *Mol. Cell. Proteomics*, **1**, 845–867.

33 Ramstrom, M., Hagman, C., Mitchell, J.K., Derrick, P.J., Hakansson, P. and Bergquist, J. (2005) Depletion of high-abundant proteins in body fluids prior to liquid chromatography Fourier transform ion cyclotron resonance mass spectrometry. *J. Proteome Res.*, **4**, 410–416.

34 Constantopoulos, T.L., Jackson, G.S. and Enke, C.G. (1999) Effects of salt concentration on analyte response using electrospray ionization mass

spectrometry. *J. Am. Soc. Mass Spectrom.*, **10**, 625–634.

35 Wang, H. and Hanash, S. (2003) Multi-dimensional liquid phase based separations in proteomics. *J. Chromatogr. B Analyt. Technol. Biomed. Life Sci.*, **787**, 11–18.

36 Jens Schindler, H.G.N. (2006) Aqueous polymer two-phase systems: Effective tools for plasma membrane proteomics. *Proteomics* 9999, NA.

37 Macfarlane, D.E. (1989) Two dimensional benzyldimethyl-n-hexadecylammonium chloride–sodium dodecyl sulfate preparative polyacrylamide gel electrophoresis: a high capacity high resolution technique for the purification of proteins from complex mixtures. *Anal. Biochem.*, **176**, 457–463.

38 Hartinger, J., Stenius, K., Hogemann, D. and Jahn, R. (1996) 16-BAC/SDS-PAGE: A two-dimensional gel electrophoresis system suitable for the separation of integral membrane proteins. *Anal. Biochem.*, **240**, 126–133.

39 Fournier, M.L., Gilmore, J.M., Martin-Brown, S.A. and Washburn, M.P. (2007) Multidimensional separations-based shotgun proteomics. *Chem. Rev.*, **107**, 3654–3686.

40 Mawuenyega, K.G., Kaji, H., Yamauchi, Y., Shinkawa, T., Saito, H., Taoka, M., Takahashi, N. and Isobe, T. (2003) Large-scale identification of *Caenorhabditis elegans* proteins by multidimensional liquid chromatography-tandem mass spectrometry. *J. Proteome Res.*, **2**, 23–35.

41 Motoyama, A., Xu, T., Ruse, C.I., Wohlschlegel, J.A. and Yates, J.R. (2007) Anion and cation mixed-bed ion exchange for enhanced multidimensional separations of peptides and phosphopeptides. *Anal. Chem.*, **79**, 3623–3634.

42 Villen, J., Beausoleil, S.A., Gerber, S.A. and Gygi, S.P. (2007) Large-scale phosphorylation analysis of mouse liver. *Proc. Natl. Acad. Sci. USA*, **104**, 1488–1493.

43 Gruhler, A., Olsen, J.V., Mohammed, S., Mortensen, P., Faergeman, N.J., Mann, M. and Jensen, O.N. (2005) Quantitative phosphoproteomics applied to the yeast pheromone signaling pathway. *Mol. Cell. Proteomics*, **4**, 310–327.

44 Larsen, M.R., Thingholm, T.E., Jensen, O.N., Roepstorff, P. and Jorgensen, T.J.D. (2005) Highly selective enrichment of phosphorylated peptides from peptide mixtures using titanium dioxide microcolumns. *Mol. Cell. Proteomics*, **4**, 873–886.

45 Xiong, L., Andrews, D. and Regnier, F. (2003) Comparative proteomics of glycoproteins based on lectin selection and isotope coding. *J. Proteome Res.*, **2**, 618–625.

46 Durham, M. and Regnier, F.E. (2006) Targeted glycoproteomics: Serial lectin affinity chromatography in the selection of O-glycosylation sites on proteins from the human blood proteome. *J. Chromatogr. A*, **1132**, 165–173.

47 Candiano, G., Bruschi, M., Musante, L., Santucci, L., Ghiggeri, G.M., Carnemolla, B., Orecchia, P., Zardi, L. and Righetti, P.G. (2004) Blue silver: a very sensitive colloidal Coomassie G-250 staining for proteome analysis. *Electrophoresis*, **25**, 1327–1333.

48 Nesterenko, M.V., Tilley, M. and Upton, S.J. (1994) A simple modification of Blum's silver stain method allows for 30 minute detection of proteins in polyacrylamide gels. *J. Biochem. Biophys. Methods*, **28**, 239–242.

49 Jensen, O.N., Wilm, M., Shevchenko, A. and Mann, M. (1999) Sample preparation methods for mass spectrometric peptide mapping directly from 2-DE gels. *Methods Mol. Biol.*, **112**, 513–530.

50 Palmblad, M. and Vogel, J.S. (2005) Quantitation of binding recovery and desalting efficiency of peptides and proteins in solid phase extraction micropipette tips. *J. Chromatogr. B*, **814**, 309–313.

51 Wille, S.M. and Lambert, W.E. (2007) Recent developments in extraction procedures relevant to analytical toxicology. *Anal. Bioanal. Chem.*, **388**, 1381–1391.

52 Blonder, J., Chan, K.C., Issaq, H.J. and Veenstra, T.D. (2007) Identification of membrane proteins from mammalian cell/tissue using methanol-facilitated solubilization and tryptic digestion coupled with 2D-LC-MS/MS. *Nat. Protocols*, **1**, 2784.

53 Slysz, G.W., Lewis, D.F. and Schriemer, D.C. (2006) Detection and identification of sub-nanogram levels of protein in a NanoLC-Trypsin-MS system. *J. Proteome Res.*, **5**, 1959–1966.

54 Medzihradszky, K.F. and Burlingame, A.L. (2005) In-solution digestion of proteins for mass spectrometry in: *Methods in Enzymology*, Academic Press, pp. 50–65.

55 Whitelegge, J., Halgand, F., Souda, P. and Zabrouskov, V. (2006) Top-down mass spectrometry of integral membrane proteins. *Expert Rev. Proteomics*, **3**, 585–596.

56 Reid, G.E. and McLuckey, S.A. (2002) 'Top down' protein characterization via tandem mass spectrometry. *J. Mass Spectrom.*, **37**, 663–675.

57 Wells, J.M. and McLuckey, S.A. (2005) Collision-induced dissociation (CID) of peptides and proteins. *Methods Enzymol.*, **402**, 148–185.

58 Heck, A.J. and Van Den Heuvel, R.H. (2004) Investigation of intact protein complexes by mass spectrometry. *Mass Spectrom. Rev.*, **23**, 368–389.

59 Ruotolo, B.T. and Robinson, C.V. (2006) Aspects of native proteins are retained in vacuum. *Curr. Opin. Chem. Biol.*, **10**, 402–408.

60 Benesch, J.L. and Robinson, C.V. (2006) Mass spectrometry of macromolecular assemblies: preservation and dissociation. *Curr. Opin. Struct. Biol.*, **16**, 245–251.

61 Gruber, K.A., Stein, S., Brink, L., Radhakrishnan, A. and Udenfriend, S. (1976) Fluorometric assay of vasopressin and oxytocin: a general approach to the assay of peptides in tissues. *Proc. Natl. Acad. Sci. USA*, **73**, 1314–1318.

62 Schaefer, H., Chervet, J.P., Bunse, C., Joppich, C., Meyer, H.E. and Marcus, K. (2004) A peptide preconcentration approach for nano-high-performance liquid chromatography to diminish memory effects. *Proteomics*, **4**, 2541–2544.

63 Mitulovic, G., Smoluch, M., Chervet, J.P., Steinmacher, I., Kungl, A. and Mechtler, K. (2003) An improved method for tracking and reducing the void volume in nano HPLC-MS with micro trapping columns. *Anal. Bioanal. Chem.*, **376**, 946–951.

64 Wilson, I.D., Nicholson, J.K., Castro-Perez, J., Granger, J.H., Johnson, K.A., Smith, B.W. and Plumb, R.S. (2005) High resolution 'ultra performance' liquid chromatography coupled to TOF mass spectrometry as a tool for differential metabolic pathway profiling in functional genomic studies. *J. Proteome Res.*, **4**, 591–598.

65 Mitulovic, G. and Mechtler, K. (2006) HPLC techniques for proteomics analysis – a short overview of latest developments. *Brief Funct. Genomic Proteomics*, **5**, 249–260.

66 Apffel, A., Fischer, S., Goldberg, G., Goodley, P.C. and Kuhlmann, F.E. (1995) Enhanced sensitivity for peptide mapping with electrospray liquid chromatography-mass spectrometry in the presence of signal suppression due to trifluoroacetic acid-containing mobile phases. *J. Chromatogr. A*, **712**, 177–190.

67 Funk, J., Li, X. and Franz, T. (2005) Threshold values for detergents in protein and peptide samples for mass spectrometry. *Rapid Commun. Mass Spectrom.*, **19**, 2986–2988.

68 Mihailova, A., Lundanes, E. and Greibrokk, T. (2006) Determination and removal of impurities in 2-D LC-MS of peptides. *J. Separation Sci.*, **29**, 576–581.

69 Molina, H., Horn, D.M., Tang, N., Mathivanan, S. and Pandey, A. (2007) Global proteomic profiling of

phosphopeptides using electron transfer dissociation tandem mass spectrometry. *Proc. Natl. Acad. Sci. USA*, **104**, 2199–2204.
70 Everley, P.A., Gartner, C.A., Haas, W., Saghatelian, A., Elias, J.E., Cravatt, B.F., Zetter, B.R. and Gygi, S.P. (2007) Assessing enzyme activities using stable isotope labeling and mass spectrometry. *Mol. Cell. Proteomics*, **10**, 1771–1777.
71 Liu, H., Sadygov, R.G. and Yates, J.R. III (2004) A model for random sampling and estimation of relative protein abundance in shotgun proteomics. *Anal. Chem.*, **76**, 4193–4201.
72 Ishihama, Y., Oda, Y., Tabata, T., Sato, T., Nagasu, T., Rappsilber, J. and Mann, M. (2005) Exponentially modified protein abundance index (emPAI) for estimation of absolute protein amount in proteomics by the number of sequenced peptides per protein. *Mol. Cell. Proteomics*, **4**, 1265–1272.
73 Rappsilber, J., Ryder, U., Lamond, A.I. and Mann, M. (2002) Large-scale proteomic analysis of the human spliceosome. *Genome Res.*, **12**, 1231–1245.
74 Colinge, J., Masselot, A., Cusin, I., Mahe, E., Niknejad, A., Argoud-Puy, G., Reffas, S., Bederr, N., Gleizes, A., Rey, P.A. and Bougueleret, L. (2004) High-performance peptide identification by tandem mass spectrometry allows reliable automatic data processing in proteomics. *Proteomics*, **4**, 1977–1984.
75 Perkins, D.N., Pappin, D.J.C., Creasy, D.M. and Cottrell, J.S. (1999) Probability-based protein identification by searching sequence databases using mass spectrometry data. *Electrophoresis*, **20**, 3551–3567.
76 Sadygov, R.G., Cociorva, D. and Yates, J.R. III (2004) Large-scale database searching using tandem mass spectra: looking up the answer in the back of the book. *Nat. Methods*, **1**, 195–202.
77 Stephan, C., Reidegeld, K.A., Hamacher, M., van Hall, A., Marcus, K., Taylor, C., Jones, P., Muller, M., Apweiler, R., Martens, L., Korting, G., Chamrad, D.C., Thiele, H., Bluggel, M., Parkinson, D., Binz, P.A., Lyall, A. and Meyer, H.E. (2006) Automated reprocessing pipeline for searching heterogeneous mass spectrometric data of the HUPO Brain Proteome Project pilot phase. *Proteomics*, **6**, 5015–5029.
78 Elias, J.E. and Gygi, S.P. (2007) Target-decoy search strategy for increased confidence in large-scale protein identifications by mass spectrometry. *Nat. Methods*, **4**, 207–214.
79 Elias, J.E., Haas, W., Faherty, B.K. and Gygi, S.P. (2005) Comparative evaluation of mass spectrometry platforms used in large-scale proteomics investigations. *Nat. Methods*, **2**, 667–675.
80 Higdon, R., Hogan, J.M., Van Belle, G. and Kolker, E. (2005) Randomized sequence databases for tandem mass spectrometry peptide and protein identification. *Omics*, **9**, 364–379.
81 Blackler, A.R., Klammer, A.A., MacCoss, M.J. and Wu, C.C. (2006) Quantitative comparison of proteomic data quality between a 2D and 3D quadrupole ion trap. *Anal. Chem.*, **78**, 1337–1344.
82 Wanders, R.J. and Waterham, H.R. (2006) Biochemistry of Mammalian peroxisomes revisited. *Annu. Rev. Biochem.*, **75**, 295–332.
83 Wanders, R.J. and Waterham, H.R. (2006) Peroxisomal disorders: The single peroxisomal enzyme deficiencies. *Biochim. Biophys. Acta*, **1763**, 1707–1720.
84 Ofman, R., Speijer, D., Leen, R. and Wanders, R.J. (2006) Proteomic analysis of mouse kidney peroxisomes: identification of RP2p as a peroxisomal nudix hydrolase with acyl-CoA diphosphatase activity. *Biochem. J.*, **393**, 537–543.
85 Wanders, R.J., van Roermund, C.W., Schor, D.S., ten Brink, H.J. and Jakobs, C. (1994) 2-Hydroxyphytanic acid oxidase activity in rat and human liver and its deficiency in the Zellweger syndrome. *Biochim. Biophys. Acta*, **1227**, 177–182.

86 Wiese, S., Gronemeyer, T., Ofman, R., Kunze, M., Grou, C.P., Almeida, J.A., Eisenacher, M., Stephan, C., Hayen, H., Schollenberger, L., Korosec, T., Waterham, H.R., Schliebs, W., Erdmann, R., Berger, J., Meyer, H.E., Just, W., Azevedo, J.E., Wanders, R.J. and Warscheid, B. (2007) Proteomics characterization of mouse kidney peroxisomes by tandem mass spectrometry and protein correlation profiling. *Mol. Cell. Proteomics*, **12**, 2045–2057.

87 Olsen, J.V., Ong, S.E. and Mann, M. (2004) Trypsin cleaves exclusively C-terminal to arginine and lysine residues. *Mol. Cell. Proteomics*, **3**, 608–614.

88 Wessel, D. and Flugge, U.I. (1984) A method for the quantitative recovery of protein in dilute solution in the presence of detergents and lipids. *Anal. Biochem.*, **138**, 141–143.

89 Laemmli, U.K. (1970) Cleavage of structural proteins during the assembly of the head of bacteriophage T4. *Nature*, **227**, 680–685.

4.2
Sample Preparation for the Application of MALDI Mass Spectrometry in Proteome Analysis

Andreas Tholey, Matthias Glückmann, Kerstin Seemann, and Michael Karas

4.2.1
Introduction

Since its introduction during the late 1980s [90], matrix-assisted laser desorption/ionization mass spectrometry (MALDI MS) has become one of the most valuable tools for the investigation of biomolecules such as peptides, proteins, and oligonucleotides. Other fields of applications include the analysis of technical polymers, small organic molecules and low molecular-weight compounds of biological interest such as amino acids, lipids, and carbohydrates [91].

The basic principle of MALDI MS (Figure 4.6) is based on mixing of the analyte with a high excess (up to 100 000 : 1, mol:mol) of a low molecular-weight matrix consisting of small organic compounds (the first successful matrix was nicotinic acid) which exhibit a strong resonance absorption at the laser wavelength used. Most matrices are organic, aromatic compounds that are able to absorb the laser light applied in MALDI instruments. The matrix–analyte mixture is then applied on a metal plate (target) where the solvent evaporates, leaving behind a matrix–analyte cocrystallite. Within the high vacuum of the mass spectrometer, pulsed laser shots (typical nanosecond pulse lengths) are applied onto the cocrystallite. This controllable energy transfer into the condensed phase of matrix and analyte results both in desorption into the gas phase (plume) and ionization of the analyte. The mechanisms of the MALDI process itself are complex and, although far from final clarification, are beyond the scope of this chapter. A good overview over the fundamental processes can be found in several recent reviews [92–94].

The ions formed in this way are accelerated in an electric field and separated according to their mass-to-charge (m/z) ratio. In contrast to electrospray ionization (ESI) MS, mainly singly charged ions are formed. Protons are most often the charge

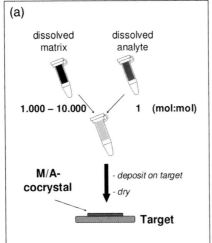
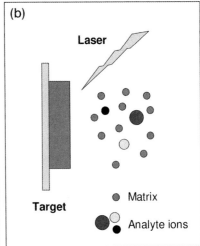

Figure 4.6 Basic scheme of a MALDI experiment. (a) Sample preparation, here shown as the dried droplet method. The matrix and analyte are mixed and applied to the target. After evaporation of the solvent, the matrix–analyte–cocrystallite is formed. (b) The target is shot in high vacuum by short laser pulses, leading to the desorption of matrix and analyte ions. (c) Structures of three commonly used MALDI matrices: 2,5-dihydroxybenzoic acid (DHB); α-cyano-4-hydroxy-cinnamic acid (CHCA); sinapinic acid (SA) (from left to right).

carriers leading to analyte ions of the form of $[M+H]^+$ or $[M-H]^-$ ions. Because of the pulsed ion generation in MALDI the combination with a time-of-flight (TOF) analyzer is widely used. Developments such as time lag focusing or delayed extraction have greatly improved the robustness of the MALDI-TOF technology [95,96].

The most striking features of MALDI MS are the relatively simple sample preparation, low sample consumption and high sensitivity, the high tolerance against impurities such as salts and detergents, the speed and throughput encountered with the possibility of multiplexed analyses, and the relatively straightforward analysis of the spectra due to the predominance of singly charged analyte ions. These unique properties make MALDI MS one of the most powerful tools for protein identification in proteomics.

Sample preparation in MALDI-based proteome analysis consists of two steps: (1) preparation of the biological sample; and (2) preparation of the MALDI sample itself. The MS experiment itself is the final step, which is carried out offline.

4.2 Sample Preparation for the Application of MALDI Mass Spectrometry in Proteome Analysis

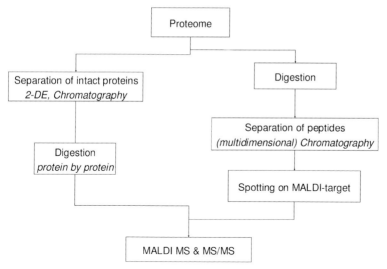

Figure 4.7 Possible workflows for the proteome analysis. Left side: gel-based. Right side: LC-MALDI-based approach.

For protein identification based on MALDI MS, either by peptide mass fingerprint analysis (MS) or in combination with peptide fragment analysis (MS/MS), two different workflows are amenable (Figure 4.7). The first workflow (which in the following text is referred to as gel-based workflow) comprises the separation of a complex protein mixture (the proteome) by means of either chromatographic or (more frequently) one- or two-dimensional (2-D) electrophoretic separation followed by digestion of the protein and MALDI MS or MS/MS analysis. The second workflow encompasses digestion of the complete proteome, separation of the resulting peptide mixture by chromatography coupled with offline measurement of the separated peptides by MALDI MS (liquid chromatography (LC)-MALDI coupling).

In the following section some basic aspects of the MALDI based analysis of complex protein mixtures are discussed. The selection of the matrix substance and different methods for sample preparation, as well as strategies suited to the removal of common contaminants are discussed. Two exemplary experiments using a gel-based and a LC-MALDI workflow are described

4.2.2
Sample Preparation for MALDI-Based Protein Identification

4.2.2.1 Selection of the MALDI Matrix
The matrix fulfils three essential functions in the MALDI process:

- It absorbs the laser light via electronic (ultraviolet (UV)-MALDI) or vibrational (infrared (IR)-MALDI) excitation.

- Due to the high molar excess of the matrix over the analyte, the intermolecular interactions of analyte molecules are reduced, thus facilitating transfer into the gas phase.
- Matrix–analyte interactions play an active role both in the ionization of the analyte as well as in its desorption [97].

Numerous substances have been tested and applied as matrices. The choice of matrix influences the ionization behavior, the formation of adducts, the stability (or fragmentation) of the analytes, and has also practical impacts on the performance of the experiments. With the exception of the absorption behavior, there are no general rules for the prediction of a substance to be suited as matrix, neither based on chemical nor on physical properties. The search for new matrices applied therefore in most cases an empirical approach, taking into account a few practical rules, such as capability of solubility with the analyte, absorption behavior at the laser wavelength applied, inertness of the matrix and vacuum stability.

Most commercially available MALDI mass spectrometers are equipped with UV-lasers; hence, the following discussion concentrates on MALDI matrices suited for the wavelength range between $\lambda = 266$ nm and $\lambda = 355$ nm. Today, many commercial MALDI instruments are equipped either with nitrogen gas lasers emitting at a wavelength of $\lambda = 337$ nm, or with Nd:YAG solid-state lasers, which are frequency tripled, emitting at a wavelength of $\lambda = 355$ nm. Nevertheless, there are some interesting applications of infrared MALDI MS also in the field of proteome analysis, such as the detection of proteins directly from sodium dodecylsulfate (SDS) gels [98].

The majority of the UV-compatible matrices are small organic compounds absorbing in the range of $\lambda = 266$ to 355 nm. Many aromatic matrices possess hydroxyl- or amino groups in the *ortho*- or *para*-position and acidic groups or carbonyl functions (carboxylic groups, amides, ketones), although the role of the functional groups is not yet clear. Moreover, constitutional isomers of a compound can show different properties in MALDI MS [99]. An important fact is that most of the matrix substances are not suited for the analysis of all classes of analyte.

For the analysis of peptides in a proteomics environment, derivatives of cinnamic acid [100], especially α-cyano-4-hydroxycinnamic acid (CHCA) [101], are the most frequently applied matrices; another cinnamic acid derivative, sinapinic acid (SINA) [100], is primarily used for the analysis of intact proteins (Table 4.3; see Figure 4.6c). Derivatives of benzoic acid, such as 2,5-dihydroxybenzoic acid (DHB) [102], are especially suited for the analysis of low molecular-weight compounds, and of proteins and glycoproteins [103]. Polyhydroxyacetophenones (e.g., 2,3,4-trihydroxyacetophenone [104] or 2,6-dihydroxyacetophenon [105,106]) contain only weakly acidic hydroxyl functions, and are preferentially used for the analysis of acid-labile molecules and of oligonucleotides. Heteroaromatic and condensed aromatic compounds, for example nitrogen-containing compounds such as derivatives of pyridine (e.g., nicotinic acid), 3-amino-quinoline and derivatives of picolinic acid (e.g., 3-hydroxypicolinic acid; 3HPA), are preferentially used for the measurement of oligonucleotides and acid-labile

Table 4.3 Matrices, matrix properties, and main area of application.

Common abbreviation of matrix/Reference	Name	Elemental formula	Molecular weight/[g mol^{-1}]	Color	Melting point/°C	Main application
3-HPA [107–109]	3-Hydroxy-picolinic acid	$C_6H_5NO_3$	139.11	White	219–221 (decomp.)	Oligonucleotides
ATT [136]	6-Aza-2-thiothymine	$C_4H_5N_3OS$	143.17	White	218–219	Oligonucleotides
DHAP [105,106]	2,6-Dihydroxyacetophenone	$C_8H_8O_3$	152.15	White to pale yellow	156–158	Oligonucleotides
DHB [102]	2,5-Dihydroxy-benzoic acid	$C_7H_6O_4$	154.12	White	205 (decomp.)	Different analytes: Peptides, proteins, glycoproteins, phosphopeptides, organic molecules
Ferulic acid [100,137]	4-Hydroxy-3-methoxycinnamic acid	$C_{10}H_{10}O_4$	194.19	White	168–171	Proteins
HABA [103]	2-(4′-Hydroxyphenylazo)- benzoic acid	$C_{13}H_{10}N_2O_3$	242.23	Orange	205–207	Peptides and polymers
CHCA [101]	α-Cyano-4-hydroxycinnamic acid	$C_{10}H_7NO_3$	189.17	Yellow	250–253 (decomp.)	Peptide Mixtures
HMBA [111,138]	2-Hydroxy-5-methoxy benzoic acid	$C_8H_8O_4$	168.15	White	141–143	As additive to DHB forming DHBs
SA [139]	Salicylic acid	$C_7H_6O_3$	138.12	White	158–161	Not vacuum stable
SINA [100,137]	Sinapic acid 3,5-Dimethoxy-4-hydroxycinnamic acid	$C_{11}H_{12}O_5$	224.21	White	203–205	Proteins
THAP [104,140,141]	2,4,6-Trihydroxycetophenone	$C_8H_8O_4 \cdot H_2O$	186.17	White to pale yellow	219–221	Oligonucleotides

Decomp.: decomposition.

substances [107–109]. In some cases, the use of recrystallized matrices has been shown to lead to improved spectral quality. Possible solvents and concentrations of commonly used matrices in MALDI mass spectrometry are listed in Table 4.4.

Today. most commercial mass spectrometers use lasers with wavelengths of either $\lambda = 337$ (nitrogen lasers) or $\lambda = 355$ nm (frequency tripled Nd-YAG solid-state lasers). Both laser are very well suited for the application of all cinnamic acid derivatives as matrices; when using DHB in case of $\lambda = 355$ nm lasers, higher laser energies/fluences must be applied for desorption and ionization due to the low absorption coefficient of this matrix at this wavelength.

An important factor characterizing a matrix is the stability of the ions formed, or the extent of fragmentation, which is induced by the complex laser–matrix–analyte interactions that occur during or after the ionization and desorption processes. Fragmentation is influenced by several factors which determine the internal energy of analytes, such as the starting velocities of the plume, the acceleration voltages applied, and the laser energy/fluence used. The initial velocities observed for matrices allow differentiation between fragmentation-favoring matrices (hot or hard matrices, e.g., CHCA) and matrices with high initial velocities leading to less-pronounced fragmentation (cold or soft matrices, e.g., DHB, 3-HPA). The analytical problem to be solved thus determines also the choice of matrix. In cases of MS/MS experiments, such as for peptide sequence analysis, hot matrices such as HCCA are more favorable, whereas for the analysis of post-translational modifications which easily undergo metastable fragmentations, the choice of cold matrices (e.g., DHB for phosphorylated peptides, 3-HPA for glycopeptides [110]) can be recommended.

Several studies have been reported regarding the use of matrix additives. Such additives may include other matrices; for example, the addition of about 10% 5-methoxysalicylic acid to DHB ("super-DHB") showed an improvement in the measurement of proteins and glycoproteins [111,112]. An improved performance of MALDI MS for peptide mass fingerprinting upon the application of a mixture of CHCA and DHB was also reported [113].

Approaches based on the addition of sugars (e.g., fucose [114]), amines [115], or the combination of two matrices together with ammonium salts [116], were described to facilitate the analysis of oligonucleotides, whereas for the measurement of peptides the addition of nitrocellulose [117] or monoammonium phosphate or citrate [118] were found to be beneficial (see section 4.3). For the analysis of phosphorylated peptides, the addition of 1% phosphoric acid to DHB was found to decrease the limits of detection [119].

4.2.3
Sample Preparation

The quality of MALDI experiments is greatly influenced by the preparation techniques and the conditions applied for sample preparation [120]. A wide variety of methods and protocols for sample preparations have been described, amongst

Table 4.4 Matrices and possible preparation protocols.

Matrix	Name	Possible preparation protocol	Possible matrix solvents and concentrations
3-HPA	3-Hydroxypicolinic acid	Cocrystallization	50 mg mL^{-1} in acetonitrile/water (1:1) 25 mg mL^{-1} in methanol/water (1:1) 25 mg mL^{-1} in ethanol/water (1:1) 25 mg mL^{-1} in water
ATT	6-Aza-2-thiothymine	Cocrystallization	10 mg mL^{-1} in acetonitrile/water (1:1)
DHAP	2,6-Dihydroxyacetophenone	Cocrystallization or Surface preparation	10 mg mL^{-1} in acetonitrile/water (1:1)
DHB	2,5-Dihydroxybenzoic acid	Cocrystallization	10 mg mL^{-1} in ATW
DHBs	"super DHB2" (mixture of DHB with 10% HMBA)	Surface preparation	20 mg mL^{-1} DHB with 10% HMBA in ATW
Ferulic acid	4-Hydroxy-3-methoxycinnamic acid	Cocrystallization	8 mg mL^{-1} in water 8 mg mL^{-1} in TFA/acetonitrile/water (1:1:2)
HABA	2-(4'-Hydroxy-phenylazo)-benzoic acid	Surface preparation	2.5 mg mL^{-1} in water/acetone (1:1)
CHCA	α-Cyano-4-hydroxycinnamic acid	Surface preparation or Cocrystallization	40 mg mL^{-1} in acetone 5 mg mL^{-1} in ATW
SA	Salicylic acid	Cocrystallization	20 mg mL^{-1} in ATW
SINA	Sinapinic acid 3,5-Dimethoxy-4-hydroxycinnamic acid	Surface preparation oder Cocrystallization	10 mg mL^{-1} in ATW
THAP	2,4,6-Trihydroxyacetophenone hydrate	Cocrystallization or Surface preparation	10 mg mL^{-1} in acetonitrile/water (1:1)

ATW, 0.1% TFA in water/acetonitrile (1:1); DHB, 2,5-dihydroxybenzoic acid; HMBA, 2-hydroxy-5-methoxy-benzoic acid; TFA, trifluoroacetic acid.

which the most frequently used are the dried droplet method, the thin-layer or fast evaporation method, and the sandwich method.

The *dried droplet method* is the original sample preparation method described in the earliest publications about MALDI MS [90]. The sample solution and matrix solution are either mixed directly on target or premixed in an Eppendorff tube and then deposited and cocrystallized on the MALDI target. The protocol works well for the majority of matrices used, and even tolerates the presence of small amounts of salts (see below). For more challenging samples such as hydrophobic peptides, the addition of 5 to 30% formic acid rather than 0.1% trifluoroacetic acid (TFA)/isopropanol can be recommended when CHCA is applied as matrix.

In the *thin-layer method*, the matrix and sample are handled separately, which simplifies sample preparation and has also been shown to improve the sensitivity and resolution attainable in peptide analysis, especially for the peptide mass fingerprint workflow [121]. The matrix is dissolved in a highly volatile solvent (e.g., acetone) and then pipetted onto the target. After evaporation of the solvent, a very homogeneous crystalline matrix layer is formed on the target. The aqueous analyte solution is placed on top of this layer and allowed to dry. Of note, the matrix should be insoluble (or only moderately soluble) in the solvent of the analyte. This procedure allows the production of prespotted and long-lived matrix preparations, which are also commercially available on disposable MALDI targets. A further advantage is that contaminants such as salts preferentially crystallize on the surface of the preparations, and thus can be depleted by simple washing of the preparation with a small amount of acidified (e.g., 0.1% TFA) aqueous solution. This method is frequently used for preparations applying CHCA, SINA and ferulic acid. In TOF/TOF-type instruments, the thin layer method is typically not used, because of the limited stability of the samples exposed to the higher laser fluences needed for MS/MS acquisitions. In a modification of the fast evaporation method – the sandwich method [122] – a second layer of matrix is placed on a thin-layer preparation.

The addition of nitrocellulose to the matrix leads to a binding of the peptides, thus allowing for a facilitated washing of the samples on target [117]. Further, in many cases a significant enhancement of the signal-to-noise ratio has been observed upon application of this method. Briefly, a solution of $20\,\text{mg mL}^{-1}$ nitrocellulose in acetone is mixed 1 : 1 with isopropanol, and $20\,\text{mg mL}^{-1}$ dry, recrystallized CHCA is added to this solution. The mixture is pipetted onto the target; when the analyte mixture is not acidified, the matrix should be acidified with 0.1% TFA prior to application of the analyte onto the matrix layer.

Factors such as solvent composition, pH or temperature can influence the rates of matrix–analyte cocrystallization, and thus the quality of the preparations. The analyte must be soluble in the final matrix–analyte mixture, especially when the dried droplet method is applied. The most commonly used solvent for the matrix are aqueous mixtures with organic solvents (e.g., methanol, isopropanol or acetonitrile); in most cases, the solvent is acidified with 0.1% TFA trifluoroacetic acid or other volatile organic acids. Non-volatile solutes, such as glycerol, polyethylene glycol and N,N-dimethylformamide, interfere strongly with matrix crystallization and must therefore be avoided.

When the solvents cannot be removed prior to MALDI sample preparation, a possible alternative sample preparation method is the *crushed crystal method*. Here, a saturated solution of the matrix (DHB, SINA or CHCA) is applied onto the target. After evaporation of the solvent, the crystals are mechanically crushed using a spatula or a microscope glass slide; the plate is then cleaned from loose crystals by applying pressurized air. Subsequently, a mixture of analyte with the same matrix is pipetted on top of the crushed crystals [123].

The amount of organic solvent used, as well as its volatility, determine the speed of crystallization and thus the shape of the crystals formed on target. Slow crystallization favors separation of the analyte and contaminants, thus leading to a cleaning of samples; however, a drawback here is the formation of so-called "hot-spots" typical for inhomogeneous sample preparations. Heating of the sample should be avoided, but the solvent evaporation process can be supported by a gentle stream of dry nitrogen or air.

In order to guarantee optimal crystallization and function, matrix stock solutions should be prepared freshly or stored at maximum for only a few days at 4 °C in the dark, and should be centrifuged prior to use. Dried MALDI samples are quite stable and can be stored in the dark and under dry conditions; under inert gas they are stable for months and thus allow the storage of important samples. Typical matrix concentrations are in the range of a few mg mL^{-1} (see Table 4.4 for possible preparation protocols). For most matrices the use of saturated solutions cannot be recommended due to the production of very inhomogeneous preparations and possible contamination of the ion source in the mass spectrometer. The matrix concentration to be applied also depends on the analyte amount to be measured, thus the molar matrix-to-analyte ratio (M/A) is an important factor. For medium to high molecular-weight analytes (e.g., peptides, proteins, polynucleotides), the best results are obtained at M/A of about 1000 : 1 to 100 000 : 1, but with larger amounts of analyte peak suppression effects can be observed [124]. Although the optimal sample amounts for peptides are in the femtomol range, the limits of detection depend on a number of factors, including the mass spectrometer and the measurement conditions used, the quality of sample preparation, the presence of contaminants, and also (due to the phenomenon of peak suppression) on the complexity of the sample. Under optimal conditions, peptide mass fingerprint analyses can be performed down to the low attomol range (Table 4.5).

Biological samples frequently carry high loads of contaminants such as buffers, salts and detergents, which are important factors for degrading the results of MALDI MS experiments. Nevertheless, MALDI MS shows a moderate tolerance against a

Table 4.5 Common analyte concentrations applied for MALDI mass spectrometry.

Compound(s)	Concentration
Peptides and proteins	0.1 to 10 pmol µL^{-1}
Oligonucleotides	10 to 100 pmol µL^{-1}
Polymers	~ 100 pmol µL^{-1}

Table 4.6 Typical contaminants in protein/peptide samples and their impact on MALDI mass spectrometry [142].

No interference	Tolerable (<50 mM)[a]	Avoid
• TFA	• HEPES	• Glycerol
• Formic acid	• MOPS	• Sodium azide
• β-Mercaptoethanol	• Tris	• DMSO
• DTT	• NH_4OAc	• SDS
• Volatile organic solvents	• Octyl glucoside	• Phosphate
• HCl		• NaCl
• NH_4OH		• 2 M urea
• Acetic acid		• 2 M guanidine hydrochloride

[a] Minimizing buffer concentrations improves performance.
Use the minimum needed to control pH.
DTT, dithiothreitol; SDS, sodium dodecylsulfate.

number of common substances used in proteomics (Table 4.6). Typical contamination effects appear as rings of contaminants that form around the crystallized matrix, that the matrix forms a glossy, non-crystalline layer, or that the droplet of spreads over a wide area.

At high contaminant concentrations, or in presence of non-tolerable substances, a sample clean-up may be necessary. In some preparation methods, a separation of salts and analytes during sample crystallization can be observed, thus leading to a purification of the sample. The main sample clean-up techniques focus on the removal of the buffer salts, urea, guanidine, EDTA, glycerol, dimethyl sulfoxide (DMSO), detergents, etc., that are commonly observed as contaminants of peptide samples for MALDI analysis. In the following section, attention is focused on the clean-up of peptide and protein samples that form the analytes in proteomics workflows.

Possible steps to eliminate or reduce the effect of contaminants include dilution of the sample, washing procedures, dialysis, cation exchange and pipette-tip column chromatography for the desalting of the samples.

The simplest way to minimize effects caused by contaminants is to dilute the sample. The goal is to dilute contaminants to the point where they no longer interfere with the analysis of the sample. However, the dilution requires that a sufficiently high analyte concentration be maintained in the sample so as to provide acceptable data when diluted out.

The application of ammonium-containing buffers as washing media for crystallized preparations leads to a significant reduction of both alkali metal adducts and matrix clusters. The best results were obtained using 5 to 10 mM diammonium citrate buffer. The application of ammonium salts (especially 4–10 mM monobasic ammonium phosphate) as matrix additive induced a similar effect. A combination of both steps – that is, washing of the sample and the use of a matrix additive – can significantly improve spectral quality and also lead to improved signal-to-noise-ratios [118].

Drop dialysis is performed to remove low molecular-weight contaminants such as detergents or salts [125]. For example, in dried droplet preparations using DHB as matrix, peptide analytes are frequently found in the crystalline rim of the preparation, whereas salts are concentrated in the polycrystalline inner zone. In thin-layer preparations, a post-preparation cleaning by washing the samples with acidified aqueous solutions can be performed [126]. For this purpose, a droplet of 0.1% TFA is placed on the dried sample and, after a few seconds, is withdrawn carefully by using a paper tissue.

In the analysis of oligonucleotides, cation-exchange beads (e.g., Dowex cation exchange resin 50WX8-200) are used for the removal of alkali metal ions. The resin is prepared in the NH_4^+ form to allow the exchange of alkali metal cations versus ammonium for enhanced ion signals in MALDI mass spectrometry. The application of cation exchange should be kept to a minimum in protein workflows due to their affinity also for peptides.

Alternative sample preparations include the use of special pipette tips (e.g., Millipore ZipTip, Millipore, Billerica, USA) [127], or stage tips [128] which contain a bed of chromatography media (e.g., hydrophobic C4 or C18 phases), which allow a straightforward desalting and concentration of samples prior to MALDI sample preparation. A direct elution of the desalted, bound peptides with matrix solution onto the target is possible. The most effective sample clean-up can be reached, when samples are eluted directly from chromatographic column on the target (LC-MALDI-coupling). Further, tips are available (e.g., using C4 material) for the chromatographic clean-up of protein samples or metal chelators for the concentration of phosphopeptides [129]. The contamination of samples by keratins is a well-known problem in proteome research. Indeed, the presence of such additional analytes (as keratin peptides) may not only contribute to peak-suppression effects but also introduce the danger of overlap with real analytes, and so should be minimized. Therefore, all experiments which involve the manual handling of samples should be performed with great care; typically, gloves should be worn, and all manipulations performed under keratin-reduced conditions. For this the use of a protected working place is recommended. Keratin signals can be added to exclusion lists in the software packages of most MALDI instruments used for data interpretation, and are thus no longer used for database searches. Yet, keratin signals can also be used as internal standards for recalibration to provide increased mass accuracy.

Different forms and materials of sample plates (targets) are commercially available, the most common being polished steel plates, which can be applied to both batch and LC-MALDI experiments. Several suppliers also offer special plates which possess a hydrophobic surface containing hydrophilic holes of between 200 and 600 µm diameter (so-called "anchor plates"). After pipetting the matrix/analyte mixture onto these anchors, the droplets shrink and are concentrated in the hydrophilic region; this subsequently leads to a concentration not only of the analyte/matrix mixture but also of the contaminants, which can then be removed by washing procedures. The size – and hence the capacity – of samples to be applied on a particular target vary between suppliers, but the target form of 384 microtiterplate formats is now the most widely used formats.

4.2.4
LC-MALDI

The coupling of chromatographic separations and MALDI MS (Figure 4.8) is one of the most effective approaches for the bottom-up strategy in proteomics. Since the introduction of MALDI tandem mass spectrometers (e.g., MALDI TOF/TOF™; Applied Biosystems), the high separation efficiency of state-of-the-art chromatographic techniques has been combined with the potential and throughput capability of modern MALDI instrumentation. In addition to reducing the complexity of the sample, the most striking arguments for this technique are the simultaneous removal of contaminating substances, as well as the offline-handling of the MALDI samples, which allows for an easy, data-dependent reproduction of the experiments. As MALDI samples may be stored over longer time ranges, an archiving of analysis runs becomes possible. A typical workflow of an LC-MALDI experiment starts with the deposition of the chromatographic run on the target, after which the complete sample plate is generally analyzed in MS mode. Although a single analyte may be distributed over more than one spot or fraction on the MALDI target, modern mass spectrometric software recognizes this and subsequently

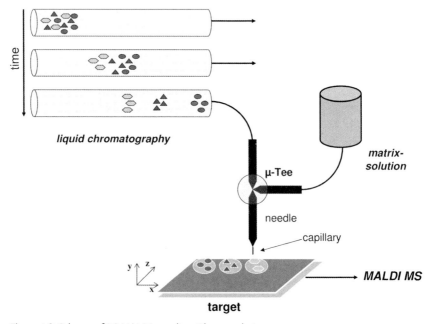

Figure 4.8 Scheme of LC-MALDI coupling. The sample is separated by liquid chromatography, the eluting analyte is then mixed with matrix solution in an µ Tee and codeposited on the target, which is placed on a table movable in the xyz-direction. Following this, offline MALDI MS and MSMS can be performed.

performs the MS/MS experiments (the final stage of this LC-MALDI experiment) only on those positions where this signal is the most intensive.

The major challenges in coupling these two orthogonal techniques are to mix the eluting sample with the matrix, and the deposition on the MALDI target, thereby retaining the chromatographic separation. In principle, two different approaches for sample preparation are possible in LC-MALDI MS. In the first method, a thin layer of matrix is placed on the target and the sample is spotted automatically onto this layer. One problem here is that the sample is dispersed due to capillary effects; hence, the most common approach is to use a variation of the dried droplet method where the eluting fractions are automatically mixed with the matrix solution and the mixture is then spotted onto the target. It is also possible to spot the analyte prior to the matrix, or vice versa. Several automated interfaces (spotting robots) are available commercially for this purpose.

A number of parameters must be optimized prior to proteome analysis. The optimal matrix concentration must be adapted to the analyte amounts and thus to the chromatographic set-up chosen. The use of diluted matrix solutions is recommended, first to avoid source contamination, and second to prevent unwanted crystallization of the solvent on the spotting device. The success of analyte transfer onto a MALDI target depends on the flow from the LC column and the spotting frequency. In the case of a highly efficient separation showing low peak widths, the spotting frequency may be in the range of only a few seconds.

In order to increase the accuracy of the measurement, it is common practice to add small amounts (e.g., 5–10 fmol μL^{-1}) of a known standard peptide to the matrix solution as an internal standard. The concentration of this internal standard should be kept to a minimum in order to avoid any danger of additional peak suppression. One peptide which is well-suited to this purpose is Glu1-Fibrinopeptide B, with a monoisotopic m/z of 1570.6774. As mentioned earlier for gel-based approaches, the MALDI targets can be stored after sample preparation. The best storage conditions are dry and dark environments, and the use of an inert gas is recommended. Plates stored under these conditions will remain stable for months, without any significant loss of spectral quality.

4.2.5
Application Example

4.2.5.1 Gel-Based Workflow

The gel-based approach was used for the analysis of the cytosolic proteome of the wild-type (strain ATCC13032) of the bacterium *Corynebacterium glutamicum*. This organism is of major biotechnological importance, and is used in the industrial-scale production of amino acids such as glutamate and lysine. This non-pathogenic organism also serves as a model for other, pathogenic, strains.

After cultivation in batch culture, the bacterial cells were disrupted by use of a swinging mill, and the cytosolic proteome was then separated using 2-D gel electrophoresis. By using colloidal-Coomassie staining, more than 600 proteins could be detected; a small region of this gel is illustrated in Figure 4.9. The spots were excised

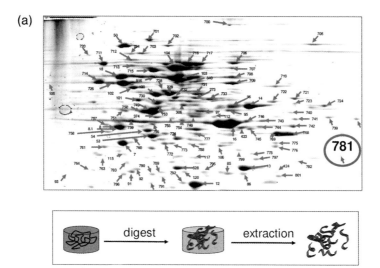

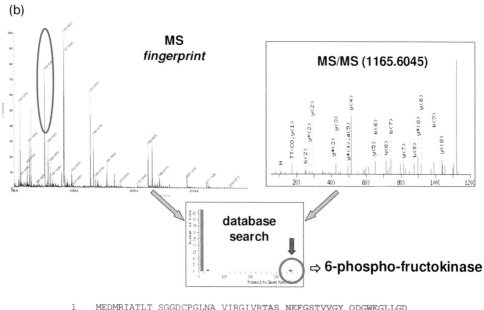

```
  1   MEDMRIATLT  SGGDCPGLNA  VIRGIVRTAS  NEFGSTVVGY  QDGWEGLLGD
 51   RRVQLYDDED  IDRILLRGGT  ILGTGRLHPD  KFKAGIDQIK  ANLEDAGIDA
101   LIPIGGEGTL  KGAKWLSDNG  IPVVGVPKTI  DNDVNGTDFT  FGFDTAVAVA
151   TDAVDRLHTT  AESHNRVMIV  EVMGRHVGWI  ALHAGMAGGA  HYTVIPEVPF
201   DIAEICKAME  RRFQMGEKYG  IIVVAEGALP  REGTMELREG  HIDQFGHKTF
251   TGIGQQIADE  IHVRLGHDVR  TTVLGHIQRG  GTPTAFDRVL  ATRYGVRAAR
301   ACHEGSFDKV  VALKGESIEM  ITFEEAVGTL  KEVPFERWVT  AQAMFG
```

Figure 4.9 (legend see p. 87)

manually, destained and the protein digested in-gel using trypsin as protease. The resultant digest was mixed with CHCA as matrix and measured both in MS and MS/MS mode. For the spot no. 781 shown here, 17 tryptic peptides could be detected; by using MS/MS analysis in a MALDI TOF/TOF™ mass spectrometer, the protein was identified unambiguously as 6-phospho-fructokinase in the MASCOT database search.

4.2.5.2 Application Example: LC-MALDI Workflow

A common laboratory animal model of chemical hepatocarcinogenesis was used successfully to demonstrate the early detection of potential biomarkers related to the classical toxicological endpoint, cancer. Liver carcinoma and adenoma were induced in male Wistar rats after 25 weeks by treatment with the liver carcinogen N-nitrosomorpholine (NNM). Then, by using 2-D-PAGE in combination with MS-based protein identification, statistically significantly deregulated proteins were visualized in NNM-exposed rat liver tissue at both late and early time-points. These proteins are considered to have predictive power for chemically induced liver carcinogenicity in rats, and the biological relevance of these early safety biomarkers has been proven [130]. However, in order to ensure reliable results, independent technology platforms must be used to prevalidate these protein expression patterns.

For this purpose, a quantitative LC-MALDI approach was applied using the iTRAQ™ reagent technology (Applied Biosystems) [131]. Following peptide labeling and MS/MS-based identification and quantitation, the regulation of 26 2-D-PAGE/MS-derived proteins could be confirmed (Figure 4.10). Proteins such as heat shock protein 90-beta, annexin A5, ketohexokinase, N-hydroxyarylamine sulfotransferase, ornithine aminotransferase and adenosine kinase showed very close changes using both proteomic quantitation strategies. Furthermore, 11 of the potential early safety biomarkers, which might have predictive power for chemically induced liver carcinogenicity, were prevalidated. The iTRAQ™ reagent analysis delivered further potential safety biomarkers with biological relevance to the processes of hepatocarcinogenicity, including the placental form of glutathione S-transferase (GST-P), carbonic anhydrase, and aflatoxin B1 aldehyde reductase. These results demonstrate the power of this new quantitation strategy for the verification of protein regulation.

Figure 4.9 Analysis of the cytosolic proteome of *Corynebacterium glutamicum*. (a) Separation of the proteins by 2-D gel electrophoresis (only a part of the gel is shown); staining: colloidal Coomassie. The spots labeled with arrows were excised and analyzed by peptide mass fingerprint and peptide fragment-fingerprint analysis. (b) MS fingerprint (left) and MS/MS spectrum of the peak at m/z 1165.6045 (circle in MS-spectrum) measured using an Applied Biosystems 4800 MALDI ToF/ToF analyzer mass spectrometer. The combination of both MS and MS/MS data allowed the identification of the protein as 6-phospho-fructokinase in a MASCOT database search. The sequence coverage (bold residues) was 70%. The underlined sequence represents the peptide sequenced by MS/MS in this example. (Experiments performed by Maria Lasaosa, Technische Biochemie, Universität des Saarlandes, Germany).

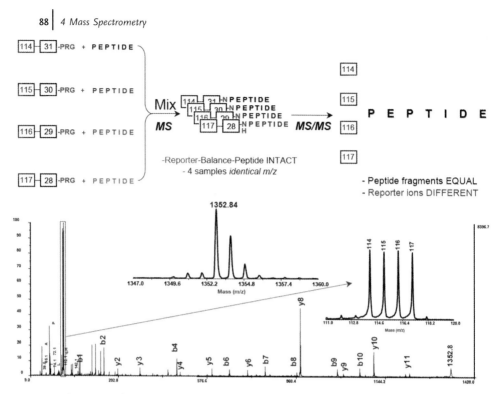

Figure 4.10 The iTRAQ™ reagents (Applied Biosystems) consist of four unique charged reporter groups, a unique neutral balance group, and a peptide reactive group to label all N-termini and lysine amino acids on the level of the peptides. Due to a total mass of 145 Da of the reagent the labeled peptides are identical in mass and, therefore, also identical in single MS mode. In MS/MS mode, low-mass reporter ion signals allow the peptide quantitation while peptide fragment ion signals overlap again and allow protein identification.

4.2.6
Summary

Despite great success having been achieved in the development of the hardware for modern MALDI mass spectrometers, sample preparation remains one of the most critical issues influencing the results of the analysis. New users are encouraged to exercise their sample preparation by measuring standard substances prior to the measurement of real samples of biological interest.

Compared to electrospray MS, MALDI shows a slightly increased tolerance against impurities, although these should be kept to a minimum whenever possible. A number of techniques are available which allow a fast and straightforward purification of the samples. The best results, in terms of accuracy, resolution and ultimately identification of the analytes, are obtained whenever the optimized sample preparations can be guaranteed. Therefore, this should be the focal point in all proteome analyses.

4.2.7
Perspectives

The high sample throughput that may be achieved with state-of-the-art MALDI instruments makes this technique the almost perfect tool for proteome analysis. Unfortunately, the methods of sample preparation prior to MALDI analysis are increasingly becoming bottlenecks in these analyses. Therefore, the development of new methods for sample preparation represents a major step towards higher sampling rates. In terms of sample preparation, the advent of multidimensional chromatography should lead, at least potentially, to a new era of proteome analysis.

A second analytical bottleneck is related to the area of data interpretation and the handling of huge datasets. Here, it is not only the mass spectrometric data that are important, but how these data are interpreted within a biological context, for example in biological networks in the sense of systems biology.

One major question in proteomics relates not only to the qualitative but also to the quantitative analysis of protein expression. Here, new developments, such as quantification on the level of MS/MS spectra (as realized in the iTRAQ reagent method [131]) promise great potential for the future. Another important issue is the analysis of post-translational modifications, and a potential key for this purpose is the increase in sequence coverage achievable in MALDI MS experiments. In addition to identification, determining the degree of modification will also be an important – but nonetheless challenging – step for the elucidation of biological impact.

For a deeper understanding of cellular processes, the simple identification and quantification of proteins will not be sufficient. Rather, it will be necessary to obtain reliable data concerning the molecular function of proteins, the molecular processes involved, and the interactions, not only with other proteins but also with other molecules in the cells. The development of mass spectrometric methods suited to determining the functions either of complete metabolic pathways, or at least of selected classes of enzymes within cells, is envisaged. The same holds true for the detection of macromolecular complexes. The analysis of non-covalent complexes of proteins with other proteins, DNA, RNA, lipids or metabolites by MALDI MS is not straightforward, and successful applications are more the exception than the rule. Nevertheless, studies on crosslinking [132,133], intensity fading [134] or hydrogen–deuterium (H/D)-exchange [135], followed by analysis on the peptide level, have been shown to possess the potential for further progress in this important field of functional proteomics.

The analysis of intact proteins also promises some interesting features, notably in the case of detecting post-translational modifications. Whilst MALDI MS shows clear potential in this area, the major drawbacks have until now been due to extensive broadening of the protein signals by fragmentation, to adduct formation, and to limited sensitivity. Potential ways in which these problems might be minimized include the development of new sample preparation methods or the introduction of new matrix systems. The formation of multiply charged ions, as well as the development of new detector systems, may also (potentially) increase the as-yet relatively poor sensitivity of MALDI MS for the measurement of intact proteins.

4.2.8
Recipes for Beginners

4.2.8.1 Sample Spotting Techniques

Dried Droplet Preparation

This is the most commonly method used applicable for all matrices:

- Premix sample and matrix solution (5 mg mL^{-1}, 70% acetonitrile/30% water/ 0.1% TFA) in 200 µL Eppendorf tube; spot 0.5–1.0 µL onto sample plate and air-dry.

- For small sample amounts, deposit 0.5 µL of sample on plate, then add 0.5 µL matrix on top; mix and allow to dry.

Thin-Layer Preparation

- Prepare a 5 mg mL^{-1} solution of α-Cyano-4-hydroxycinnamic acid (CHCA) in acetone. Apply 0.5 µL to the sample plate and allow to spread and dry.

- Place 0.5 µL sample in 0.1% TFA on top of the matrix layer and leave to dry on air.

- If the sample is not acidified, then 0.5 µL 0.1% TFA can be added to the plate prior to application of the sample.

- On-plate washing of sample with cold water can be performed after drying.

4.2.8.2 Sample Cleaning Procedures

Drop Dialysis

- Use a Millipore membrane, type VS, pore size 25 µm, diameter 25 mm.
- Fill a 250–400 mL container with deionized water.
- Float the membrane on the water (shiny side up).
- Place about 10 µL of sample solution on the membrane.
- Add 1 µL acetonitrile to the sample spot to increase surface area.
- Allow to sit for ~45 min.
- Remove an aliquot with pipette, add matrix, and spot plate.

Cation Exchange Beads for Removal of Alkali Metal Ions

- Preparation of resin in the NH_4^+ form:
  - Use Dowex cation exchange resin 50WX8-200, 8% crosslinked, H^+ loaded (can be purchased from Sigma).
  - Stir resin beads in 2×volume of 1 M ammonium acetate overnight.
  - Filter, wash with deionized water, acetone and hexane.
  - Dry and store for use.

- Use of the resin:
  - Place ~0.1 mg of beads on a clean piece of Parafilm.
  - Add 5 µL of sample and an equal amount of matrix to the beads to make a slurry of approx. 50% beads.
  - Slowly mix up and down with the pipette 10 to 15 times.
  - Allow the beads to settle for 15–30 s.
  - Pipette supernatant onto the sample plate.
  - Change tip to avoid carrying over beads to sample plate.

Note: Do not use with positively charged species!

Use of ZipTip C18 for Desalting of Peptides and Proteins

- *Condition* the ZipTip with 10 µL of acetonitrile, then with 10 µL of 50% acetonitrile/0.1% TFA, then 2×10 µL of 0.1% TFA.

- *Load* the sample onto the ZipTip by pipetting 5–10 µL sample up and down several times and discarding the liquid.

- *Wash* C_{18}-tip with 3×10 µL of 0.1% TFA to remove salts.

- *Elute* the sample from the ZipTip with 30–70% acetonitrile or elute directly into the matrix (e.g., CHCA in 50% acetonitrile/0.1% TFA); a minimal volume of ~3 µL can be used.

References

90 Karas, M. and Hillenkamp, F. (1988) Laser desorption ionization of proteins with molecular masses exceeding 10,000 daltons. *Anal. Chem.*, **60**, 2299–2301.

91 Cohen, L.H. and Gusev, A.I. (2002) Small molecule analysis by MALDI mass spectrometry. *Anal. Bioanal. Chem.*, **373**, 571–586.

92 Dreisewerd, K. (2003) The desorption process in MALDI. *Chem. Rev.*, **103**, 395–426.

93 Karas, M. and Kruger, R. (2003) Ion formation in MALDI: the cluster ionization mechanism. *Chem. Rev.*, **103**, 427–440.

94 Knochenmuss, R. and Zenobi, R. (2003) MALDI ionization: the role of in-plume processes. *Chem. Rev.*, **103**, 441–452.

95 Colby, S.M., King, T.B., Reilly, J.P. and Lubman, D.M. (1994) Improving the resolution of matrix-assisted laser desorption/ionization time-of-flight mass spectrometry by exploiting the correlation between ion position and velocity. *Rapid Commun. Mass Spectrom.*, **8**, 865–868.

96 Brown, R.S. and Lennon, J.J. (1995) Mass resolution improvement by incorporation of pulsed ion extraction in a matrix-assisted laser desorption/ionization linear time-of-flight mass spectrometer. *Anal. Chem.*, **67**, 1998–2003.

97 Horneffer, V., Gluckmann, M., Kruger, R., Karas, M., Strupat, K. and Hillenkamp, F. (2006) Matrix-analyte-interaction in MALDI MS: pellet and nano-electrospray preparations. *Int. J. Mass Spectrom.*, **249/250**, 426–432.

98 Schleuder, D., Hillenkamp, F. and Strupat, K. (1999) IR-MALDI-mass

analysis of electroblotted proteins directly from the membrane: comparison of different membranes application to on-membrane digestion, and protein identification by database searching. *Anal. Chem.*, **71**, 3238–3247.

99 Ehring, H., Karas, M. and Hillenkamp, F. (1992) Role of photoionization and photochemistry in ionization processes of organic molecules and relevance for matrix-assisted laser desorption ionization mass spectrometry. *Org. Mass Spectrom.*, **27**, 472–480.

100 Beavis, R.C. and Chait, B.T. (1989) Cinnamic acid derivatives as matrices for ultraviolet laser desorption mass spectrometry of proteins. *Rapid Commun. Mass Spectrom.*, **3**, 432–435.

101 Beavis, R.C., Chaudhary, T. and Chait, B.T. (1992) Alpha-cyano-4-hydroxycinnamic acid as a matrix for matrix assisted laser desorption mass spectrometry. *Org. Mass Spectrom.*, **27**, 156–158.

102 Strupat, K., Karas, M. and Hillenkamp, F. (1991) 2,5-Dihydroxy benzoic acid – a new matrix for laser desorption/ionization mass-spectrometry. *Int. J. Mass Spectrom. Ion Proc.*, **111**, 89–102.

103 Juhasz, P., Costello, C.E. and Biemann, K. (1993) Matrix assisted laser desorption ionization mass spectrometry with 2-(4-hydroxyphenylazo)benzoic acid matrix. *J. Am. Soc. Mass Spectrom.*, **4**, 399–409.

104 Pieles, U., Zurcher, W., Schar, M. and Moser, H.E. (1993) Matrix-assisted laser desorption ionization time-of-flight mass spectrometry: a powerful tool for the mass and sequence analysis of natural and modified oligonucleotides. *Nucleic Acids Res.*, **21**, 3191–3196.

105 Gorman, J.J., Ferguson, B.L. and Nguyen, T.B. (1996) Use of 2,6-dihydroxy-acetophenone for analysis of fragile peptides, disulphide bonding and small proteins by matrix-assisted laser desorption/ionization. *Rapid Commun. Mass Spectrom.*, **10**, 529–536.

106 Pitt, J.J. and Gorman, J.J. (1996) Matrix-assisted laser desorption/ionization time-of-flight mass spectrometry of sialylated glycopeptides and proteins using 2,6-dihydroxyacetophenone as a matrix. *Rapid Commun. Mass Spectrom.*, **10**, 1786–1788.

107 Wu, K.J., Steding, A. and Becker, C.H. (1993) Matrix-assisted laser desorption time-of-flight mass spectrometry of oligonucleotides using 3-hydroxy-picolinic acid as an ultraviolet-sensitive matrix. *Rapid Commun. Mass Spectrom.*, **7**, 142–146.

108 Wu, K., Shaler, T.A. and Becker, B.T. (1994) Time-of-flight mass spectrometry of underivatized single-stranded DNA oligomers by matrix-assisted laser desorption. *Anal. Chem.*, **66**, 1637–1645.

109 Tang, K., Taranenko, N.I., Allman, S.L., Chen, C.H., Chang, L.Y. and Jacobson, K.B. (1994) Picolinic acid as a matrix for laser mass spectrometry of nucleic acids and proteins. *Rapid Commun. Mass Spectrom.*, **8**, 673–677.

110 Karas, M., Bahr, U., Strupat, K., Hillenkamp, F., Tsarbopoulos, A. and Pramanik, B.N. (1995) Matrix dependence of metastable fragmentation of glycoproteins in MALDI TOF mass-spectrometry. *Anal. Chem.*, **67**, 675–679.

111 Karas, M., Ehring, H., Nordhoff, E., Stahl, B., Strupat, K., Hillenkamp, F., Grehl, M. and Krebs, B. (1993) Matrix-assisted laser-desorption ionization mass-spectrometry with additives to 2,5-dihydroxybenzoic acid. *Org. Mass Spectrom.*, **28**, 1476–1481.

112 Bahr, U., Stahl, B., Zeng, J., Gleitsmann, E. and Karas, M. (1997) Delayed extraction time-of-flight MALDI mass spectrometry of proteins above 25,000 Da. *J. Mass Spectrom.*, **32**, 1111–1116.

113 Laugesen, S. and Roepstorff, P. (2003) Combination of two matrices results in improved performance of MALDI MS for peptide mass mapping and protein analysis. *J. Am. Soc. Mass Spectrom.*, **14**, 992–1002.

114 Distler, A.M. and Allison, J. (2001) Improved MALDI-MS analysis of oligonucleotides through the use of fucose as a matrix additive. *Anal. Chem.*, **73**, 5000–5003.

115 Distler, A.M. and Allison, J. (2001) 5-Methoxysalicylic acid and spermine: a new matrix for the matrix-assisted laser desorption/ionization mass spectrometry analysis of oligonucleotides. *J. Am. Soc. Mass Spectrom.*, **12**, 456–562.

116 Zhang, L.K. and Gross, M.L. (2000) Matrix-assisted laser desorption/ionization mass spectrometry methods for oligodeoxynucleotides: improvements in matrix detection limits, quantification, and sequencing. *J. Am. Soc. Mass Spectrom.*, **11**, 854–865.

117 Kussmann, M., Nordhoff, E., Rahbek Nielsen, H., Haebel, S., Rossel Larsen, M., Jakobsen, L., Gobom, J., Mirgorodskaya, E., Kroll Kristensen, A., Palm, L. and Roepstorff P. (1997) Matrix-assisted laser desorption/ionization mass spectrometry sample preparation techniques designed for various peptide and protein analytes. *J. Mass Spectrom.*, **32**, 593–601.

118 Smirnov, I.P., Zhu, X., Taylor, T., Huang, Y., Ross, P., Papayanopoulos, I.A., Martin, S.A. and Pappin, D.J. (2004) Suppression of alpha-cyano-4-hydroxycinnamic acid matrix clusters and reduction of chemical noise in MALDI-TOF mass spectrometry. *Anal. Chem.*, **76**, 2958–2965.

119 Kjellstrom, S. and Jensen, O.N. (2004) Phosphoric acid as a matrix additive for MALDI MS analysis of phosphopeptides and phosphoproteins. *Anal. Chem.*, **76**, 5109–5117.

120 Cohen, S.L. and Chait, B.T. (1996) Influence of matrix solution conditions on the MALDI-MS analysis of peptides and proteins. *Anal. Chem.*, **68**, 31–37.

121 Vorm, O., Roepstorff, P. and Mann, M. (1994) Improved resolution and very high-sensitivity in MALDI TOF of matrix surfaces made by fast evaporation. *Anal. Chem.*, **66**, 3281–3287.

122 Dai, Y., Whittal, R.M. and Li, L. (1999) Two-layer sample preparation: a method for MALDI-MS analysis of complex peptide and protein mixtures. *Anal. Chem.*, **71**, 1087–1091.

123 Xiang, F. and Beavis, R.C. (1994) A method to increase contaminant tolerance in protein MALDI by the fabrication of thin protein-doped polycrystalline films. *Rapid Commun. Mass Spectrom.*, **8**, 199–204.

124 Chan, T.W.D., Colburn, A.W., Derrick, P.J., Gardiner, D.J. and Bowden, M. (1992) Suppression of matrix ions in ultraviolet-laser desorption – scanning electron-microscopy and Raman-spectroscopy of the solid samples. *Org. Mass Spectrom.*, **27**, 188–194.

125 Gorisch, H. (1988) Drop dialysis: time course of salt and protein exchange. *Anal. Biochem.*, **173**, 393–398.

126 Kussmann, M., Nordhoff, E., Rahbek Nielsen, H., Haebel, S., Rossel Larsen, M., Jakobsen, L., Gobom, J., Mirgorodskaya, E., Kroll Kristensen, A., Palm, L. and Roepstorff, P. (1997) Matrix-assisted laser desorption/ionization mass spectrometry sample preparation techniques designed for various peptide and protein analytes. *J. Mass Spectrom.*, **32**, 593–601.

127 Beranova-Giorgianni, S. and Desiderio, D.M. (2000) Mass spectrometry of the human pituitary proteome: identification of selected proteins. *Rapid Commun. Mass Spectrom.*, **14**, 161–167.

128 Rappsilber, J., Ishihama, Y. and Mann, M. (2003) Stop and go extraction tips for matrix-assisted laser desorption/ionization nanoelectrospray, and LC/MS sample pretreatment in proteomics. *Anal. Chem.*, **75**, 663–670.

129 Majors, R.E. and Shukla, A. (2005) Micropipette tip-based sample preparation for bioanalysis. *LCGC North America*, 23.

130 Fella, K., Gluckmann, M., Hellmann, J., Karas, M., Kramer, P.J. and Kroger, M. (2005) Use of two-dimensional gel electrophoresis in predictive toxicology:

identification of potential early protein biomarkers in chemically induced hepatocarcinogenesis. *Proteomics*, **5**, 1914–1927.

131 Ross, P.L., Huang, Y.N., Marchese, J.N., Williamson, B., Parker, K., Hattan, S., Khainovski, N., Pillai, S., Dey, S., Daniels, S., Purkayastha, S., Juhasz, P., Martin, S., Bartlet-Jones, M., He, F., Jacobson, A. and Pappin, D.J. (2004) Multiplexed protein quantitation in Saccharomyces cerevisiae using amine-reactive isobaric tagging reagents. *Mol. Cell. Proteomics*, **3**, 1154–1169.

132 Kuhn-Holsken, E., Lenz, C., Sander, B., Luhrmann, R. and Urlaub, H. (2005) Complete MALDI-ToF MS analysis of cross-linked peptide-RNA oligonucleotides derived from nonlabeled UV-irradiated ribonucleoprotein particles. *Rna*, **11**, 1915–1930.

133 Sinz, A., Kalkhof, S. and Ihling, C. (2005) Mapping protein interfaces by a trifunctional cross-linker combined with MALDI-TOF and ESI-FTICR mass spectrometry. *J. Am. Soc. Mass Spectrom.*, **16**, 1921–1931.

134 Yanes, O., Villanueva, J., Querol, E. and Aviles, F.X. (2005) Functional screening of serine protease inhibitors in the medical leech *Hirudo medicinalis* monitored by intensity fading MALDI-TOF MS. *Mol. Cell. Proteomics*, **4**, 1602–1613.

135 Shi, J., Koeppe, J.R., Komives, E.A. and Taylor, P. (2006) Ligand-induced conformational changes in the acetylcholine-binding protein analyzed by hydrogen-deuterium exchange mass spectrometry. *J. Biol. Chem.*, **281**, 12170–12177.

136 Juhasz, P., Papayannopulos, I.A., Zeng, C., Papov, V. and Biemann, K. (1992) The Utility of Matrix-assisted Laser Desorption for the direct Analysis of enzymatic Digests of Proteins. *Proceedings of the 40th ASMS Conference on Mass Spectrometry and Allied Topics Washington, DC, USA*, 1913–1914.

137 Beavis, R.C. and Chait, B.T. (1989) Matrix-assisted laser-desorption mass spectrometry using 355 nm radiation. *Rapid Commun. Mass Spectrom.*, **3**, 436–439.

138 Karas, M., Nordhoff, E., Stahl, B., Strupat, K. and Hillenkamp, F. (1992) Matrix-Mixtures for a Superior Performance of Matrix-Assisted Laser Desorption Ionization Mass Spectrometry. *Proceedings of the 40th ASMS Conference on Mass Spectrometry and Allied Topics Washington, DC, USA*, pp. 368–369.

139 Karas, M., Bahr, U. and Stahl-Zeng, J. (1996) Large ions: their vaporization, detection and structural analysis. In: *Steps Towards a More Refined Picture of the Matrix Function in UV MALDI* (eds T. Baer C.Y. Ng and I. Powis), John Wiley, pp. 27.

140 Papac, D.I., Wong, A. and Jones, A.J. (1996) Analysis of acidic oligosaccharides and glycopeptides by matrix-assisted laser desorption/ionization time-of-flight mass spectrometry. *Anal. Chem.*, **68**, 3215–3223.

141 Zhu, Y.F., Chung, C.N., Taranenko, N.I., Allman, S.L., Martin, S.A., Haff, L. and Chen, C.H. (1996) The study of 2,3, 4-trihydroxyacetophenone and 2,4,6-trihydroxyacetophenone as matrices for DNA detection in matrix-assisted laser desorption/ionization time-of-flight mass spectrometry. *Rapid Commun. Mass Spectrom.*, **10**, 383–388.

142 Swiderek, K., Alpert, A., Heckendorf, A., Nugent, K. and Patterson, S. (1997) Structural analysis of proteins and peptides in the presence of detergents: Tricks of the trade. *ABRF: News Methods and Reviews, December*, 17–25.

4.3
Sample Preparation for Label-Free Proteomic Analyses of Body Fluids by Fourier Transform Ion Cyclotron Mass Spectrometry

Cloud P. Paweletz, Nathan A. Yates, and Ronald C. Hendrickson

4.3.1
Introduction

Measurements of one or two proteins in serum or plasma have long been used as a diagnostic tool in patient health care. For example, B-type natriuretic peptide (BNP) is now routinely used to risk-stratify patients with heart failure and acute coronary syndromes [143]. The underlying concept, in this case, is the relationship between the concentration of each protein in the circulation and presumed damage to cardiac myocytes. Thus, while protein markers useful for certain types of heart disease are clearly established in the medical community, protein biomarkers are also notably absent for emerging health care issues, specifically neurological or lung diseases.

Given that only a fraction of the estimated 100 000 protein isoforms have been identified [144] in the blood circulation, let alone quantitated, it is believed that the human plasma proteome represents a largely untapped source of novel target engagement and diagnostic biomarkers [145]. Many biomarker measurement tools, such as ELISA, are "closed": each test must be designed to measure the concentration of a specific target analyte, and inherently limited as the identity of that analyte must be known beforehand. Clearly, approaches that are "open" and not dependent on antibody reagents are needed to take full advantage of the proteomic information content of the circulation. Quantitative protein profiling by mass spectrometry (MS) is a nascent field of clinical investigation in which global patterns of protein expression in patient samples are interpreted in support of both diagnosis and efficacy of therapy [144,146]. Mass spectrometry provides an "open" approach to identify, measure, and quantitate protein changes without prior knowledge of which specific protein to measure. This chapter is structured to provide the reader with a snapshot of current approaches for label-free liquid chromatography (LC)-MS profiling and, where warranted, offers a perspective on the future for clinical MS profiling.

Central to label-free LC-MS profiling is reproducible sample preparation [146], which has proven challenging for plasma and serum. Lower-abundance proteins are especially difficult to identify as these are embedded in the matrix of albumin ($\sim 10^8$ pg mL^{-1}), immunoglobins ($\sim 10^7$ pg mL^{-1}), lipoproteins ($\sim 10^6$ pg mL^{-1}) and other proteins spanning 10 orders of magnitude in concentrations. One solution has been to develop pre-fractionation protocols to eliminate the high-abundance proteins of plasma, thus leaving the lower-abundance proteins available for analysis. The use of commercially available albumin and IGg-depleting columns as a sample preparation for plasma and cerebral spinal fluid (CSF) prior to liquid chromatography Fourier transform ion cyclotron resonance (LC-FT-ICR) analysis was investigated by Ramstroem *et al.* [146]. In this analysis, reduced and alkylated tryptic digests of depleted and de-salted plasma samples were separated on fused silica capillaries

(10 cm; i.d. 200 μm) packed with 5 μm C18 particles coupled directly to a 9.4 T FT-ICR on-line. The investigators identified twice as many proteins in depleted samples as compared to neat (i.e., non-depleted) samples. The analytic challenge with this approach has been that this strategy necessarily leads to losses among the least abundant of the protein components. Furthermore, proteins such as albumin have been described as having their own system of interacting low-abundance proteins [146,149]. The challenge is, therefore, that the preliminarily removal of albumin or other high-abundance proteins from serum or plasma may necessitate corrections for yields of some low-abundance proteins in the depleted sample. An alternative albumin-depletion strategy was demonstrated by Zhang et al. [150]. Instead of immunodepleting albumin, Zhang et al. focused on capturing N-linked glycoproteins by selective conjugation of the sugar moiety and specific release with PNGase F. In initial proof of concept experiments, these authors showed that albumin was effectively removed and that their approach was amenable to the analysis of secreted serum proteins.

A massive depletion strategy on plasma that utilized both immunodepletion of the six most abundant proteins and the selection of N-linked glycopeptides coupled to FT-ICR was employed by Liu et al. [151]. Initially, this group depleted the six highest abundant proteins from 800 μL of crude plasma by using the multi-affinity removal column available from Agilent technologies. The plasma proteome was reduced even further by enriching for N-linked glycopeptides after oxidation with sodium periodate, coupling to hydrazide beads, digestion with trypsin, and release with PNGase F. Following the release of captured N-linked glycopeptides, the digests were separated by strong cation-exchange chromatography, and peptide identification was performed by data-dependent LC-MS/MS. The accuracy of these peptides was confirmed by subsequent mass measurement by FT-ICR. A total 2053 peptides belonging to 303 N-glycoproteins was identified. The increased mass resolution provided by FT-MS improved the ability to characterize specific isolated peptides in biological samples, whereas the observation of lower-abundance ions was aided by the reduction in chemical noise inherent in FT-MS. These initial proof-of-concept experiments utilized "Shotgun Proteomics" methods [152] that aim to produce large lists of peptides observed from thousands of tandem mass spectra (MS/MS) that are often recorded for a given sample. These qualitative results provide information about the components of a complex mixture, but do not compare between groups of patients (e.g., diseased versus normal, or treated versus non-treated).

LC-MS profiling methods that provide a quantitative measurements of abundance have been recently described. These methods differ from shotgun proteomics approaches in that they measure and compare the ion intensities recorded in the full-scan mass spectra, and are not dependent on the acquisition or interpretation of tandem mass spectra. This allows quantitative measurements to be made on all ions observed by the mass spectrometer, and not just those that are subject to an MS/MS experiment. This difference is of practical importance because the MS/MS acquisition rate of current mass spectrometers lags far behind the speed that would be required to collect an MS/MS spectrum for every ion in a full-scan spectrum. This MS/MS duty cycle limitation becomes acutely apparent when profiling complex mixtures such as plasma or CSF. In the past, quantitative (i.e., comparative) MS required some sort of label, either by incorporating them metabolically (e.g., stable incorporation of essential amino acids; SILAC [153]) or via chemical conjugation of stable isotopes (e.g., acetylation, or isotope-coded affinity tags, ICAT [154]). Some of the protein manipulations required are 30 years old, and therefore have been proven to be quite dependable and reliable. However, a whole new set of "label-free" mass spectrometric approaches are on the horizon, and are becoming readily the mainstay for direct comparative analyses of biofluids. The important elements of the label-free proteomic workflows are:

- the preservation of individual sample integrity by analyzing samples without the need to pool patient cohorts,
- reproducible biochemical sample preparation,
- interwoven and block experimental design to eliminate systematic bias,
- efficient and reproducible LC coupled to,
- high-resolution mass spectrometers [155,156].

Recently, statistically tools have been developed to enable label-free comparative proteomics [157–160]. At their core, label-free data analysis methods convert raw mass spectrometric data files into three-dimensional (3-D) images (x axis time, y axis = m/z, z axis = intensity), extract peptide features, aligned and normalize ions across data sets, and measure relative abundances of peptides based on peak areas (Figure 4.11). The result is a rank order list that shows a statistically significant ion characterized by its m/z, charge state, ion intensity, and elution time. Not using mass-tag labeling [153,154] simplifies sample preparation, and allows individual samples to be kept separate rather being pooled. This allows the quantitation of many features arising from peptides, and detailed analysis to be conducted of how well these features distinguish individual samples (rather than pools).

In subsequent studies, Wang *et al.* show that it was possible to identify as little as 100 fmol of horse myoglobin spiked into human plasma, based on relative ion abundance signals using statistical approaches on de-isotoped peptides [161]. However, their analysis assumed that all peptides could be de-isotoped by a single run which, depending on the cut-off variables, may miss lower-abundance peptides. Later, Aebersold's group introduced a software suite that aligns multiple LC runs across experiments and identifies differences on aligned data sets; the efficacy was demonstrated by identifying protein differences in serum between five male and five female mice [157]. Both of these approaches rely on the manipulation of raw mass spectrometric data before first finding the peptides. Prakash *et al.* [158] also performed comparisons from LC-MS runs directly from the raw spectra level by first generating "signal maps" that associate different signal from different runs across experiments, before aligning multiple data set. However, their algorithm was not tested on "real-time data". An alternative approach for label-free FT-ICR profiling was demonstrated by Fang and coworkers in Smith's laboratory, when label-free proteomic analyses were combined with accurate mass and time tags [159]. In this approach the first set of samples to be compared is digested with trypsin and

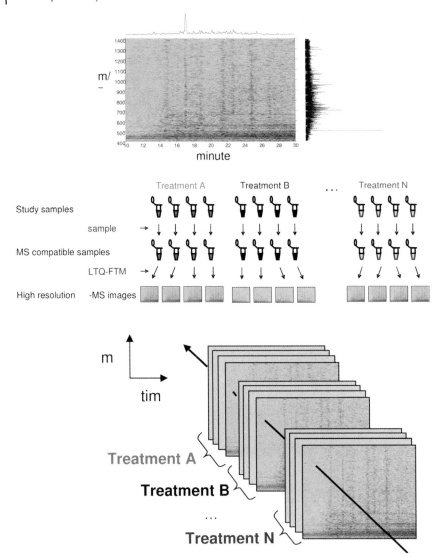

Figure 4.11 General strategy for label-free differential mass spectrometry (dMS). (a) Mass spectrometric spectra are converted to three-dimensional data cubes (x axis: time, y axis: m/z, and z axis: intensity). (b) Individual samples from different groups (i.e., treatment, time course, disease status) are run in an interwoven block design to eliminate statistical bias and individually without pooling of samples. (c) For each m/z and retention time pair, dMS determines the statistical significance (using the non-parametric Kruskal test) of the difference between the sets of intensities from samples in different conditions. dMS then looks for consecutive statistically significant points that persist in time at a particular m/z-value. This helps to reduce the problem of false-positives that can arise from performing a large number of statistical tests. Charge states are assigned, and signals that appear to arise from different charge states and isotopes of a single analyte are grouped.

analyzed by data-dependent LC-MS/MS to generate an initial list of peptides each of which contains an accurate mass and normalized LC time tags. "Correct" identifications (based on predefined cut-off values) and accurate mass and time tags were stored in a relational database, after which a second set of samples was analyzed by LC-FT-ICR. Subsequently, area ratios from both groups were compared if the same mass and time tag was identified based on the accurate mass alone. The analytical disadvantage here, however, is that this label-free approach is still dependent on data-dependent acquisition.

The present authors have developed a novel workflow – termed "differential mass spectrometry (dMS)" – which finds differences directly from the raw full spectrum mass spectrometric data files [155,160]. dMS determines the statistical significance (using the non-parametric Kruskal test) of the difference between the sets of intensities from samples in different conditions on a point-by-point basis from raw mass spectrometric data. Because LC-MS data is acquired as a function of time, dMS then looks for runs of statistically significant points that persist in time at a particular m/z. Once the user has a ranked list of differences, targeted MS/MS runs are acquired for subsequent protein identification. In this workflow, proteins are usually digested proteolytically to peptides with an enzyme such as trypsin, which cleaves the protein at the C-terminal side of lysine and arginine, prior to profiling by LC-MS. In this manner, information at the protein level is no longer retained. Information is then generated at the feature or peptide level. In theory, post-translational modifications that shift the mass of a peptide should be identifiable by this technique.

4.3.2
Perspective

Although extensive chromatography has shown promise for cataloging lower-abundant ions in plasma [162–165], the reproducible identification of lower-abundance ions is still difficult for quantitative shotgun LC-MS/MS methods. Data-dependent ion selection is governed by the number of peptide ions present in samples at any particular moment, and how many of those ions exceed a predetermined threshold. For complex samples, although the MS/MS acquisition rates of mass spectrometers has increased by orders of magnitude during the past decade, the MS/MS sampling rate is still by far exceeded and hence an undersampling error occurs. Fortunately, investigators have been able to overcome this limitation by generating dynamic or static exclusion list and repeating the data-dependent acquisition multiple time, although for large biomarker studies this approach neither scaleable nor practicable. By performing quantification on ion intensity generated in the full-scan mass spectra in the dMS approach, and by executing biomarker experiments on high-resolution instruments that maintain high mass accuracy, it is possible to profile every species detected in the full-scan mass spectrum. The quantitative utility of differential mass spectrometry on an FT-MS has been demonstrated recently by Meng et al. [155].

The application of FT-MS (resolution >100 000, high mass accuracy <2 ppm, attomole sensitivity, increase in dynamic range) offers the potential to probe increasingly complex biological samples, without the requirement for multistage chromatographic separations. In this way, the variance introduced by intricate biochemical manipulation of the samples is eliminated. Principally, when profiling complex mixtures such as CSF or plasma, peptides of similar mass and hydrophobicity may coelute, which makes it difficult to detect the relative amounts of each individual species. The increased mass resolution provided by FT-MS improves the ability to characterize – and therefore also to quantitate – specific peptides that may be degenerate in retention time but distinct in mass. These advantages are readily observed when one compares the resolution and ion detection between LTQ and LTQ-FT-MS directly, side-by-side, as shown in Figure 4.12.

The observation of lower-abundant ions is aided by the reduction in chemical noise for FT-MS-based protein profiling. Once the feature or peptide of interest has been identified by m/z and retention time, then that sequence can be targeted in separate 1-D-μLC-MS/MS runs. This approach enables a workflow which is designed to identify statistically significant changes in abundance among the thousands of features profiled in an experiment.

Today, dMS is sufficiently mature today to support adaptive design clinical studies. The consequence will be to bring the diagnostic and prognostic benefits to clinical proteomics, in real time, and ultimately to translate these findings to individual patients in hospitals and outpatient clinics worldwide, as biomarkers identified by global protein profiling are incorporated in the fit for purpose paradigm of exploration, demonstration, and characterization [169]. Future technologies such as ion mobility mass spectrometry and innovations in Orbitrap™ instrumentation promise to extend these capabilities even further. Collectively, these technologies will impact science in general by enhancing our ability to probe elusive and profound biological problems at the protein level.

4.3.3
Recipe for Beginners

The following is designed to allow for a direct comparison of ion trap spectra (unit resolution spectra) and FT-spectra (60 000 resolution). This example demonstrates the effect of increased resolution, decrease in chemical noise, and increase in dynamic range in the context of a complex biological sample. The second goal is to demonstrate the reproducible chromatography of complex digests.

The materials needed include:

- Rat plasma (Charles River Laboratory)
- Multiple affinity spin cartridge against top three mouse sera proteins (Agilent Technologies)
- Buffers A and B (Agilent)
- 5000 MWCO ultrafiltration spin concentrators (VivaSpin)

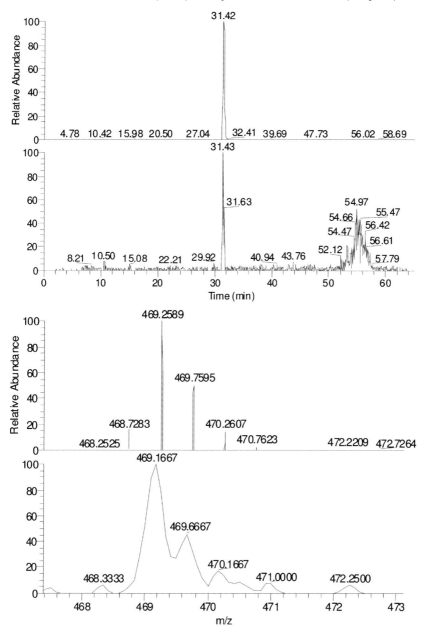

Figure 4.12 Selective ion chromatogram of LTQ-FTMS (60000 resolution) and LCQ (unit resolution) run on the exact 60-min gradient.

- TCEP (Sigma)
- LTQ-FT(ICR/Orbitrap)

4.3.3.1 Step-By-Step Instructions

- Centrifuge the MAR cartridge (place cartridge in 2-mL Eppendorf tube) for 2 min at $700 \times g$.
- Remove any residual liquid from the cartridge bed.
- Equilibrate cartridge by adding 200 μL of Agilent buffer A onto cartridge and spin for 2 min at $700 \times g$. (The bed should be dry, but if not the residual liquid must be removed.)
- Dilute 10 μL of plasma into 90 μL of Buffer A.
- Apply 100 μL of the diluted plasma onto the column.
- Place the cartridge in a new 2-mL Eppendorf tube. Centrifuge for 2 min at $700 \times g$.

Retain the flowthrough.

- Add an additional 400 μL of Buffer A. Centrifuge at $700 \times g$ for 2 min and collect the flowthrough. Combine the flowthrough from step 6 with that from step 7.
- Add the combined fractions to the MWCO 5000 spin filter and concentrate to ca. 50 μL. (The column can be regenerated by applying 200–L of buffer B, then centrifuge at $700 \times g$ for 2 min, and re-equilibrate as described in step 3.)
- Reconstitute to 100 μL with 25 mM ammonium bicarbonate.
- Add 5 μL of 4 mM TCEP to solution, and then incubate for 30 min.
- Add 10 mM iodoacetamide, and incubate for 30 min in the dark.
- Add trypsin and incubate at 37°C overnight.
- Quench the reaction with 10 μL acetic acid.
- Concentrate to dryness.
- Reconstitute the dried residue to 20 μL with 0.1% acetic acid.

The peptides are then analyzed using a LTQ-FT-MS with a binary 65-min gradient with three gradient sections: (1) binary gradient from 0% solvent B (90% acetonitrile, 0.1 M acetic acid) to 30% B from 4.01 min to 29 min at a flow rate of $1 \mu L \, min^{-1}$; (2) 29.01 min to 39 min to 90% B at $1 \mu L \, min^{-1}$; and (3) 40 to 65 min to 100% A at $1 \mu L \, min^{-1}$. The acquisition should consist of a cycle of a full-scan ion trap mass spectrum, full-scan FT mass spectrum (resolution @ 60 000) and (optionally) seven MS/MS spectra recorded sequentially on the most abundant ions detected in the initial ion trap scan. A general configuration as shown in Figure 4.13 can be used for LC-FT-MS for sample lading and analysis. Samples are loaded at $3 \mu L \, min^{-1}$ onto a

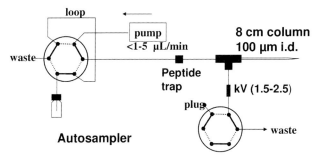

Figure 4.13 Suggested configuration for nano-LC interface. Loading and elution of samples into the spray column can be achieved with a single pump.

trap column, washed for 3 min at 3 µL min^{-1}, and than eluted into the spray column at the gradient described above.

References

143 Sabatine, M.S., Morrow, D.A., de Lemos, J.A., Gibson, C.M., Murphy, S.A., Rifai, N., McCabe, C., Antman, E.M., Cannon, C.P. and Braunwald, E. (2002) Multimarker approach to risk stratification in non-ST elevation acute coronary syndromes: simultaneous assessment of troponin I C-reactive protein, and B-type natriuretic peptide. *Circulation*, **105**, 1760–1763.

144 Anderson, N.L. and Anderson, N.G. (2002) The human plasma proteome: History character and diagnostic prospects. *Mol. Cell. Proteomics*, **1**, 845–867.

145 Polanski, M. and Anderson, L.N. (2006) A list of cancer biomarkers for targeted proteomics. *Biomarker Insights*, **2**, 1–48.

146 Petricoin, E.F., Belluco, C., Araujo, B.P. and Liotta, L.A. (2006) The blood peptidome: a higher dimension of information content for cancer biomarker discovery. *Nat. Rev. Cancer*, **6**, 961–967.

147 Mann, M., Hendrickson, R.C. and Pandey, A. (2001) Analysis of proteins and proteomes by mass spectrometry. *Annu. Rev. Biochem.*, **70**, 437–473.

148 Ramstroem, M., Hagman, C., Mitchell, J.K., Derrick, P.J., Hakansson, P. and Bergquist, J. (2005) Depletion of high abundant proteins in body fluids prior to liquid chromatography Fourier transform ion cyclotron resonance mass spectrometry. *J. Proteome Res.*, **4**, 410–416.

149 Lowenthal, M.S., Mehta, A.L., Frogale, K., Bandle, R.W., Arauja, R.P., Hood, B.L., Veenstra, T.D., Conrads, T.P., Goldsmith, P., Fischman, D., Petricoin, E.F. and Liotta, L.A. (2005) Analysis of albumin-associated peptides and proteins from ovarian cancer patients. *Clin. Chem.*, **51**, 1933–1945.

150 Zhang, H., Li, X.J., Martin, D.B. and Aebersold, R. (2003) Identification and quantification of N-linked glycoproteins using hydrazide chemistry stable isotope labeling and mass spectrometry. *Nat. Biotech.*, **21**, 660–666.

151 Liu, W., Qian, W.-J., Gritsenko, M.A., Camp, D.G., Monroe, M.E., Moore, R.J. and Smith, R.D. (2005) Human plasma N-glycoproteome analysis by immunoaffinity subtraction hydrazide

152 Washburn, M.P., Wolters, D. and Yates, J.R. (2001) Large-scale analysis of the yeast proteome by multidimensional protein identification technology. *Nat. Biotechnol.*, **19**, 242–247.

153 Ong, S.E., Blagoev, B., Kratchmarova, I., Kristensen, D.B., Steen, H., Pandey, A. and Mann, M. (2002) Stable isotope labeling by amino acids in cell culture SILAC, as a simple and accurate approach to expression proteomics. *Mol. Cell. Proteomics*, **1**, 376–386.

154 Gygi, S.P., Rist, B., Gerber, S.A., Turecek, F., Gelb, M.H. and Aebersold, R. (1999) Quantitative analysis of complex protein mixtures using isotope-coded affinity tags. *Nat. Biotechnol.*, **19**, 994–999.

155 Meng, F., Wiener, M.C., Sachs, J.R., Burns, C., Verma, P., Paweletz, C.P., Mazur, M.T., Deyanova, G.D., Yates, N.A. and Hendrickson, R.C. (2007) Quantitative analysis of complex peptide mixtures using FTMS and differential mass spectrometry. *J. Am. Soc. Mass Spectrom.*, **18**, 226–233.

156 Wang, G., Wu, W.W., Zheng, W., Chou, C.L. and Shen, R.F. (2005) Label-free protein quantification using LC-coupled ion trap or FT mass spectrometry: Reproducibility linearity, and application with complex proteomes. *J. Proteome Res.*, **5**, 1214–1223.

157 Prakash, A., Mallick, P., Whiteaker, J., Zhang, H., Paulovich, A., Flory, M., Lee, H., Aebersold, R. and Schwikowski, B. (2006) Signal maps for mass spectrometry-based comparative proteomics. *Mol. Cell. Proteomics*, **5**, 423–432.

158 Li, X.J., Yi, E.C., Kemp, C.J., Zhang, H. and Aebersold, R. (2005) A software suite for the generation and comparison of peptide arrays from sets of data collected by liquid chromatography-mass spectrometry. *Mol. Cell. Proteomics*, **4**, 1328–1340.

159 Fang, R., Elias, D.A., Monroe, M.E., Shen, Y., McIntosh, M. et al. (2006) Differential label-free quantitative proteomic analysis of *Shewanella oneidensis* cultured under aerobic and suboxic conditions by accurate mass and time tag approach. *Mol. Cell Proteomics*, **5**, 714–725.

160 Wiener, M.C., Sachs, J.R., Deyanova, E.G. and Yates, N.A. (2004) Differential mass spectrometry: a label-free LC-MS method for finding significant differences in complex peptide and protein mixtures. *Anal. Chem.*, **76**, 6085–6096.

161 Wang, W., Zhou, H., Lin, H., Roy, S., Shaler, T.A., Hill, L.R., Norton, S., Kumar, P., Anderle, M. and Becker, C.H. (2003) Quantification of proteins and metabolites by mass spectrometry without isotopic labeling or spiked standards. *Anal. Chem.*, **75**, 4818–4826.

162 Strittmatter, E.F., Ferguson, P.L., Tang, K. and Smith, R.D. (2003) Proteome analyses using accurate mass and elution time peptide tags with capillary LC time-of-flight mass spectrometry. *J. Am. Soc. Mass Spectrom.*, **14**, 980–991.

163 Pieper, R., Gatlin, C.L., Makusky, A.J., Russo, P.S. and Schatz, C.R. et al. (2003) The human serum proteome: display of nearly 3700 chromatographically separated protein spots on two-dimensional electrophoresis gels and identification of 325 distinct proteins. *Proteomics*, **7**, 1345–1364.

164 Adkins, N.J., Varnum, S.M., Auberry, K.J., Moore, R.J., Angell, N.H. et al. (2002) Toward a human blood serum proteome: analysis by multidimensional separation coupled with mass spectrometry. *Mol. Cell. Proteomics*, **1**, 947–955.

165 Jin, W.H., Dai, J., Xia, Q.C., Zou, H.F. and Zeng, R. (2005) Human plasma proteome analysis by multidimensional chromatography prefractionation and linear ion trap mass spectrometry identification. *J. Proteome Res.*, **4**, 613–619.

166 Syka, J.E.P., Marto, J.A., Bai, D.L., Stevan, H., Senko, M.W., Schwartz, J.C., Ueberheide, B., Garcia, B., et al. (2004)

Novel linear quadrupole ion trap/FT mass spectrometer: Performance characterization and use in the comparative analysis of histone H3 post-translational modifications. *J. Proteome Res.*, **3**, 621–626.
167 Hardman, M. and Makarov, A.A. (2003) Interfacing the Orbitrap mass analyzer to an electrospray ion source. *Anal. Chem.*, **75**, 1699–1705.
168 Yates, J.R., Cociorva, D., Liao, L.J. and Zabrouskov, V. (2006) Performance of a linear ion trap-Orbitrap hybrid for peptide analysis. *Anal. Chem.*, **78**, 493–500.
169 Lee, J.W., Devanaarayan, V., Barrett, Y.C., Weiner, R., Allinson, J., Fountain, S., Keller, S., Weinryb, I., Green, M., Duan, L., Roger, J.A., Millham, R., O'Brien, P.J., Sailsatd, J., Khan, M., Ray, C. and Wagner, J.A. (2006) Fit-for-purpose method development and validation of successful biomarker measurement. *Pharm. Res.*, **23**, 312–328.

4.4
Sample Preparation for Differential Proteome Analysis: Labeling Technologies for Mass Spectrometry

Josef Kellermann

4.4.1
Introduction

One key focus in proteome research is the determination of changes in protein expression and their modifications. Today, two-dimensional electrophoresis (2-DE) followed by mass spectrometry (MS)-based techniques such as peptide mapping or MS-sequencing (MS/MS) is still regarded as a standard tool for the relative quantification and identification of proteins within complex mixtures [170,171]. Despite its high resolving power, it is clear that this strategy suffers from significant analytical limitations that confine high-throughput proteome analysis.

Limited sample capacity and detection sensitivity, which are especially apparent during the analysis of unfractionated complex samples, result in the identification of only relatively high-abundant proteins. The separation of membrane-spanning proteins still remains a challenge. Likewise, the comigration of different proteins complicates the quantification of the single protein species. Labeling of proteins with fluorescent dyes, as introduced during the difference gel electrophoresis (DIGE) technology, partly overcomes the limitations of the 2-DE [172].

The rapid improvements in multidimensional separation technologies, together with faster and more sensitive mass spectrometers and bioinformatics tools for data analysis, have provided the basics for high-throughput approaches to study complex proteomes. The inclusion of chemical modification reactions has a longstanding history in protein chemistry [173], and this led to mass spectrometry – together with those "chemical tags" – becoming a central technology in proteomics.

For many years, metabolic studies have used isotope labeling combined with mass spectrometry as a powerful means of quantification [174]. Analogues of the metabolites to be tested were synthesized as non-radioactive versions containing ^{13}C, ^{15}N

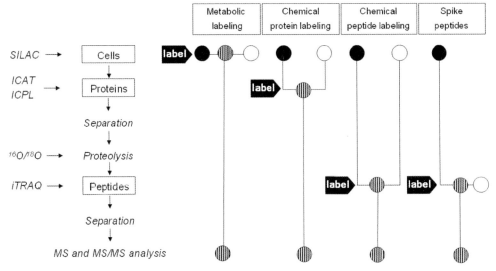

Figure 4.14 Labeling strategies for stable isotope incorporation. The workflows differ in their time point when stable isotopes are introduced into proteins or peptides. Sample (●) and control (○) are labeled separately and combined afterwards. Before the labeling step, samples are processed independently in parallel. This can result in quantification errors caused by non-reproducible purification steps.

or ^{2}H, and spiked in defined amounts into the biological sample. As isotopic variants of all molecules behave identically and exhibit the same ionization behavior during one experiment, quantification by signal intensity is highly accurate.

During the past few years this strategy was recognized as being valid also for proteomics, with the development of several methods [175–177] that differed in the way – and when during the proteome analysis – the stable isotope labels were introduced into the peptides or proteins.

In this respect, four different workflow categories (Figure 4.14) can be distinguished:

- the *in-vivo* incorporation of stable isotope-containing amino acids (metabolic labeling) which are added to cell culture media,
- the introduction of chemical tags to isolated protein mixtures,
- the labeling of peptides after or during enzymatic cleavage of proteins,
- the spiking of labeled peptides, for absolute quantification.

All these strategies have in common, that two or more protein or peptide samples are differentially labeled, one with an isotopically "light" and the others with isotopically "heavy" tags. The samples are then combined, thereby "freezing" the relative amount of proteins or peptides. The complexity of the samples is then reduced by using one or more separation steps. Peptides resulting from corresponding

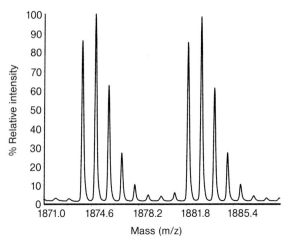

Figure 4.15 Typical isotope cluster of a duplexed peptide; differentially labeled peptides are detected as duplet peaks (two proteome samples) during mass spectrometry only differing by the mass introduced by the isotopic label.

proteins of both samples retain the same chemical properties despite being differentially labeled, and can be detected as a peptide pair or triplet peaks (two or three proteome samples) during mass spectrometry, differing only by the mass introduced by the isotopic label (Figure 4.15). Corresponding peak heights or areas are then compared to calculate the relative abundance of corresponding peptides of the different samples.

4.4.2
Isotopic Labeling of Peptides and/or Proteins

4.4.2.1 Stable Isotope Labeling of Proteins in Cell Culture

One principal difference between the four above-mentioned strategies is the time point of incorporation of the isotopic label. The main advantage of the metabolic labeling strategy is that the label is introduced into living cells, and therefore cells from different states, if differentially labeled, can be mixed before lysis. The following steps of fractionation and purification do not affect the accuracy of quantification. Consequently, stable isotope labeling by amino acids in cell culture (SILAC) [178] became one of the most widely used strategies in quantitative proteomics. Two or more cell populations are simply grown in different media, each containing a light or one or more heavy versions of a suitable amino acid. Several amino acids are described as being used in the SILAC approach. The labeling of arginine and lysine, followed by tryptic digestion, results in labeling of almost every peptide except the C-terminal of each protein. Other amino acids such as tyrosine [179] or methionine [180] have also been described. The combination with bioorthogonal non-canonical amino acid tagging (BONCAT) allows for the selective isolation of newly synthesized proteins as a response to internal or external cues. Azidohomoalanine (AHA) is thus incorporated instead of methionine when serving as reactant for a biotin-FLAG-alkyne tag which can be purified using a Neutravidine affinity matrix [181].

SILAC is mainly used for cell culture-based proteomics approaches. Oda and coworkers [182] have described a culture-derived isotope tag (CDIT) method which can be applied also to compare two tissue samples, using SILAC-labeled cells as a linking internal standard. The *in-vivo* incorporation of stable isotopes has been demonstrated even in whole animals [183]. One significant limitation of SILAC is that it cannot be used for samples which are not grown in culture. Samples obtained from patients (e.g., body fluids or tissue) can only be processed by chemical or enzymatic labeling.

4.4.2.2 Chemical Isotopic Labeling of Peptides or Proteins

When using a chemical-based labeling approach, stable isotope-bearing reagents react with the reactive sites of a protein or peptide. In 1999, Gygi and colleagues [184] introduced this new approach based on chemical stable isotope labeling of proteins using isotope-coded affinity tags (ICAT). Since then, many groups have adopted the principle of this strategy, generating different approaches (see Table 4.7), each with its own strengths and weaknesses.

These techniques are based on the differential isotopic labeling of proteins or peptides derived from two states (e.g., healthy and tumor tissues) with either light or heavy tags. In the latter case, hydrogen atoms can be exchanged by deuterium, thereby increasing the M_r by one Dalton per incorporated isotope. More recently, other isotopes such as ^{13}C or ^{15}N have also been employed, avoiding different elution times (isotopic shift) during reverse-phase liquid chromatography (LC) and assuring more accurate quantification [185,186]. The most frequently used methods are described in the following sections.

Isotope-Coded Affinity Tags (ICAT) The ICAT reagent [184] consists of a SH-reactive group that is cysteine-directed. A polyether linker region with eight hydrogen atoms, which are replaced by deuterium atoms in the heavy version and a biotin-containing tag, which allows for the isolation of modified peptides by avidin affinity chromatography. During an average ICAT experiment, solubilized proteins from both experiment and control are denatured, reduced and modified with a light or heavy reagent, respectively. Subsequently, the combined protein mixtures are enzymatically cleaved and the labeled peptides isolated by avidin chromatography, fractionated by ion-exchange chromatography, and quantified by MS. The avidin chromatography step reduces the complexity by a factor of 10, thereby simplifying the separation effort. On the other hand, the sequence coverage of proteins is reduced by almost the same factor. Recent 2-DE/MS studies have demonstrated that single-gene products result in an average of 10 to 15 different spots, mostly derived from post-translational modifications [187–189]. In order to characterize these modifications, which have a major influence on protein activity and biological function, a high sequence coverage of the modified proteins is essential. The disadvantage of

Table 4.7 Chemical labeling technologies for MS-based proteomics.

Reactive group	Method or reagent	Reference(s)
Amines	Isotope-coded protein label (ICPL)	[196]
	Isobaric tag for relative and absolute quantification (iTRAQ)	[194]
	Acetic anhydride	[200]
	Formaldehyde	[201]
	Propionic anhydride	[202]
	Succinic anhydride	[203]
	Nicotinoyloxy succinimide	[204]
	N-acetoxysuccinimide	[205]
	Phenylisocyanate	[206]
	Sulfo-NHS-biotin and ^{13}C, D_3-methyl iodide	[207]
	4-Sulfophenyl isothiocyanate	[208]
	Tandem Mass tags	[209]
Lysines	Guanidination(O-methyl-isourea) mass-coded abundance tagging (MCAT)	[210]
	Guanidination(O-methyl-isourea)	[211]
	Quantitation using enhanced sequence tags (QUEST)	[212]
	2-Methoxy-4,5-1-H-imidazole	[213]
Sulfhydryl	Isotope-coded affinity tag (ICAT)	[184]
	Cleavable ICAT	[190,191]
	Solid-phase ICAT	[214]
	HysTag	[215]
	Acrylamide	[216]
	Isotope-coded reduction of chromatographic support (ICROC)	[217]
	2-Vinylpyridine	[218]
	Iodacetanilide	[219]
	N-t-butyliodoacetamide	[219]
	Acid-labile isotope-coded extractants (ALICE)	[220]
	Solid-phase mass tagging	[221,222]
Carboxyl	Ethyl esterification	[223]
	Methyl esterification	[224]
Others	COFRADIC	[225,226]
	Phosphoprotein isotope-coded affinity tag (PhIAT)	[227]
	Phosphoprotein isotope-coded solid-phase tag (PhIST)	[228]
	Beta elimination and Michael addition with dithiothreitol (BEMAD)	[229–231]
	2-Nitrobenzenesulfenyl chloride (NBSCI)	[232]

this less complex mixture is that those proteins which contain no cysteine cannot be quantified at all, and must be quantified and identified on the basis of only a single peptide. Additionally, the ICAT tag influences the fragmentation spectra, thereby making the identification step of these peptides more difficult. Although the method became more practicable after substituting a cleavable and coeluting tag [190,191], it still suffers from low sequence coverage.

$^{16}O/^{18}O$ **Exchange** The isotope-labeling technique termed global internal standard technology (GIST) overcomes the problem of low sequence coverage, thus allowing the isotopic tagging of all peptides [192] obtained after or during separate enzymatic cleavage of the two protein samples.

The enzymatic labeling of peptides with heavy oxygen during the cleavage process is widely used in comparative proteomic studies. The technique involves the proteolysis of proteins in the presence of light ($H_2^{16}O$) and heavy ($H_2^{18}O$) water. Two oxygen atoms from the carboxy terminus of the proteolyzed peptide are exchanged with two oxygen atoms of the surrounding water molecules. The exchange from light to heavy water introduces a mass shift of 4 Da in the mass spectrum. Although the method has shown to provide good quantitative information on relative protein abundances [193], it suffers from some basic problems. For example, as the proteomes to be compared are cleaved before any separation step, the method struggles with the tremendous complexity on the peptide level. The limited separation space of two- ore even more-dimensional LC steps makes the quantification in MS crucial.

Isobaric Tag for Relative and Absolute Quantification (iTRAQ) The above-discussed technologies are based on the relative quantification of stable isotope labeled peptides prior to MS/MS analysis. During the analysis of complex biological samples, many peptides coelute during chromatography, causing signal suppression and potentially making interpretation difficult. To allow for higher multiplexing, and thereby avoid an increase in peptide complexity, a higher multiplexing GIST technology was recently introduced. The core of this methodology is a multiplexed set of four isobaric reagents (iTRAQ) [194]. The label consists of an N-hydroxysuccinimide moiety, reacting with amino groups, and two isotope-coded regions, a balance group and a reporter group, which is cleaved off during MS/MS yielding MS signals between 114 and 117 Da. The combination of different isotopic forms ($^{12}C/^{13}C$, $^{14}N/^{15}N$, $^{16}O/^{18}O$) allows the generation of four isobaric tags. Routinely, protein mixtures of four samples are reduced, alkylated and cleaved into peptides, after which each sample is labeled separately with one isobaric tag. Corresponding isobaric peptides of four proteomes coelute during chromatography and are indistinguishable in MS, but exhibit low-mass MS/MS signature ions that support relative peptide quantitation. This comprises one severe drawback of this technology. Since for quantification, MS/MS spectra of each peptide are needed, then ten thousands of MS/MS spectra per analysis are necessary. The second disadvantage is that the reduction of complexity is, as with all GIST approaches, limited to the peptide level.

Besides those two technical drawbacks, all GIST approaches are unable to address an important biological feature, namely that they cannot distinguish the different regulation of different protein isoforms, or post-translational modifications such as proteolytic processing. If a gene product is present in two or more proteolytically processed variants, it is not possible to distinguish between those species after enzymatic cleavage. A different regulation of protein isoforms can only be quantified if the labeling step is performed on the protein level. Therefore, although the iTRAQ approach was recently adopted for protein labeling, it continued to show minor problems during electrospray ionization (ESI)-MS identification [195].

Isotope-Coded Protein Label The isotope-coded protein label (ICPL) was designed to overcome the drawbacks of the above-described methods by labeling all free amino groups in proteins [196]. This method provides highly accurate and reproducible quantitation, high protein sequence information, including post-translational modifications and isoforms, and is compatible with all commonly used protein and peptide separation techniques. A N-nicotinoyloxy-succinimide (Nic-NHS) label allows for the replacement of six ^{12}C by ^{13}C isotopes. For higher multiplexing, a third label can be introduced, where four H are replaced by their deuterium isotope.

Two or more protein mixtures, obtained from different states, are individually reduced and alkylated to denature the proteins and to ensure easier access to free amino groups that are subsequently derivatized with the ^{12}C (light), ^{13}C (heavy), or ^{2}H (medium) -containing form, respectively, of the ICPL reagent. After combining the mixtures, any separation method can be adopted to reduce the complexity of the sample on the protein level. Even 2-D gel electrophoresis can be used as a pure, high-resolving separation technology, where especially protein isoforms can be well distinguished. After digestion, the peptides are quantified by MS; thus, identical peptides derived from the differentially labeled protein samples differ in mass, and appear as triplets in the acquired MS-spectra. The ratios of the ion intensities of these sister peptide triplets allow for the determination of the relative abundance of their parent proteins in the original samples. After relative quantification, only those proteins which are differently regulated have to be identified either by peptide mass fingerprint (PMF) or collision-induced dissociation (CID) of single peptides.

4.4.2.3 Spiking of Labeled Peptides

Some 17 years ago, in 1990, Desiderio and coworkers [197] first described an isotope dilution method to quantify neuropeptides by FAB-MS, and in 2001 this absolute quantitation strategy (AQUA) was combined with modern MS-technologies [198]. This represents the simplest way to introduce stable isotopes to a protein mixture by spiking labeled peptides in exactly known quantities. The strategy has been used to quantify a limited number of proteins, for example to validate candidate biomarkers in the plasma or urine of a large number of specimens. In order to reduce the background during MS, and thereby achieve higher sensitivity, specific peptide fragments are quantified using selective- or multiple-reaction monitoring (SRM or MRM). The different isotopes can be introduced either during synthesis or by several of the above-described methods. For each protein some suitable peptides must be selected and synthesized. This approach could, in principle, be extended to all proteins of a proteome [199].

4.4.3
Summary

The labeling of proteins or peptides with stable isotopes introduces a defined mass difference in peptides derived from two or more biological states, thereby allowing for a highly accurate and sensitive quantification during MS-analysis. A large number of methods in which an isotopic tag is introduced have been described. As the quality and accuracy of the analysis depends heavily on the time point at which the isotopic tag is introduced during the analytical workflow, only those methods that introduce the label directly to proteins and/or before extensive handling steps, promise to fulfill the necessary requirements. At present, SILAC represents the method of choice when an *in-vivo* introduction of isotopes can be achieved. For all samples of body fluids and tissues, or for any pathological sample, the ICPL technology provides highly accurate results. However, for targeted proteome analyses, where only a limited number of proteins must be quantified, iTRAQ allows for higher multiplexing and monitoring a large number of samples.

4.4.4
Perspectives

In order to understand the biological response of proteins and proteomes, it is absolutely necessary to recognize quantitative information about their changes in expression level. Isotopic labeling strategies, in combination with MS, have not only enabled a rapid evolution in proteomics but have also driven the science from a qualitative to a quantitative basis. However, in order to achieve proteome-wide quantitative analyses within a reasonable time, the technologies used for protein separation must be improved and even more sensitive and faster MS instrumentation must be developed to address the enormous dynamic range in the abundance of proteins. Despite these improvements, the main bottleneck in proteomics remains that of informatics; consequently, not only the instrument-based analytical software but also the interpretational tools capable of handling the vast amounts of data acquired during proteomics experiments must be improved.

4.4.5
Recipe for Beginners

In the following example of isotopic labeling, two protein mixtures – each containing four standard proteins in different proportions – are labeled using the ICPL-technology [196].

- Dissolve equal amounts (100 µg) of each sample separately in 6 M guanidine HCl, 0.1 M HEPES, pH 8.5, to obtain a total protein concentration of 5 mg mL^{-1} (100 µg 20 µL).

- Add 1 µL of 0.2 M Tris-2-carboxyethyl phosphine and reduce the disulfide bonds for 30 min at 60°C.

- After cooling, the samples are alkylated by adding 1 µL of 0.4 M iodoacetamide. The reaction is allowed to proceed for 30 min at room temperature in the dark.

- Excess iodoacetamide is quenched by adding 0.5 µM N-acetylcysteine. For nicotinoylation a 10-fold molar excess of ^{12}C-Nic-NHS (nicotinoyl-N-hydroxysuccinimide, light label) and ^{13}C-Nic-NHS (Serva electrophoresis, Heidelberg or Bruker Daltonics, Bremen) is added individually to each sample.

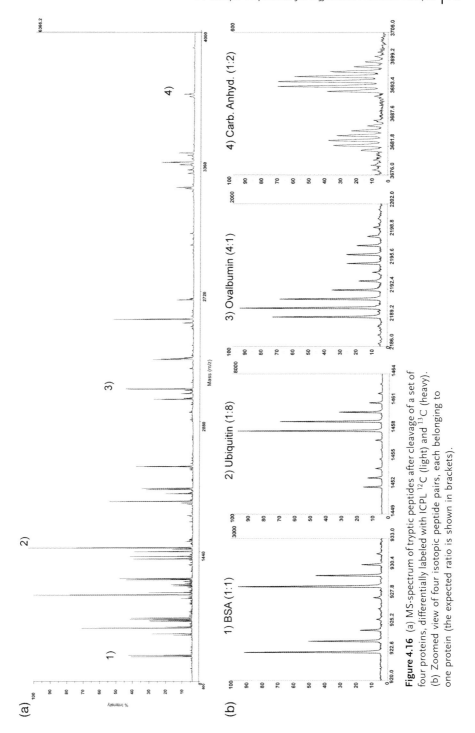

Figure 4.16 (a) MS-spectrum of tryptic peptides after cleavage of a set of four proteins, differentially labeled with ICPL ^{12}C (light) and ^{13}C (heavy). (b) Zoomed view of four isotopic peptide pairs, each belonging to one protein (the expected ratio is shown in brackets).

- A third sample may be processed in parallel with ^{2}H-Nic-NHS.

- Overlay with argon.

- The reaction is carried out at room temperature, at pH 8.3. Adjust the pH carefully to 8.3 using NaOH before labeling.

- After destroying any excess of reagents with 4 µL 1.5 M hydroxylamine, the two samples are combined, diluted with 25 mM Tris, pH 8.5, to a final guanidine/HCl concentration of 0.5 M, and digested overnight at 37°C with trypsin (substrate: enzyme ratio, 50:1).

The samples may then be directly analyzed by mass spectrometry. An example of a MS analysis of the applied protein mixture is shown in Figure 4.16.

The complex protein mixtures, which must be processed with any other separation technique, are combined and precipitated:

- Add an equal volume of distilled water to the sample.

- Add a fivefold excess (in relation to the total volume of sample and water) of ice-cold acetone to the sample and leave at –20°C overnight.

- Centrifuge at 20 000 × g for 30 min at 4°C to sediment the precipitated proteins.

- Overlay the precipitated proteins with ca. 100–200 µL ice-cold 80% acetone, shake carefully by hand, and re-centrifuge at 20 000 × g for 5 min at 4°C.

- Discard the supernatant and remove the remaining acetone at room temperature by evaporation (allow the cup to stand open).

The samples can either be stored, or dissolved directly in appropriate buffers for any further separation process (chromatography or electrophoresis).

References

170 Klose, J., Nock, C., Herrmann, M., Stuhler, K., Marcus, K., Bluggel, M., Krause, E., Schalkwyk, L.C., Rastan, S., Brown, S.D., Bussow, K., Himmelbauer, H. and Lehrach, H. (2002) Genetic analysis of the mouse brain proteome. *Nat. Genet.*, **30**, 385–393.

171 Wildgruber, R., Reil, G., Drews, O., Parlar, H. and Gorg, A. (2002) Web-based two-dimensional database of *Saccharomyces cerevisiae* proteins using immobilized pH gradients from pH 6 to pH 12 and matrix-assisted laser desorption/ionization-time of flight mass spectrometry. *Proteomics*, **2**, 727–732.

172 Unlu, M., Morgan, M.E. and Minden, J.S. (1997) Difference gel electrophoresis: a single gel method for detecting changes in protein extracts. *Electrophoresis*, **18**, 2071–2077.

173 Lundblad, R.L. (2004) *Chemical Reactions for Protein Modification*, 3rd edn. CRC Press, Boca Raton.

174 Deleenheer, A.P. and Thienpont, L.M. (1992) Applications of isotope-dilution mass-spectrometry in clinical-chemistry pharmacokinetics, and toxicology. *Mass Spectrom. Rev.*, **11**, 249–307.

175 Roe, M.R. and Griffin, T.J. (2006) Gel-free mass spectrometry-based high

throughput proteomics: Tools for studying biological response of proteins and proteomes. *Proteomics*, **6**, 4678–4687.

176 Leitner, A. and Lindner, W. (2006) Chemistry meets proteomics: The use of chemical tagging reactions for MS-based proteomics. *Proteomics*, **6**, 5418–5434.

177 Ong, S.E. and Mann, M. (2005) Mass spectrometry-based proteomics turns quantitative. *Nat. Chem. Biol.*, **1**, 252–262.

178 Ong, S.E., Blagoev, B., Kratchmarova, I., Kristensen, D.B., Steen, H., Pandey, A. and Mann, M. (2002) Stable isotope labeling by amino acids in cell culture SILAC, as a simple and accurate approach to expression proteomics. *Mol. Cell. Proteomics*, **1**, 376–386.

179 Ibarrola, N., Molina, H., Iwahori, A. and Pandey, A. (2004) A novel proteomic approach for specific identification of tyrosine kinase substrates using [C-13] tyrosine. *J. Biol. Chem.*, **279**, 15805–15813.

180 Ong, S.E., Mittler, G. and Mann, M. (2004) Identifying and quantifying *in vivo* methylation sites by heavy methyl SILAC. *Nat. Methods*, **1** (2), 119–126.

181 Dieterich, D.C., Link, A.J., Graumann, J., Tirrell, D.A. and Schuman, E.M. (2006) Selective identification of newly synthesized proteins in mammalian cells using bioorthogonal noncanonical amino acid tagging (BONCAT). *Proc. Natl. Acad. Sci. USA*, **103**, 9482–9487.

182 Ishihama, Y., Sato, T., Tabata, T., Miyamoto, N., Sagane, K., Nagasu, T. and Oda, Y. (2005) Quantitative mouse brain proteomics using culture-derived isotope tags as internal standards. *Nat. Biotechnol.*, **23**, 617–621.

183 Wu, C.C., MacCoss, M.J., Howell, K.E., Matthews, D.E. and Yates, J.R. (2004) Metabolic labeling of mammalian organisms with stable isotopes for quantitative proteomic analysis. *Anal. Chem.*, **76**, 4951–4959.

184 Gygi, S.P., Rist, B., Gerber, S.A., Turecek, F., Gelb, M.H. and Aebersold, R. (1999) Quantitative analysis of complex protein mixtures using isotope-coded affinity tags. *Nat. Biotechnol.*, **17**, 994–999.

185 Krijgsveld, J., Ketting, R.F., Mahmoudi, T., Johansen, J., Artal-Sanz, M., Verrijzer, C.P., Plasterk, R.H. and Heck, A.J. (2003) Metabolic labeling of *C. elegans* and *D. melanogaster* for quantitative proteomics. *Nat. Biotechnol.*, **21** (8), 927–931.

186 Zhang, R., Sioma, C.S., Thompson, R.A., Xiong, L. and Regnier, F.E. (2002) Controlling deuterium isotope effects in comparative proteomics. *Anal. Chem.*, **74**, 3662–3669.

187 Fountoulakis, M., Juranville, J.F., Berndt, P., Langen, H. and Suter, L. (2001) Two-dimensional database of mouse liver proteins. An update. *Electrophoresis*, **22**, 1747–1763.

188 Fountoulakis, M., Berndt, P., Langen, H. and Suter, L. (2002) The rat liver mitochondrial proteins. *Electrophoresis*, **23**, 311–328.

189 Mann, M. and Jensen, O.N. (2003) Proteomic analysis of post-translational modifications. *Nat. Biotechnol.*, **21**, 255–261.

190 Hansen, K.C., Schmitt-Ulms, G., Chalkley, R.J., Hirsch, J., Baldwin, M.A. and Burlingame, A.L. (2003) Mass spectrometric analysis of protein mixtures at low levels using cleavable C-13-isotope-coded affinity tag and multidimensional chromatography. *Mol. Cell. Proteomics*, **2**, 299–314.

191 Li, J.X., Steen, H. and Gygi, S.P. (2003) Protein profiling with cleavable isotope-coded affinity tag (cICAT) reagents – The yeast salinity stress response. *Mol. Cell. Proteomics*, **2**, 1198–1204.

192 Chakraborty, A. and Regnier, F.E. (2002) Global internal standard technology for comparative proteomics. *J. Chromatogr. A*, **949**, 173–184.

193 Yao, X.D., Freas, A., Ramirez, J., Demirev, P.A. and Fenselau, C. (2001) Proteolytic O-18 labeling for comparative proteomics: Model studies with two sero-

types of adenovirus. *Anal. Chem.*, **73**, 2836–2842.
194 Ross, P.L., Huang, Y.N., Marchese, J.N., Williamson, B., Parker, K., Hattan, S., Khainovski, N., Pillai, S., Dey, S., Daniels, S., Purkayastha, S., Juhasz, P., Martin, S., Bartlet-Jones, M., He, F., Jacobson, A. and Pappin, D.J. (2004) Multiplexed protein quantitation in *Saccharomyces cerevisiae* using amine-reactive isobaric tagging reagents. *Mol. Cell. Proteomics*, **3** (12), 1154–1169.
195 Wiese, S., Reidegeld, K.A., Meyer, H.E. and Warscheid, R. (2006) Protein labeling by iTRAQ: A new tool for quantitative mass spectrometry in proteome research. *Proteomics*, **7** (3), 340–350.
196 Schmidt, A., Kellermann, J. and Lottspeich, F. (2005) A novel strategy for quantitative proteomics using isotope-coded protein labels. *Proteomics*, **5**, 4–15.
197 Kusmierz, J.J., Sumrada, R. and Desiderio, D.M. (1990) Fast-atom-bombardment mass-spectrometric quantitative-analysis of methionine-enkephalin in human pituitary tissues. *Anal. Chem.*, **62**, 2395–2400.
198 Stemmann, O., Zou, H., Gerber, S.A., Gygi, S.P. and Kirschner, M.W. (2001) Dual inhibition of sister chromatid separation at metaphase. *Cell*, **107** (6), 715–726.
199 Aebersold, R. (2003) Constellations in a cellular universe. *Nature*, **422**, 115–116.
200 Che, F.Y. and Fricker, L.D. (2002) Quantitation of neuropeptides in Cpe (fat)/Cpe(fat) mice using differential isotopic tags and mass spectrometry. *Anal. Chem.*, **74**, 3190–3198.
201 Hsu, J.L., Huang, S.Y., Chow, N.H. and Chen, S.H. (2003) Stable-isotope dimethyl labeling for quantitative proteomics. *Anal. Chem.*, **75** (24), 6843–6852.
202 Zhang, X., Jin, Q.K., Carr, S.A. and Annan, R.S. (2002) N-Terminal peptide labeling strategy for incorporation of isotopic tags: a method for the determination of site-specific absolute phosphorylation stoichiometry. *Rapid Commun. Mass Spectrom.*, **16** (24), 2325–2332.
203 Wang, S.H. and Regnier, F.E. (2001) Proteomics based on selecting and quantifying cysteine containing peptides by covalent chromatography. *J. Chromatogr.*, **924**, 345–357.
204 Munchbach, M., Quadroni, M., Miotto, G. and James, P. (2000) Quantitation and facilitated de novo sequencing of proteins by isotopic N-terminal labeling of peptides with a fragmentation directing moiety. *Anal. Chem.*, **72**, 4047–4057.
205 Ji, J.Y., Chakraborty, A., Geng, M., Zhang, X., Amini, A., Bina, M. and Regnier, F. (2000) Strategy for qualitative and quantitative analysis in proteomics based on signature peptides. *J. Chromatogr. B*, **745**, 197–210.
206 Mason, D.E. and Liebler, D.C. (2003) Quantitative analysis of modified proteins by LC-MS/MS of peptides labeled with phenyl isocyanate. *J. Proteome Res.*, **2**, 265–272.
207 Hoang, V.M., Conrads, T.P., Veenstra, T.D., Blonder, J., Terunuma, A., Vogel, J.C. and Fisher, R.J. (2003) Quantitative proteomics employing primary amine affinity tags. *J. Biomol. Techniques*, **14** (3), 216–223.
208 Lee, Y.H., Han, H., Chang, S.B. and Lee, S.W. (2004) Isotope-coded N-terminal sulfonation of peptides allows quantitative proteomic analysis with increased de novo peptide sequencing capability. *Rapid Commun. Mass Spectrom.*, **18** (24), 3019–3027.
209 Thompson, A., Schafer, J., Kuhn, K., Kienle, S., Schwarz, J., Schmidt, G., Neumann, T. and Hamon, C. (2003) Tandem mass tags: A novel quantification strategy for comparative analysis of complex protein mixtures by MS/MS. *Anal. Chem.*, **75**, 4942.
210 Cagney, G. and Emili, A. (2002) De novo peptide sequencing and quantitative profiling of complex protein mixtures

using mass-coded abundance tagging. *Nat. Biotechnol.*, **20**, 163–170.

211 Brancia, F.L., Montgomery, H., Tanaka, K. and Kumashiro, S. (2004) Guanidino labeling derivatization strategy for global characterization of peptide mixtures by liquid chromatography matrix-assisted laser desorption/ionization mass spectrometry. *Anal. Chem.*, **76**, 2748–2755.

212 Beardsley, R.L. and Reilly, J.P. (2003) Quantitation using enhanced signal tags: A technique for comparative proteomics. *J. Proteome Res.*, **2**, 15–21.

213 Peters, E.C., Horn, D.M., Tully, D.C. and Brock, A. (2001) A novel multifunctional labeling reagent for enhanced protein characterization with mass spectrometry. *Rapid Commun. Mass Spectrom.*, **15**, 2387–2392.

214 Zhou, H.L., Ranish, J.A., Watts, J.D. and Aebersold, R. (2002) Quantitative proteome analysis by solid-phase isotope tagging and mass spectrometry. *Nat. Biotechnol.*, **20**, 512–515.

215 Olsen, J.V., Andersen, J.R., Nielsen, P.A., Nielsen, M.L., Figeys, D., Mann, M. and Wisniewski, J.R. (2004) HysTag – A novel proteomic quantification tool applied to differential display analysis of membrane proteins from distinct areas of mouse brain. *Mol. Cell. Proteomics*, **3**, 82–92.

216 Sechi, S. (2002) A method to identify and simultaneously determine the relative quantities of proteins isolated by gel electrophoresis. *Rapid Commun. Mass Spectrom.*, **16**, 1416–1424.

217 Shen, M., Guo, L., Wallace, A., Fitzner, J., Eisenman, J., Jacobson, E. and Johnson, R.S. (2003) Isolation and isotope labeling of cysteine- and methionine-containing tryptic peptides – Application to the study of cell surface proteolysis. *Mol. Cell. Proteomics*, **2**, 315–324.

218 Sebastiano, R., Citterio, A., Lapadula, M. and Righetti, P.G. (2003) A new deuterated alkylating agent for quantitative proteomics. *Rapid Commun. Mass Spectrom.*, **17**, 2380–2386.

219 Pasquarello, C., Sanchez, J.C., Hochstrasser, D.F. and Corthals, G.L. (2004) N-t-butyliodoacetamide and iodoacetanilide: two new cysteine alkylating reagents for relative quantitation of proteins. *Rapid Commun. Mass Spectrom.*, **18**, 117–127.

220 Qiu, Y.C., Sousa, E.A., Hewick, R.M. and Wang, J.H. (2002) Acid-labile isotope-coded extractants: A class of reagents for quantitative mass spectrometric analysis of complex protein mixtures. *Anal. Chem.*, **74**, 4969–4979.

221 Shi, Y., Xiang, R., Crawford, J.K., Colangelo, C.M., Horvath, C. and Wilkins, J.A. (2004) A simple solid phase mass tagging approach for quantitative proteomics. *J. Proteome Res.*, **3**, 104–111.

222 Shi, Y., Xiang, R., Horvath, C. and Wilkins, J.A. (2005) Quantitative analysis of membrane proteins from breast cancer cell lines BT474 and MCF7 using multi-step solid phase mass tagging and 2D LC/MS. *J. Proteome Res.*, **4**, 1427–1433.

223 Goodlett, D.R., Keller, A., Watts, J.D., Newitt, R., Yi, E.C., Purvine, S., Eng, J.K., von Haller, P., Aebersold, R. and Kolker, E. (2001) Differential stable isotope labeling of peptides for quantitation and de novo sequence derivation 3. *Rapid Commun. Mass Spectrom.*, **15**, 1214–1221.

224 Syka, J.E., Marto, J.A., Bai, D.L., Horning, S., Senko, M.W., Schwartz, J.C., Ueberheide, B., Garcia, B., Busby, S., Muratore, T., Shabanowitz, J. and Hunt, D.F. (2004) Novel linear quadrupole ion trap/FT mass spectrometer: performance characterization and use in the comparative analysis of histone H3 post-translational modifications. *J. Proteome Res.*, **3** (3), 621–626.

225 Gevaert, K., Van Damme, J., Goethals, M., Thomas, G.R., Hoorelbeke, B., Demol, H., Martens, L., Puype, M., Staes, A. and Vandekerckhove, J. (2002) Chromatographic isolation of methionine-containing peptides for gel-free proteome analysis: identification of

more than 800 *Escherichia coli* proteins. *Mol. Cell. Proteomics*, **1**, 896–903.

226 Gevaert, K., Damme, P.V., Martens, L. and Vandekerckhove, J. (2005) Diagonal reverse-phase chromatography applications in peptide-centric proteomics: Ahead of catalogue-omics? *Anal. Biochem.*, **345**, 18–29.

227 Goshe, M.B., Conrads, T.P., Panisko, E.A., Angell, N.H., Veenstra, T.D. and Smith, R.D. (2001) Phosphoprotein isotope-coded affinity tag approach for isolating and quantitating phosphopeptides in proteome-wide analyses. *Anal. Chem.*, **73**, 2578–2586.

228 Qian, W.J., Gosche, M.B., Camp, D.G., Yu, L.R., Tang, K.Q. and Smith, R.D. (2003) Phosphoprotein isotope-coded solid-phase tag approach for enrichment and quantitative analysis of phosphopeptides from complex mixtures. *Anal. Chem.*, **75**, 5441–5450.

229 Amoresano, A., Marino, G., Cirulli, C. and Quemeneur, E. (2004) Mapping phosphorylation sites: a new strategy based on the use of isotopically-labeled dithiothreitol and mass spectrometry. *Eur. J. Mass Spectrom.*, **10**, 401–412.

230 Vosseller, K., Hansen, K.C., Chalkley, R.J., Trinidad, J.C., Wells, L., Hart, G.W. and Burlingame, A.L. (2005) Quantitative analysis of both protein expression and serine/threonine post-translational modifications through stable isotope labeling with dithiothreitol. *Proteomics*, **5**, 388–398.

231 Wells, L., Vosseller, K., Cole, R.N., Cronshaw, J.M., Matunis, M.J. and Hart, G.W. (2002) Mapping sites of O-GlcNAc modification using affinity tags for serine and threonine post-translational modifications. *Mol. Cell. Proteomics*, **1** (10), 791–804.

232 Kuyama, H., Watanabe, M., Toda, C., Ando, E., Tanaka, K. and Nishimura, O. (2003) An approach to quantitative proteome analysis by labeling tryptophan residues. *Rapid Commun. Mass Spectrom.*, **17**, 1642–1650.

4.5
Determining Membrane Protein Localization Within Subcellular Compartments Using Stable Isotope Tagging

Kathryn S. Lilley, Tom Dunkley, and Pawel Sadowski

4.5.1
Introduction

In this chapter, attention is focused on sample preparation for the subsequent use of the LOPIT (Localization of Organelle Proteins by Isotope Tagging) technique in assigning proteins to the plant Golgi apparatus and endoplasmic reticulum (ER).

Within eukaryote cells, proteins are organized spatially according to their function, and consequently knowledge of their location is an important step towards assigning function to the thousands of proteins predicted by genome-sequencing projects which have yet to be characterized. Moreover, the localization of novel proteins to organelles will also improve the present understanding of the functions of these compartments. Proteomic approaches can provide powerful tools for characterizing the proteins resident within organelles in a high-throughput manner. To enable confident protein localization, however, it is prerequisite that organelle preparations are free of contaminants, or that techniques are utilized which discriminate between genuine organelle residents and contaminating proteins [233]. In the case of some organelles, such as mitochondria, it is relatively straightforward to produce a largely homogeneous preparation of the organelle. The components of the endomembrane system, however, such as the Golgi apparatus and ER are impossible to produce in a purified form as the organelles within the system are of similar size and density, which makes them difficult to separate one from one another [234,235]. Furthermore, proteins associated with this class of organelle are in a constant state of flux as they traffic through the endomembrane system en route to their final destination. This leads to additional complications when assigning proteins to a subcellular location. A final confounding factor is that proteins within the endomembrane system cycle between compartments; for example, ER residents continuously escape to the Golgi and are retrieved in COPI (coat protein complex) vesicles [236]. These factors necessitate measurement of the steady-state distributions of proteins within the whole endomembrane system in order to obtain a realistic insight into the subcellular localization of endomembrane proteins.

Recently, several techniques have emerged which enable the assignment of proteins to organelles. All of these methods use quantitative proteomics approaches to characterize the distribution pattern of organelles amongst partially enriched fractions generated by various separation technologies, and can discriminate between genuine organelle residents and contaminants, without the preparation of pure organelles [233,237–240]. Here, details are discussed of the sample preparation necessary for use with one of these methods – Localization of Organelle Proteins by Isotope Tagging (LOPIT) [239]. LOPIT relies on two assumptions: (1) that organelles exhibit distinct distribution patterns within a density gradient; and (2) that proteins residing in the same membrane co-fractionate. The LOPIT work-flow involves the partial separation of organelles using equilibrium density gradient centrifugation, measuring the relative abundance of proteins amongst the fractions along the length of the gradient, and comparing the profiles of proteins with the profiles of known organelle markers. The measurement of relative abundance is achieved in a robust manner using stable isotope tagging procedures and tandem mass spectrometry (MS/MS), while the comparison of profiles employs the use of multivariate statistical techniques. Currently, several methods are available for the incorporation of stable isotope labels into the proteins within fractions in a differential manner. Hence, LOPIT utilizes one of these approaches in which primary amine groups present on peptides generated from proteins are derivatized with iTRAQ (isotope tags for relative and absolute quantitation) reagents [241].

At all stages of the LOPIT protocol, the use of an appropriate sample preparation protocol is essential if successful organelle enrichment, isotope labeling, and subsequent mass spectrometric analysis is to be achieved.

The LOPIT technology is applicable to the study of protein subcellular localization in many sample types, including cultured cells, tissues, and whole organisms. The example provided here focuses on the assignment of membrane proteins to the plant endomembrane system, and in particular on the Golgi apparatus from *Arabidopsis* liquid callus culture.

4.5.2
Preparation and Treatment of Samples in the Early-Stage LOPIT Protocol

The procedure starts with the collection of crude membranes from *Arabidopsis* callus and partial separation of organelles through dense medium (for details of the LOPIT schema, see Figure 4.17). Here, LOPIT utilizes iodixanol (Optiprep) because of its low osmolarity and ability to produce self-generating gradients in a highly reproducible manner. To ensure the distinct distribution of organelles that need to be resolved, the degree of organelle separation is monitored by use of one-dimensional SDS–PAGE and Western blotting employing antibodies raised against marker proteins with already known organelle location. Alternatively, when certain antibodies are not available, enzyme assays measuring the activity characteristic for a particular organelle can be utilized. If necessary, the separation of organelles can be further optimized by manipulation of the iodixanol gradient profile. On the basis of Western blotting results, the gradient fractions enriched in specific organelles of interest are selected for quantitative labeling. This entails isotope tagging of peptides that have been generated from proteins by proteolysis using trypsin with the iTRAQ reagents. This system enables the labeling of up to eight separate fractions from a given gradient per experiment. The soluble proteins (and also some peripheral proteins) are removed from membranes by use of a high-pH wash with sodium carbonate prior to labeling. The penultimate step involves, first, the separation of peptides by strong cation-exchange chromatography, and further separation of peptides within fractions eluted from cation-exchange columns by reverse-phase chromatography coupled to tandem mass spectrometry (LC-MS/MS).

The final steps of the LOPIT protocol entail the identification and quantitation of peptides using fragmentation data generated by MS/MS using a tailor-made open access suite of programs [242], and analysis of the resulting data with unsupervised and supervised multivariate statistical approaches.

In this chapter, attention is focused on the important factors that influence sample preparation during the early stages of the LOPIT protocol. For an overall description of the protocol, the reader is referred to the review of Sadowski *et al.* [243]. Methods required during the later stages of the procedure, such as strong cation-exchange chromatography and LC-MS/MS, are described in details elsewhere in this book (see Section 4.1.2.1).

4.5.2.1 Preparation and Fractionation of Organelles

The procedure detailed here is based on the membrane preparation from *Arabidopsis* callus liquid culture grown according to Sherrier *et al.* [244]. This protocol can be optimized for membranes from various sources, however. Irrespective of the nature of the starting tissue/cell culture, it is important that there is limited disruption of the organelles during the first stages of sample preparation. The integrity of cellular compartments after homogenization can be checked by a variety of methods, including latent enzyme activity (where a high level of luminal enzyme activity would suggest lysis of the organelle) or electron microscopy of the homogenate.

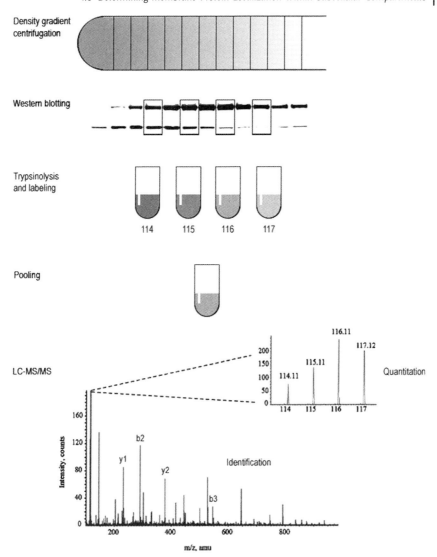

Figure 4.17 An overview of the LOPIT workflow. Organelles from *Arabidopsis* callus are partially separated through a self-generating iodixanol density gradient, and their enrichment patterns across the gradient are visualized by Western blotting with antibodies specific to known organelle residents. Four fractions which demonstrate distinct enrichment patterns of organelles under study, are next selected for quantitative labeling with iTRAQ reagents that is performed on the peptide level after trypsinolysis. Once labeled, samples are pooled and analyzed by tandem mass spectrometry (MS/MS), where iTRAQ tags fragment to release diagnostic reporter ions detected at 114 to 117 m/z. Areas under reporter ion peaks are proportional to peptide distribution across four density gradient fractions. The remaining part of the spectrum containing b, y, and other types of fragmention is used for peptide, and hence protein, identification. The localization of a protein is achieved when its distribution matches that of known organelle markers.

A typical preparation would involve the homogenization of 60 g of *Arabidopsis* callus tissue in an equal volume of homogenization medium (HM; for details, see Section 5), using a Polytron homogenizer with a 94 PTA 10EC head attachment (Kinematica) for two 7-s pulses at 57 000 r.p.m.

The amount of HM used and the time of homogenization are crucial to minimize organelle disruption during this stage of the preparation. Sufficient starting material should be used to yield at a minimum 100 µg of membrane proteins in each gradient fraction, which ultimately will be labeled with iTRAQ tags.

Concentration of Membranes on Iodixanol Cushion After homogenization, the next stage involves the removal of intact cells, cell wall and nuclei using low-speed centrifugation. Crude membranes are concentrated from the homogenate in the additional centrifugation step in the interphase that forms between the supernatant and an iodixanol cushion of typically 18% (v/v). The centrifugation conditions in this step are selected in such a way that the iodixanol does not form a gradient but rather acts as a filter that results in the partial pelleting of abundant organelles such as mitochondria and plastids, which are more dense than the cushion, and otherwise would dominate the analysis. Organelles less dense than the cushion form an interphase between the cushion and the supernatant. The iodixanol cushion concentration can be modified to enrich the interphase with different membranes [245,246]. A typical strategy would be as follows:

- Centrifuge the homogenate in four 50-mL centrifuge tubes to pellet the intact cells, nuclei and cell wall fragments for 5 min at $2200 \times g$ at $4°C$.

- Decant the supernatants into four clean centrifuge tubes, and re-centrifuge as above.

- Aliquot the supernatants into a total of six SW28 centrifuge tubes and underlay each supernatant with 6 mL of 18% iodixanol cushion solution. This underlaying can be achieved by using a 10-mL syringe fitted with a blunt-ended needle.

- Centrifuge at $100\,000 \times g$ at 4 C for 2 h in SW28 rotor, using an Optima L-XP ultracentrifuge (Beckman Coulter).

- Collect the crude membranes from the interphases (typically 6–8 mL).

Partial Fractionation of Organelles using Equilibrium Density Gradient Centrifugation
The next step in the protocol is membrane fractionation using equilibrium density gradient centrifugation. The example given here utilizes a 16% iodixanol gradient that enables separation of the Golgi apparatus, ER, vacuoles, plasma membrane, and mitochondria/plastids. Under appropriate centrifugation conditions, iodixanol generates the density gradient such that the membranes are fractionated according to their densities. The shape of the iodixanol gradient will determine the distribution of organelles, and hence the extent of their separation. It is important to optimize the separation of organelles of interest by manipulating the centrifugation time, centrifugal force, or iodixanol concentration, hence allowing different

membranes to be resolved. If a vertical type of centrifuge rotor is used, the slowest deceleration must be used to keep the gradient undisturbed. The procedure is as follows:

- Dilute membranes with 16% iodixanol gradient solution and adjust to give a final percentage of iodixanol of 16% using iodixanol working solution or HM accordingly.
- The density should be measured using a refractometer (Leica) according to the refractive index table of iodixanol-sucrose gradient solutions (available from Axis Shield).
- Decant the adjusted solution into two Vti65.1 polycarbonate Opti Seal tubes (Beckman Coulter) and centrifuge at $350\,000 \times g$ at $4\,°C$ for 3 h in a Vti65.1 rotor using an Optima L-XP ultracentrifuge, with the slowest deceleration rate selected.

Fraction Collection When the centrifugation step is complete, it is necessary to collect the gradient fractions such that the relative abundance of proteins in each can be determined. One approach is to use an Auto Densi-flow collection device (Labconco Corporation), which comprises a liquid-sensing probe connected to a peristaltic pump. The probe automatically detects the meniscus at the top of the gradient and moves downwards with the meniscus whilst the fractions are being collected. The use of such a precision instrument is crucial to minimize mixing of the gradient during fraction collection.

- Harvest 20 fractions, each of 0.5 mL, from the top of each gradient into TLA 100.3 polycarbonate tubes.

Preliminary Analysis of Organelle Separation by Western Blotting At this stage it is important to check the degree of membrane separation by performing Western blot analysis of gradient fractions with antibodies specific to different organelle marker proteins. It is essential that the organelles of interest exhibit distinct distributions. If this is not the case, then further optimization of the density gradients is necessary. Fractions should be chosen for further analysis, and quantitation based on whether they are likely to be the peak fractions of a particular organelle distribution and whether they contain sufficient protein for labeling.

4.5.2.2 Carbonate Washing of Fractions to Lyse Organelles, and Removal of Soluble and Peripheral Proteins

If membrane proteins are of interest, this step is necessary to remove any contaminating cytoplasmic proteins from the organelles, as well as some weakly interacting peripheral membrane proteins. The high pH of the carbonate buffer used at this stage disrupts electrostatic interactions between the peripheral proteins and the membrane. In addition, the high pH results in closed vesicles being transformed into membrane sheets. Consequently, soluble proteins contained within organelles are released [245].

- Add 800 mL of 162.5 mM Na_2CO_3 to 0.5 mL of each density fraction to give a final concentration of 100 mM Na_2CO_3.

- Incubate for 30 min on ice, and then centrifuge for 25 min at $100\,000 \times g$ at 4°C in a TLA 100.3 rotor using an Optima TLX ultracentrifuge (Beckman Coulter); discard the supernatants.

- Wash the pellets with 1 mL of deionized water at 4°C, centrifuge for 10 min as above, and discard the resulting supernatants.

4.5.2.3 iTRAQ Labeling

This stage in the LOPIT protocol involves choosing up to eight carbonate-washed fractions along the length of the density gradient. The fractions chosen should exhibit different distributions of specific organelles based on the distribution of markers as visualized by Western blotting. For example, the peak Golgi apparatus and peak ER fractions may be chosen along with two others from areas within the equilibrium density gradient which do not have a great abundance of these organelles. The carbonate-washed membranes from each of the chosen samples must be solubilized in a suitable buffer system, typically containing urea, and stringent detergents such as sodium dodecyl sulfate (SDS) and Triton X-100. Care must be taken when using such buffers. First, the iTRAQ reagents are supplied as N-hydroxysuccinimide esters which react with primary amine moieties. It is important, therefore, that the samples are not heated above 30°C as the iTRAQ labeling buffer contains urea, and heating may result in the carbamylation of amino groups; these would then not be available to react with the iTRAQ reagents. The second step of the protocol involves reduction and alkylation of the proteins (to avoid the formation of disulfide bridges during downstream processing) and the generation of peptides by proteolysis. Trypsin is the most commonly used protease, and its use is incompatible with the high concentrations of detergent and urea in the solubilization buffer.

- Solubilize pellets of chosen fractions in 100 µL of iTRAQ labeling buffer. Vortex at 20°C until the pellets are completely redissolved.

- Determine the protein concentration of each fraction using the BCA Protein Assay kit, according to the manufacturer's instructions, and using a bovine serum albumin standard in iTRAQ labeling buffer.

- Reduce 100 µg protein by adding 2 µL of 50 mM Tris(carboxyethyl) phosphine (TCEP) and incubate at 20°C for 1 h.

- Block cysteine sulfhydryls by the addition of 1 µL of 200 mM methyl methanethiosulfonate (MMTS), followed by incubation at 20°C for 10 min.

- Dilute each of the four protein samples after reduction and blocking with 50 mM triethylammonium bicarbonate (TEAB), such that the urea concentration is below 1 M, before adding 2.5 µg trypsin; incubate for 1 h at 37°C.

- Add a further 2.5 µg trypsin to each sample, followed by incubation overnight at 37°C.

- Lyophilize the resulting tryptic digests using a Speed Vac centrifugal vacuum concentrator, without heating.

- Resuspend the peptides in 100 µL of 250 mM TEAB, 75% (v/v) ethanol, and then add to one unit of the corresponding iTRAQ reagent followed by incubation for 1 h at 20°C. Of the four sets of peptides from the four starting fractions, label the least-dense fraction with iTRAQ reagent 114; the second least dense fraction with 115; the third least dense fraction with 116; and the most dense fraction with 117.

- Hydrolyze the residual iTRAQ reagents by adding 100 µL of deionized water and incubating for 15 min at 20°C. Pool all four samples together and lyophilize in a Speed Vac centrifugal vacuum concentrator.

The above procedure concludes the sample preparation for the LOPIT protocol. The next steps entail peptide separation using a combination of strong cation-exchange and reverse-phase chromatography coupled to tandem mass spectrometry. This results in identification and quantification of peptides and the assignment of proteins to organelles based on analysis using multivariate statistical approaches. Unfortunately, however, a detailed description of this part of the protocol is beyond the scope of this chapter.

4.5.3
Application of LOPIT to Map the Organelle Proteome of *Arabidopsis*

To date, the LOPIT technology has been used to best effect in the identification of novel proteins within the plant endomembrane system [233]. By using this approach, the density gradient distributions of 689 proteins from *Arabidopsis* callus cells were determined, enabling the confident and simultaneous localization of 527 proteins to the ER, Golgi apparatus, vacuole, plasma membrane, or to the mitochondria and plastids. The parallel analysis of endomembrane components enabled protein steady-state distributions to be measured, and this resulted in a "snapshot" of the protein localization within a cell. Figure 4.18 is a schematic representation of a Principal Components Analysis scores plot showing the clustering of proteins according to the similarity in their density gradient distributions, and therefore localizations. The verification of some of the novel assignments was carried out by green fluorescent protein (GFP) fusion to selected proteins, and confocal microscopy to allow their visualization within the subcellular compartments. This LOPIT study has resulted in the first proteomic analysis in which genuine organelle residents have been distinguished not only from contaminating proteins but also from proteins in transit through the plant secretory pathway.

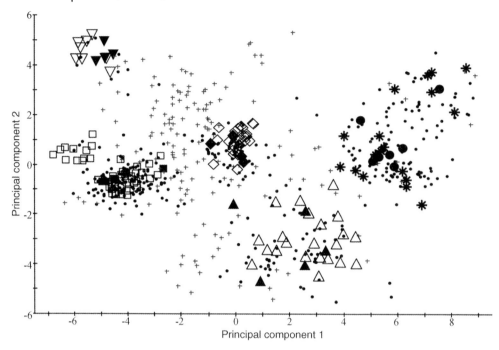

Figure 4.18 Statistical analysis of the dataset collected from the LOPIT experiment. Each symbol represents a protein characterized by its iTRAQ quantitation information that was projected onto first two principal components axes to draw the Principal Components Analysis (PCA) plot (shown on the figure). PCA clustered proteins according to the similarities in their gradient profiles and, hence, localization. Known and predicted organelle marker proteins are indicated, using larger filled or open shapes, respectively. Proteins to which localization was assigned on the basis of subsequent Partial Least Square Discriminant Analysis are represented by corresponding smaller shapes, whereas all remaining proteins that were not localized by this study are shown as crosses. The PCA plot represents the "snapshot" of the cell interior where proteins are "captured" at their steady-state distributions. Inverted triangles: vacuolar membrane; squares: endoplasmic reticulum; diamonds: plasma membrane; circles: mitochondria-plastids; triangles: Golgi apparatus.

4.5.4
Summary

The knowledge of the localization of protein to a particular subcellular structure or organelle is an important step towards assigning function to proteins, and also enhances the understanding of how organelles function. Many organelles cannot be purified, and only partially enriched preparations can be obtained. In many previous studies, where the degree of contamination by organelles with similar physical parameters to the organelle being studied has gone unchecked, the mis-localization of proteins has been observed. Localization of Organelle Proteins by Isotope Tagging (LOPIT) is a technique which does not rely on the ability to obtain pure organelle preparations. Instead, the approach traces the enrichment of organelles along the length of density gradients using stable isotope-coded tags and mass spectrometry.

The subcellular localization of proteins is then achieved by comparing their enrichment patterns to those of previously localized proteins by assuming that proteins which belong to the same organelle will fractionate in a similar way. In order to ensure the successful application of the LOPIT protocol, the use of an appropriate sample preparation is essential. Care should be taken with the enrichment of organelle fractions, particularly with respect to the homogenization of cells, the preparation of crude membranes, and the subsequent careful optimization of density gradients to maximize resolution of the subcellular compartments being studied.

4.5.5
Recipe for Beginners

- Homogenization buffer (HM): 250 mM sucrose, 10 mM HEPES-NaOH (pH 7.4), 1 mM EDTA (pH 8.0), 1 mM dithiothreitol (DTT). This solution must be prepared freshly before use and kept at 4°C.

- Optiprep stock solution (Axis Shield, Huntingdon, UK): 60% solution of iodixanol in water.

- Working solution buffer (WSB): 60 mM HEPES-NaOH (pH 7.4), 6 mM EDTA (pH 8.0), 6 mM DTT. This solution must be prepared freshly before use and kept at 4°C.

- Iodixanol working solution (IWS): prepared by diluting five parts of Optiprep stock solution with one part of working solution buffer. This solution must be prepared freshly before use and kept at 4 °C.

- iTRAQ labeling buffer: 25 mM TEAB, 8 M urea, 2% Triton X-100, 0.1% SDS.

- Iodixanol cushion solution: for an 18% (23.25 °Brix) cushion, mix 14.4 ml of IWS with 25.6 ml of HM. Measure the iodixanol concentration using a refractometer and the refractive index table of iodixanol-sucrose gradient solutions available from axis shield. Use IWS and HM media to increase or decrease the refractive index, respectively.

- Iodixanol gradient solution: for a 16% (22 °Brix) gradient, mix 16.4 ml of IWS and 13.6 ml of HM. Adjust the refractive index with IWS or HM.

References

233 Dunkley, T.P.J., Hester, S., Shadforth, I., Runnions, J., Weimar, T., Hanton, S., Griffin, J.L., Bessant, C., Brandizzi, F., Hawes, C., Watson, R., Dupree, P. and Lilley, K.S. (2006) Mapping the *Arabidopsis* organelle proteome. *Proc. Natl. Acad. Sci. USA*, **103** (17), 6518–6523.

234 Gabaldon, T. and Huynen, M.A. (2004) Shaping the mitochondrial proteome. *Biochim. Biophys. Acta - Bioenergetics*, **1659**, 212–220.

235 Peck, S.C. (2005) Update on proteomics in *Arabidopsis*. Where do we go from here? *Plant Physiol.*, **138**, 591–599.

236 Hanton, S.L., Bortolotti, L.E., Renna, L., Stefano, G. and Brandizzi, F. (2005) Crossing the divide – transport between the endoplasmic reticulum and Golgi apparatus in plants. *Traffic*, **6**, 267–277.

237 Andersen, J.S., Lam, Y.W., Leung, A.K., Ong, S.E., Lyon, C.E., Lamond, A.I. and Mann, M. (2005) Nucleolar proteome dynamics. *Nature*, **433**, 77–83.

238 Foster, L.J., de Hoog, C.L., Zhang, Y., Zhang, Y., Xie, X., Mootha, V.K. and Mann, M. (2006) A mammalian organelle map by protein correlation profiling. *Cell*, **125** (1), 187–199.

239 Dunkley, T.P.J., Watson, R., Griffin, J.L., Dupree, P. and Lilley, K.S. (2004) Localization of organelle proteins by isotope tagging (LOPIT). *Mol. Cell. Proteomics*, **3**, 1128–1134.

240 Gilchrist, A., Au, C.E., Hiding, J., Bell, A.W., Fernandez-Rodriguez, J., Lesimple, S., Nagaya, H., Roy, L., Gosline, S.J., Hallett, M., Paiement, J., Kearney, R.E., Nilsson, T. and Bergeron, J.J. (2006) Quantitative proteomics analysis of the secretory pathway. *Cell*, **127** (6), 1265–1281.

241 Ross, P.L., Huang, Y.L.N., Marchese, J.N., Williamson, B., Parker, K., Hattan, S., Khainovski, N., Pillai, S., Dey, S., Daniels, S. *et al.* (2004) Multiplexed protein quantitation in *Saccharomyces cerevisiae* using amine-reactive isobaric tagging reagents. *Mol. Cell. Proteomics*, **3**, 1154–1169.

242 Shadforth, I., Dunkley, T., Lilley, K. and Bessant, C. (2005) i-Tracker: for quantitative proteomics using iTRAQ. *BMC Genomics*, **6**, 145.

243 Sadowski, P.G., Dunkley, T.P., Shadforth, I.P., Dupree, P., Bessant, C., Griffin, J.L. and Lilley, K.S. (2006) Quantitative proteomic approach to study subcellular localization of membrane proteins. *Nat. Protoc.*, **1** (4), 1778–1789.

244 Sherrier, D.J., Prime, T.A. and Dupree, P. (1999) Glycosylphosphatidylinositol-anchored cell-surface proteins from *Arabidopsis*. *Electrophoresis*, **20**, 2027–2035.

245 Ford, T., Graham, J. and Rickwood, D. (1994) Iodixanol: a nonionic iso-osmotic centrifugation medium for the formation of self-generated gradients. *Anal. Biochem.*, **220**, 360–366.

246 Graham, J., Ford, T. and Rickwood, D. (1994) The preparation of subcellular organelles from mouse liver in self-generated gradients of iodixanol. *Anal. Biochem.*, **220**, 367–373.

5
Electrophoresis

5.1
Sample Preparation for Two-Dimensional Gel Electrophoresis

Walter Weiss and Angelika Görg

5.1.1
Introduction

Two-dimensional gel electrophoresis (2-DGE) with immobilized pH gradients (IPGs) is currently the most commonly used protein separation technology in proteome analysis [1,2]. The procedure is used to deliver a map of intact proteins that reflects changes in protein expression level, isoforms or post-translational modifications. The method is technically challenging, due not only to the high dynamic range of protein abundance in cell or tissue extracts but also to the highly diverse physico-chemical properties of proteins.

Sample preparation is the first step and, hence, a key factor for successful 2-DGE-based proteome analysis. Although a universal procedure for sample preparation would be desirable, there is no single method that can be applied to all types of sample, and most often the optimal procedure must be found empirically. Whilst the application of unfractionated samples to 2-DGE gels is preferred with regards to simplicity and reproducibility, there are instances in which subfractionation methods are required. Typical examples include the reduction of sample complexity, or the depletion of highly abundant proteins (e.g., albumin or immunoglobulins in plasma) that dominate the sample and obscure less-abundant proteins. Likewise, the solubilization of a subset of proteins by applying sequential extraction procedures with increasingly denaturing buffers, or the isolation of organelles by differential centrifugation, represent powerful methods for the enrichment of low-abundance proteins. The latter procedures, however, are beyond the scope of this chapter, and the reader is referred elsewhere for details [3,4]. The same applies to the solubilization of membrane proteins, which still remains a challenge for proteome analysis despite recent progress in this field [5,6]. In this chapter, the most important sample preparation guidelines and the major constituents of sample solubilization/lysis buffers for 2-DGE will be discussed, and the most relevant sample preparation procedures and protein solubilization cocktails exemplified.

5.1.2
General Aspects of Sample Preparation for 2-DGE

The treatment of samples prior to 2-DGE typically involves: (1) cell disruption; (2) sample clean-up (i.e., removal and/or inactivation of interfering protein and non-protein components); and (3) protein solubilization (for reviews, see [7–9]).

5.1.2.1 Cell Disruption

Except for liquid samples such as serum, plasma, urine or cerebrospinal fluid, cell disruption (cell lysis) is the first step in sample preparation for 2-DGE. Whereas in animal cells the *plasma membrane* alone separates the cell contents from the environment, in bacteria, fungi and plants the plasma membrane is surrounded by a *rigid cell wall*. This lack of a cell wall makes animal cells relatively easy to disrupt, in contrast to most microbial and plant cells that require harsher conditions. Depending on the precise requirements of the cell/tissue homogenate, different procedures and apparatus are available for cell disruption, ranging from mechanical and physical (e.g., homogenization, sonication, hypotonic lysis) to chemical (i.e., detergent) and enzyme-based methods, which may be used either individually or in combination. The technique(s) chosen must take into consideration not only the origin and type of cells and tissues with respect to the ease or difficulty of cell disruption, but also the amount of material available for processing (e.g., biopsy samples). Another important aspect to be considered is compatibility with downstream applications such as isoelectric focusing (IEF), which is susceptible to ionic contaminants, and in particular to salt or ionic detergents. Moreover, the procedure must be robust and effective to obtain reproducible results and a high rate of yield, without introducing any charge or size modifications in proteins. Finally, it should be borne in mind that gentle cell disruption methods are required for the preparation of intact organelles (e.g., ribosomes, mitochondria, chloroplasts) for subproteome analysis.

Mechanical Cell Disruption One of most widely used mechanical cell disruption methods to process fairly large amounts of soft tissue relies on the use of stainless steel blades rotating at 5000 to 10 000 rpm, and with a shearing action (e.g., Waring blender). Typically, the blender is first cooled by the addition of liquid nitrogen or dry-ice. The cells or tissue to be disrupted are suspended in lysis buffer (see Section 5.16), frozen in liquid nitrogen, added to the blender, and then disrupted by pulsing the blender for several minutes until the tissue resembles a fine powder.

Alternatively, the tissue sample may be ground manually using a mortar and pestle, typically at liquid nitrogen temperature. Liquid nitrogen and the deep-frozen cells or tissue are placed into the pre-cooled mortar and crushed with a pestle until a fine powder is obtained. The powder is then suspended (while still deep-frozen) either in lysis buffer for subsequent protein solubilization, or in ice-cold trichloroacetic acid (TCA)/acetone for protein precipitation (see Section 5.1.3). Thorough cleaning of the device is mandatory to avoid cross-contamination between different samples; however, this is valid for all cell disruption methods based on non-disposable equipment. Whilst this method is the most efficient procedure for disrupting plant tissue such as leaves, it is also applicable for animal tissue (e.g., liver). Dry plant seeds may simply be disrupted by crushing with a hammer and subsequent grinding in a mortar, without cooling with liquid nitrogen.

Liquid-Based Homogenization This is one of the most commonly used cell disruption methods for small amounts (1–10 mL) of cells cultured *in vitro*, soft tissues, and prokaryotes. The *Potter homogenizer* uses the shearing forces generated between a motorized or manually driven polytetrafluoroethylene (PTFE)-coated plunger and a glass cylinder to disrupt the samples. By cooling the glass vessel and suspending the cells or tissue in denaturing lysis buffer containing protease inhibitors, protein degradation during disruption is minimized. The *French press* has been recommended for the disintegration of blood cells, unicellular organisms, and minced animal tissues. The sample (typically suspended in lysis buffer) is placed in the bottom chamber of the device, and the pressure on the sample raised to several hundred MPa. Hence, the cells are forced through a tiny hole into a second chamber, so that the rapid change in pressure between the two chambers causes their disruption. Although the French press is expensive, it is often the method of choice for the mechanical disruption of bacterial cells.

Sonication Sonication is very popular for the disruption of bacteria or yeast cells and minced animal tissue. The sound waves are usually delivered by an ultrasonic probe that is immersed into the sample already suspended in lysis buffer. In order to prevent excessive heating, ultrasonic waves are transmitted into the solution in multiple short bursts, with intermittent cooling in an ice bath. Sonication is often combined with an upstream cell disruption method such as grinding in a mortar with pestle or liquid homogenization or other mechanical force, such as vortexing the cells with glass beads. The system is capable of almost complete disruption of yeast cells in 10 to 15 min, of prokaryotes such as *Escherichia coli* in approximately 5 min, and of mammalian tissues in 1 to 2 min.

Other Methods In contrast to the aforementioned procedures, which employ rather harsh conditions for cell lysis, several more gentle methods are available that facilitate the isolation of intact organelles such as ribosomes, mitochondria, or chloroplasts. These techniques are usually based on osmolysis, freeze/thawing, enzymatic lysis, or cell lysis using detergents. As proteolysis may be a problem, protease inhibitors (*see* Section 5.1.2.3) should always be added.

Circulating mammalian cells (e.g., erythrocytes) or cells grown *in vitro* in suspension culture are harvested by centrifugation and then lysed by suspending the cells in a *hypoosmotic solution* (or deionized water), which causes them to swell and burst.

Freeze/thawing has been recommended for the lysis of mammalian (and also certain bacterial) cells. The cell suspension is usually deep-frozen in an alcohol/dry ice bath or in a freezer ($-70\,^\circ$C), and then thawed; the cell disruption is achieved by ice crystal formation during freezing. For higher yields, several freeze/thaw cycles may be performed. For cells surrounded by a cell wall, osmotic shock or freeze/thawing may be insufficient for cell disruption, and access to organelles is complicated by the requirement of a homogenization method sufficiently strong to disrupt the cell wall but gentle enough to preserve the intactness of organelles. A popular strategy for the isolation of intact organelles is to prepare spheroplast cells, the walls of which have been digested by polysaccharide-cleaving enzymes. Depending on the type of cell wall, different *hydrolytic enzymes* are used, such as lysozyme for Gram-negative bacteria, 1,3-glucanases for yeast cells, or cellulases for plant cells. The content of the spheroblast is then released by mild hypotonic treatment, and the subcellular components are fractionated by centrifugation in a sucrose gradient.

Detergent lysis is often combined with mechanical cell disruption or homogenization, but may also be applied solely. The choice of detergent depends primarily on the type of cell wall/membrane. Whereas, ionic detergents such as SDS or CTAB (see Figure 5.1) are strongly denaturing and totally disrupt cell membranes, non-ionic or zwitterionic detergents are milder. However, ionic detergents have deleterious effects on subsequent IEF, so that precautions must be taken to ensure that these detergents have been removed (or diluted below the critical concentration) prior IEF (see Section 5.1.2.3).

5.1.2.2 Sample Clean-Up

During or after cell disruption, interfering components such as proteolytic enzymes, salts, lipids, nucleic acids, polysaccharides, plant phenols, insoluble material and/or

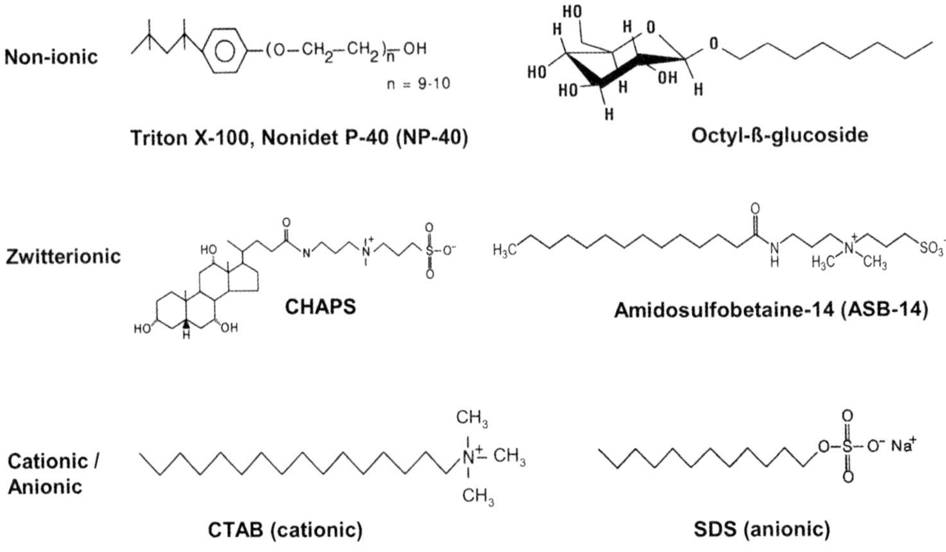

Figure 5.1 Commonly used detergents for protein solubilization.

highly abundant proteins must either be removed or inactivated, as they may complicate subsequent protein solubilization or IEF. Typically, proteases and salt cause the most severe problems.

Proteases All living organisms contain *proteolytic enzymes* (proteases). Whereas protease activity in the cell is strictly regulated by compartmentalization, proteases are liberated after disruption and may react with cellular proteins, thereby introducing charge- and/or mass modifications. Hence, rapid and complete inactivation of endogenous proteases is mandatory to prevent damage to cellular proteins that otherwise would result in artifactual spots on the 2-DGE gel. If proteases are present in the sample, the cells should be disrupted in a deep-frozen state (either at $-70\,^\circ\mathrm{C}$ or in liquid nitrogen), and protease inhibitors should be added. The use of a correct protease inhibitor(s) is important as several different types of proteases exist, such as serine (e.g., trypsin), thiol (e.g., papain), acid (e.g., pepsin), and metalloproteases (e. g., carboxypeptidase A). Hence, different types of inhibitor may be required to protect the proteins against proteolysis. Several cocktails containing up to ten different protease inhibitors are commercially available. Alternatively, the cells may be disrupted in the presence of strongly denaturing agents such as (boiling) SDS buffer or TCA [10]. Protease activity can vary greatly between different organisms, and as certain proteases of microbial or plant origin are rather difficult to inactivate, the efficiency of protease inhibition should always be verified, for example by one-dimensional SDS-PAGE. If the high molecular mass (M_r) proteins are missing, the sample preparation protocol must be improved until protein degradation has been eliminated. Care should also be taken to inactivate hydrolytic enzymes other than proteases, such as phosphatases and oxidoreductases (that may covalently crosslink proteins), by adding inhibitors of these enzymes and/or denaturing agents to cell-disruption and protein-solubilization buffers.

Salt High concentrations (>100 mM) of *salt ions* interfere with IEF and must be removed to avoid problems such as low current, zones of dehydration and "empty lanes" within the IEF gel. Salt increases the conductivity of the IEF gel, thereby prolonging the time required to reach the steady state, so that in extreme cases IEF may virtually stop due to salt fronts. The high concentration of salts, when combined with a relatively low protein content in certain biological samples such as urine or cerebrospinal fluid, makes these samples less suitable for IEF. Salt removal and/or up-concentration of proteins can be achieved by ultrafiltration, spin dialysis and/or precipitation of proteins with TCA and organic solvents (e.g., cold acetone). Sample clean-up kits based on TCA precipitation are available commercially from different suppliers. The disadvantage of protein precipitation lies in the fact that not all proteins may precipitate, or some of the precipitated proteins will not resolubilize, so that a different set of proteins may be obtained compared to, for example, ultrafiltered samples. On the other hand, this effect can be exploited for the enrichment of particular protein classes, such as ribosomal or nuclear proteins, from total cell lysates [11]. If the salt concentration is not excessively high, the sample may also be desalted during IEF with little processing by simply diluting the sample extract with lysis buffer and applying a larger volume onto the IPG gel. Salt is removed by the application of rather low voltages (50–100 V) at the start of the IEF run for up to several hours, and replacing the filter paper pads beneath the electrodes where the salt ions have collected [1,2]. High concentrations (>0.2%) of *ionic detergents* such as SDS also have a deleterious effect on IEF, and must be removed prior to IEF by dilution with a non-ionic or zwitterionic detergent (see Section 5.1.2.3).

Other Interfering Components   *Nucleic acids* may have various adverse effects on 2-DGE gel. Typically, they may: (i) increase sample viscosity; (ii) block the pores of the IEF and SDS gels, resulting in horizontal and vertical streaks; and (iii) interact with nucleic acid-binding proteins that will be depleted and are then missing, or underrepresented, on the gel. Nucleic acids may be removed by digestion with nucleases or by sonication. Although not generally a problem, large amounts of *lipids* interact with hydrophobic proteins, "consume" detergents, or alter the pI of these proteins if the lipids are charged (e.g., phospholipids). Delipidation using organic solvents (e.g., cold ethanol or acetone) has been reported, but this may lead to severe and unpredictable loss of protein [8]. In some cases, lipids form a layer on top of the sample after ultracentrifugation that can easily be removed. Uncharged *polysaccharides* (e.g., starch, glycogen) increase the viscosity of the sample and obstruct the pores of the polyacrylamide gels, thereby inhibiting migration of the sample proteins. Unless present at low concentrations, these should be removed by ultracentrifugation. Charged polysaccharides (e.g., dextrans, mucins) can interact with sample proteins, and result in poor focusing or even depletion of some proteins. Removal is possible by, for example, TCA/acetone precipitation, similar to the removal of nucleic acids.

Phenols present in plants (especially in green tissues such as leaves) may bind to sample proteins, giving rise to horizontal streaks in 2-DGE patterns. Polyphenolic compounds can be removed by binding to polyvinylpolypyrrolidone (PVPP), or by TCA/acetone precipitation of proteins and subsequent rinsing of the pellet with ice-cold acetone [12–15].

On occasion, *highly abundant proteins* (e.g., albumin in plasma) cause problems as they impair the separation and detection of less-abundant proteins by limiting the amount of these proteins to be loaded onto the 2-DGE gel, or by masking them on the 2-DGE pattern. At present, several albumin-removal kits are available commercially, but proteins other than albumin may also be removed due to non-specific binding (for a review, see Ref. [4]).

Insoluble material (cell debris, nucleic acids, polysaccharides, etc.) which may cause horizontal and/or vertical smears and block the pores of the polyacrylamide gels is usually removed by high-speed centrifugation (30 min at $40\,000 \times g$ at 15 °C). Even if no insoluble material appears to be present, it is generally recommended that sample extracts are centrifuged (using at least 20 min at $15\,000 \times g$ at 20 °C before being applied to the IEF gel.

5.1.2.3 Protein Solubilization

In order to take full advantage of the high resolution of 2-DGE, the sample proteins must be reduced, denatured, and disaggregated prior to solubilization so that all non-covalently (ionic and hydrophobic interactions, hydrogen bonds, van der Waals forces) and disulfide-bonded interactions are completely disrupted. This will ensure that, after 2-DGE, each spot represents an individual polypeptide. If such molecular interactions are not blocked, the proteins may aggregate such that they are poorly resolved on the gels (seen as streaking), or there is a loss of proteins due to precipitation.

Protein solubilization is usually carried out in a buffer (see Section 5.1.6) containing *chaotropes*, *surfactants (detergents)*, *reducing agent(s)*, *carrier ampholytes*, and/or *enzyme (protease/phosphatase) inhibitors*. Protein denaturation and solubilization each require some time to complete (30–90 min, depending on the type of sample), although the process may be accelerated by using sonication.

Chaotropes   *Urea* and *thiourea* are the chaotropic agents of choice for disrupting hydrogen bonds and hydrophobic interactions both between and within proteins. As these compounds are uncharged, they have no effect on the intrinsic charge of the proteins, which makes them ideal for IEF. Urea is typically used at concentrations ranging from 7 to 9 M, often in combination with thiourea at concentrations up to 2 M for improved solubilization of hydrophobic proteins [16]. Unfortunately, however, urea exists in equilibrium with ammonium (iso)cyanate, which can react with α-amino groups of the N-terminus and ε-amino groups of lysine residues of proteins, thereby forming artifacts such as blocked N-termini or introducing charge heterogeneities (i.e., altered pI-values). In order to prevent this so-called *carbamylation reaction*, urea-containing solutions must not be heated above 37 °C (samples dissolved in urea buffer should be sonicated with cooling on ice!), and carrier ampholytes are usually added as they act as cyanate scavengers.

Surfactants   *Surfactants* (detergents) are used to prevent hydrophobic interactions between the hydrophobic protein domains within and between proteins, in order to avoid loss of proteins due to aggregation and precipitation. They are particularly useful for the solubilization of otherwise insoluble proteins. The term *surfactant* is a contraction of "surface active agent". Surfactants are amphipathic compounds, which means they contain both hydrophobic "tails" and hydrophilic "heads". The ionic character of the polar head group forms the basis for a broad classification of surfactants: they may be ionic (charged, either anionic or cationic), non-ionic (uncharged), or zwitterionic (having both positively and negatively charged groups, but with a net charge of zero).

Detergents used in sample preparation for 2-DGE must be soluble in urea/thiourea buffer, even at higher concentrations, and they should not carry a net charge to be compatible with subsequent IEF. Detergents are typically used in concentrations of between 2 and 4%. Uncharged detergents include NP-40, Triton X-100, octyl-β-glucoside, and dodecyl maltoside. CHAPS and ASB-14 are the most commonly used zwitterionic detergents (see Figure 5.1). Ionic detergents are highly efficient for breaking protein–protein interactions, while SDS is unparalleled in its ability to solubilize proteins efficiently and rapidly. However, as these detergents are charged, they can only be applied for the initial protein solubilization, and must be removed from the proteins prior to IEF by dilution with an at least eightfold excess of nonionic and/or zwitterionic detergent(s) to ensure that they do not interfere with the IEF separation.

Reducing Agents The reduction and prevention of reoxidation of intra- and intermolecular disulfide bonds (i.e., within and between protein subunits) are necessary steps for complete protein unfolding. This is usually achieved with a free-thiol-containing reductant such as *dithiothreitol* (DTT) [17], which is converted into a stable cyclic disulfide, or with *trialkylphosphines* (see Figure 5.2). The DTT-based thiol–disulfide interchange process is an equilibrium reaction, so that quite high DTT concentrations ($\sim$100 mM) are required. Moreover, DTT becomes negatively charged at alkaline pH and migrates towards the anode during IEF, thereby depleting the basic end of the IEF gel. This allows disulfide bonds to reform randomly both intra- and inter-molecularly, and this results in protein aggregation and precipitation due to oxidative crosslinking. The prevention of aggregation/precipitation can be accomplished by alkylation (e.g., with iodoacetamide) of the reduced proteins so as to block, irreversibly, the free protein sulfhydryl groups. Alternatively, cysteines can be stabilized as mixed disulfides by using hydroxyethyldisulfide in the IPG strip rehydration solution instead of the reductant DTT [18]. Trialkylphosphines [19,20] are non-ionic and do not migrate during IEF, and therefore maintain reducing conditions during the first-dimension

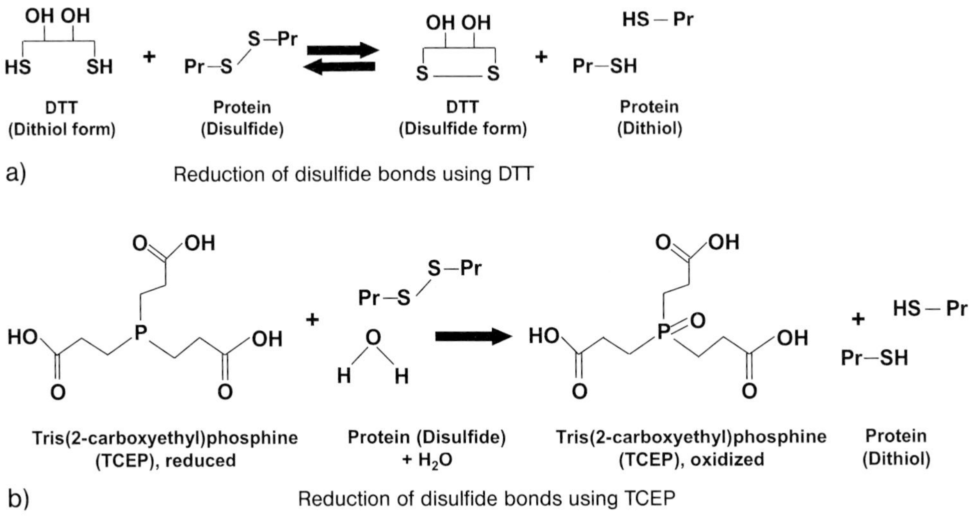

Figure 5.2 Cleavage of disulfide bonds by the reducing agents (a) dithiothreitol (DTT) and (b) tris-(2-carboxyethyl)phosphine (TCEP).

separation. Moreover, these reducing agents can be used at lower concentrations than the sulfhydryl reductant DTT. The most prominent examples are tributylphosphine (TBP) and Tris-(2-carboxyethyl)phosphine (TCEP). Due to its low solubility and short half-life in water, the use of TBP is, however, somewhat restricted, for example to the reduction of proteins containing a high number of cysteines (e.g., wool proteins) [21]. Although these obstacles were overcome by TCEP, which is non-volatile, odorless, and resistant to air oxidation, the use of TCEP in 2-DGE is not widespread. TCEP is primarily employed in the saturation-labeling procedure [22] of fluorescent difference gel electrophoresis (DIGE) [23].

Carrier Ampholytes Carrier ampholytes (1–2%) are commonly added to sample preparation buffers because they increase buffering capacity, improve protein solubilization, and prevent the carbamylation of protein amino groups by scavenging cyanate ions, without interfering with subsequent IEF. The carrier ampholyte mixture used need not necessarily reflect the pH of the IEF gel to be run, and satisfying results are generally obtained with a carrier ampholyte mix pH 3 to 10.

Protease and Phosphatase Inhibitors Inhibitors of proteases and/or phosphatases are usually low molecular-weight compounds that function by reversibly or irreversibly binding to the (active center of the) enzyme. However, enzyme inhibitors should be used with some caution because they may covalently modify proteins and cause charge artifacts. Moreover, the full inhibition of all enzymatic activity is not always easy to accomplish, even if a combination of different inhibitors is applied. Typical examples of protease inhibitors include phenylmethylsulfonylfluoride (PMSF) and the (less toxic) 4-(2-aminoethyl)-benzenesulfonylfluoride hydrochloride (Pefabloc) serine- and cysteine-protease inhibitors, EDTA (which inactivates metalloproteases), or the acid-protease inhibitor Pepstatin. For display of the phosphoproteome, phosphatase activity must be inhibited, for example with a cocktail composed of sodium fluoride, sodium molybdate, and/or sodium orthovanadate.

5.1.3
Application Samples

Sample preparation should be as simple as possible in order to increase reliability and reproducibility. The growth conditions of microorganisms or cultured cells must be strictly standardized, whilst tissue samples should be as fresh as possible and always kept frozen to avoid protein degradation. Protein modifications during sample preparation must be minimized, because these might result in artifactual spots on 2-DGE gels. In particular, proteolytic enzymes in the sample must be inactivated, while samples which contain urea must not be heated in order to avoid protein carbamylation. Samples must be kept sufficiently long in solubilization buffers for the complete reduction, denaturation and solubilization of proteins, and this is best carried at room temperature. In addition, clean glassware and protective gloves must be used, or contaminants (e.g., keratin) may produce artifactual spots on

2-DGE. The final protein concentration of the extracts should approximate 5 to 10 mg mL^{-1}, and the extracts should be stored deep-frozen in small aliquots at $-78\,°C$, or under liquid nitrogen at $-196\,°C$.

5.1.3.1 Mammalian Tissues

Mammalian tissue samples are collected as soon as possible after death of the "donor" and immediately frozen at $-196\,°C$ in liquid nitrogen. Ideally, the samples are disrupted while still deep-frozen. Small tissue specimens (e.g., biopsy samples) are wrapped in aluminum foil, frozen in liquid nitrogen, and crushed with a precooled hammer, whereas larger tissue pieces are ground under liquid nitrogen using a mortar and pestle or in a French press. The resulting deep-frozen powder is transferred into an appropriate amount of denaturing lysis buffer, left for approximately 1 h at room temperature with occasional vortexing or sonication, and then centrifuged at $40\,000 \times g$ at $15\,°C$. For the enrichment of alkaline proteins [11], the tissue is ground under cooling with liquid nitrogen in a mortar and pestle, suspended in 20% TCA in ice-cold ($-20\,°C$) acetone containing 0.2% DTT, and kept at $-20\,°C$ overnight in order to ensure complete protein precipitation. Following centrifugation (60 min, $40\,000 \times g$ at $-15\,°C$), the supernatant is discarded and the pellet resuspended in ice-cold acetone containing 0.2% DTT. The sample is recentrifuged, and the pellet dried under vacuum and then solubilized in lysis buffer. Treatment in a waterbath sonicator helps to improve solubilization, particularly of material that is otherwise difficult to resuspend.

5.1.3.2 Microbial Cell Cultures

Microbial cell cultures such as bacteria or yeast must be standardized and optimized for the growth conditions, and the growth phase determined from which the sample is taken, as this has an enormous impact upon the biochemical state of the cells. Because the cells may excrete proteases and other extracellular enzymes into the growth media, they must first be washed with an isotonic buffer such as phosphate-buffered saline or sucrose (at the same temperature as the culture, so as not to cold- or heat-shock the cells) prior to harvesting by centrifugation. As bacteria and yeast cells are surrounded by a rigid cell wall, the protein extraction cannot be effected simply by osmotic shock, as for mammalian cells. Instead, extensive disruption of the cells is required, either by vigorously shaking them in the presence of glass beads, by sonicating them on ice in the presence of urea/thiourea lysis buffer, or by heating the sample in the presence of SDS [24]. Protease-free DNAse and RNAse may be added to digest nucleic acids.

5.1.3.3 Plant Cells

Dry plant seeds which contain fairly high amounts of protein (10–20%), and where proteases are usually not active, are first crushed with a hammer and then ground with a mortar and pestle, with or without cooling by liquid nitrogen. The powder is then suspended in lysis buffer, sonicated, and centrifuged ($40\,000 \times g$, 60 min at $15\,°C$. It is also possible to extract specific protein fractions only, such as water-soluble or alcohol-soluble proteins from cereals [25,26].

Plant leaves contain not only proteases, which are liberated immediately after cell disruption, but also high concentrations of interfering compounds such as plant phenols, which may adsorb proteins and cause horizontal and/or vertical streaks on the 2-DGE gel. To counter this, the cells are first disrupted with a mortar and pestle in the presence of liquid nitrogen, and the proteins then precipitated with 20% TCA in ice-cold acetone, similar to the procedure described above for the enrichment of alkaline proteins from mammalian tissue. Following the removal of phenols by rinsing the pellet with ice-cold acetone, the proteins are solubilized in lysis buffer [12,15].

5.1.4
Summary

Two-dimensional gel electrophoresis (2-DGE), combined with protein identification by mass spectrometry, is currently the "workhorse" of proteomics. Moreover, sample preparation represents the first step in the process and is, therefore, a key factor for successful 2-DGE-based proteome analysis. Although a universal procedure for sample preparation would be highly desirable, there is no single method that can be applied to all types of sample. However, some general guidelines can be applied to increase efficiency and reproducibility:

- Sample preparation should be as simple as possible.
- Protein modifications during sample preparation must be minimized, because they might result in artifactual spots on 2-DGE gels.
- Samples should be as fresh as possible, and kept deep-frozen to avoid protein degradation.
- Proteolytic enzymes in the sample must be inactivated, and urea-containing solutions must not be heated in order to avoid the carbamylation of protein amino groups.

The three fundamental steps in sample preparation for 2-DGE are: (1) cell disruption; (2) sample clean-up (i.e., the removal and/or inactivation of interfering protein and non-protein components); and (3) protein solubilization. Depending on the type of cell wall, lysis is achieved by gentle cell disruption procedures such as freeze/thaw cycling, osmotic, enzymatic or detergent lysis of the cell wall, or with harsher procedures such as homogenization with glass beads in a bead beater, liquid homogenization, or grinding the sample with (or without) liquid nitrogen. During or after cell disruption, interfering compounds such as proteolytic enzymes, salts, lipids, nucleic acids, polysaccharides, plant phenols, insoluble material or highly abundant proteins must be removed or inactivated, as they may impair the subsequent solubilization and/or isoelectric focusing of proteins. After cell lysis and sample clean-up, the proteins of the sample must be reduced, denatured and disaggregated in order to disrupt all non-covalently (e.g., hydrophobic or ionic) and disulfide-bonded interactions, and solubilized while maintaining their inherent charge properties. Sample solubilization is usually carried out in a buffer containing chaotropes (urea/thiourea), non-ionic or zwitterionic surfactants, reducing agent, carrier ampholytes, and protease inhibitors. After protein solubilization, the extracts are centrifuged, aliquoted and stored deep-frozen at $-78\,°C$ (or in liquid nitrogen) until further use.

5.1.5
Perspective

Current research focuses on: (1) improved solubilization of highly hydophobic proteins; (2) sample prefractionation procedures; and (3) pre-electrophoretic labeling for differential display (i.e., multiplexing). The solubilization of *highly hydrophobic proteins* has been improved by the introduction of chaotropes such as thiourea [16], novel sulfobetaine detergents [16], and organic solvents. Regardless of these recent advances, there does not currently exist a single solution for the complex solubility problem of membrane proteins [27], and empirically testing and optimizing the sample solubilization buffer remains of major importance.

Due to the high number, wide dynamic range and diversity of proteins expressed in higher organisms, *prefractionation* steps are often carried out to reduce sample complexity, enrich for low-abundance proteins, or to deplete highly abundant proteins that obscure the less-abundant proteins. This can be accomplished by: (1) the isolation of specific cell types, for example by laser capture microdissection or fluorescence-activated cell sorting; (2) by the isolation of cell compartments or organelles, for example by sucrose gradient centrifugation; (3) by the selective precipitation or sequential extraction of proteins with increasingly powerful solubilizing buffers; (4) with immunoaffinity binding and related methods; and (5) with chromatographic or electrophoretic procedures (for reviews, see [3,4]). Although, these sample prefractionation procedures have many advantages, they also demonstrate limitations, such as a lower reproducibility due to cross-contamination between individual fractions. Moreover, these techniques are often time-consuming, complicated to use, require concentration steps due to adsorption/desorption effects, and/or do not permit the simultaneous handling of more than a few samples in parallel.

In order to enhance the reproducibility of electrophoretic separations, to simplify image analysis, and/or to improve the quantitation of 2-DGE separated proteins, *differential display* methods for the visualization of multiple samples separated on a single 2-DGE gel have been recently introduced, the most important of which are based on *pre-electrophoretic in-vitro* labeling of proteins from different samples with fluorescent dyes that differ in their excitation and emission wavelengths (DIGE technology based on cyanine dyes [23]). After consecutive excitation at the corresponding wavelength(s), the detected spot patterns are overlaid, normalized and subtracted, whereby only differences (e.g., up- or down-regulated proteins) between the samples are visualized. Methodical variations in spot positions are excluded due to the comigration of two (or three) samples on a single 2-DGE gel, and image analysis is significantly accelerated. For improved quantitation, an internal standard (typically a pooled mixture of all samples), which is run on all gels within a series of experiments, can be included [28].

5.1.6
Recipes for Beginners

Most sample solubilization buffers are based on O'Farrell's lysis buffer [29] and modifications thereof. SDS buffer is sometimes recommended for initial solubilization of proteins; however, the extract must be diluted with an excess of urea or urea/thiourea lysis buffer prior to IEF to displace the anionic detergent SDS from the proteins and replace it with an uncharged detergent to decrease the amount of SDS below a critical concentration of 0.2%. Both, the urea and urea/thiourea lysis buffers can be prepared in fairly large quantities, but should be stored deep-frozen at −78 °C in small aliquots (for up to several months) and not exposed to repeated thawing and freezing.

Urea Lysis Buffer [29]

The urea lysis buffer contains 9.5 M urea, 2% CHAPS, 1% DTT, and 2% carrier ampholytes.

- To prepare 40 mL of urea lysis buffer, dissolve 29.0 g urea (Merck, Darmstadt, Germany) in ca. 25 mL deionized water and make up to 50 mL.
- Add 0.5 g of mixed ion-exchange resin (e.g., Serdolit MB-1; Serva, Heidelberg, Germany), stir for 15 min, and then filter.
- Add 0.8 g CHAPS (Roche, Mannheim, Germany), 400 mg DTT (Sigma-Aldrich, Munich, Germany), and 0.8 mL Pharmalyte pH 3–10 (40%, w/v) (GE Healthcare, Freiburg, Germany) to 40 mL of the filtrate.
- If necessary, add (immediately before use!) 1.0 mg Pefabloc® proteinase inhibitor (Merck) per mL of lysis buffer.

Urea/Thiourea Lysis Buffer [16]

The urea/thiourea lysis buffer contains 7 M urea, 2 M thiourea, 4% CHAPS, 2% DTT, and 2% carrier ampholytes.

- To prepare 40 mL of urea/thiourea lysis buffer, dissolve 21.5 g urea and 7.7 g thiourea (Fluka, Munich, Germany) in ca. 25 mL of deionized water and make up to 50 mL.
- Add 0.5 g of mixed ion-exchange resin (e.g., Serdolit MB-1), stir for 15 min, and then filter.
- Then add 1.6 g CHAPS, 800 mg DTT, and 0.8 mL Pharmalyte pH 3–10 (40%, w/v) to 40 mL of the filtrate.
- If necessary, add (immediately before use!) 1.0 mg of Pefabloc® proteinase inhibitor per mL of lysis solution.

SDS Buffer (1% SDS)

- To prepare 50 mL of SDS lysis solution, dissolve 500 mg of SDS (Serva) in 50 mL deionized water, and then filter. Store in the refrigerator.

Abbreviations

ASB14	tetradecanoylamidopropyldimethylammoniopropanesulfonate
CHAPS	3-[(3-cholamidopropyl)dimethylammonio]-1-propanesulfonate
CTAB	cetyltrimethyl-ammonium bromide
EDTA	ethylenediaminetetraacetic acid
M_r	relative molecular mass
NP-40	Nonidet P-40
pI	isoelectric point
PAGE	polyacrylamide gel electrophoresis
SB3-10	decyldimethylammoniopropanesulfonate
SDS	sodium dodecylsulfate
TBP	tributyl-phosphine

References

1 Görg, A., Obermaier, C., Boguth, G., Harder, A., Scheibe, B., Wildgruber, R. and Weiss, W. (2000) The current state of two-dimensional electrophoresis with immobilized pH gradients. *Electrophoresis*, **21**, 1037–1053.

2 Görg, A., Weiss, W. and Dunn, M.J. (2004) Current two-dimensional electrophoresis technology for proteomics. *Proteomics*, **4**, 3665–3685.

3 Righetti, P.G., Castagna, A., Antonioli, P. and Boschetti, E. (2005) Prefractionation techniques in proteome analysis: The mining tools of the third millennium. *Electrophoresis*, **26**, 297–319.

4 Simpson, R.E. (2004) International proteomics initiatives: technological challenges. *Eur. Pharm. Rev.*, **9**, 25–36.

5 Chevallet, M., Santoni, V., Poinas, A., Rouquie, D., Fuchs, A., Kieffer, S., Rossignol, M., Lunardi, J., Garin, J. and Rabilloud, T. (1998) New zwitterionic detergents improve the analysis of membrane proteins by two-dimensional electrophoresis. *Electrophoresis*, **19**, 1901–1909.

6 Molloy, M.P. (2000) Two-dimensional electrophoresis of membrane proteins using immobilized pH gradients. *Anal. Biochem.*, **280**, 1–10.

7 Dunn, M.J. (1993) *Gel Electrophoresis: Proteins*, Bios, Oxford, UK.

8 Rabilloud, T. (1999) Solubilization of proteins in 2-D electrophoresis. *Methods Mol. Biol.*, **112**, 9–19.

9 Shaw, M.M. and Riederer, B.M. (2003) Sample preparation for two-dimensional gel electrophoresis. *Proteomics*, **3**, 1408–1417.

10 Harder, A., Wildgruber, R., Nawrocki, A., Fey, S.J., Larsen, P.M. and Görg, A. (1999) Comparison of yeast cell protein solubilization procedures for two-dimensional electrophoresis. *Electrophoresis*, **20**, 826–829.

11 Görg, A., Boguth, G., Obermaier, C. and Weiss, W. (1998) Two-dimensional electrophoresis of proteins in an immobilized pH 4–12 gradient. *Electrophoresis*, **19**, 1516–1519.

12 Damerval, C., de Vienne, D., Zivy, M. and Thiellement, H. (1986) Technical improvements in two-dimensional electrophoresis increase the level of genetic variation detected in wheat-seedling proteins. *Electrophoresis*, **7**, 52–54.

13 Cremer, F. and Van de Walle, C. (1985) Method for extraction of proteins from green plant tissues for two-dimensional polyacrylamide gel electrophoresis. *Anal Biochem.*, **147**, 22–26.

14 Flengsrud, R. and Kobro, G. (1989) A method for two-dimensional electrophoresis of proteins from green plant tissues. *Anal. Biochem.*, **177**, 33–36.

15 Méchin, V., Consoli, L., Le Guilloux, M. and Damerval, C. (2003) An efficient solubilization buffer for plant proteins in immobilized pH gradients. *Proteomics*, **3**, 1299–1302.

16 Rabilloud, T. (1998) Use of thiourea to increase the solubility of membrane proteins in two-dimensional electrophoresis. *Electrophoresis*, **19**, 758–760.

17 Cleland, W.W. (1964) Dithiothreitol, a new protective reagent for SH groups. *Biochemistry*, **3**, 480–482.

18 Olsson, I., Larsson, K., Palmgren, R. and Bjellqvist, B. (2002) Organic disulfides as a means to generate streak-free two-dimensional maps with narrow range IPG strips as first dimension. *Proteomics*, **2**, 1630–1632.

19 Ruegg, U.T. and Gattner, H.G. (1975) Reduction of S-sulpho groups by tributylphosphine: an improved method for the recombination of insulin chains. *Hoppe Seyler's Z. Physiol. Chem.*, **356**, 1527–1533.

20 Burns, J.A., Butler, J.C., Moran, J. and Whitesides, G.M. (1991) Selective reduction of disulfides by tris(2-carboxyethyl)phosphine. *J. Org. Chem.*, **56**, 2648–2650.

21 Herbert, B.R., Molloy, M.P., Gooley, A.A., Walsh, B.J., Bryson, W.G. and Williams, K.L. (1988) Improved protein solubility in two-dimensional electrophoresis using tributyl phosphine as reducing agent. *Electrophoresis*, **19**, 845–851.

22 Shaw, J., Rowlinson, R., Nickson, J., Stone, T., Sweet, A., Williams, K. and Tonge, R. (2003) Evaluation of saturation labelling two-dimensional difference gel electrophoresis fluorescent dyes. *Proteomics*, **3**, 1181–1195.

23 Ünlü, M., Morgan, M.E. and Minden, J.S. (1997) Difference gel electrophoresis: a single gel method for detecting changes in protein extracts. *Electrophoresis*, **18**, 2071–2077.

24 Görg, A., Drews, O. and Weiss, W. (2004) Separation of proteins using two-dimensional gel electrophoresis. in *Purifying Proteins for Proteomics* (ed. R.E. Simpson), Cold Spring Harbor Laboratory Press, New York, pp. 391–430.

25 Weiss, W., Postel, W. and Görg, A. (1992) Application of sequential extraction procedures and glycoprotein blotting for the characterization of the 2-D polypeptide patterns of barley seed proteins. *Electrophoresis*, **13**, 770–773.

26 Weiss, W., Vogelmeier, C. and Görg, A. (1993) Electrophoretic characterization of wheat grain allergens from different cultivars involved in bakers' asthma. *Electrophoresis*, **14**, 805–816.

27 Santoni, V., Molloy, M. and Rabilloud, T. (2000) Membrane proteins and proteomics: un amour impossible? *Electrophoresis*, **21**, 1054–1070.

28 Alban, A., David, S.O., Bjorkesten, L., Andersson, C., Sloge, E., Lewis, S. and Currie, I. (2003) A novel experimental design for comparative two-dimensional gel analysis: two-dimensional difference gel electrophoresis incorporating a pooled internal standard. *Proteomics*, **3**, 36–44.

29 O'Farrell, P.J. (1975) High resolution two-dimensional electrophoresis of proteins. *J. Biol. Chem.*, **250**, 4007–4021.

5.2
Sample Preparation for Native Electrophoresis

Ilka Wittig and Hermann Schägger

5.2.1
Introduction

A number of native electrophoresis systems have been described to separate native water-soluble proteins and membrane protein complexes. Examples include: native Tris/glycine PAGE using the Lämmli buffer system [30]; native isoelectric focusing (native IEF) [31]; blue-native electrophoresis (BNE) [32]; and clear-native electrophoresis (CNE) [33].

In contrast to the denaturing Lämmli SDS–PAGE, Triton X-100 is used instead of SDS for sample preparation and for gel buffers in native Tris/glycine PAGE. Native Tris/glycine PAGE separates proteins with isoelectric point pI <9.5, since the running pH in the gel at room temperature is around 9.5 (higher than the pH 8.8 of the gel buffer!) and, in contrast to the anionic SDS, the neutral detergent Triton X-100 does not induce a charge shift on the proteins. Native Tris/glycine PAGE separates native water-soluble proteins but offers rather low resolution for membrane proteins [34].

In contrast to native Tris/glycine PAGE, IEF also separates water-soluble proteins with acidic and basic pI. IEF that is not specified as "native" commonly means denaturing IEF, since high urea concentrations are typically used that denature proteins. Native IEF means that no urea is added to the sample and gel; hence, as urea is missing, protein aggregation is favored and proteins are able to precipitate, especially when they reach areas in the gel where the gel pH approaches the intrinsic pI of the protein. Especially, the resolution of hydrophobic proteins by native and denaturing IEF also still poses considerable problems due to protein aggregation in the aqueous media. Native Tris/glycine PAGE is mostly used for water-soluble proteins because the running pH of Tris/glycine PAGE (pH 9.5) is higher than in CNE (pH 7.5) and more proteins – even those that are basic in nature – migrate to the anode.

Currently, BNE is mostly preferred for membrane proteins and complexes [32,33]. Protocols for BNE, two-dimensional (2-D) BNE/BNE, and 2-D BNE/SDS electrophoresis systems, native electroblotting, native electroelution, and other associated protocols have been recently summarized in detail [35].

BNE has initially been developed for the separation of enzymatically active mitochondrial membrane proteins and complexes in the mass range 10 kDa to 10 MDa (Figure 5.3) [32,33,36]. The essential component of BNE is the anionic dye Coomassie Blue G-250 (Serva Blue G, abbreviated as Coomassie dye in the following) that is added to the sample and cathode buffer, and binds to the surface of membrane proteins and many water-soluble proteins. Coomassie dye shifts the pI even of extremely basic proteins into the acidic range [33]. The charge shift causes native proteins to migrate to the anode at the neutral pH 7–7.5 of the gel. Proteins must be

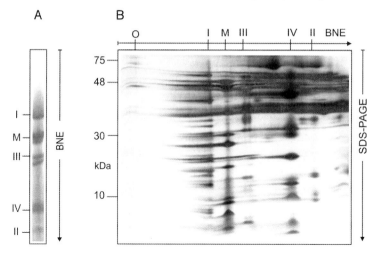

Figure 5.3 Separation of bovine heart respiratory chain complexes by blue-native electrophoresis (BNE). (A) Bovine heart mitochondria were solubilized by dodecyl-maltoside ($2.5\,g\,g^{-1}$) and the five mitochondrial complexes (complex I–V) were separated by BNE on a 4 → 13% acrylamide gradient gel. The five mitochondrial complexes can be used for mass calibration. I, complex I: ca. 1000 kDa; M, monomeric ATP synthase: ca. 700 kDa; III, complex III: 480 kDa; IV, complex IV: 200 kDa; II, complex II: 130 kDa. (B) One gel strip from the 1-D BNE gel was cut out and placed on the top of a 2-D SDS gel using a 16% acrylamide gel for Tricine-SDS–PAGE which separates the subunits of the membrane protein complexes. O, oxoglutarate dehydrogenase complex: 2500 kDa.

negatively charged to allow for migration towards the anode, but they are separated according to their size and not according to their charge/mass ratio. During electrophoresis on acrylamide gradient gels, proteins migrate into gel areas with gradually decreasing gel pore size. Once a protein complex approaches the size-dependent specific pore-size limit it will cease to migrate. BNE can, therefore, be used to determine native protein masses. As BNE is especially suited to separate and isolate large enzymatically active membrane protein complexes, this technique has predominantly been used to study the components of the oxidative phosphorylation system from bacterial membranes, mammalian, yeast and plant mitochondria [37–41] as well as total cell and tissue homogenates [42–46], including patient biopsies for the clinical diagnosis of human mitochondrial disorders [43,44,46–48]. Recently, BN–PAGE was also used for the separation of plasma membrane protein complexes [49], soluble protein complexes (e.g., the proteasome complex) [45], and nuclear protein complexes [50]. BNE is used for in-gel enzyme activity assays [51–54], for native electroblotting and immunodetection [48], to study physiological protein–protein interactions [55–60], to isolate supramolecular physiological protein assemblies [55–64], and many more tasks. Typically, microgram quantities of protein can be isolated on analytical gels, while up to milligram quantities can be recovered by native electroelution from preparative gels for immunization, Edman degradation [65], 2-D crystallization [66], and electron microscopic single-particle analyses [67].

The experimental set-up for CNE is essentially the same as for BNE, except that no Coomassie dye is used and protein migration therefore depends on the protein's intrinsic charge [68]. CNE separates only proteins with pI <7, and offers considerably lower resolution compared to the charge shift method BNE. However, CNE exhibits advantages whenever Coomassie dye interferes with further analyses, as in-gel enzyme activity assays or in-gel fluorescent dye detection [68–70]. Furthermore, CNE acts more mildly to protein complexes than does BNE, and preserves higher-order supramolecular protein assemblies such as oligomeric forms of the ATP synthase [68].

5.2.2
Sample Preparation: General Considerations

BNE and CNE have been elaborated to generally applicable protocols, and specific protocols for the crucial step – solubilization and sample preparation – are available for many specific biological samples. This solubilization and sample preparation step is crucial, as it depends on many different factors that can affect, for example, the solubility and the functional and structural integrity of specific proteins and protein assemblies. The most important factors are: the choice of detergent; the detergent/protein ratio, ionic strength, and pH.

5.2.2.1 Choice of Detergent and Detergent/Protein Ratio
Biological membranes are mostly solubilized by neutral and mild detergents for both BNE and CNE. To date, three neutral detergents have been used preferentially, namely dodecyl-β-D-maltoside (DDM), Triton X-100, and digitonin:

- Digitonin – the mildest detergent – can preserve supramolecular associations of multiprotein complexes and has been essential for the identification of dimeric mitochondrial ATP synthase and specific respiratory supercomplexes (Figure 5.4) [55–57,68].

- DDM is mild, but seems to have stronger delipidating properties compared to digitonin. The solubilization of mammalian mitochondria by DDM resulted in the isolation of individual membrane protein complexes – that are, the building blocks of respiratory supercomplexes (see Figure 5.3).

- Triton X-100 seems to hold an intermediate position between DDM and digitonin. Low amounts of Triton X-100 preserved the supramolecular associations of multiprotein complexes, as did digitonin. Higher levels of Triton X-100 solubilized single respiratory chain complexes from mitochondria, similar to DDM.

Prior to adding any detergent, the samples (membranes) should be carefully suspended in detergent-free solubilization buffer. This facilitates the access of detergent to well-dispersed membranes. Detergent is then added (from 1 to 20% stock solutions in water) to set a specific detergent/protein ratio. Membrane solubilization conditions must be optimized empirically by varying the detergent/protein ratio in the range

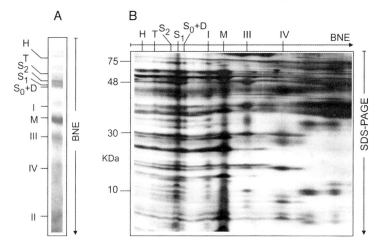

Figure 5.4 Solubilization by digitonin preserves supramolecular assemblies during separation on blue-native electrophoresis (BNE) gels. (A) Digitonin-solubilized bovine heart mitochondria ($6 g g^{-1}$) were separated by 1-D BNE. Supercomplexes S_{0-2} are associates of complex I, dimeric complex III, and zero (S_0) to two (S_2) copies of complex IV. ATP synthase was retained in various oligomeric forms (M, monomer; D, dimer; T, tetramer; and H, hexamer). I–V, mitochondrial complexes I to V. (B) 2-D Tricine-SDS–PAGE was used to separate the subunits of the membrane protein complexes.

from 0.5 to 4 g detergent per gram protein (DDM/protein or Triton X-100/protein 0.5, 1, 2, and $4 g g^{-1}$). Solubilization by digitonin requires higher detergent/protein ratios (e.g., 2, 4, 6, and $8 g g^{-1}$). These series of experiments are used to test the solubilization efficiency, and to identify the conditions that preserve catalytically activity and/or physiological protein associations. Specific protocols for biological membranes are summarized in Table 5.1.

5.2.2.2 Choice of Ionic Strength and pH for Sample Solubilization

Many membrane proteins are not dissolved by neutral detergents using low ionic strength buffers. A certain minimal salt concentration is prerequisite in these cases, for example 50 mM NaCl. Higher salt concentrations improve solubilization, and commonly do not dissociate proteins from complexes using concentrations as high as 500 mM NaCl. In spite of that, salt concentrations should be kept low in the samples for native electrophoresis (<50 mM NaCl) for several reasons: the higher the salt concentration in the sample for BNE or CNE, the higher is the extent of protein stacking (protein concentration to very thin bands). This high (membrane) protein concentration in the sample strongly favors protein aggregation and may prevent proteins from entering the gel. The counterions – especially divalent cations or potassium ions – can precipitate Coomassie dye and the bound proteins with it. Higher salt concentrations, as are often observed with chromatographically prepared samples (especially after ion-exchange chromatography) should therefore be removed by

Table 5.1 Solubilization of mitochondria and bacterial membranes for blue-native electrophoresis (BNE).

	Yeast mitochondria	Mammalian mitochondria	Bacterial membranes
Pelleted protein	400 µg	400 µg	400 µg
Add solubilization buffer A	40 µL	40 µL	40 µL
Homogenize by twirling a tiny spatula before adding detergent.			
Add dodecylmaltoside (20%)	2.0 µL (1.0 g g^{-1})	5 µL (2.5 g g^{-1})	2 µL (1.0 g g^{-1})
or Triton X-100 (20%)	4.8 µL (2.4 g g^{-1})	6 µL (3.0 g g^{-1})	4 µL (2.0 g g^{-1})
or digitonin (20%)	6.0 µL (3.0 g g^{-1})	12 µL (6.0 g g^{-1})	8 µL (4.0 g g^{-1})
Solubilization is complete within several minutes. Then centrifuge for 15 min at 100 000×g (4 °C) Proceed with the supernatant.			
Add glycerol (50%)	5 µL	5 µL	5 µL
Add 5% Coomassie dye to set a detergent/dye ratio of 8/1	1 or 2.4 or 3 µL	2.5 or 3 or 6 µL	1 or 2 or 4 µL
Load around 20 µL each to two 0.15×0.5 cm sample wells for BNE.			

Detergent/protein ratios (in brackets) are empirically determined for yeast mitochondria (*Saccharomyces cerevisiae* and *Yarrowia lipolytica*), mammalian mitochondria (bovine, rat, mouse, pig), and bacterial membranes (*Paracoccus denitrificans*).

dialysis, desalting on columns, or by using centrifuge filter units with an appropriate cut-off limit.

Sample pellets are preferentially homogenized in solubilization buffer "A" (50 mM NaCl, 50 mM imidazole/HCl, 5 mM 6-aminohexanoic acid, 1 mM EDTA, pH 7) before adding detergent for solubilization. If the sample suspension contains salt, or if low ionic strength conditions are required, an alternative solubilization buffer "B" should be used (50 mM imidazole, 500 mM 6-aminohexanoic acid, 1 mM EDTA, pH 7). 6-Aminohexanoic acid can assist membrane solubilization by detergents similar to NaCl but, as a zwitterionic compound, it can avoid certain disadvantages of salts (the zwitterions do not migrate in the electric field, do not contribute to protein stacking and protein aggregation, and do not precipitate Coomassie dye and associated proteins, as described above for salts).

BNE and CNE are performed at neutral physiological pH 7.5 and at low temperature (4–7 °C). Buffer pH values are also adjusted at 4 to 7 °C to avoid temperature-dependent pH shifts.

5.2.2.3 Storage of Biological Membranes

Biological membranes as isolated by differential cell fractionation protocols [57] and density gradient centrifugations [68,71,72] are usually dispersed in small volumes of carbohydrate or glycerol-containing buffers (e.g., 250 mM sucrose, or 400 mM sorbitol, 10% glycerol) to avoid dissociation of multiprotein complexes upon freezing. Samples should be shock-frozen, for example in liquid nitrogen, and then stored at −80 °C for several months.

5.2.2.4 Effects of Adding Coomassie Dye to Sample and/or Cathode Buffer for BNE

The binding of a large number of negatively charged Coomassie dye molecules to proteins has several beneficial effects:

- Coomassie dye shifts the pI of proteins to more negative values and allows even basic proteins to migrate to the anode during electrophoresis ("running pH" = 7.5).
- Negatively charged surfaces repel each other and prevent membrane protein aggregation in the sample well and during electrophoresis.
- Bound Coomassie dye keeps the membrane complexes soluble, thus converting membrane proteins into water-soluble proteins. Therefore, no addition of further detergent to the gels for BNE is required, and the risk of denaturation of detergent-sensitive membrane protein complexes is considerably reduced.
- Proteins and complexes migrate as blue bands to the anode. The blue-stained bands of native proteins can be immediately detected and directly excised for electroelution or for transfer of the gel piece to 2-D protein separation.
- Lipid–detergent micelles, as formed during solubilization by detergent, incorporate added Coomassie dye and migrate rapidly to the running front. Once pulled to the running front, they are no longer interfering with the protein resolution.

Coomassie dye is added immediately after centrifugation and before loading the sample onto the gel wells of the gradient acrylamide gel. The amount of Coomassie dye added to the samples depends on the detergent used for membrane solubilization. Coomassie dye is added from a suspension of 5% dye in 500 mM 6-aminohexanoic acid (w/v) to set a Coomassie/detergent ratio of 8/1, as shown in Tables 5.1 and 5.2.

Chromatographically prepurified membrane proteins mostly contain reduced amounts of boundary lipid and an excess of neutral detergent from the last buffer used. Adding anionic Coomassie dye to this sample would form detergent/anionic dye micelles that can mimic some unfavorable properties of anionic detergents. As detergent-labile subunits would easily dissociate upon direct addition of Coomassie dye, a different approach to avoid dissociation has been proposed, namely to add 5% glycerol, 0.01% Ponceau S dye (from a 50% glycerol, 0.1% Ponceau S stock) but no Coomassie dye to the sample before loading it onto the BNE gel. The hydrophilic dye Ponceau S serves to facilitate sample application but does not bind to the proteins. The Coomassie dye that, of course, is required for BNE, joins the applied protein later, as it requires some time to migrate from the cathode buffer to the protein applied to the sample well. (For more detailed explanation, see Ref. [35], trouble shooting, step 11.)

Table 5.2 Solubilization of mammalian tissue and cell lines.

	Cells	Heart	Muscle	Kidney	Brain	Liver
Aliquots	20 mg	5 mg	15 mg	10 mg	10 mg	10 mg
Add solubilization buffer A	35 µL	35 µL	35 µL	35 µL	35 µL	35 µL
	Homogenize pellet containing mitochondria, nuclei and large fragments					
Add dodecylmaltoside (20%) or	5 µL	2 µL	5 µL	8 µL	10 µL	10 µL
digitonin (20%)	10 µL	4 µL	10 µL	16 µL	20 µL	20 µL
	Solubilization is completed within 5 min. Centrifuge for 15 min at 100 000 × g (4 °C) Proceed with the supernatant.					
50% glycerol	5 µL	5 µL	5 µL	5 µL	5 µL	5 µL
Add 5% Coomassie dye	2.5 µL	1 µL	2.5 µL	5 µL	5 µL	5 µL
Sample loading	Load around 20 µL each to two 0.16 × 0.5 cm sample wells for BNE.					

5.2.3
Applications

5.2.3.1 Solubilization of Bacterial Membranes, Yeast and Mammalian Mitochondria

The maximum protein load applicable to gel wells depends on the DNA content of the sample. DNA blocks the pores of the sample gel and prevents proteins from entering the gel. The maximum load onto 0.15 × 1 cm sample wells is 200 to 400 µg. Concentrated protein suspensions (>10 mg mL^{-1}) can be used directly for solubilization. Aliquots of diluted samples (400 µg) are pelleted by centrifugation (10 min 10 000 × g for mitochondria, 30 min 100 000 × g for bacterial membranes). Water must be added to high-density samples to reduce the density sufficiently for sedimentation by centrifugation. Pelleted membranes are then resuspended in solubilization buffer "A". Table 5.1 provides a step-by-step protocol for the solubilization of bacterial membranes, yeast and mammalian mitochondria. The sample preparation for BNE and CNE is identical, except that Coomassie dye is added to samples for BNE, and Ponceau S/glycerol for CNE.

5.2.3.2 Homogenization and Solubilization of Mammalian Cells and Tissues

The following sample preparation scheme was optimized to investigate respiratory chain complexes from patient skeletal muscle biopsies, animal tissues, and human cell lines. The first step aims at the preparation of a crude mitochondria fraction, avoiding the loss of mitochondrial protein. Biopsy specimens (10–50 mg) from skeletal muscle, heart, kidney, brain, and liver are homogenized in 500 µL sucrose buffer (250 mM sucrose, 20 mM imidazole/HCl, pH 7.0) using a motor-driven, tightly fitting 0.5- to 1-mL glass/Teflon Potter–Elvehjem homogenizer (500 rpm; 10–20 strokes). The sedimented cells (e.g., mouse embryonic stem cells, human fibroblast and lymphoblast cell lines, 143B osteosarcoma and derived rho zero cell lines) are homogenized with 20 to 40 strokes in diluted sucrose buffer (83 mM sucrose, 6.6 mM imidazole/HCl, pH 7). Aliquots (see Table 5.2 for appropriate amounts) are centrifuged for 10 min at 10 000 to 20 000×g to obtain "crude mitochondrial pellets" containing nuclei, mitochondria, and larger cell fragments. The pellets can be stored for several months at −80 °C after shock freezing in liquid nitrogen. Examples of solubilization protocols for mammalian cells and tissues are shown in Table 5.2.

5.2.3.3 Recipe for Beginners: Mass Calibration Ladder for BNE

Common protocols to test buffers and gels and the normal performance of BNE use bovine heart specimens (see Table 5.2, second column) or isolated bovine heart mitochondria (see Table 5.1, center column), because these samples are easily accessible and contain high amounts of membrane protein complexes. These five mitochondrial oxidative phosphorylation complexes (I–V) cover the mass range 130 kDa to 1 MDa, and seem ideal for molecular mass calibration. The apparent masses of complexes I to V on BN-gels are: complex I, ca. 1 MDa; complex II, 130 kDa; complex III, 480 kDa; complex IV, 200 kDa; and complex V, ca. 700 kDa. Acrylamide gradient gels (4 → 13%) are commonly used for this mass range, as exemplified by Figure 5.3A. Supercomplexes (associates of mitochondrial complexes with masses even larger than 2 MDa) are best resolved using 3 → 13% acrylamide gradient gels, as shown in Figure 5.3B.

The isolation of mammalian mitochondria requires fresh, non-frozen tissue, as the use of frozen or aged, non-frozen tissue results in mitochondria of very low purity and yield. However, using non-frozen heart samples from the slaughterhouse or butcher directly, without prior preparation of mitochondria, seems ideal for beginners, as even aged, non-frozen tissue did not show any considerable decay of complexes (as tested using rat brain and skeletal muscle, but not bovine heart).

Some commercially available proteins and protein complexes (mostly water-soluble, some of which appear as monomers and dimers in the BN-gel) have also been used as molecular mass markers, and these may also be used by beginners to test the performance of BNE, the buffers, and the gel. These proteins are: pyruvate dehydrogenase complex from porcine heart (10 MDa), 2-oxoglutarate dehydrogenase from porcine heart (2.5 MDa), ferritin from horse spleen (880 kDa and 440 kDa), catalase (230 kDa and 460 kDa) from bovine liver, lactate dehydrogenase (140 kDa and 280 kDa) from rat muscle, hexokinase from yeast (99 kDa), and bovine serum albumin (66 kDa and 132 kDa).

5.2.4
Summary and Perspectives

Both, BNE and CNE have been successfully applied to study single membrane protein complexes [32,33,37–41] and supramolecular assemblies, such as respiratory supercomplexes and oligomeric ATP synthase [55–60,68]. The native electrophoresis techniques offer some advantages compared to co-immunoprecipitation:

- No specific antibody is required for the isolation of protein complexes whenever the available protein amounts are sufficient for detection by silver-staining and subsequent mass spectrometric analysis. The use of reactive fluorescent dyes (e.g., Cy dyes), as currently used for differential gel electrophoresis (DIGE), will certainly further reduce the minimal protein load.

- Detergents must only be added once, namely for the solubilization of membranes. Recurrent washings with detergent-containing buffers, that may dissociate detergent-labile protein components from multiprotein complexes, are common in co-immunoprecipitation but are avoided in BNE.

- Protein complexes are separated according to size and are easily characterized by 2-D Tricine/SDS–PAGE [32,33,73–75].

CNE has a more mild action towards protein complexes than BNE, and offers advantages whenever Coomassie dye interferes with further analyses, for example, with the in-gel detection of fluorescent-tagged proteins and in-gel catalytic activity assays [68]. Furthermore, the isolation of large physiological protein assemblies by mild electrophoresis techniques such as CNE offers the chance to better understand structural and functional networks within cells [76]. However, current protocols for CNE mostly separate protein complexes with considerably lower resolution than BNE. Recent attempts to retain the advantages of both BNE and CNE led to the development of high-resolution clear native electrophoresis (hrCNE) which seems especially useful for functional proteomics [77].

Abbreviations

M_r relative molecular mass
PAGE   polyacrylamide gel electrophoresis
p*I*   isoelectric point
SDS   sodium dodecylsulfate

References

30 Laemmli, U.K. (1970) Cleavage of structural proteins during assembly of the head of bacteriophage T4. *Nature*, **227**, 680–685.

31 Tsirogianni, I. and Georgios, T. (2003) Preparative isoelectric focusing. in *Membrane Protein Purification and Crystallization. A Practical Guide* (eds C. Hunte, G. von Jagow and H. Schägger), Academic Press, San Diego.

32 Schägger, H. and von Jagow, G. (1991) Blue native electrophoresis for isolation of membrane protein complexes in enzymatically active form. *Anal. Biochem.*, **199**, 223–231.

33 Schägger, H., Cramer, W.A. and von Jagow, G. (1994) Analysis of molecular masses and oligomeric states of protein complexes by blue native electrophoresis and isolation of membrane protein complexes by two-dimensional native electrophoresis. *Anal. Biochem.*, **217**, 220–230.

34 Castro, O.A., Zorreguieta, A., Ielmini, V., Vega, G. and Ielpi, L. (1996) Cyclic beta-(12)-glucan synthesis in Rhizobiaceae:

roles of the 319-kilodalton protein intermediate. *J. Bacteriol.*, **178**, 6043–6048.

35 Wittig, I., Braun, H.P. and Schägger, H. (2006) Blue native PAGE. *Nat. Protocols*, **1**, 418–428.

36 Schägger, H. (2003) Blue native electrophoresis. in *Membrane Protein Purification and Crystallization. A Practical Guide* (eds. C. Hunte, G. von Jagow and H. Schägger), Academic Press, San Diego.

37 Kügler, M., Jänsch, L., Kruft, V., Schmitz, U.K. and Braun, H.P. (1997) Analysis of the chloroplast protein complexes by blue-native polyacrylamide gel electrophoresis. *Photosynth. Res.*, **53**, 35–44.

38 Schägger, H. (2001) Blue native gels to isolate protein complexes from mitochondria. *Methods Cell Biol.*, **65**, 231–244.

39 Pfeiffer, K., Gohil, V., Stuart, R.A., Hunte, C., Brandt, U., Greenberg, M.L. and Schägger, H. (2003) Cardiolipin stabilizes respiratory chain supercomplexes. *J. Biol. Chem.*, **278**, 52873–52880.

40 Ludwig, J., Kerscher, S., Brandt, U., Pfeiffer, K., Getlawi, F., Apps, D.K. and Schägger, H. (1998) Identification and characterization of a novel 9.2 kDa membrane sector associated protein of vacuolar proton-ATPase from chromaffin granules. *J. Biol. Chem.*, **273**, 10939–10947.

41 Schägger, H. (1995) Native electrophoresis for isolation of mitochondrial oxidative phosphorylation protein complexes. *Methods Enzymol.*, **260**, 190–202.

42 Vahsen, N., Cande, C., Briere, J.J., Benit, P., Joza, N., Larochette, N., Mastroberardino, P.G. and Pequignot, M.O. et al.((2004) AIF deficiency compromises oxidative phosphorylation. *EMBO J.*, **23**, 4679–4689.

43 Acin-Perez, R., Bayona-Bafaluy, M.P., Fernandez-Silva, P., Moreno-Loshuertos, R., Perez-Martos, A., Bruno, C., Moraes, C.T. and Enriquez, J.A. (2004) Respiratory complex III is required to maintain complex I in mammalian mitochondria. *Mol. Cell*, **13**, 805–815.

44 Schägger, H., deCoo, R., Bauer, M.F., Hofmann, S., Godinot, C. and Brandt, U. (2004) Significance of respirasomes for the assembly/stability of human respiratory chain complex I. *J. Biol. Chem.*, **279**, 36349–36353.

45 Camacho-Carvajal, M., Wollscheid, B., Aebersold, R., Steimle, V. and Schamel, W.A. (2004) Two-dimensional blue native/SDS gel electrophoresis of multi-protein complexes from whole cellular lysates. *Mol. Cell. Proteomics*, **3**, 176–182.

46 Schägger, H., Bentlage, H., Ruitenbeek, W., Pfeiffer, K., Rotter, S., Rother, C., Bottcher-Purkl, A. and Lodemann, E. (1996) Electrophoretic separation of multiprotein complexes from blood platelets and cell lines: Technique for the analysis of diseases with defects in oxidative phosphorylation. *Electrophoresis*, **17**, 709–714.

47 Schägger, H. (1995) Quantification of oxidative phosphorylation enzymes after blue native electrophoresis and two-dimensional resolution: Normal complex I protein amounts in Parkinson's disease conflict with reduced catalytic activities. *Electrophoresis*, **16**, 763–770.

48 Carrozzo, R., Wittig, I., Santorelli, F.M., Bertini, E., Hofmann, S., Brandt, U. and Schägger, H. (2006) Subcomplexes of human ATP synthase mark mitochondrial biosynthesis disorders. *Ann. Neurol.*, **59**, 265–275.

49 Zapata, D.A., Schamel, W.W., Torres, P.S., Alarcon, B., Rossi, N.E., Navarro, M.N., Toribio, M.L. and Regueiro, J.R. (2004) Biochemical differences in the alphabeta T cell receptor-CD3 surface complex between CD8+ and CD4+ human mature T lymphocytes. *J. Biol. Chem.*, **279**, 24485–24492.

50 Novakova, Z., Man, P., Novak, P., Hozak, P. and Hodny, Z. (2006) Separation of nuclear protein complexes by blue native polyacrylamide gel electrophoresis. *Electrophoresis*, **27**, 1277–1287.

51 Zerbetto, E., Vergani, L. and Dabbeni-Sala, F. (1997) Quantification of muscle mitochondrial oxidative phosphorylation enzymes via histochemical staining of blue native polyacrylamide gels. *Electrophoresis*, **18**, 2059–2064.

52 Jung, C., Higgins, C.M.J. and Xu, Z. (2000) Measuring the quantity of mitochondrial electron transport chain complexes in tissues of central nervous system using Blue native polyacrylamide gel electrophoresis. *Anal. Biochem.*, **286**, 214–223.

53 Eubel, H., Heinemeyer, J., Sunderhaus, S. and Braun, H.P. (2004) Respiratory chain supercomplexes in plant mitochondria. *Plant Physiol. Biochem.*, **42**, 937–942.

54 Sunderhaus, S., Dudkina, N.V., Jansch, L., Klodmann, J., Heinemeyer, J., Perales, M., Zabaleta, E., Boekema, E.J. and Braun, H.P. (2006) Carbonic anhydrase subunits form a matrix-exposed domain attached to the membrane arm of mitochondrial complex I in plants. *J. Biol. Chem.*, **281**, 6482–6488.

55 Arnold, I., Pfeiffer, K., Neupert, W., Stuart, R.A. and Schägger, H. (1998) Yeast mitochondrial F_1F_O-ATP synthase exists as a dimer: identification of three dimer-specific subunits. *EMBO J.*, **17**, 7170–7178.

56 Schägger, H. and Pfeiffer, K. (2000) Supercomplexes in the respiratory chains of yeast and mammalian mitochondria. *EMBO J.*, **19**, 1777–1783.

57 Stroh, A., Anderka, O., Pfeiffer, K., Yagi, T., Finel, M., Ludwig, B. and Schägger, H. (2004) Assembly of respiratory chain complexes I III, and IV into NADH oxidase supercomplex stabilizes Complex I in *Paracoccus denitrificans*. *J. Biol. Chem.*, **279**, 5000–5007.

58 Eubel, H., Jänsch, L. and Braun, H.P. (2003) New insights into the respiratory chain of plant mitochondria: supercomplexes and a unique composition of complex II. *Plant Physiol.*, **133**, 274–286.

59 Eubel, H., Heinemeyer, J. and Braun, H.P. (2004) Identification and characterization of respirasomes in potato mitochondria. *Plant Physiol.*, **134**, 1450–1459.

60 Paumard, P., Vaillier, J., Coulary, B., Schaeffer, J., Soubannier, V., Mueller, D. M., Brethes, D., di Rago, J.P. and Velours, J. (2002) The ATP synthase is involved in generating mitochondrial cristae morphology. *EMBO J.*, **21**, 221–230.

61 Dudkina, N.V., Heinemeyer, J., Keegstra, W., Boekema, E.J. and Braun, H.P. (2005) Structure of dimeric ATP synthase from mitochondria: An angular association of monomers induces the strong curvature of the inner membrane. *FEBS Lett.*, **579**, 5769–5772.

62 Dudkina, N.V., Eubel, H., Keegstra, W., Boekema, E.J. and Braun, H.P. (2005) Structure of a mitochondrial supercomplex formed by respiratory-chain complexes I and III. *Proc. Natl. Acad. Sci. USA*, **102**, 3225–3229.

63 Minauro-Sanmiguel, F., Wilkens, S. and Garcia, J.J. (2005) Structure of dimeric mitochondrial ATP synthase: novel F_O bridging features and the structural basis of mitochondrial cristae biogenesis. *Proc. Natl. Acad. Sci. USA*, **102**, 12356–12358.

64 Schamel, W.W., Arechaga, I., Risueno, R. M., van Santen, H.M., Cabezas, P., Risco, C., Valpuesta, J.M. and Alarcon, B. (2005) Coexistence of multivalent and monovalent TCRs explains high sensitivity and wide range of response. *J. Exp. Med.*, **202**, 493–503.

65 Schägger, H., Brandt, U., Gencic, S. and von Jagow, G. (1995) Ubiquinol-cytochrome c-reductase from human and bovine mitochondria. *Methods Enzymol.*, **260**, 82–96.

66 Poetsch, A., Neff, D., Seelert, H., Schägger, H. and Dencher, N.A. (2000) Dye removal catalytic activity and 2D-crystallization of chloroplast H^+-ATP synthase purified by blue native electrophoresis. *Biochim. Biophys. Acta*, **1466**, 339–349.

67 Schäfer, E., Seelert, H., Reifschneider, N.H., Krause, F., Dencher, N.A. and Vonck, J. (2006) Architecture of active mammalian respiratory chain super-complexes. *J. Biol. Chem.*, **281**, 15370–15375.

68 Wittig, I. and Schägger, H. (2005) Advantages and limitations of clear native polyacrylamide gel electrophoresis. *Proteomics*, **5**, 4338–4346.

69 Gavin, P.D., Devenish, R.J. and Prescott, M. (2003) FRET reveals changes in the F_1-stator stalk interaction during activity of F_1F_O-ATP synthase. *Biochim. Biophys. Acta*, **1607**, 167–179.

70 Gavin, P.D., Prescott, M. and Devenish, R.J. (2005) Yeast F_1F_O-ATP synthase complex interactions in vivo can occur in the absence of the dimer specific subunit e. *J. Bioenerg. Biomembr.*, **37**, 55–66.

71 Meisinger, C., Pfanner, N. and Truscott, K.N. (2006) Isolation of yeast mitochondria. *Methods Mol. Biol.*, **313**, 33–39.

72 Heinemeyer, J., Eubel, H., Wehmhöner, D., Jänsch, L. and Braun, H.P. (2004) Proteomic approach to characterize the supramolecular organization of photo-systems in higher plants. *Phytochemistry*, **65**, 1683–1692.

73 Schägger, H. and von Jagow, G. (1987) Tricine-sodium dodecyl sulfate poly-acrylamide gel electrophoresis for the separation of proteins in the range from 1–100 kDalton. *Anal. Biochem.*, **166**, 368–379.

74 Schägger, H. (2003) SDS electrophoresis techniques. in *Membrane Protein Purification and Crystallization. A Practical Guide* (eds C. Hunte, G. von Jagow and H. Schägger), Academic Press, San Diego.

75 Schägger, H. (2006) Tricine–SDS-PAGE. *Nat. Protocols*, **1**, 16–22.

76 Wittig, I., Carrozzo, R., Santorelli, F.M. and Schägger, H. (2006) Supercomplexes and subcomplexes of mitochondrial oxidative phosphorylation. *Biochim. Biophys. Acta*, **1757**, 1066–1072.

77 Wittig, I., Karas, M. and Schägger, H. (2007) High resolution clear native electrophoresis for in-gel functional assays and fluorescence studies of membrane protein complexes. *Mol. Cell Proteomics*, **6**, 1215–1225.

5.3
Sample Preparation for LC-MS/MS Using Free-Flow Electrophoresis

Mikkel Nissum, Afsaneh Abdolzade-Bavil, Sabine Kuhfuss, Robert Wildgruber, Gerhard Weber, and Christoph Eckerskorn

5.3.1
Introduction

Continuous free-flow electrophoresis (FFE) is a liquid-based separation technique first described by Hannig some 35 years ago [78]. Subsequently, Zeiller *et al.* [79,80] used the technique to separate cells such as lymphocytes and bone marrow cells at constant pH, performing zone electrophoresis. More recently, isoelectric focusing (IEF) using FFE has been enabled by the introduction of a new concept of separation media, and also through improvements in FFE methodology [81]. Since then, FFE has been employed to separate complex mixtures of proteins or peptides according to their isoelectric points (pI) using various linear pH gradients [82–86].

The absence of a solid matrix for separation in the FFE separation system is in contrast to the situation in gel electrophoresis, where acrylamide is used for the solid separation phases that are essential for chromatographic resolution. Instead, in FFE the proteins or peptides migrate according to their charge and electrophoretic mobility in a laminar flow of buffer solutions in an electric field applied perpendicular to the flow direction. The laminar flow is collected into 96 capillaries at the end of the separation chamber, thus enabling continuous fractionation into a collection plate (Figure 5.5). In the IEF mode each fraction has a specific pH value determined by the chosen composition of media, and will contain the proteins or peptides with corresponding pI-values.

In FFE, in contrast to gel electrophoresis or chromatography, the application of sample is continuous. The sample is injected continuously into the running laminar flow, thereby allowing sample volumes ranging from 50 µL in a few minutes to 100 mL in a 24-h run to be processed. Sample quantities can be separated at a rate of

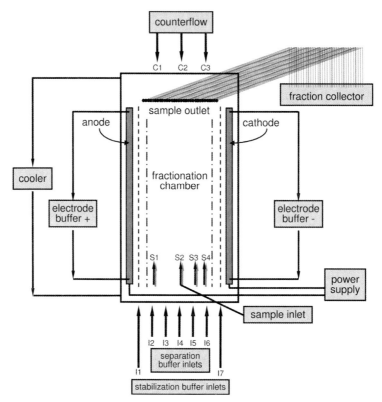

Figure 5.5 Schematic representation of the continuous free-flow electrophoresis (FFE) apparatus. Dimensions of the separation chamber are 500 × 100 × 0.4 mm. The inlets I1 to I7, S1 to S4 and C1 to C3 represent ports for stabilization and separation media, for sample delivery and for counterflow, respectively.

up to more than 30 mg h^{-1}. The separation time – and hence the residence time of the sample in the electric field – is dependent on the flow rate, and is approximately 15 to 25 min, depending on the applied separation protocol.

The compatibility of FFE media with downstream analytical methods has been demonstrated in the past, especially with regards to reversed-phase liquid chromatography (RPLC) and tandem mass spectrometry (MS/MS). FFE has been adapted to different workflows in proteomics as the first mode of separation for proteins with subsequent chromatographic peptide separation for detailed proteome mapping [82,87]. Further, FFE has also been used successfully for separating peptides in conjunction with RPLC-MS/MS, where the high loading capacity of the FFE system and the pI information obtained from the IEF separation were of advantage when performing database searches [85,86,88]. Despite these findings, debate has persisted within the literature as to what extent components of the separation media would interfere with mass spectrometric measurements, and how such interfering components might be removed [89].

In this chapter, the use of separation media that can easily be removed after FFE separation, and prior to RPLC-MS/MS, is described. Following the sample purification and concentration steps, no interfering components were observed in the mass spectrometric analyses. The separation of peptides generated from a whole cellular extract from rat liver was demonstrated, including subsequent purification and mass spectrometric analysis. The approach was not limited to peptides but also was applicable to the separation of proteins.

5.3.2
The Problems of Sample Preparation: The Pros and Cons

The identification of proteins from complex mixtures is typically a multi-step procedure involving separation, enzymatic digestion, and mass spectrometric analysis.

5.3.2.1 Separation
Separation is performed either at the protein or peptide level after digestion with an appropriate enzyme, typically trypsin. The generated peptides from the selected sample are analyzed using tandem MS, and the proteins identified by a search against a database containing protein sequences of the selected or homologous organism. The successful analysis of complex protein mixtures, such as cellular extracts or body fluids, requires more than one separation dimension. This is because more than ten orders of magnitude in concentration may separate the high-abundance proteins from the least-abundant ones, and the restricted dynamic range of currently available analytical techniques is not compatible with this wide range [90].

Initially, two-dimensional gel electrophoresis (2-DGE) was applied to resolve the protein pattern from complex mixtures. Proteins in the sample were separated by isoelectric focusing (IEF) in the first dimension and by molecular weight in the second, resulting in a two-dimensional map of proteins [91,92]. Although 2-DGE provides a high-resolution separation of proteins, the technique suffers from limited applicability to small and very large proteins, proteins with extreme p*I*-values, and hydrophobic proteins [93]. Alternatively, multidimensional chromatographic methods were combined and coupled online to tandem MS in the "shotgun" proteomics approach for the separation of peptides. Especially, the combination of strong cation-exchange chromatography followed by RPLC proved useful in the mapping of the proteome [94,95]. This approach greatly facilitated the proteomics workflow as a direct online coupling of the two separation dimensions was achieved. The highest resolution of a multidimensional separation system is obtained when orthogonal separation techniques are combined [96]. Malmstrom *et al.* [85] showed that there is no relationship between p*I* and hydrophobicity; hence, IEF and RPLC represent a 2-D separation system that provides an orthogonal separation with high peak capacity. These two techniques were combined by performing in-gel IEF of peptides with immobilized pH gradient (IPG) gels as the first separation dimension, followed by RPLC-MS/MS [97–99]. Another approach for IEF-based peptide separation has been described using off-gel IEF [100,101]. Although the peptides were focused in an IPG gel into 15 fractions, the peptides were extracted directly into a liquid medium in the focusing device, which facilitated coupling to RPLC-MS/MS.

5.3.2.2 Extraction

The extraction of peptides and proteins from a gel matrix suffers from the effect of non-specific adsorption to the gel matrix and surfaces that, in both cases, can be substantial. A further complication is that only a limited part of the protein sequence is normally recovered for analysis. This is mainly due to difficulties in the extraction of large peptides from the gel matrix and to variable accessibility of the proteolytic enzyme to the protein [102]. Ideally, it should be possible to reduce these losses using solution-based IEF. Indeed, this was shown by Moritz *et al.* [83], who performed in-solution IEF-FFE. A high sample recovery was obtained when separating proteins from a total cellular extract of a colon carcinoma cell line.

5.3.2.3 Media Composition

The composition of the solution-based media is important for further analysis of the generated FFE fractions. Initially, the FFE media contained hydroxypropyl methyl cellulose (HPMC) to reduce electro-endosmosis and increase the viscosity of the medium. Although these media were used successfully for protein and peptide separations in conjunction with RPLC [86,87], some difficulties have been encountered using this type of media [89]. Cho *et al.* [103] described a cumbersome procedure using acetone or trichloroacetic acid precipitation followed by SDS–PAGE to remove interfering media constituents. Similar media were observed to cause an increased background of the MS spectra in a RPLC-MS/MS run, even after purification using ultrafiltration [85]. Recently, the present authors developed a new type of HPMC-free IEF-FFE media which contained mannitol as a low molecular-weight alternative to HPMC, in addition to well-defined low molecular-weight components used to build the pH gradient called Prolytes (BD Diagnostics, Munich, Germany) and urea. All components of the mannitol-based media were efficiently removed

5.3 Sample Preparation for LC-MS/MS Using Free-Flow Electrophoresis

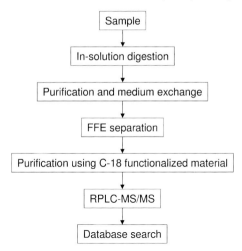

Figure 5.6 Workflow including free-flow electrophoresis for shotgun proteomics.

prior to RPLC-MS/MS; subsequently, no interference with the MS measurements was observed and there was no discernable background in the MS spectra [84,85].

5.3.3
Application Example

The use of FFE in shotgun proteomics was demonstrated where FFE provided an alternative separation dimension to reduce sample complexity prior to RPLC-MS/MS analysis. A total cellular protein extract from rat liver was digested using trypsin. The generated peptides were separated in a pH gradient from 3 to 9 according to their pI, using FFE. Prior to further separation and mass spectrometric analysis using RPLC-MS/MS, the peptide-containing FFE fractions were purified and concentrated. All constituents of the separation media were efficiently removed, and no interference from the media was observed during subsequent analysis of the FFE fractions. The complete workflow is shown in Figure 5.6.

5.3.3.1 Reagents
IEF Prolyte Buffer 3-9 and SPADNS (2-(4-sulfophenylazo)1,8-dihydroxy-3,6-naphthalenedisulfonic acid) were obtained from BD Diagnostics (Munich, Germany). Sulfuric acid, sodium hydroxide and mannitol were purchased from Riedl-de-Haen (St. Louis, MO, USA). Ethanolamine, taurine, betaine and n-acetylglycine were from Fluka (St. Louis, MO, USA). HEPES and AMPSO [N-(1,1-dimethyl-2-hydroxyethyl)-3-amino-2-hydroxypropanesulfonic acid] were purchased from Roth (Karlsruhe, Germany), TAPS [N-tris(hydroxymethyl)methyl-3-aminopropanesulfonic acid] from CalBiochem (Darmstadt, Germany), and urea from Serva (Heidelberg, Germany). Sequencing-grade modified trypsin was purchased from Promega (Madison, WI, USA). All other reagents and chemicals were from either

Sigma-Aldrich or Merck (Darmstadt, Germany). SepPak cartridges for solid-phase extraction were purchased from Waters (Milford, MA, USA), and ultrafiltration spin columns (MWCO 5 kDa) from Vivascience (Hannover, Germany).

5.3.3.2 Sample Preparation for FFE

Tissue from rat liver which had been frozen in liquid nitrogen was ground using a pestle and mortar, and solubilized in lysis buffer containing 7 M urea, 2 M thiourea, 2% CHAPS, 1% dithiothreitol and 5 mM of the protease inhibitor Pefabloc (Serva). The disrupted cell suspension was centrifuged at $16\,000 \times g$ for 45 min at 4 °C. The supernatant was collected (cytosolic fraction) and the protein concentration determined using the 2-D Quant Kit from GE Healthcare (Munich, Germany). A 2-mg portion of protein in 500 µL lysis buffer was reduced in 10 mM TCEP for 60 min at 25 °C and alkylated in the dark in 50 mM iodoacetic acid for 60 min at 25 °C. The pH was adjusted to 7.8 with 100 mM ammonium bicarbonate and the urea concentration was diluted to approximately 2 M using an ultrafiltration spin column. Trypsin (80 µg) was added to the sample which was then incubated overnight at 37 °C. Complete digestion was verified using SDS–PAGE (results not shown). Peptides were purified using SepPak™ C18 reversed-phase cartridges according to the manufacturer's instructions. The eluted peptides where dried down using SpeedVac centrifugation and stored at −20 °C until further use.

5.3.3.3 FFE Separation of Peptides

FFE was performed in the isoelectric focusing mode using a BD™ FFE System (BD Diagnostics). For a detailed description of the principles, see Ref. [104]. A schematic illustration of the current version of the apparatus is shown in Figure 5.5. All media contained 8 M urea and 250 mM mannitol. Except for the counterflow medium, Prolyte (BD Diagnostics) or other electrolytes (acid or base) were added to create the pH gradient. The anodic stabilization medium contained 100 mM sulfuric acid, 30 mM N-acetylglycine, 200 mM taurine and 50 mM betaine anhydrous; the separation medium contained 26% IEF Prolyte Buffer 3-9; and the cathodic stabilization medium contained 150 mM sodium hydroxide, 75 mM ethanolamine, 150 mM AMPSO, 75 mM TAPS and 30 mM HEPES. The electrolyte anode circuit was 100 mM sulfuric acid, and the electrolyte cathode circuit 100 mM sodium hydroxide. All media were prepared fresh daily. The media were pumped into the FFE separation chamber via the media inlets, as shown in Figure 5.5. The flow rates for the separation media were set at $60\,\text{mL}\,\text{h}^{-1}$. The flow rate for delivering the sample solution was $1\,\text{mL}\,\text{h}^{-1}$, corresponding to $2\,\text{mg}\,\text{h}^{-1}$. A Tecan (Maennedorf, Switzerland) liquid-handling system equipped with a pH electrode was used to measure the pH value of each fraction.

Immediately prior to FFE separation the peptides were dissolved in 1 mL separation medium and 15 µL of SPADNS, and introduced into the separation chamber as described above. Electrophoresis was performed at 400 V. A volume of approximately 1 mL was collected of each of the 96 fractions. The total collection time was 60 min. SPADNS was used as marker to indicate the start and end of sample collection [85]. Peptides from the FFE fractions were purified using C18 SepPak™

cartridges using 600 μL of each fraction, respectively. After purification, the eluted peptides where dried down and dissolved in approximately 50 μL of 0.1% trifluoroacetic acid.

5.3.3.4 RPLC-MS/MS Analysis

Electron spray ionization-based LC-MS/MS (HCTultra, Bruker, Bremen, Germany) analyses were carried out using an Agilent 1100 series NanoPump (Agilent Technologies, Waldbronn, Germany) on a 75 μm × 15 cm fused silica microcapillary reversed-phase column (Agilent). Sample volumes of 1 μL were loaded onto the pre-column (300 μm × 0.5 cm reversed-phase (C18) column from Agilent) at a flow rate of $10\,\mu L\,min^{-1}$ for 5 min using a microflow CapPump (Agilent). After sample loading, the sample was separated and analyzed at a $200\,nL\,min^{-1}$ flow rate with a gradient of 2% B to 50% B over 35 min. The column was directly coupled to the spray needle from New Objective (Woburn, MA, USA). Mobile phase A was 0.1% formic acid, and mobile phase B was 100% acetonitrile and 0.1% formic acid. Peptides eluting from the capillary column were selected for collision-induced dissociation (CID) by the mass spectrometer using a protocol that alternated between one MS scan (300–1500 m/z) and three MS/MS scans. The three most abundant precursor ions in each survey scan were selected for CID, if the intensity of the precursor ion peak exceeded 10 000 ion counts. The electrospray voltage was set to 1.8 kV, and the specific m/z value of the peptide fragmented by CID was excluded from reanalysis for 2 min.

5.3.3.5 Data Processing

Each MS/MS spectrum was searched against the IPI rat database (www.ebi.ac.uk), release no. 3.21, using the Mascot Software (Matrix Science Ltd., London, UK). The probability score calculated by the software was used as the criterion for correct identification search. An expectation value of less than 0.01 was required for identification. In addition, peptides were required to have a minimum sequence length of seven amino acids and to be fully tryptic with one internal missed cleavage site allowed. Carbamidomethylation was assigned as fixed modification and methionine oxidation as variable. The mass tolerances were set to 1.5 Da and 0.5 Da for MS and MS/MS, respectively. Proteins with at least one peptide passing these criteria were accepted as an identification. A random database search was performed to estimate the false-positive rate, although the data sets were relatively small in this study.

5.3.4
Summary

A basic requirement to a sample preparation system is *reproducibility*. In the case of FFE, this was demonstrated by performing a protein separation of cellular extract from rat liver using a slightly different gradient in the pH range from 4 to 8. Triplicate separations were obtained on two consecutive days, respectively. All six separations provided the same protein pattern when SDS–PAGE was used for the analysis (Figure 5.7).

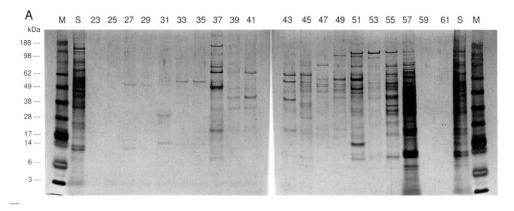
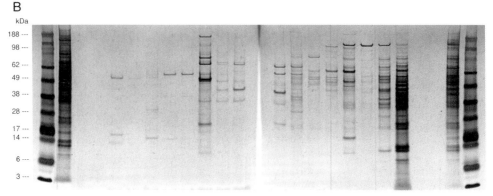

Figure 5.7 Reproducibility of free-flow electrophoresis (FFE) separations. A liver cell extract prepared as described in Section 5.3.3.2 was separated using a pH 4–8 gradient in triplicate on two consecutive days, respectively. The separation medium contained 1% CHAPS to increase protein solubility. The FFE fractions were analyzed by SDS–PAGE. Representative gel images are shown from day 1, experiment one (A) and day 2, experiment one (B), respectively. Numbers above the gel images represent the individual FFE fractions. M, protein marker; S, starting material.

Recovery is another important parameter for a sample preparation system. In order to measure recovery, a sample of the peptide angiotensin II was prepared in the FFE separation medium. Equal amounts of the sample were subsequently either separated using the FFE system and then purified and analyzed using RPLC-MS/MS as described above, or simply purified and analyzed. A full recovery from the FFE system of the angiotensin II peptide was demonstrated (Figure 5.8), with angiotensin II being distributed over three fractions. The fact that approximately 80% of the peptide load was located in the central fraction illustrated the focusing capacity and thereby the potential resolution of the system.

The FFE system was used in the application example to separate a complex mixture of peptides according to the workflow shown in Figure 5.6. Sample purification prior to FFE is not mandatory as long as the salt concentration is below 25 mM, but it is

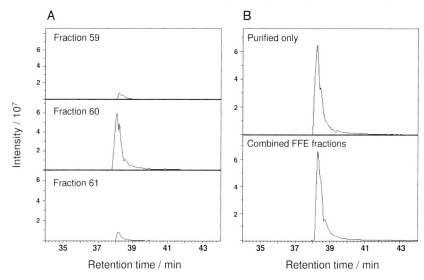

Figure 5.8 Extracted ion chromatograms of angiotensin II, demonstrating the recovery and resolution of the FFE system. (A) A 2-µg quantity of angiotensin II was processed in a pH 4–8 gradient. Angiotensin II was identified in the fractions 59 to 61, with approximately 80% of peptide content in fraction 60. (B) The extracted ion chromatogram of the three combined fractions was compared to the extracted ion chromatogram of the same amount of angiotensin II that was purified, only, using a solid-phase extraction cartridge.

recommended in order to provide consistent and reproducible results. An optimized pH gradient (Figure 5.9) provided additional fractions in the abundant peptide region from pH 3.5 to 5 compared to the pH 4–8 gradient used by Malmstrom et al. [85]. Increased resolution in this region is desirable to prevent under-sampling in the mass spectrometer that, potentially, could lead to missed peptide identifications. The simultaneous removal of media constituents and up-concentration of the sample were performed before RPLC-MS/MS, as only a few microliters were loaded onto the chromatographic column. However, this procedure was performed in a consistent and reproducible manner using standard solid-phase extraction (SPE) cartridges. No interference from the media constituents was observed in the MS measurements. An alternative to the cartridges are 96-well SPE plates in which all the FFE fractions may be purified and concentrated in parallel [84]. Such a process significantly reduces the sample preparation time prior to RPLC-MS/MS.

In order to obtain a comprehensive coverage of the proteome, all 32 peptide fractions across the whole pH range from 3 to 9 should be analyzed with RPLC-MS/MS [85]. In the present study, the principles of solution-based peptide separation were demonstrated by analyzing three fractions in the abundant peptide region. The base peak chromatograms of the fractions 30, 32 and 34 with the measured pH-values of 4.5, 4.8 and 5.1, respectively, are shown in Figure 5.10. Significant differences are observed in the chromatograms, indicating that the peptides from the cellular liver extract were resolved in the pH gradient. The number of proteins identified from the three fractions are listed in Table 5.3. As only the number of

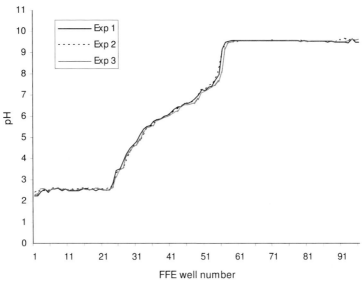

Figure 5.9 The pH gradient used for peptide separation. Graphs from three independent experiments were overlaid. The two flat regions, one below pH 3 and the other above pH 9, represent the pH of the anodic stabilization medium and the pH of cathodic stabilization medium, respectively.

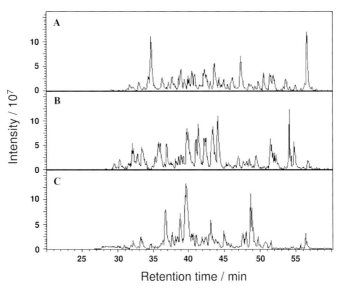

Figure 5.10 Base peak chromatograms of selected FFE fractions. Peptides from a liver cell extract were separated in a pH 3–9 gradient using FFE. Fractions 30 (A), 32 (B) and 34 (C) were purified and analyzed using RPLC-MS/MS.

Table 5.3 Numbers of proteins identified from the FFE fractions.

Fraction no.	Fraction pH	Protein IDs Forward	Protein IDs Random[a]
30	4.5	146	0
32	4.8	131	2
34	5.1	100	0

[a] Hits to the random database provide an approximate measure of the false-positive rate.

true positives should be counted, the rate of false-positive assignments was estimated. In addition to the normal forward IPI rat database, a decoy database consisting of randomized sequences of the same length and amino acid distributions as the original database was searched. The number of protein hits to the randomized database was used as a proxy measure of the number of false protein hits to the IPI rat database. The number of true positive protein hits was then determined by subtracting this estimated number of false-discoveries from the number of hits to the normal database. In the present case, the false-positive rate was estimated to be below 1%. A total of 277 distinct gene products was identified from the three fractions. In 81 cases the same protein was identified in more than one fraction, and for only four proteins was the same peptide identified in more than one fraction. This not only demonstrated the high resolution of the FFE system but also was in agreement with the focusing properties of the system shown in Figure 5.8. In addition, the analysis of more FFE fractions increased the protein sequence coverage, and thereby the confidence with which several proteins was identified.

5.3.5
Perspective

A common occurrence in the sequence database searching of MS/MS data is that protein identifications based on only a single peptide sequence match. In many cases, these peptides contribute to a significant proportion of the identified proteins. However, several of these matches are determined with low confidence and require manual evaluation [94,95]. One way of increasing confidence is to include the pI information obtained from the FFE fractions in the database search. Xie et al. [86] used a combination of pI filtering and probability scoring to minimize false-negative and false-positive identifications of proteins, and thereby reduce manual evaluation. The predicted pI of peptide sequences was calculated according to Bjellqvist [105], and database search results were processed using the PeptideProphet program [106] modified to include the pI information [85,86].

Separation techniques with high capacity are essential in a proteomics workflow for the detection of low-abundance proteins. Considering the large dynamic range of protein concentrations in the cell of up to nine orders of magnitude [90], processing of sample material in the milligram range is needed to access proteins in the picogram range, based on the sensitivity of currently available mass spectrometers. Detailed mass spectrometric analyses involving electron transfer dissociation for the identification of post-translational modifications require even higher sample amounts if

Table 5.4 FFE working conditions.

Operation Mode	Horizontal	Sample flow rate	1 mL h^{-1}
Spacer	0.4 mm	Sample injection	S2
Filter	0.6 mm	Sample tube	I.D. 0.51 mm
Temperature	10 °C	Voltage	Set: 400 V
Counterflow inlets	I.D. C1: 0.76 mm		Actual: 395 V
	I.D. C2: 1.42 mm	Current	Set: 50 mA
	I.D. C3: 0.76 mm		Actual: 19 mA
Media flow inlets 1–7	I.D. 0.64 mm	Power	Set: 60 W
Media flow rate	60 mL h^{-1}		Actual: 8 W

low-abundance proteins are to be analyzed [107]. In the present study, the separation of peptides in the low milligram range was demonstrated using the continuous sample application of the FFE system. With extended separation time, quantities of 20 mg or more may be separated within a normal working day. Thus, the FFE system may be used in a preparative manner at an initial stage of the proteomics workflow to access low-abundance proteins, as demonstrated by Moritz et al. [87].

5.3.6
Recipe for Beginners

5.3.6.1 FFE Set-Up Procedure

- Switch on the instrument. Switch on the cooling circuit and set appropriate experiment temperature.

- Set the working conditions as described in Table 5.4.

- Place the separation chamber in the upright position and open the separation chamber.

- Clean the interior surfaces of the separation chamber.

- Position spacer (Table 5.4) on the interior surface of the front part of separation chamber.

- Insert membranes and filters (Table 5.4) and close separation chamber.

- Place the fractionation plate on the fraction collector housing. Place counterflow valves in "OPEN" position. Immerse the media inlet tubes 1–7 as well as the counterflow tubes 1–3 in a reservoir filled with H_2O. Fill the separation chamber with H_2O. Ensure that all air bubbles are displaced from the chamber.

- Check that all 96 fractionation tubes are dripping consistently.

- Place the sample tube inlet in a small empty beaker. Relax the sample pump until the inlet tube is filled with H_2O and dripping. Ensure that all air bubbles are pumped out of sample inlet tube. Finally, fix the sample inlet tube until dripping is stopped.

- Calibrate the media and sample pump.

- To verify that the system has a consistent laminar flow over the length of the separation chamber without changes in laminar flow and thickness, a colored red dye (SPADNS) is used to obtain three stripes from the buffer inlets 2, 4, and 6 to the collection ports of the FFE chamber.

- The separation chamber is coated with HPMC by immersing the media and counterflow inlets in a beaker filled with 300 mL 0.2% HPMC solution. Run the media pump at 180 mL h^{-1} for 20 min to equilibrate the FFE system.

5.3.6.2 Pre-Experimental Quality Control (QC)

- Stop the media pump and immerse the media inlets in a beaker filled with H$_2$O to wash the tube surfaces.

- Relocate the media inlet in the following order:
  - media inlets 1 and 2 in the anodic stabilization buffer (Table 5.5),
  - media inlets 5–7 in the cathodic stabilization buffer (Table 5.5),

Table 5.5 FFE-IEF buffers pH 3–9.

IEF-buffer	Composition[a]
Counter flow (CF) buffer Inlets C 1–3	600.0 g H$_2$O 432.0 g urea 40.8 g mannitol
Separation buffer pH 5.60 Conductivity: 460 µS cm^{-1} Inlets I3 and I4	52.5 g IEF Prolyte buffer 3–9 147.5 g H$_2$O 144.0 g urea 13.6 g mannitol
Stabilization buffer anode pH 2.2 Conductivity: 7790 µS cm^{-1} Inlets I1 and I2	4.0 g 5 M H$_2$SO$_4$ 196.0 g CF buffer 0.7 g N-acetylglycine 5.0 g taurine 1.2 g betaine anhydrous
Stabilization buffer cathode pH 9.5 Conductivity: 5000 µS cm^{-1} Inlets I5–I7	11.3 g 4 M NaOH 288.7 g CF buffer 5.5 g TAPS 10.2 g AMPSO 2.2 g HEPES 1.4 g ethanolamine
Electrode buffer anode	5.0 g 5 M H$_2$SO$_4$ 245.0 g H$_2$O
Electrode buffer cathode	25.0 g 1 M NaOH 225.0 g H$_2$O

[a] It is strongly recommend that the buffer components are measured by weight, not only to facilitate preparation but also to ensure accurate preparation.

- media inlets 3 and 4 and the counterflow inlets in the counter flow buffer (Table 5.5).

- Introduce media at 240 mL h^{-1} for 10 min. Turn off the media pump and relocate the media inlets to the required media (Table 5.5). Equilibrate system at 180 mL h^{-1} for 10 min.

- Relocate sample inlet to S2 and run the sample pump in the reverse direction to remove any air bubbles in the sample inlet and fill the sample inlet tube with separation buffer.

- Prepare the IEF test solution for QC. Weigh 360 mg urea and 34.5 mg mannitol into a 2-mL Eppendorf tube and add 500 µL pI-Markers (BD Diagnostics) and 500 µL separation buffer.

- Turn on high voltage and wait approximately 10 min. When the FFE current conditions are stable, perform the IEF Test (QC).

- Collect the p*I*-Markers at 35 min after injection in the separation chamber and verify that p*I*-Markers are sharply focused.

5.3.6.3 Experiment

- Relocate sample inlet to a small beaker and rinse the sample inlet using reverse flow of the sample pump. When the sample inlet has been rinsed, switch off the sample pump.

- Place sample inlet in sample and apply sample.

- Collect the fractionated sample after 40 min in appropriate sample plate.

- Proceed to sample downstream analysis.

Abbreviations

AMPSO	[N-(1,1-dimethyl-2-hydroxyethyl)-3-amino-2-hydroxypropanesulfonic acid]
HPMC	hydroxypropyl methyl cellulose
MWCO	molecular weight cut-off
SDS–PAGE	sodium dodecylsulfate polyacrylamide gel electrophoresis
SPE	solid-phase extraction
TAPS	[N-tris(hydroxymethyl)methyl-3-amino-propanesulfonic acid]

References

78 Zeiller, K., Pascher, G. and Hannig, K. (1970) Free-flow electrophoresis: a new method for the elucidation of immunological events at the cellular level, II. The formation of 19S hemolysin-producing cells in intestinal lymph nodes of the rat. *Hoppe Seylers Z. Physiol. Chem.*, **351**, 435–447.

79 Zeiller, K., Hannig, K. and Pascher, G. (1971) Free-flow electrophoretic separation of lymphocytes. Separation of graft versus host reactive lymphocytes of rat spleens. *Hoppe Seylers Z. Physiol. Chem.*, **352**, 1168–1170.

80 Zeiller, K., Schubert, J.C., Walther, F. and Hannig, K. (1972) Free flow electrophoretic separation of bone marrow cells. Electrophoretic distribution analysis of in vivo colony forming cells in mouse bone marrow. *Hoppe Seylers Z. Physiol. Chem.*, **353**, 95–104.

81 Weber, G. and Bocek, P. (1998) Recent developments in preparative free flow isoelectric focusing. *Electrophoresis*, **19**, 1649–1653.

82 Hoffmann, P., Ji, H., Moritz, R.L., Connolly, L.M., Frecklington, D.F., Layton, M.J., Eddes, J.S. and Simpson, R.J. (2001) Continuous free-flow electrophoresis separation of cytosolic proteins from the human colon carci-noma cell line LIM 1215: a non two-dimensional gel electrophoresis-based proteome analysis strategy. *Proteomics*, **1**, 807–818.

83 Moritz, R.L., Ji, H., Schutz, F., Connolly, L.M., Kapp, E.A., Speed, T.P. and Simpson, R.J. (2004) A proteome strategy for fractionating proteins and peptides using continuous free-flow electrophoresis coupled off-line to reversed-phase high-performance liquid chromatography. *Anal. Chem.*, **76**, 4811–4824.

84 Immler, D., Greven, S. and Reinemer, P. (2006) Targeted proteomics in biomarker validation: detection and quantification of proteins using a multi-dimensional peptide separation strategy. *Proteomics*, **6**, 2947–2958.

85 Malmstrom, J., Lee, H., Nesvizhskii, A.I., Shteynberg, D., Mohanty, S., Brunner, E., Ye, M., Weber, G., Eckerskorn, C. and Aebersold, R. (2006) Optimized peptide separation and identification for mass spectrometry based proteomics via free-flow electrophoresis. *J. Proteome Res.*, **5**, 2241–2249.

86 Xie, H., Bandhakavi, S. and Griffin, T.J. (2005) Evaluating preparative isoelectric focusing of complex peptide mixtures for tandem mass spectrometry-based proteomics: a case study in profiling chromatin-enriched subcellular fractions in *Saccharomyces cerevisiae*. *Anal. Chem.*, **77**, 3198–3207.

87 Moritz, R.L., Clippingdale, A.B., Kapp, E.A., Eddes, J.S., Ji, H., Gilbert, S., Connolly, L.M. and Simpson, R.J. (2005) Application of 2-D free-flow electrophoresis/RP-HPLC for proteomic analysis of human plasma depleted of multi high-abundance proteins. *Proteomics*, **5**, 3402–3413.

88 Xie, H., Rhodus, N.L., Griffin, R.J., Carlis, J.V. and Griffin, T.J. (2005) A catalogue of human saliva proteins identified by free flow electrophoresis-based peptide separation and tandem mass spectrometry. *Mol. Cell. Proteomics*, **4**, 1826–1830.

89 Lee, H.J., Lee, E.Y., Kwon, M.S. and Paik, Y.K. (2006) Biomarker discovery from the plasma proteome using multidimensional fractionation proteomics. *Curr. Opin. Chem. Biol.*, **10**, 42–49.

90 Anderson, N.L. and Anderson, N.G. (2002) The human plasma proteome: history character, and diagnostic prospects. *Mol. Cell. Proteomics*, **1**, 845–867.

91 O'Farrell, P.Z. and Goodman, H.M. (1976) Resolution of simian virus 40 proteins in whole cell extracts by two-dimensional electrophoresis: heterogeneity

of the major capsid protein. *Cell*, **9**, 289–298.

92 Klose, J. (1975) Protein mapping by combined isoelectric focusing and electrophoresis of mouse tissues. A novel approach to testing for induced point mutations in mammals. *Humangenetik*, **26**, 231–243.

93 Richard, D.S. (2002) Advanced mass spectrometric methods for the rapid and quantitative characterization of proteomes. *Comp. Funct. Genomics*, **3**, 143–150.

94 Washburn, M.P., Wolters, D. and Yates, J.R. (2001) Large-scale analysis of the yeast proteome by multidimensional protein identification technology. *Nat. Biotech.*, **19**, 242–247.

95 Peng, J., Elias, J.E., Thoreen, C.C., Licklider, L.J. and Gygi, S.P. (2003) Evaluation of multidimensional chromatography coupled with tandem mass spectrometry (LC/LC-MS/MS) for large-scale protein analysis: the yeast proteome. *J. Proteome Res.*, **2**, 43–50.

96 Giddings, J.C. (1987) Concepts and comparisons in multidimensional separation. *J. High-Res. Chromatogr.*, **10**, 319–323.

97 Cargile, B.J., Bundy, J.L., Freeman, T.W. and Stephenson, J.L. (2004) Gel-based isoelectric focusing of peptides and the utility of isoelectric point in protein identification. *J. Proteome Res.*, **3**, 112–119.

98 Cargile, B.J., Talley, D.L. and Stephenson, J.L., Jr. (2004) Immobilized pH gradients as a first dimension in shotgun proteomics and analysis of the accuracy of pI predictability of peptides. *Electrophoresis*, **25**, 936–945.

99 Krijgsveld, J., Gauci, S., Dormeyer, W. and Heck, A.J.R. (2006) In-gel isoelectric focusing of peptides as a tool for improved protein identification. *J. Proteome Res.*, **5**, 1721–1730.

100 Heller, M., Ye, M., Michel, P.E., Morier, P., Stalder, D., Junger, M.A., Aebersold, R., Reymond, F. and Rossier, J.S. (2005) Added value for tandem mass spectrometry shotgun proteomics data validation through isoelectric focusing of peptides. *J. Proteome Res.*, **4**, 2273–2282.

101 Ros, A., Faupel, M., Mees, H., Oostrum, J., Ferrigno, R., Reymond, F., Michel, P., Rossier, J.S. and Girault, H.H. (2002) Protein purification by off-gel electrophoresis. *Proteomics*, **2**, 151–156.

102 Jonsson, A.P., Aissouni, Y., Palmberg, C., Percipalle, P., Nordling, E., Daneholt, B., Jornvall, H. and Bergman, T. (2001) Recovery of gel-separated proteins for in-solution digestion and mass spectrometry. *Anal. Chem.*, **73**, 5370–5377.

103 Cho, S.Y., Lee, E.Y., Lee, J.S., Kim, H.Y., Park, J.M., Kwon, M.S., Park, Y.K., Lee, H.J., Kang, M.J., Kim, J.Y., Yoo, J.S., Park, S.J., Cho, J.W., Kim, H.S. and Paik, Y.K. (2005) Efficient prefractionation of low-abundance proteins in human plasma and construction of a two-dimensional map. *Proteomics*, **5**, 3386–3396.

104 Moritz, R.L. and Simpson, R.J. (2005) Liquid-based free-flow electrophoresis-reversed-phase HPLC: a proteomic tool. *Nat. Methods*, **2**, 863–873.

105 Bjellqvist, B., Hughes, G.J., Pasquali, C., Paquet, N., Ravier, F., Sanchez, J.C., Frutiger, S. and Hochstrasser, D. (1993) The focusing positions of polypeptides in immobilized pH gradients can be predicted from their amino acid sequences. *Electrophoresis*, **14**, 1023–1031.

106 Keller, A., Nesvizhskii, A.I., Kolker, E. and Aebersold, R. (2002) Empirical statistical model to estimate the accuracy of peptide identifications made by MS/MS and database search. *Anal. Chem.*, **74**, 5383–5392.

107 Gunawardena, H.P., Emory, J.F. and McLuckey, S.A. (2006) Phosphopeptide anion characterization via sequential charge inversion and electron-transfer dissociation. *Anal. Chem.*, **78**, 3788–3793.

5.4
Sample Preparation for Capillary Electrophoresis

Ross Burn and David Perrett

5.4.1
Introduction

Since the goal of proteomics is to identify every protein expressed in a cell or tissue, high resolution is the essential prerequisite of an analytical method, more so than high sensitivity. In addition, proteomes are complex, dynamic mixtures of proteins with diverse physicochemical properties that can span over tens of orders of magnitude and therefore require suitable sample collection and pretreatment [108]. Current separation technologies offering high resolution are two-dimensional electrophoresis (2DE), high-performance liquid chromatography (HPLC), and capillary electrophoresis (CE).

Capillary electrophoresis is applicable to the rapid and high-resolution separation of macromolecules, such as proteins and peptides, and includes a family of techniques that employ narrow-bore capillaries to perform electrophoretic separations [109]. A CE instrument consists of the capillary, a detector, a high-voltage power supply, two buffer reservoirs, a control system (PC) and a recording device (Figure 5.11). The capillary is made of fused silica and is coated externally with a layer of polyimide to provide flexibility. Typical capillary dimensions are 25 to 100 μm internal diameter, 375 μm outer diameter, and between 20 and 90 cm length. In order to allow detection, a few millimeters of the polyimide coat is removed near one end, thus producing a UV-transparent window (although this also becomes a weak point in the capillary).

In operation, the capillary's ends are placed in reservoirs filled with background electrolyte (BGE) and the capillary is flushed by applying an external pressure. The sample is loaded into the capillary using either a short pressure pulse or a transient high voltage. The high voltage (typically 20–30 kV) is then applied, and the migrating analytes (proteins and/or peptides) are recorded as they pass the detector. Today, modern commercial instruments are robust, automated, temperature-controlled

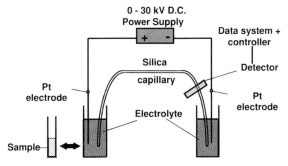

Figure 5.11 Schematic representation of the capillary electrophoresis instrument.

and have high data acquisition rates to ensure effective data capture. The most common detectors are ultraviolet (UV), fluorescence, and mass spectrometry (MS).

Capillary electrophoresis has five fundamental modes separation; capillary zone electrophoresis (CZE); capillary isoelectric focusing (CIEF); capillary gel electrophoresis (CGE); capillary isotachophoresis (cITP); and micellar electrokinetic chromatography (MEKC). The fundamentals of CE differ little from those of traditional electrophoresis, in that the separation of charged analytes result from their different mobilities (i.e., speed plus direction of movement) when placed in a conducting medium under the influence of an applied direct current (DC) electric field. In electrophoresis, negative analytes (anions) move towards the oppositely charged electrode (anode), and positively charged analytes (cations) to the negative cathode. Neutral compounds cannot move under the influence of the electric field alone, but may diffuse from the load position. However, in CE they are also carried by the electroosmotic flow (EOF), which also enhances the mobility of both cations and anions. These fundamentals are reviewed in detail elsewhere [110,111].

In proteomics, CZE is commonly used due to its high resolution, speed and simplicity. In CZE, although the analytes separate depending on their charge to size ratio, the EOF helps to move all analytes in discrete zones towards the detector. The cations travel fastest, followed by neutrals, and then the anions (Figure 5.12). Capillary isoelectric focusing is a popular equilibration technique, and permits whole capillary injection; it is also the highest-resolving of all the electrophoretic modes, resolving proteins and peptides to 0.005 of a pI unit [112]. At equilibrium, the analytes focus within the capillary at their isoelectric point (pI) (Figure 5.13). The EOF can be used to mobilize the analytes beyond the detector, but if the EOF is

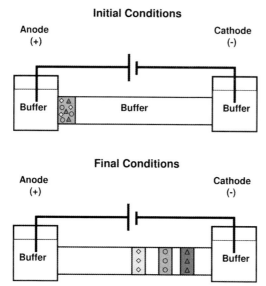

Figure 5.12 Schematic representation of a capillary zone electrophoresis (CZE) separation.

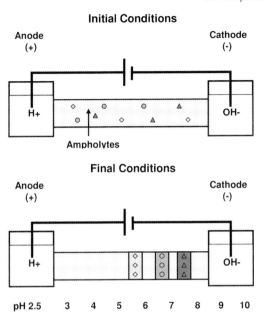

Figure 5.13 Schematic representation of a capillary isoelectric focusing (CIEF) separation.

suppressed then the addition of salt or the application of pressure/vacuum is used for mobilization.

Other features make CE ideal for proteome analysis. For example, the speed and high efficiency of CE allows the resolution of complex mixtures of proteins with minimal sample loss. A sample volume of less than 10 µL is required, although <20 nL is typically analyzed, and automated quantitation is easily performed. For CE fingerprinting, as well as MS identification, it is common to produce peptides by digesting a protein or protein mixture with a proteolytic enzyme such as trypsin. With only minor modification, CE is able to separate peptides with very high efficiency, and can be directly linked with MS for protein identification [113,114]. Finally, in an attempt to resolve more proteins or peptides in a separation and to increase the dynamic range of detection, CE capillaries may be coupled together with HPLC or other modes of CE [115–117]. To date, many proteomes have been mapped using CE, including those from human plasma, yeast, and *Escherichia coli*; details of these have been reviewed elsewhere [118–120].

5.4.2
Sample Preparation

Before any sample preparation, it is important to define the problem at hand and to select which technique(s) should be used. Many factors must be considered, including: How much sample is available? Is the sample going to be analyzed globally or will a targeted approach be required? Is MS detection required? Is sample digestion with trypsin required and, if so, at what point in the workflow will be best?

When such questions have been answered the best analytical platform can be selected. In CE methods, the preparation of the capillary, the BGE to be used for CZE, or the ampholyte mixture for CIEF, are equally as important as preparation of the sample.

5.4.2.1 Sample Collection and Storage

These aspects are no different for CE than for other proteomics techniques. It is pivotal to ensure that the samples are collected properly, aliquoted into workable volumes, and stored at $-50\,°C$ (or colder) until required, taking care to avoiding unnecessary freeze–thaw cycles. After thawing, the samples should be kept chilled whenever possible, although CE systems are not fitted with cooled sample compartments. If sample preparation takes several hours, the addition of a protease inhibitor cocktail (e.g., Sigma-Aldrich, P2714) is recommended. For human plasma and serum analysis, the Human Proteome Organisation (HUPO) has published specific guidelines for sample collection, storage and preparation [121,122], and these should be adhered to whenever possible.

5.4.2.2 Sample Preparation for CE

An ideal sample preparation procedure will do all of the following:

- Prepare analyte(s) at optimum concentration with sufficient volume for complete analysis.
- Improve analyte(s) selectivity.
- Prepare analyte(s) within a buffer solution at a desired pH and concentration that permits sample stacking and also provides a good peak shape.
- Stabilize the analyte(s).
- Remove all interfering components from the matrix.
- Ensure high recoveries of the analyte(s).
- Remove particulate matter that might block the capillary.

One "obvious" point, especially with peptides and proteins, is to ensure that all analytes of interest are in solution (preferably in aqueous buffer) prior to injection. If the sample is in an organic solvent and then injected, the components may precipitate in the BGE, blocking the capillary. In order to aid dissolution, the sample may be sonicated or heated gently, but great care must be taken so as not to overheat the sample and degrade proteins. All samples should be filtered through a 0.45-µm membrane filter or, if the volumes are limited, centrifuged at $13\,000 \times g$ for 5 min at $4\,°C$ to remove particulates. Normally, samples are prepared in ca. 20 µL for injection. Sample preparation is achieved on-line via sample injection, or off-line prior to sample injection. Although off-line methods are readily available, inexpensive and easy to perform, the on-line methods are faster in operation.

5.4.2.3 Sample Concentration

The limit of detection (LOD) for UV/Visible detection in CE is approximately 10^{-13} to 10^{-16} moles of analyte injected [123], which corresponds to a sample concentration LOD of approximately 1 mM to 10 µM. This LOD is restricted due to the minute sample volumes injected (typically 10–50 nL) and the short pathlength of the detector cell. It is therefore important to enrich low-abundant analytes to concentrations above the LOD, or to induce derivatization methods in order to reduce the LOD. When enriching a protein sample it is important not to cause any degradation and to ensure maximum recovery. The protein concentration can be assayed using one of many protein assays, for example Lowry [124], Bradford [125], and OPA/NAC fluorescence [126]. In CIEF, the initial total protein concentration should be <0.5 mg mL^{-1} in order to prevent protein band precipitation, as the localized protein concentration may increase 100-fold [127].

5.4.2.4 Off-Line Preconcentration

As proteins and peptides are relatively stable, off-line concentration methods such as evaporation, liquid–liquid extraction and freeze-drying can be used. Protein precipitation can be achieved using ammonium sulfate or organic solvents (e.g., acetone, ethanol), in conjunction with acids (e.g., trichloroacetic acid). The process of evaporation can be speeded up by using a rotary centrifugal evaporator (Speedvac) or a dry block with nitrogen supply. Other off-line methods include ultrafiltration (UF), protein precipitation, and chromatographic extraction. Ultrafiltration devices are supplied in a range of sizes and porosities and, most importantly for CE, do not dilute the sample. UF-centrifugation through a <50 kDa membrane is useful for removing large proteins, especially albumin. Ultrafiltration and protein precipitation can also be used to desalt samples, but generally provide poor protein recoveries.

Chromatographic preconcentration techniques include solid-phase extraction (SPE), affinity chromatography, and immunoaffinity chromatography. SPE using C_2–C_{18} phases is used to bind hydrophobic analytes, followed by elution with acetonitrile or methanol. *Affinity chromatography* is used to enrich specific classes of analytes such as phosphorylated proteins via immobilized metal affinity chromatography [128], glycoproteins via lectin affinity chromatography [129], and albumin and kinases via dye affinity chromatography [130]. *Antibodies* are used in immunoaffinity chromatography to enrich a target protein or to deplete interfering proteins from the matrix [131,132]. As very small sample volumes are required for CE, the extraction technologies may be miniaturized. Examples are the use of solid-phase micro-extraction (SPME) and ZipTips (Millipore); these chromatographic methods clean the sample and permit sample elution in a solvent of choice.

5.4.2.5 On-Line Preconcentration

A simple form of on-capillary concentration is achieved electrophoretically by sample stacking, isotachophoresis (ITP) or field-amplified injection (FAI) [133]. Sample stacking works by injecting the sample in a solvent with conductivity lower than the BGE. The mobility of the analyte ions is higher than that of the BGE ions, and so they will focus between the running buffer zones. Stacking is readily produced using organic solvents, a low buffer concentration, or by injecting a short plug of water prior to the sample. A more than tenfold increase in analyte concentration is achieved.

Isotachophoresis is similar to sample stacking, except that two different buffers are employed. The leading electrolyte (LE) contains ions of higher mobility than the analytes, while the terminating buffer (TE) contains ions of lower mobility than the analytes. Consequently, the analyte is concentrated between the LE and TE. In the second step, CZE is performed with the LE as the BGE.

Field-amplified injection is also similar to sample stacking, but here a voltage is applied to the capillary whilst injecting the sample, after which sample stacking is performed as normal. As a consequence, the whole sample can be processed, with preconcentration factors >100 having been reported [134].

Several on-line capillary methods are based on chromatography such as solid-phase extraction (SPE), immunoaffinity chromatography, and molecular imprinted polymer (MIP) technologies. In capillary SPE, the chromatographic phase is packed into the injection end of the capillary and kept in place by the use of frits. After sample loading and washing, the analytes are eluted into the CZE region of the capillary using typically 50 to 100 nL of solvent. Original volumes of 25 to 100 µL are possible, with a resultant improvement in LOD of up to four orders of magnitude, but the loading sequence may be protracted. On-line immunoaffinity chromatography is similar to SPE (except that analyte-specific antibodies are employed for enrichment), and a 5000-fold increase in signal has been reported [135]. Although the use of MIP with proteins and peptides is described as being of limited use, the technique has clear potential for future use.

5.4.2.6 Desalting

As salts interfere with CE they should be removed from the samples, with ultrafiltration, precipitation and SPE as the main procedures. The classical method for desalting samples is that of dialysis, although special care must be taken when dialyzing sparingly soluble or concentrated protein solutions as they may precipitate. Dialysis may require long periods of time, and therefore it must be performed at 4 °C. In order to prevent protein precipitation, the sample should be dialyzed against dilute buffer (buffer exchange), for example, 10% BGE. It is also important to desalt samples when MS detection is employed, since certain ions (e.g., Na^+, PO_4^{3-}, NH_4^+, BO_3^{3-}) can suppress the ionization of the analyte [136].

5.4.2.7 Analyte Modification

Proteins can be chemically modified prior to CE to improve the selectivity, sensitivity, or efficiency of the analysis. Prior to trypsin digestion, proteins may be reduced and alkylated to remove any intramolecular disulfide bridges, thus increasing digestion efficiency. Proteins may also be prepared with chaotropes and detergents to unfold the protein and remove any tertiary structures; the classic example is to use urea and sodium dodecylsulfate (SDS). In order to improve LODs, protein and peptide samples may be derivatized with a fluorophoric or fluorogenic reagent; subsequent LODs of 100 to 300 zmol have been reported for peptides [137]. These derivatization procedures are performed both off-line and on-line, either before or after separation. For further information, the reader should consult Refs. [138–140].

5.4.3
Background Electrolyte

In CZE, the BGE is the most important variable in developing a successful separation [141], as it controls the pH, which in turn controls the mobility of the analytes and magnitude of the EOF. As the BGE carries the electrical current for the separation, its concentration or ionic strength is very important: a too-high concentration will generate too much heat, while a too-low concentration may causes poor peak shapes due to insufficient buffering of the sample. An efficient BGE should possess the following properties: good buffering capacity at the pH of choice (e.g., within one pH unit of the pK_a for the buffer); low absorbance at the detection wavelength; low conductivity to minimize current generation; and mobilities matching those of the analytes, thus conserving peak shape. Some commonly used CZE buffers are listed in Table 5.6.

The BGE must be prepared using only high-grade reagents and good quality, deionized water (≥ 18 MΩ), and be accurately adjusted to the required pH using an appropriately calibrated pH meter. The BGE should then be degassed and filtered through a 0.45-μm membrane to remove any particulate matter. The BGE can be

Table 5.6 Common buffers used in CE, and their physico-chemical properties.[a]

Buffer salt[b]	pK_a[d]	Native pH[e]	Mobility/ $\times 10^5$ cm^2 V^{-1}	Absorbance at 210 nm
Phosphate (pK_{a1})	2.12	4.61	-7.76×10^{-3}	0.026
Glycine	2.35	6.25	-3.62×10^{-3}	0.839
Citrate (pK_{a1})[c]	3.06	2.56	-3.43×10^{-3}	1.103
Formate	3.75	6.77	-5.18×10^{-4}	1.201
Citrate (pK_{a2})	4.74	2.56	-7.77×10^{-3}	1.57
Acetate	4.76	7.16	-16.6	0.95
Citrate (pK_{a3})	5.40	2.47	-3.62×10^{-3}	1.72
Bicarbonate (pK_{a2})	6.15	8.58		0.15
Phosphate (pK_{a2})	7.21	4.52	-7.77×10^{-3}	0.176
TRIS	8.06	10.32	-3.63×10^{-3}	0.647
Tetraborate	9.23	5.42	-1.10×10^{-5}	0.085
Glycine (pK_{a2})	9.78	5.85	-3.62×10^{-3}	2.25
Carbonate (pK_{a2})	10.25	10.7		2.56
Phosphate (pK_{a3})	12.35	4.58	-7.66×10^{-3}	2.44

[a] All measurements obtained using AnalaR grade salts and usually the sodium salt.
[b] All determinations performed using 25 mM solution at 25 °C.
[c] All determinations performed using 5 mM solution at 25 °C.
[d] All values measured at the pK_a adjusted from the native pH with 25 mM HCl or 25 mM NaOH, as appropriate.
[e] Native pH = pH of the simple solution of the salt water.

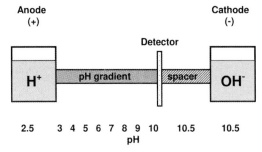

Figure 5.14 The use of spacers in capillary isoelectric focusing (CIEF).

"cleaned" by passing it through a C_{18} SPE cartridge, thus removing any traces of UV-absorbing contaminants. It is important to change the BGE buffer reservoirs regularly in order to prevent the elution and re-injection of sample analytes. The present authors recommend changing BGE reservoirs every 10 runs, whilst the ionic strength of the sample should be between 10- and 100-fold that of the BGE.

It should be noted that BGEs are not used in CIEF; rather, ampholyte mixtures are used to create a pH gradient. Typical ampholyte mixtures span pH 3 to 10, although mixtures as narrow as one pH unit may be used. The final ampholyte concentration should be between 1 and 2% [142], and the ionic strength should be as low as possible in order to avoid disruption of the gradient and the precipitation of proteins. Samples with a salt concentration >50 mM should be desalted using one of the above-mentioned methods. To permit detection of all focused proteins, the capillary beyond the detector must be "blocked" prior to mobilization; this may be achieved using a spacer such as TEMED (N,N,N',N'-tetramethylethylenediamine), or by back-flushing the catholyte (Figure 5.14).

5.4.4
Capillary Preparation

5.4.4.1 Capillary Dimensions
Capillaries can be obtained pre-cut with a prepared detection window or in bulk, for example, in a 10-m roll that can be cut to the required length using a capillary cutter. The ends should be inspected with a magnifier to ensure a straight cut, or trimmed until they are cut straight. Once cut, the detection window is best made using an electrical burner; any charred remains should be removed using methanol. The capillary total length (TL) and effective length (EL), the distance between the injection end and the detection window, should be noted.

5.4.4.2 Capillary Conditioning
After preparation, the capillary must be cleaned and conditioned. Following an initial rinse with 1 M NaOH at 50 °C to any remove adsorbed contaminants, the capillary is equilibrated with BGE. A less-aggressive clean with 0.1 M NaOH is used

at start of each analysis, or when the separation becomes poor. This routine wash should improve the reproducibility of a separation. If the separation is performed at low pH, a phosphoric acid wash may be required prior to equilibration in order to protonate the capillary wall silanols. Coated capillaries should be washed only with BGE in order to prevent damage to the coating.

5.4.4.3 Capillary Coating

The internal surface of a capillary can be modified to suppress EOF and increase efficiency by reducing the number and magnitude of interactions between the analyte and the capillary wall. Two fundamental approaches have been considered: (i) permanent modification of the inner wall surface; and (ii) dynamic deactivation by using BGE additives, such as celluloses, polyvinylalcohol (PVA), SDS, and CHAPS (3-[(3-cholamidopropyl)dimethylammonio]-1-propanesulfonate). When MS detection is required care must be taken when selecting dynamic modifiers, as many polymers and detergents are not MS-compatible. Details of permanent wall modifications have been reviewed elsewhere [143,144]. After capillary modification, the EOF should be determined to ensure good deactivation; this is achieved by injecting a neutral compound such as acetone or dimethyl sulfoxide (DMSO).

5.4.5
Summary

It is essential that proteomic samples are prepared at the optimum concentration and in the required matrix prior to CE separation, whilst ensuring minimal sample loss. Failure to do so can result in poor separation and reduced detection. Classical sample preconcentration and desalting methods are used, as well as more specialized online methods. In order to increase the LOD, protein or peptide samples are derivatized with a fluorogenic reagent. CE is a versatile, high-efficiency technique that is used for the comprehensive global analysis of whole proteomes. Whilst the direct linking to modern MS allows rapid, automated identification of all detectable sample proteins, for more targeted approaches CE may be coupled, either offline or online, to enrichment techniques such as SPE and affinity chromatography.

5.4.6
Perspective

Capillary electrophoresis has developed into an established technology for proteomic analysis, especially prior to MS. Recent advances have enabled direct hyphenation with high-performance liquid chromatography (HPLC), and with other modes of CE and MS. The orthogonal, high-resolution nature of CZE, CIEF and reverse-phase liquid chromatography (RPLC) allow upwards of several hundreds – and even thousands – of proteins to be identified in just one multidimensional analysis [145,146]. Moreover, the incorporation of stable isotope peptide standards will permit the quantitation of multiple proteins in one analysis. One notable advantage of CIEF is the intrinsic capability to calculate the isoelectric points of detectable proteins, and these can be used as an extra dimension of data for purposes of identification. Likewise, if in future a more selective analysis is required, then the method is easily tailored such that a particular pH range can be targeted, even within a single pH unit.

5.4.7
Recipe for Beginners

5.4.7.1 Method 1: Analysis of Human Serum/Plasma by CZE

An automated CE system, integration and reporting software is used, in this case an Agilent ^{3D}CE system and Chemstation V.02.01. However, if unavailable, alternative instrumentation and software may be used.

Capillary Preparation

- A 40-cm length of 50 μm i.d. capillary is cut using a ceramic tile. A window is made by burning the polyimide coating 8.5 cm from one end of the capillary using an electrical capillary electrical burner. (A match or Bunsen burner can be used.) Any charred polyimide is removed using a methanol-soaked tissue.

- The capillary is placed into the cassette with the window properly aligned with the detector interface. Care must be taken to ensure that the injection and detection ends are both at the same level, to ensure no siphoning of BGE and sample.

- Wash the capillary for 30 min at 45 °C with 1 M NaOH, followed by a 10-min wash at 30 °C with deionized water.

BGE Preparation

- Dissolve 0.5 g sodium tetraborate in 75 mL 18 MΩ H_2O, adjust to pH 9.25 with NaOH, and make up to 100 mL with 18 MΩ H_2O. Filter through a 0.45-μm membrane filter and store at 4 °C until use. This buffer should be re-prepared every 3 months.

Sample Preparation

- Dilute 100 μL of human plasma or serum to 1 mL and filter through a 0.45-μm membrane filter. Using a Bradford or alternative protein assay, measure the protein concentration of the diluted plasma; a typical value is 6–8 mg mL^{-1}.

- Pipette a minimum of 20 μL into a sample vial (Agilent: 5188-2788) and remove any air bubbles by tapping on the bench. Place vial on the instrument carousel.

- Aliquot the residual sample into 50-μL volumes and store at −80 °C until required.

CE Analysis

- Pipette 700 μL of BGE into two buffer vials (Agilent: 5182-0567). These vials should be changed every 10 runs.

- Create a method file for the analysis as follows:
  - Preconditioning: Flush 0.1 M NaOH for 2 min and BGE for 2 min
  - Temperature: 30 °C
  - Injection: 150 mbar s^{-1}
  - Detection: 215 nm
  - Voltage: 25 kV
  - Analysis time: 10 min
- Start the acquisition.

Data Analysis

- In ChemStation, select the electropherogram required and process the data. If incorrect integration events are assigned, manual integration may be required.

- Display/print report. Peak areas correlate to the concentration of each protein in the sample.

- A typical electropherogram for the analysis is displayed in Figure 5.15.

Optional Further Analysis

- Peak-fraction collection is used for protein identification.

- External/internal standards are used for protein quantitation.

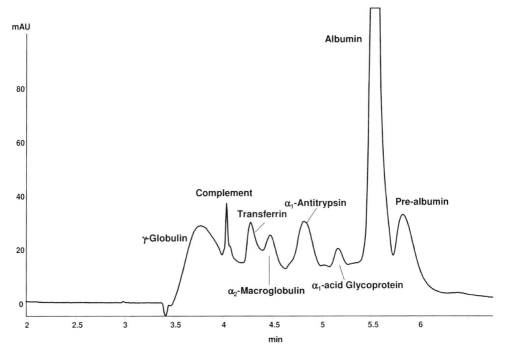

Figure 5.15 A typical electropherogram for capillary zone electrophoresis (CZE) of human plasma.

- Direct coupling to mass spectrometry permits rapid identification and quantitation.

Notes

- The 0.1 M NaOH and deionized water should be replaced every week.
- Clean the capillary for 30 min at 45 °C with 1 M NaOH, followed by a 10-min wash at 30 °C with deionized water every week.
- If the baseline is noisy, add 1 mM EDTA to BGE and apply solid-phase extraction. Ensure no contamination from pipetting, sample vials or stock solutions. The lamp may also need to be changed. If spikes are present throughout the analysis, remove particulates from sample.
- If reduced signal intensity, check injection mechanism and that capillary end is free from blockage. If a new capillary has been installed, check correct alignment of window.
- If negligible current, check electrode contacts, no air in capillary, and capillary integrity. Small cracks may form in capillary and prevent conduction.
- Consult instrument user manual for further troubleshooting tips and advice.

5.4.7.2 Method 2: Analysis of Tryptic Digests by CZE

Many excellent separations of tryptic digests have been performed using CZE. The following example will work well with most medium to large proteins. A typical electropherogram of a 1 mg mL^{-1} alcohol dehydrogenase digest is shown in Figure 5.16.

Instrument Conditions, Capillary, and Sample Preparation

- As for Method 1, except where stated otherwise
- BGE: 50 mM NaH_2PO_4, pH 2.5
- Analysis time: 20 min

Digest Preparation

- A protein is dissolved in water at ca. 10 mg mL^{-1}; 0.1 vol. of trypsin (2 mg mL^{-1} in water) is added, followed by 0.2 vol. of 0.5 M NH_4HCO_3 (39.5 g NH_4HCO_3 L^{-1}) to give a final pH of ca. 8.5. The mixture is stirred to redissolve any precipitate and incubated at 37 °C for a minimum of 3 h.
- The pH is then lowered to 6.4 by addition of 0.5 M CH_3COOH (29 mL glacial acetic acid L^{-1}), and the solution is immersed in ice to stop further enzyme reaction and precipitate insoluble peptides.
- The mixture is centrifuged and the supernatant withdrawn and divided into suitable aliquots; these are then evaporated to dryness or freeze-dried, and may be reconstituted in water or dilute buffer prior to analysis.

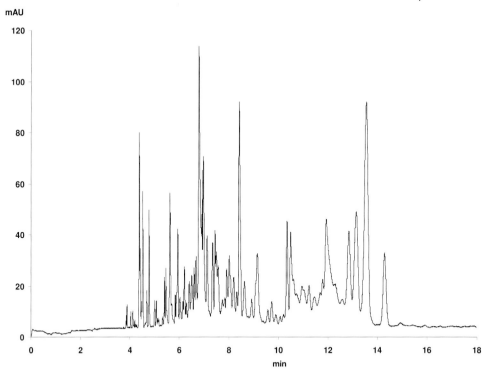

Figure 5.16 A typical electropherogram for capillary zone electrophoresis (CZE) of a protein digest.

5.4.7.3 Method 3: Analysis of Proteomes by CIEF

CIEF is suited to the analysis of whole proteomes, either as the parent proteins or as a global digest of peptides. The following example will work well with most proteomic samples.

Instrument Conditions, Capillary, and Sample Preparation

- As in Method 1, except where stated otherwise

Capillary
- A PVA-coated capillary, 100 μm i.d. × 64.5 cm TL (Agilent: G1600-61419). Do NOT wash with 1 M NaOH; follow manufacturer's instructions.

Sample Preparation
- 2% (w/w) solution of ampholytes, pH 3–10 (Sigma-Aldrich)
- 0.1 mg mL^{-1} solution of sample in 2% ampholytes
- pH 10.5, NH$_4$OH (~1%) solution (catholyte)
- pH 2.5, acetic acid (~1%) solution (anolyte)

CE Analysis

- Preconditioning:
  - Flush 2 min with 18 MΩ H$_2$O and 1 min with 2% ampholytes
  - Flush 2 min with proteomic sample in 2% ampholytes
  - Backflush for ca. 40 s with catholyte

- Injection: No injection

- Home values: The catholyte at capillary outlet and the anolyte at the inlet

- Temperature: 30 °C

- Detection: 280 nm

- Voltage: 25 kV

- Mobilization: +10 mbar pressure to inlet at 15 min

- Analysis time: 60 min

Additional Notes

- Resolution and analysis time are optimized by changing the pressure for elution.

- If spikes in electropherogram, reduce mobilization pressure and ensure capillary ends and solutions prepared properly. If persists, reduce protein concentration and ramp voltage at start.

- If high background noise and poor peak shapes, monitor the regularity of pH gradient at lower wavelengths e.g., 215 nm. If poor, try reducing ampholyte concentration, using blends of ampholytes and a different coating for the capillary.

Abbreviations

2DE	two-dimensional electrophoresis
BGE	background electrolyte
CE	capillary electrophoresis
CGE	capillary gel electrophoresis
CHAPS	3-[(3-cholamidopropyl)dimethylammonio]-1-propanesulfonate hydrate
cIEF	capillary isoelectric focussing
cITP	capillary isotachophoresis
CZE	capillary zone electrophoresis
DC	direct current
DMSO	dimethyl sulfoxide
EDTA	ethylenediaminetetraacetic acid
EL	effective length
EOF	electroosmotic flow
FIA	field-amplified injection

HPLC	high performance liquid chromatography
HUPO	human proteome organisation
ITP	isotachophoresis
LOD	limit of detection
MEKC	micellar electrokinetic chromatography
MIP	molecular imprinted polymer
MS	mass spectrometry
OPA/NAC	phthaldialdehyde/*N*-acetyl-L-cysteine
pI	isoelectric point
PVA	polyvinyl alcohol
RPLC	reversed phase liquid chromatography
SDS	sodium dodecyl sulfate
SPE	solid phase extraction
SPME	solid phase micro-extraction
TEMED	N,N,N',N'-tetramethylethylenediamine
TL	total length
Tris	tris(hydroxymethyl)aminomethane
UF	ultrafiltration
UV	ultraviolet
Vis	visible

References

108 Anderson, N.L. and Anderson, N.G. (2002) *Mol. Cell. Proteomics*, **1**, 845–867.

109 Jorgenson, J.W. and Lukacs, K.D. (1981) *Clin. Chem.*, **27**, 1551–1553.

110 Mosher, R.A., Saville, D.A. and Thormann, T. (1992) *The Dynamics of Electrophoresis*, VCH, Weinheim.

111 Perrett, D. (2000) *Encyclopedia of Separation Science* (eds. I.D. Wilson, M. Cooke and C.F. Poole), Academic Press, London. pp. 103–118.

112 Shen, Y., Berger, S.J., Anderson, G.A. and Smith, R.D. (2000) *Anal. Chem.*, **72**, 2154–2159.

113 Stutz, H. (2005) *Electrophoresis*, **26**, 1254–1290.

114 Simpson, D.C. and Smith, R.D. (2005) *Electrophoresis*, **26**, 1291–1305.

115 Bushey, M.M. and Jorgenson, J.W. (1990) *Anal. Chem.*, **62**, 161–167.

116 Mohan, D. and Lee, C.S. (2002) *Electrophoresis*, **23**, 3160–3167.

117 Chen, J., Lee, C.S., Shen, Y., Smith, R.D. and Baehrecke, E.H. (2002) *Electrophoresis*, **23**, 3143–3148.

118 Cooper, J.W., Wang, Y. and Lee, C.S. (2004) *Electrophoresis*, **25**, 3913–3926.

119 Shen, Y. and Smith, R.D. (2002) *Electrophoresis*, **23**, 3106–3124.

120 Dolnik, V. (2006) *Electrophoresis*, **27**, 126–141.

121 Rai, A.J., Gelfand, C.A., Haywood, B.C., Warunek, D.J., Yi, J., Schuchard, M.D., Mehigh, R.J., Cockrill, S.L., Scott, G.B., Tammen, H., Schulz-Knappe, P., Speicher, D.W., Vitzthum, F., Haab, B.B., Siest, G. and Chan, D.W. (2005) *Proteomics*, **5**, 3262–3277.

122 Haab, B.B., Geierstanger, B.H., Michailidis, G., Vitzthum, F., Forrester, S., Okon, R., Saviranta, P., Brinker, A., Sorette, M., Perlee, L., Suresh, S., Drwal, G., Adkins, J.N. and Omenn, G.S. (2005) *Proteomics*, **5**, 3278–3291.

123 Perrett, D. (2004) *Clarke*'s Analysis of Drugs and Poisons, Pharmaceutical Press, UK, pp. 535–549.
124 Lowry, O.H., Rosebrough, N.J., Farr, A.L. and Randall, R.J. (1951) *J. Biol. Chem.*, **193**, 265–275.
125 Bradford, M.M. (1976) *Anal. Biochem.*, **72**, 248–254.
126 Roth, M. (1971) *Anal. Chem.*, **43**, 880–882.
127 Chen, J., Gao, J. and Lee, C.S. (2003) *J. Proteome Res.*, **2**, 249–254.
128 Ueda, E.K., Gout, P.W. and Morganti, L. (2003) *J. Chromatogr. A*, **988**, 1–23.
129 Yang, Z., Hancock, W.S., Chew, T.R. and Bonilla, L. (2005) *Proteomics*, **5**, 3353–3366.
130 Denizli, A. and Piskin, E. (2001) *J. Biochem. Biophys. Methods*, **49**, 391–416.
131 Yocum, A.K., Yu, K., Oe, T. and Blair, I.A. (2005) *J. Proteome Res.*, **4**, 1722–1731.
132 Greenough, C., Jenkins, R.E., Kitteringham, N.R., Pirmohamed, M., Park, B.K. and Pennington, S.R. (2004) *Proteomics*, **4**, 3107–3111.
133 Stroink, T., Paarlberg, E., Waterval, J.C., Bult, A. and Underberg, W.J. (2001) *Electrophoresis*, **22**, 2375–2383.
134 Veraart, J.R., Lingeman, H. and Brinkman, U.A. (1999) *J. Chromatogr. A*, **856**, 483–514.
135 Heegaard, N.H., Nilsson, S. and Guzman, N.A. (1998) *J. Chromatogr. B Biomed. Sci. Appl.*, **715**, 29–54.
136 Eriksson, J.H., Mol, R., Somsen, G.W., Hinrichs, W.L., Frijlink, H.W. and de Jong, G.J. (2004) *Electrophoresis*, **25**, 43–49.
137 Baars, M.J. and Patonay, G. (1999) *Anal. Chem.*, **71**, 667–671.
138 Bardelmeijer, H.A., Waterval, J.C., Lingeman, H., van't Hof, R., Bult, A. and Underberg, W.J. (1997) *Electrophoresis*, **18**, 2214–2227.
139 Waterval, J.C., Lingeman, H., Bult, A. and Underberg, W.J. (2000) *Electrophoresis*, **21**, 4029–4045.
140 Underberg, W.J. and Waterval, J.C. (2002) *Electrophoresis*, **23** 3922–3933.
141 Hows, M., Alfazema, L.N. and Perrett, D. (1997) *LCGC*, 656–666.
142 Wehr, T., Rodriguez, R. and Zhu, M. (1999) *Capillary Electrophoresis of Proteins*, Marcel Dekker, New York, pp. 131–227.
143 Horvath, J. and Dolnik, V. (2001) *Electrophoresis*, **22**, 644–655.
144 Liu, C.Y. (2001) *Electrophoresis*, **22**, 612–628.
145 Wang, Y., Rudnick, P.A., Evans, E.L., Li, J., Zhuang, Z., Devoe, D.L., Lee, C.S. and Balgley, B.M. (2005) *Anal. Chem.*, **77**, 6549–6556.
146 Wang, Y., Balgley, B.M., Rudnick, P.A., Evans, E.L., Devoe, D.L. and Lee, C.S. (2005) *J. Proteome Res.*, **4**, 36–42.

6
Optical Methods

6.1
High-Throughput Proteomics: Spinning Disc Interferometry (SDI)

Patricio Espinoza Vallejos, Greg Lawrence, David Nolte, Fred Regnier, and Joerg Schreiber

6.1.1
Proteomics as a Tool for Health Assessment

Changes in the concentration of plasma and serum proteins associated with a specific disease have been used clinically for more than 50 years to assess health and monitor therapy [1]. Blood is used widely in diagnostics because it is easily obtained and contains proteins from most tissues in the body. Fewer than 4000 proteins have been identified in plasma based on literature references and listings from the HUPO Plasma Proteome Project [2], although it is certain there are many thousands of proteins and their isoforms in this proteome. Of these vast numbers, only a few proteins are approved for diagnostic use in clinical laboratories. With the advent of mass spectrometry-based proteomics, and the ability to identify and quantify large numbers of proteins in tissue and plasma proteomes, many proteins are being found that might be connected biochemically to disease states and thus serve as candidates for disease biomarkers. Given the accumulation of large numbers of disease marker candidates, there is great excitement that, in the future, disease progression and therapy will be defined by the quantitative signature of ten to a few hundred proteins that are mechanistically associated with specific aspects of a disease [3]. These protein biomarkers may be of enormous value in screening, early diagnosis, and prediction of response to therapy. However, the concept in general, the utility of individual protein markers, and the methods and tools used in clinical measurements must be validated in a large population of genetically diverse patients in order for this to happen [4].

Proteomics Sample Preparation. Edited by Jörg von Hagen
Copyright © 2008 WILEY-VCH Verlag GmbH & Co. KGaA, Weinheim
ISBN: 978-3-527-31796-7

6.1.2
Translational Proteomics

The organized effort to carry out the large-scale validation of protein biomarkers and clinically relevant tools has come to be known as "translational proteomics" [5]. Unfortunately, the tools and methods used in discovery proteomics are not well suited for the requisite high-throughput quantification of specific proteins in thousands of patient samples. The problem arises from the fact that mass is being targeted in discovery proteomics, so that peptide sequence can be compared with peptide masses generated *in silico* from DNA databases [6]. Because large numbers of polypeptides have the same mass, discrimination must be achieved through: (1) sample proteolysis and extensive liquid-phase separation, followed by; (2) a gas-phase separation of peptides by mass; (3) further gas-phase fragmentation of peptide ions; (4) additional mass analysis of the product fragment ions; and (5) extensive computational analysis of the data. The power of mass spectrometry-based discovery proteomics is in the non-routine identification of hundreds to thousands of proteins in a small number of samples. The Achilles' heel of mass spectrometry-based proteomics in the case of the plasma proteome is that abundant proteins interfere with the analysis of low-abundance marker proteins.

Translational proteomics differs from this in that a small number of specific proteins and peptides are being targeted for validation, they are of known structure, and the objective is to determine their concentration in thousands of samples. Moreover, it is unnecessary that the analytical platform being used in validation studies is capable of determining the structure of large numbers of other proteins.

There is a long history in clinical chemistry of using antibodies for the purpose of targeting specific proteins in plasma and serum for selective purification and quantification [7]. The strength of immunological assay methods is that they are robust, sensitive, rapid, inexpensive, scalable, and require little skill to use. The limitations are that frequent calibration is required, and inter-laboratory reproducibility can be poor. The careful selection of high-specificity antibodies is also a necessity. Although differentiation between closely related forms of a protein can be achieved, it is totally dependent on the ability of the antibody to discriminate between the structural variants. Based on the microtiter plate ELISA systems of a half-century ago, a generation of automated clinical analyzers have evolved that are capable of analyzing hundreds of samples per day, with high sensitivity and precision. One limitation of these systems in proteomics is that they were not designed to analyze large numbers of proteins in the same sample. A solution to this problem has been to use immunological microarrays capable of simultaneously analyzing multiple analytes [8]. At present, antibody array systems are capable of examining more than a hundred analytes at a time, using small volumes of sample and still showing detection sensitivities that range down to $1\,\text{pg}\,\text{mL}^{-1}$ [9]. Clearly, immunological arrays are a powerful alternative to mass spectrometry-based proteomics in clinical medicine [10]. However, even immunological arrays have limitations in translational proteomics. One such problem is that single sample array plates are used in a serial fashion, whereas the need in translational proteomics is for integrated, parallel analysis of large numbers of samples simultaneously. A much better approach would be to analyze many samples at a time using integrated sets of arrays. A second limitation is the manner in which quantification is achieved along with array element read time, as huge numbers of array elements must be read during the analysis of hundreds of proteins in hundreds to thousands of samples. Ideally, all of the array elements relating to a sample could be read individually, multiple times, and within a few milliseconds. Although this can be achieved in a CCD format, individual, direct reading of array elements multiple times provides a greater sensitivity of detection.

In this chapter, a new type of immunological array is described that uses spinning disc interferometry (SDI) to achieve molecular assays with high sensitivity, integrated sample analysis, high throughput, high reliability, and low costs.

6.1.3
The Principles of Spinning Disc Interferometry

6.1.3.1 The Spinning Disc

The Spinning Disc is a biological compact disc that spins at high speed, bringing large numbers of protein spots (generally antibody array elements) in rapid succession to a focused laser spot for interferometric detection of bound protein [11]. Interferometry is the most sensitive optical direct detection method for the direct detection of intermolecular association, as in protein–protein interactions. The technique surpasses surface-plasmon resonance (SPR) in surface mass sensitivity, and is highly quantitative. Perhaps more importantly, it can be massively multiplexed over large-array formats. This high sensitivity across large areas is achieved because it is non-resonant. In the search for high signal-to-noise (S/N) ratios, there are two approaches. One approach is to introduce gain in the detection to boost the signal; this is the approach used in resonant techniques such as SPR. However, resonance requires tight constraints on operating conditions because of the narrow bandwidth of the resonant enhancement. Small drifts in device thicknesses or other dimensions can cause the operating wavelength to shift. The constraints on the fabrication of the system become tighter as the area of the system becomes larger. This is one reason why highly-multiplexed SPR systems have been slow to develop. Furthermore, in the presence of optical gain, noise also tends to increase; consequently, an engineering-enhanced signal gain into a detection system often does not yield a proportional improvement in the S/N ratio.

With the Spinning Disc, the opposite approach is taken such that no system gain is used at all, but instead the noise floor is suppressed. This approach is easier to accomplish than engineering gain into an optical system, because noise can be suppressed by simple high-speed data averaging.

Interferometric Quadrature Interferometry compares two light waves: one light wave is the reference wave, and acts like a high-precision meter stick; the other light wave is the signal wave that is identical to the reference, with the single exception that it has passed through a thin layer of material, such as biomolecules bound to a surface. The peaks and troughs of the signal wave field are compared with the peaks and troughs of the reference wave. Relative shifts of the peaks and troughs (known as the phase of the wave) can be measured with picometer resolution. These shifts provide the information about how many molecules are on the surface-bound layer.

The key to interferometric sensitivity is the condition known as "phase quadrature", a condition where the signal and reference waves have a 90° relative phase shift. In this situation, a shift in the phase of the signal is transduced linearly to a change in intensity at the detector. Furthermore, in phase quadrature, the conversion factor of phase-to-intensity takes on its maximum value.

Clearly, for high-precision interferometry, it is necessary to stabilize the phase between the signal and the reference waves to high accuracy. In many interferometers, this phase stabilization is very difficult, requiring expensive vibration isolation systems to shield the interferometer from mechanical disturbances. However, there is a class of interferometers known as "common-path interferometers" in which the signal and the reference waves share common paths through the entire system: only in a microscopic portion of the system – the sample region – do their paths vary. Because of the common path, these interferometers require no path length stabilization and hence are ultra-stable and inexpensive.

The Spinning Disc assay system described here uses common-path interferometric detection. The common-path architecture enables high-precision surface height measurements down to a picometer on the surface of the Spinning Disc, even though the disc surface may be wobbling by many microns. Over the past few years, several different ways have been identified of achieving common-path quadrature conditions that are applicable to the Spinning Disc assay format.

Spinning Disc Quadrature Classes There are currently four different quadrature classes of the Spinning Disc, each of which maintains highly-stable phase quadrature but uses very different mechanisms. Although there are relative trade-offs between these classes, the "in-line" class has recently emerged as the most practical.

The original Spinning Disc quadrature class was the micro-diffraction (MD) class [12,13]. This class is closest to the original concept of the digital compact disc. The MD-class Spinning Disc uses microfabricated gold ridges on a highly-reflecting surface, called the "land". The focused laser spot straddles the ridge so that half the intensity falls on the ridge, and half on the land. The height of the ridge is an eight-wavelength, putting the double-pass distance to the land relative to the ridge height at a quarter-wavelength, which is the quadrature condition.

A second class of Spinning Disc uses adaptive optics to stabilize the phase quadrature condition. This is accomplished using an adaptive optic (AO) mixer that phase-locks the signal and reference waves [14,15]. This AO-class of Spinning Disc [16] has the advantage of a simple disc structure, with no microfabrication, but the disadvantage of needing the adaptive mixer, which is a relatively sophisticated non-linear optical device.

A third class of Spinning Disc is the phase-contrast (PC) class [17,18]. This has the advantages of a simple disc structure, requiring only simple protein immobilization, combined with a relatively simple optical detection in the far field that uses a split detector with an inverting amplifier. This optical system is equivalent to differential phase-contrast scanning. The PC class has the advantage of a high surface mass sensitivity, but is more susceptible to disc wobble. It also only measures the slopes of the protein topology on the disc, rather than the direct topology.

The fourth class of Spinning Disc is the in-line (IL) class [17], which uses a thin layer on a reflective surface to set the quadrature condition. All of the detection is along the same line – hence the name "in-line". It has the advantage of simple disc patterning, with the additional advantage of simple intensity detection in the far field without the need for split detectors and inverters. This interferometer class is similar to a reflectometer, but uses a monochromatic light source; it measures the direct protein density rather than only the slope. In addition, it can be implemented using simple thermal oxide on silicon, or other inexpensive dielectric substrates. For these reasons, the IL-class of Spinning Disc has emerged as the most practical for commercialization, and it represents the core Quadraspec technology.

6.1.3.2 Why Spin?

The act of spinning is of fundamental importance to achieving high interferometric sensitivity. The advantage of spinning over stationary interferometric detection comes from the advantage of high-frequency sampling in the presence of $1/f$ noise. Almost all measurement systems are dominated by $1/f$ noise, in which the noise spectrum increases at lower sampling frequency. Static measurements, also known as DC measurements, occur at the peak of the $1/f$ noise spectrum and are hence the least advantageous to achieving good S/N ratios. On the other hand, if the sampling frequency is high, it significantly reduces the noise. It is not uncommon in optical systems to suppress the noise floor of the detection at high frequency by 40 to 50 dB. Conversely, it is extremely difficult to engineer 50 dB of gain into a system, and therefore, the simple act of spinning is the key to the advantage of the Spinning Disc over other systems, such as SPR.

In order to obtain a sense of the S/N advantage of Spinning Disc interferometry, it is instructive to perform a back-of-the-envelope calculation of S/N ratios, comparing static linear scanners detection to spinning detection. First, we assume equal-time measurements for this comparison to give the two cases equal signal values $S : T_m^2$, where T_m is the total measurement time for a single pixel within scan across a protein spot. The difference is that the static measurement makes a single measurement transit across the spot in the time T_m, while the spinning measurement makes nM measurements transits accumulated over nM rotations. Equal measurement time requires that $M = v_{spin}/v_{scan}$, where $v_{spin} = r\omega$ is the linear velocity of the Spinning Disc at radius r, and where ω is the angular velocity and v_{scan} is the velocity of the scanner. The power-spectrum noise of the measurement then scales according to

$$\text{Noise}: \frac{1}{f_{samp}} BW$$

where f_{samp} is the sampling frequency, and BW is the detection bandwidth. For both linear scanning static measurement and spinning, $f_{samp} = 1/T_m$ under equal-time measurements and the bandwidth is $BW = 1/T_m$. Conversely, for spinning detection $f_{samp} = n/T_m$, where n is the total number of sub-measurements that combine to equal T_m, and the bandwidth is again $BW = 1/T_m$. which is the inverse of the total acquisition time for the spot. However, for spinning $f_{samp} = v_{spin}/d$, where d is the size of the spot, whereas for linear scanning $f_{samp} = v_{scan}/d$. Therefore, the S/N ratios of these two cases are

$$\frac{\frac{S}{N}|_{spin}}{\frac{S}{N}|_{scan}} = M = \frac{v_{spin}}{v_{scan}}$$

which gives spinning an n-fold increase in the S/N ratio by the ratio of linear speeds.

As a concrete example, consider a 1-s measurement time T_m for a 130 μm-diameter protein spot. For a laser spot at a radius of 30 mm on a disc spinning at 20 Hz, this is a linear velocity of 4000 mm s^{-1}. The transit time for a spot is 35 μs. The ratio of 1 s to 35 μs is equal to $M = 29\,000$. By putting this in dB, it becomes

$$\text{Noise suppression (dB)} = 10\log_{10}(29{,}000) = 45\text{ dB}$$

The noise suppression in this case is 45 dB for spinning detection relative to an equal-time linear scanner. As stated above, it is much easier to suppress noise by 45 dB (simply spin), than to engineer 45 dB of gain (as in resonant systems), and this is why the BioCD is spun.

The same arguments apply in the case of fluorescence detection. High-speed sampling, as on a spinning disc, moves the detection off of the $1/f$ noise associated with the long-term drifts in excitation intensity. The fluorescent photon flux is unaffected by the high disc speed as long as the fluorescent lifetime is shorter than the transit time of a fluorophore across the focal spot. For typical fluorophores, excited lifetimes are in the range of nanoseconds, while transit times are in the range of microseconds for the spinning disc. Therefore, high-speed spinning does not affect the fluorescence detection, while affording the advantages of high-speed sampling. We have implemented dual-channel fluorescent/interferometric spinning disc systems that simultaneously acquire fluorescent and interferometric signals from the same location on the disc. This system combines the complementary advantages of high-sensitivity interferometry with the well-known properties of conventional fluorescence assays.

6.1.3.3 In-line Quadrature

In-line quadrature takes its name from the planar optical configuration that places a partially reflecting reference surface in the same line as the signal beam. For example, a dielectric layer on top of a reflecting surface provides a partial reflection. Furthermore, if the optical thickness of the layer is an eight-wavelength, then the wave reflected from the top surface has a 90° phase offset from the lower reflecting surface. The thin film is mechanically stable, and thus the interferometry is stable, without the need for any phase or path stabilization.

Theoretical Analysis To analyze the performance of in-line quadrature, we begin with the normalized two-dimensional (2-D) intensity distribution of the incident Gaussian beam

$$I(\rho) = \frac{1}{\pi\sigma^2} e^{-\rho^2/\sigma^2} \tag{1}$$

where $\rho^2 = x^2 + y^2$, $\sigma = \frac{\sqrt{2}}{2} w_0$, and w_0 is the focal spot radius. The corresponding dimensionless electric field is

$$g(\rho) = g(x,y) = \frac{1}{\sigma\sqrt{\pi}} e^{-\frac{\rho^2}{2\sigma^2}} \tag{2}$$

The 2-D diffraction problem is considered in the Fraunhofer regime. The reflected near-field is

$$\begin{aligned} E(x,y) &= r'(x,y)g(x,y) \\ &= r[1 + i\phi(r)h(x-vt,y)]g(x,y) \end{aligned} \tag{3}$$

where

$$\phi(r) = \left[\frac{(r_p - r)(1 - rr_p)}{r(1 - r_p^2)}\right] \frac{4\pi n_p}{\lambda} \tag{4}$$

is a complex-valued function that plays the role of phase, where n_p is the protein refractive index and λ is the free-space wavelength of the light. In this expression, r_p is the reflectivity of the air–protein interface, and r is the original surface reflectivity. The surface topology, including the motion of the disc, is contained in the real-valued height function $h(x + \eta, y)$, here $\eta = -vt$, and v is the linear speed of the disc at the radius of the probe beam. Performing the integral that propagates the near-field to the far-field yields the expression [19]

$$i^{IL}_d(\eta) = 2|r|^2 \phi_{Img} g^2(\eta) \otimes h(\eta) \tag{5}$$

which is the convolution of the beam shape with the protein profile. The in-line signal depends on the imaginary (90° phase) component of the phase. This component is maximized when the thickness of the dielectric layer is nearly an eight-wavelength, establishing the quadrature condition.

Oxide on Silicon Silicon is one of the most common materials available because of its importance to the electronics industry. It therefore is a good substrate choice for

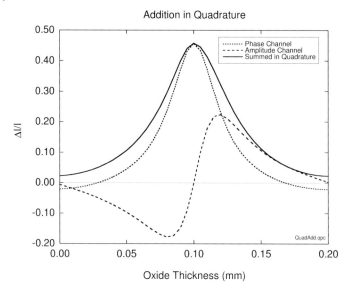

Figure 6.1 Intensity modulation in response to an 8-nm monolayer of antibody, showing the response of the phase and intensity channels and their summation in quadrature.

economic reasons, as well as for its compatibility to thermal oxide coatings. Thermally grown silicon dioxide on silicon is one of the most near-perfect dielectric structures, and it also provides a good refractive index difference between both air/oxide and oxide/silicon interfaces. The intensity response to immobilization of an 8 nm-thick monolayer of antibody is shown in Figure 6.1 as a function of the oxide thickness. Two channels are shown: differential phase contrast, and in-line, including a quadrature sum of these two channels. Differential phase contrast is a maximum for an oxide thickness near 100 nm. However, the in-line channel passes through zero at this condition. The in-line channel is maximized on either side of the phase-contrast peak, at conditions near eighth-wave. In practice, the in-line maximum on the high-thickness side is used because a protein layer leads to an increase in signal. This makes it possible to distinguish between thin layers (with positive signal) and light loss from scattering (with negative signal). For in-line Spinning Disc applications, we operate at an oxide thickness of approximately 120 nm.

6.1.3.4 Scaling Mass Sensitivity

The sensitivity of interferometry to surface-bound mass is a function of the number of statistical samples that are acquired. This means that often-quoted values of mass sensitivity in units of mass per area are not intrinsic properties of the detection system. However, it *is* possible to derive a scaling surface mass sensitivity that *is* an intrinsic property.

To derive the experimental scaling mass sensitivity of the Spinning Disc, we performed an experiment on a protein-spotted disc in which the disc was washed for 20 h in a phosphate-buffered saline (PBS) solution containing 10 ng mL^{-1} casein.

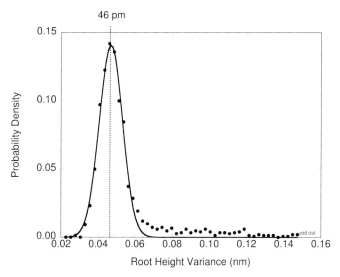

Figure 6.2 Histogram of the root height variance between two scans of the same disc before and after a 20-h buffer wash. The maximum is at 46 pm. (Reprinted from [20]).

The disc was scanned prior to and after washing, and the scans were differenced to measure the change in protein mass in addition to noise contributions. A histogram of the root variance of the data differenced between the two scans is shown in Figure 6.2 for a region containing approximately 1000 antibody spots. The maximum root variance of the height difference is 46 pm. This is the root-mean-squared height measurement error per focal spot area. It is dominated by the mass variability caused by the 20-h wash, and also by mechanical performance of the system (repositioning error between scans).

The root variance of 46 pm in the surface height corresponds to 5 fg of protein per focal spot, with a diameter of 15–20 μm. Assuming Gaussian random statistics, the surface height sensitivity at the scale of 1 mm is given by

$$\Delta h_{mm} = \Delta h_{meas} \sqrt{\frac{a_{foc}}{1mm^2}} \tag{6}$$

where a_{foc} is the area of the focused laser spot and Δh_{meas} is the root variance in the height difference. For $\Delta h_{meas} = 46$ pm and $a_{foc} = 200$ μm^2, this gives $\Delta h_{mm} = 0.65$ pm. The mass associated with this protein height is $\Delta m_{mm} = \Delta h_{mm} \rho_m 1mm^2$, which for $\Delta h_{mm} = 0.65$ pm, gives $\Delta m_{mm} = 0.25$ pg. The appropriate scaling mass sensitivity is therefore

$$S = \frac{\Delta m_A}{\sqrt{A}} = \rho_m \Delta h_{meas} w_{meas} = 0.25 \ pg/mm \tag{7}$$

which has the units of mass per length. In order to obtain the minimum detectable surface mass density, the scaling sensitivity is divided by the square-root of the sensing area. For a square millimeter this is

$$S_{mm} = \frac{S}{\sqrt{1\,mm}} = 0.25\,pg/mm^2 \tag{8}$$

This area-dependent sensitivity is better than the best values determined by SPR [21]. This sensitivity is gained without the need for resonance, and hence is much more robust and easier to manufacture than other interferometric or resonance approaches.

6.1.4
The Spinning Disc as a High-Throughput Immunological Assay Platform

The Quadraspec Assay System is comprised of four basic components, which include an assay substrate, a sample processor, a dual channel analytical instrument, and a software system for workflow, data accumulation and data analysis. These components provide technicians, scientists and clinicians with an easy-to-use solution for a wide range of immunological systems with turnkey simplicity and high-throughput capability. The non-software components of the system are depicted in Figure 6.3.

The disc is a polished 100 mm-diameter silicon wafer with a 120-nm SiO_2 layer that constitutes the surface at which biomolecules are immobilized. The rotational orientation of the surface features on the disc is defined relative to the flat of the silicon wafer. This flat section on the disc facilitates all automated disc processing, including sample application, sample processing, and reading. Hydrophobic wells define areas of sample application and containment. A single disc can contain from tens to a few hundred separate sample application wells. A barcode tracking system is employed on the disc to facilitate tracking, sample processing, automation and sample identification. Software integration of the sample and disc barcode facilitates a true turnkey immunological assay system for scientists and clinicians.

Figure 6.3 Left to right: Assay substrate (disc); sample processor; interferometric and fluorescence detector.

6.1.4.1 Immunological Assays Using a Disc Array Format

Local Array Architecture Because the SDI tool is sensitive to any kind of molecule or mass, provisions must be taken to compensate for mass that is not directly related to the analyte under consideration. This is a new and broader definition of non-specific binding (NSB). In traditional assays, NSB is defined in terms of analytes and labels binding to non-target sites. In the more general sense, NSB of mass will depend on the characteristic of the antibodies, and the properties of the substrate, including passivation and blocking steps. The non-specific binding can be measured by using reference antibodies. These antibodies are not specific to the antigen or marker under study, and are immobilized in the vicinity of the target antibodies (often also called "capture antibody"). Redundancy of the measurement is obtained by repeated spotting of the target and reference antibodies organized in a repetitive pattern, as shown in Figure 6.4. These repetitions further enhance the S/N ratio.

Multiplexing is achieved by printing numerous different antibodies in each well. Printed reference antibodies simulate the NSB of the target antibody under study, and are not specific to the target analyte. Several other factors affect the performance of the assay: number of samples, quality of the hydrophobic barrier, required detection limit, antibody activity, size of the panel, and others. As the degree of multiplexing increases, the required data analysis for assay development significantly increases in complexity, and the analysis process must be automated. New ground will have to be broken to take into account cross-reactivities, though the solution to the problem will be easier than in traditional multiplexed assays due to the smaller number of reacting molecules involved.

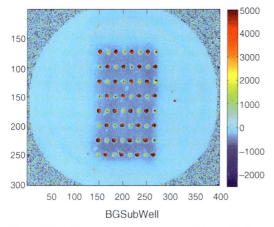

Figure 6.4 A 64-array element sample well of roughly 4 mm in diameter surrounded by a hydrophobic barrier layer. High-intensity spots (red) are the target antibodies; the low-intensity spots correspond to reference antibodies that show no specific response to the antigen. Very small dark centers in the protein spots mark the algorithms finding spots.

Figure 6.5 Hydrophobic barriers separate samples on the disc. The silicone/silicone oxide appears blue, three sample drops are visible, the remaining gray structures represent the barriers between the sample wells.

Disc Architecture Samples corresponding to multiple patients must be dispensed on the disc for incubation. These samples must be separated from each other to avoid cross contamination. Figure 6.5 shows a hydrophobic barrier that allows sample containment for the time period necessary for the immunoreactions to occur. This time varies from a few minutes to hours, depending on multiple factors such as antibody activity, antigen concentration, incubation temperature, and (if designed for) mixing.

6.1.4.2 Assay Formats

One of the most attractive characteristics of SDI is the ability to perform measurements without labels or tags. This reduces the cost and significantly facilitates multiplexing. As shown in Figure 6.6, the testing of multiple analytes is possible

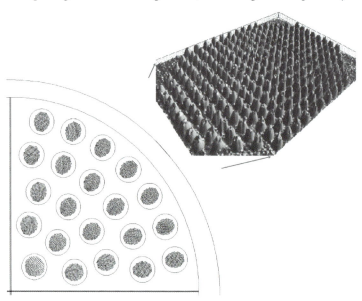

Figure 6.6 Example of a disc architecture that contains a low density of wells (84) and a high density of antibody spots per well (256).

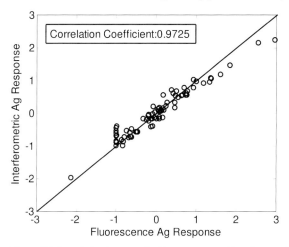

Figure 6.7 Correlation between the fluorescence and interferometric responses from a dual-channel instrument.

due to the topological separation of the target antibodies. The antibody arrangements can be customized depending on the application. The lowest noise is achieved by maximizing the redundancy.

The antibodies are arranged as arrays typically consisting of alternating target and reference spots on a 100 mm-diameter silicon wafer. For SDI, the "target" is any molecule capable of a specific biological binding event of interest. The "reference" is an immobilized protein that is chosen based on non-specific binding characteristics that mimic the target molecule. A collection of target and reference binding spots within a single well constitutes a protein array.

Each of the array elements is contained within the hydrophobic barrier that defines the application of a single sample. In a typical immunological microarray application, a sample well can contain a few hundred target and reference spots of approximately 100 µm in diameter. Assay multiplexing is possible within a given sample well through the immobilization of a multitude of target and reference molecules.

Immunological formats also can include mass amplification and fluorescent tags. The former can be achieved by adding a secondary reagent containing an antibody against the antigen under study. The second antibody can also be fluorescently tagged, which makes possible to correlate the interferometric measurements to more conventional fluorescence technology. A high degree of correlation as been achieved, as shown in Figure 6.7.

6.1.4.3 Assay Protocols

Sample and disc processing can be performed either manually or through the Inspira Sample Processor. Through a menu-driven software interface, the Inspira Sample Processor robotically tracks and applies as many as 370 samples from test tubes or microtiter plates per hour on the disc. The samples are applied to the disc in a humidity-controlled chamber to prevent biases from sample evaporation or contamination. Once the samples are applied and the appropriate sample incubation time has elapsed, the sample processor continues to process the disc by washing and drying. The sample processor can accommodate further disc processing by the addition of a secondary reagent or label if desired to any or all wells required for a given application. Software integration of the Inspira Sample Processor allows all sample and disc processing to be logged and recorded.

6.1.5
Types of Assay that Fit the Spinning Disc

6.1.5.1 Assay Structure

The assay structure of the Spinning Disc is completely analogous to other immunological assays, with the exception that secondary labeling is not specifically required for interferometric mass quantification. The assay format can be configured to facilitate multiplexing, high-throughput screening, antibody characterization including isotyping, and biomarker validation. The possible assay structures include general methodology such as direct or competitive, reagent-limited or reagent-excess assays. Additionally, the Inspira reader provides an interferometric as well as a fluorescence channel acquisition and analysis support for an even wider assay scope. One particularly useful aspect of the Inspira dual-channel instrument is the ability to use the interferometric and fluorescent results to provide an easy validation of an assay system, one of the requirements for FDA approval.

The array elements of the disc consist of target and reference spots of varied structural configurations. Literally, the target can be constructed from any system capable of a specific binding event (e.g., antibodies, oligonucleotides, or aptamers). The reference can be constructed from any system that mimics the target in non-specific binding events. The core fundamental design of the array element is to match the target and reference spots in all aspects except the specific binding characteristics of the target.

For immunological systems, the structure of the target or capture system can be constructed exactly as in other immunological assays. The target can be directly immobilized through covalent or passive adsorption or immobilized through a secondary binding molecule. For example, the target can be immobilized through a binding protein such as protein A, G, L, or fusion proteins. The target system can also be constructed using an immobilized biotin binding protein such as avidin, streptavidin or NeutrAvidin. Any biotinylated molecule can then be immobilized as the target. Alternatively, biotin can be immobilized on the disc surface to bind such proteins needed to build a target system.

The flexibility of the target system cannot be overstated. Any molecule that can be immobilized which exhibits binding specificity can be exploited in SDI. Antibodies, antibody fragments, antigens, epitopes, haptens, and signaling peptides are just a few potential target systems that can be exploited in SDI.

Construction of the reference system is equally flexible in structural motif. Fundamentally, the role of the reference spot is to mimic the target spot in all aspects except the specific binding characteristics. Generally, this is accomplished by immobilization of molecules that are structurally similar to the target system. In practice, immobilization of the proteins that make the reference system is performed in a manner identical to the target system. Care is taken to immobilize the reference proteins using the same buffer system, concentration, incubation time, temperature, humidity and protein homogeneity, as well as a host of other issues. These precautions are employed in an effort to approximate the target system non-specific binding characteristics. Additionally, the target and reference spots are in close physical proximity, which maximizes the value of the target-reference system by minimizing small local variations in the disc surface.

6.1.5.2 Assay Development Kit (ADK)

In the current implementations, the fabrication of a disc protein array requires the use of relatively sophisticated protein printing equipment for optimal interferometric assay performance. Although protein printers are becoming increasingly common in microassay fabrication, high operational costs and high capital costs exclude wide adoption of custom fabrication of desired assay systems. To address the needs of researchers that can benefit by a cost-effective microarray solution, Quadraspec Inc. prefabricates discs that allow the user to specify the target molecule in the array. This pre-fabricated disc is known as the "Assay Development Kit" or ADK.

The ADK is a pre-fabricated 96- or 264-well disc that can be customized by the end user. In the ADK, the target spots are fabricated by immobilization of a binding protein or biotin. In one implementation the fusion protein AG is immobilized in the target spot, whereas the reference spot consists of a pre-saturated protein AG-Mouse IgG complex. In this way, the end user can immobilize a wide range of antibodies of interest without the need to custom fabricate the microassay in-house.

Protein AG is the product of gene fusion of the Fc binding domain of protein A and protein G. It is a 50 449 Da protein containing approximately 10% lysine residues. Protein AG binds through the Fc region of all IgG subclass antibodies and contains two protein G and four protein A binding domains [22]. Because protein AG binds to all IgG subclasses of human and mouse IgGs, the protein AG ADK is particularly useful in the immobilization of polyclonal or monoclonal antibodies of undetermined subclasses. The ADK's use lies in research applications such as developing single plex assays (before deciding on multiplexes), determining solutions to NSB, or selecting the best antibodies for desired immunoassays.

6.1.6
Assay and Sample Processing

6.1.6.1 High-Throughput System

One of the outstanding properties of the SDI platform is the simplicity of sample preparation and handling. Performing an assay using the using the Spinning Disc immunological platform begins with the selection of the well format. As previously

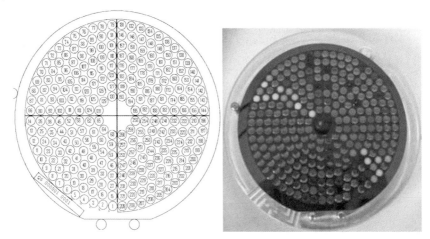

Figure 6.8 Left: The format of a 260-well Spinning Disc. Right: A 260-well Spinning Disc after the biological samples have been applied via the sample processor (note the different sample types, e.g., lipemic sera in white color).

discussed, the assay format is flexible with respect to the number and placement of the sample wells, as well as to the specific immunological system to be studied. In the first example, a 260-well Spinning Disc is selected for the assay substrate to test the effects of common immunological interfering substances. In this example, the immunological system consists of an array of antibodies specific to the canine heartworm antigen. Quadraspec Inc. prefabricates the Spinning Disc used in this example (Figure 6.8), for a veterinary diagnostic application.

For pre-developed assays, parameters for processing the assay are specified by the manufacturer. In this case, the 260 sample-well anti-canine heartworm Spinning Disc processing parameters are specified by Quadraspec Inc., but may be easily modified by the user if required.

Once the user specifies the Spinning Disc well format and immunological system of interest to meet the given needs, the samples are scanned via a barcode system for tracking (Figure 6.9). When the samples are logged into the system, the menu-driven software allows the user to specify all aspects of subsequent processing. The samples may consist of controls, biological samples, secondary reagents or any other fluid required to perform the immunological assay. Serum samples are used without further preparation.

One of the outstanding properties of the SDI platform is the simplicity of sample preparation and handling. No specific sample preparation is required, allowing the research scientist or clinician the ability to begin the assay by recording the sample identity. For the 260 sample-well anti-canine heartworm Spinning Disc, the biological samples containing the interfering substances under investigation are logged into the software system for tracking and to establish an assay workflow. The software specifies and records the relationship between a specific sample well and the samples.

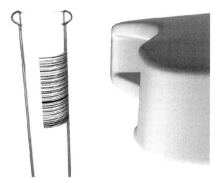

Figure 6.9 Scanning of sample for tracking.

When the user has loaded the samples, controls, secondary reagents as required for the assay, the software directs the user to specific assay processing parameters. The assay processing parameters include all aspects of processing, including the volume of the sample to be applied, the time of sample incubation including secondary reagents, the rate of flow of any required washing buffer volume and rate of spinning for washing and drying of the Spinning Disc. Prefabricated assays are accompanied by processing parameters that allow the user a turn-key assay system that can be performed with little direct interaction by the user.

After logging the samples into the system, the 260 sample-well anti-canine heartworm assay begins with a brief disc washing cycle to remove antibody stabilization agents that have been applied in disc manufacturing. In the example experiment, samples containing the interfering substances, controls and sera are applied to the disc in a humidity controlled chamber via the sample processor. After a 25-min incubation time, the disc is washed with a buffer, followed by deionized water. A fluorescently labeled reagent is then applied to the specified wells and allowed to incubate in the sample processor chamber for an additional 5 min. Following the reagent incubation, the disc is washed and dried.

The disc is removed from the sample processor chamber and placed into the dual channel reader for scanning. In this example, fluorescent scanning is performed on the 260 sample-well anti-canine heartworm Spinning Disc. If required either interferometric, fluorescent or both methods of scanning may be employed.

Following the Spinning Disc scans, the data are processed and analyzed with the aid of the integrated software. The output is expressed in an easy-to-understand format to yield the immunological response of the applied samples. All results are recorded and documented at the completion of the assay workflow.

6.1.6.2 The ADK

Often, research and clinical investigations are more basic in nature, whereupon these users have a need to interrogate a number of different immunological systems, and also require flexibility in experimental design. The ADK provides a platform for these more basic immunological assay investigations.

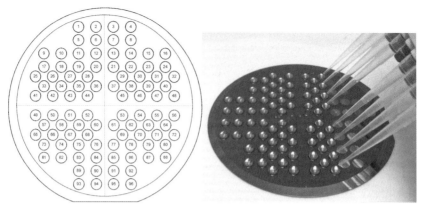

Figure 6.10 Left: Format of a 96-well Spinning Disc. Right: Example of disc dosing with a standard multi-tip pipette.

The Protein A/G Assay Development Kit (Protein A/G ADK) allows the user to build the same or different immunological system in each well of the Spinning Disc. The 96-well Spinning Disc has the additional advantage of well spacing that allows the simultaneous application of samples or reagents with standard multi-tip microtiter pipettes (Figure 6.10).

For the Protein A/G ADK, the protein A/G and reference antibodies are immobilized in an array within each well. To build an immunological system, the researcher selects target antibodies of interest that have affinity for protein AG. This allows them to screen a variety of monoclonal or polyclonal antibodies. In the ADK Spinning Disc Immunological Assay format, the user can easily develop the assay processing conditions, including the use of secondary fluorescent labels if required.

The general procedure for using the Protein A/G ADK is as follows:

- Wash the disc with a standard PBS buffer containing 0.05% Tween 20 (PBST). The typical molarity of PBS solutions is 50 mM phosphate and 150 mM NaCl. The PBST wash is followed by a second wash with deionized water.

- The excess deionized water is removed by a short spinning cycle.

- Pipette the target antibodies into the designated wells. The concentration of the antibody may range from $5\,\mu g\,mL^{-1}$ to $150\,\mu g\,mL^{-1}$.

- Allow the antibody to incubate on the Spinning Disc for 30 min in a covered container.

- Wash the disc with a standard PBST buffer solution, followed by a deionized water wash.

- Spin the disc dry.

- Load the disc into the Dual Channel Reader. Using the software program, designate the disc and scan identification. Typically, the scans are expressed in a "disc name, Ab" scan format. Initiate the disc scanning with the software.

- When the scan is complete, remove the disc and apply the antigen of interest to the designated wells.

- Allow the antigen solution to incubate on the Spinning Disc for a set time in a covered container. Antigen incubations can vary from as little as 5 min to over 1 h.

- Wash the disc with a standard PBST buffer solution, followed by a deionized water wash. This wash cycle is typically 1 min for each solution.

- Spin the disc dry.

- Load the disc into the Dual Channel Reader. Follow the previous nomenclature for disc and data identification. Typically, the antigen scans are expressed in a "disc name, AbAg" scan format. Initiate the disc scanning with the software.

6.1.7
Conclusions

Quadraspec has developed and is currently commercializing its Spinning Disc Interferometry (SDI) system to achieve molecular assays with high sensitivity, high throughput, high reliability, and low costs. Whereas, older technologies have relied on secondary reagents and labeled detection molecules, SDI is label-free, measuring directly the bound molecule of interest, such as a bound antigen in an immune reaction. A disc contains up to 264 sample locations, and each such location contains an array of up to 256 unique biomolecule spots, such as antibodies. As the disc spins under a narrow laser beam, a map is created through in-line quadrature interferometry with signals proportional to the bound mass on the disc surface. After the binding event and appropriate washes, a second measurement is taken and the differential is proportional to the mass bound in the binding event and can be used to create dose–response curves and, subsequently, analytical or diagnostic tests. SDI is self-referencing and can subtract out any inhomogeneous background. Also, if several spots in the array are printed to contain non-specific binding biomolecules, their signal can be used to subtract the non-specific portion of the specific binding reaction. If, in one sample location, several specific antibodies are arrayed, one has created a multiplexed assay. As the disc is mapped in the interferometric scan, it is always known which signal on the map comes from which specific capture molecule, thus allowing easy identification of the reactions of interest.

The foundation of proteomic and genomic information is accumulated through the measurement of highly specific biological binding events. The ability to measure biological binding events is limited by the sensitivity, accuracy and specificity of the testing method. Traditional proteomic and genomic assays involve the amplification of a generated signal through the use of a labeled reagent, biological amplification and/or complex methodologies. The limiting factor in obtaining proteomic and genomic information then becomes a function of the ability to individually construct a biological system with an appropriate labeled reagent. Because of their complexity, these systems provide limited throughput. As the labeled reagents must not exhibit cross-reactivity, the ability to build multiplexed assays is also limited.

The time-associated cost of building proteomic and genomic assays are directly related to the need to use labeled reagents. Currently, there are no high-throughput proteomic and genomic non-labeled or unamplified analytical methods available to the scientific and clinical diagnostic community. The limitation of current analytical techniques has an even more profound implication on bioinformatics. There is no efficient method to extract the vast information contained within complex protein or gene matrices and biological events. Simply put, there are no labeled analytical tools available that are capable of such analytical diversity with the required sensitivity and specificity. The direct measurement of an analyte is a prerequisite for widespread bioinformatics inspired assays.

Spinning Disc Interferometry has demonstrated the potential to directly measure analytes present within a sample matrix, without the need for labeled reagents. Because SDI does not require a reagent label, complex biological interactions can be investigated. The most fundamental issues related to genetic expression in terms of mRNA, protein expression including regulatory effects, diagnostic protein fingerprinting, and disease expression all become targets of investigation. The implication is that any biologically mediated binding event can be identified for scientific or diagnostic purposes. The potential to discover the most fundamental connection between genetic expression, protein expression and a diseased state can be realized.

Today, we have only begun to explore the capabilities of SDI as a technology and product platform. Undoubtedly, its full potential will be realized in a future where multi-analyte signatures will be examined for disease-state recognition and monitoring.

References

1 Sternberg, J., Bourcher, R. and Prolx, A. (1952) New method of clinical diagnosis determination on proteins by paper electrophoresis. *L'union medicale du Canada*, **81** (8), 908–919.

2 Muthusamy, B., Hanumanthu, G., Suresh, S., Rekha, B., Srinivas, D., Karthick, L., Vrushabendra, B.M., Sharma, S., Mishra, G., Chatterjee, P., Mangala, K.S., Shivashankar, H.N., Chandrika, K.N., Deshpande, N., Suresh, M., Kannabiran, N., Niranjan, V., Nalli, A., Prasad, T.S., Arun, K.S., Reddy, R., Chandran, S., Jadhav, T., Julie, D., Mahesh, M., John, S.L., Palvankar, K., Sudhir, D., Bala, P., Rashmi, N.S., Vishnupriya, G., Dhar, K., Reshma, S., Chaerkady, R., Gandhi, T.K., Harsha, H. C., Mohan, S.S., Deshpande, K.S., Sarker, M. and Pandey, A. (2005) Institute of Bioinformatics Bangalore, India. Plasma Proteome Database as a resource for proteomics research. *Proteomics*, **5** (13), 3531–3536.

3 Guo, Y., Fu, Z. and Van Eyk, J.E. (2007) A proteomic primer for the clinician. *Proc. Am. Thorac. Soc.*, **4** (1, Making Genomics Functional in Lung Diseases), 9–17.

4 Rifai, N., Gillette, M.A. and Carr, S.A. (2006) Protein biomarker discovery and validation: the long and uncertain path to clinical utility. *Nat. Biotechnol.*, **24** (8), 971–983.

5 Ono, M., Shitashige, M., Honda, K., Isobe, T., Kuwabara, H., Matsuzuki, H., Hirohashi, S. and Yamada, T. (2006) Label-free quantitative proteomics using large peptide data sets generated by nanoflow liquid chromatography and mass spectrometry. *Mol. Cell. Proteomics*, **5** (7), 1338–1347.

6 van der Greef, J., Martin, S., Juhasz, P., Adourian, A., Plasterer, T., Verheij, E.R. and McBurney, R.N. (2007) The art and practice of systems biology in medicine: mapping patterns of relationships. *J. Proteome Res.*, **6** (4), 1540–1559.

7 Thorpe, R. (1998) Developments in immunological standardization. *J. Immunol. Methods*, **216** (1–2), 93–101.

8 Ekins, R.P. and Chu, F.W. (1991) Multianalyte microspot immunoassay-microanalytical compact disk of the future. *Clin. Chem.*, **37** (11), 1955–1967.

9 Schweitzer, B., Roberts, S., Grimwade, B., Shao, W., Wang, M., Fu, Q., Shu, Q., Laroche, I., Zhou, Z., Tchernev, V.T., Christiansen, J., Velleca, M. and Kingsmore, S.F. (2002) Multiplexed protein profiling on microarrays by rolling-circle amplification. *Nat. Biotechnol.*, **20** (4), 359–365.

10 Haab, B.B. (2006) Applications of antibody array platforms. *Curr. Opin. Biotechnol.*, **17** (4), 415–421.

11 Nolte, D.D. and Regnier, F.E. (2004) Spinning-disk interferometry: The BioCD. *Optics and Photonics News*, 48–53.

12 Varma, M.M., Inerowicz, H.D., Regnier, F.E. and Nolte, D.D. (2004) High-speed label-free detection by spinning-disk micro-interferometry. *Biosensors Bioelectronics*, **19**, 1371–1376.

13 Varma, M.M., Nolte, D.D., Inerowicz, H.D. and Regnier, F.E. (2004) Spinning-disk self-referencing interferometry of antigen-antibody recognition. *Optics Lett.*, **29**, 950–952.

14 Nolte, D.D. (1999) Semi-insulating semiconductor heterostructures: Optoelectronic properties and applications. *J. Appl. Phys.*, **85**, 6259.

15 Nolte, D.D., Cubel, T., Pyrak-Nolte, L.J. and Melloch, M.R. (2001) Adaptive beam combining and interferometry using photorefractive quantum wells. *J. Opt. Soc. Am. B*, **18**, 195–205.

16 Peng, L., Varma, M.M., Cho, W., Regnier, F.E. and Nolte, D.D. (2007) Adaptive Interferometry of Protein on a BioCD. *Appl. Opt.*, **46**, 5384–5395.

17 Zhao, M., Nolte, D.D., Cho, W.R., Regnier, F., Varma, M., Lawrence, G. and Pasqua, J. (2006) High-speed interferometric detection of label-free immunoassays on the biological compact disc. *J. Clin. Chem.*, **52**, 2135–2140.

18 Zhao, M., Cho, W.R., Regnier, F. and Nolte, D.D. (2007) Differential phase-contrast BioCD Biosensor. *J. Appl. Opt.*, **46** (24), 6196–6209.

19 Wang, X., Zhao, M. and Nolte, D.D. Common-path interferometric detection of patterned protein on the BioCD (submitted).

20 Nolte, D.D. and Zhao, M. (2006) Scaling mass sensitivity of the BioCD at 0.25 pg/mm. Presented at Smart Medical and Biomedical Sensor Technology IV, Boston.

21 Homola, J., Yee, S.S. and Gauglitz, G. (1999) Surface plasmon resonance sensors: review. *Sensors Actuators B-Chemical*, **54**, 3–15.

22 Eliasson, M., Olsson, A., Palmcrantz, E., Wiberg, K., Inganas, M., Guss, B., Lindberg, M. and Uhlén, M. (1988) Chimeric IgG-binding receptors engineered from staphylococcal protein A and streptococcal protein G. *J. Biol. Chem.*, **263**, 4323–4327.

6.2
Optical Proteomics on Cell Arrays

Andreas Girod and Philippe Bastiaens

6.2.1
Introduction

The discovery and utilization [23] of green fluorescence protein (GFP) from the jellyfish, *Aequorea victoria*, was a milestone for biological and medical research. Since then, many color variants have either been created by genetic modifications [24,25] or found in other species [26,27], covering the entire visible spectrum. In response to excitation with light with a variant-specific wavelength, fluorescent proteins (FPs) emit red-shifted light. This intrinsic fluorescence does not require exogenous cofactors or substrates, thus allowing the expression and visualization in almost all bacterial and cellular systems, and even intact, living organisms. Fluorescent proteins are marker molecules that can be genetically fused to many other proteins of interest, without compromising their functionality. Due to spectral differences between the FP-variants, it is even possible to visualize several constructs simultaneously in a single cell. These features initiated the development of a broad range of fluorescence microscopy-based applications, such as the visualization of subcellular localization, tracking of molecules in a spatial and temporal manner, and the detection of molecular interactions.

The detection of molecular interactions addresses the dynamics of molecular systems in cell biology. However, this application of fluorescence microscopy is handicapped by the limited spatial resolution of classical light microscopy techniques: resolution limit: approx. 1/2 wavelength; $R = \lambda/2n(\sin)$.

By using the photophysical process of fluorescence resonance energy transfer (FRET), this optical limitation is overcome, allowing the detection of molecular interactions (Figure 6.11). FRET occurs via a non-radiative, dipole–dipole-coupling process. The energy of an excited donor fluorophore can be transferred to an acceptor fluorophore, if: (1) the donor emission spectrum has an overlap with the excitation spectrum of the acceptor; and (2) the fluorophores are in very close proximity, as the efficiency of FRET falls dramatically with their distance (usually >11 nm) [28,29].

This restriction guarantees that a FRET signal is only detectable when molecules are interacting. Thus, excitation of the donor produces a sensitized emission of the acceptor that, in the absence of FRET, ordinarily would not occur, while simultaneously quenching the fluorescence of the donor. Molecules of interest can either be fused to genetically encoded FPs, chemically labeled by covalent attachment of synthetic fluorophores, or indirectly marked by fluorescently labeled antibodies. The molecular interaction of the molecules in question can then be inferred by FRET between the fluorophores.

Typically, FRET is detected by donor–acceptor fluorescence ratio measurements [30,31], sensitized emission [32,33], or acceptor photobleaching [34,35]. The strength

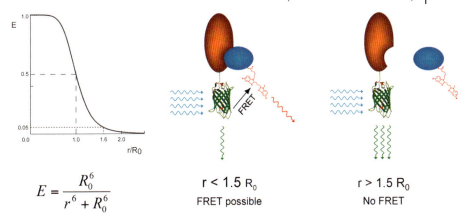

Figure 6.11 FRET efficiency, E, as a function of distance between fluorophores in units of R_0 (in nm; R_0 is the distance of a given donor/acceptor pair, at which 50% energy transfer takes place). Indicated are $E = 50\%$ at $r = R_0$, and $E \approx 5\%$ $r \approx \sim 1.6 \times R_0$, where r is a given distance between the fluorophores. Assuming a detection limit of $E \approx 5\%$, the maximum distance for reliably detecting FRET corresponds to $r \approx \sim 1.6 \times R_0$. Since R_0 will generally not exceed 7 nm, this confers an upper distance limit of ~ 11 nm for the detection of FRET (corresponding to a detection limit of $E \approx 5\%$).

of these intensity-based methods is, that they can be performed with standard fluorescence wide-field microscope set-ups. However, they require rather complex measurement and data analysis strategies in order to achieve reliable results.

Fluorescence lifetime imaging microscopy (FLIM) [28,35,36] is a more robust and sensitive technique for detecting FRET, although the instrumentation required for this method is more sophisticated. FLIM measures the dynamics of the emission decay for an excited fluorophore (Figure 6.12c and g). The measurable parameter, fluorescence lifetime (τ), is characteristic for each fluorophore. When FRET occurs, the lifetime of the donor fluorophore is reduced, indicating an interaction between observed molecules (Figure 6.12d, f and h). These features, when combined with a more reliable, easily automated, data acquisition routine, makes FLIM a very powerful tool for high-throughput screening (HTS) and high-content screening (HCS) *in vivo*, for example in the identification of intracellular networks and their dynamics.

However, until recently, optical–cellular *in-vivo* assays were based on the overexpression of recombinant proteins on a more or less gene-by-gene scale due to the established "single well–single DNA" liquid-phase transfection (LPT) techniques. As a consequence, each protein is manipulated individually and independently. This is a critical drawback especially for analysis of network dynamics, which depends on precisely performed, reproducible sets of experiments. A breakthrough to overcome this weakness was the development of the reverse transfection (RT) technique, as described by Ziauddin and Sabatini in 2001 [37].

Here, cells are cultured in glass-bottomed dishes on which different plasmid expression constructs are spotted in an array (Figure 6.12a). These constructs are transfected and expressed only by those cells growing on top of the individual spots

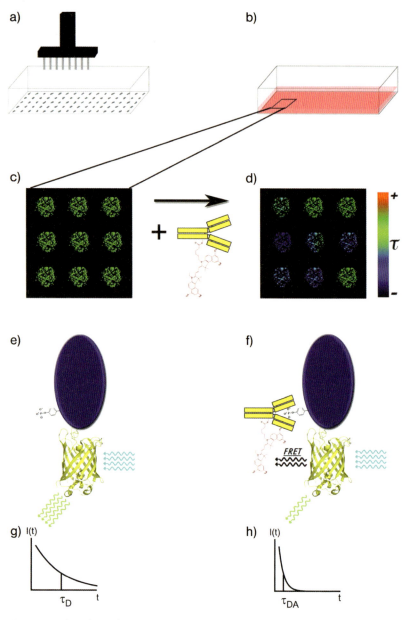

Figure 6.12 Flow sheet of FLIM-bases screen. (a) Automated spotting of nucleic acid into glass-bottomed dishes. (b) Seeding of cells, transfection and performance of the experiment. (c–h) FLIM measurement. Fluorophores have a characteristic lifetime (g). (d, f, h) In the situation of FRET this lifetime will be reduced, indicated by a change from green to blue (c, d). (c) and (d) show colonies of cells expressing EYFP-tagged proteins, treated with an antibody labeled with an acceptor fluorophore. The colors represent a measure for the lifetime and not the intensity or color of the emitted light.

(Figure 6.12b). This results in a cell array expressing hundreds of different proteins – typically 384 – in a locally well-separated manner in a single dish (e.g., Nunc™ Lab-Tek™ Chamber Slides; spotted area ca. 3.6×1.4 cm), enabling simultaneous treatment under absolutely identical conditions. The preparation of the transfection-probe and its spotting can be automated, and may provide the high numbers of samples needed for the screens. Although this transfection technique is highly suitable for the FLIM approach described above, it can also be used for many other applications. For example, a recently reported project was the life cell small interfering (si)RNA screen for the identification of proteins involved in mitosis [38].

6.2.2
A Description of the Problem with Regards to Sample Preparation

6.2.2.1 General Remarks
Unlike classical single-probe approaches, high-throughput data acquisition often does not allow interactive compensation for inhomogeneous sample properties (e.g., cell distribution and spot integrity). The significance of this limiting factor increases strongly with the degree of automation. Therefore, one of the most critical factors of HTS is the provisioning or preparation of suitable, high-quality probes, to maximize the readout of the employed technology with a minimum of compromise. In that respect, the RT array technology presents a challenge as it requires elaborate optimization of each step. This involves finding a suitable transfection reagent and the correct composition of the transfection mixture (transfectant). In addition, cell lines with a sufficient transfection rate should be used, as the efficiency of RT is lower in comparison to liquid-phase methods. These critical aspects are highlighted in the following sections.

6.2.2.2 Cell Line Selection
The cell line used is the most critical key component of an *in vivo* experiment; hence, special care should be taken in choosing an appropriate line. Depending on the design of the experiment, certain characteristics of individual cell lines finally determine whether – and under which conditions – they can be used for RT. A note of caution should be noted at this point that, under confluent cell growth conditions, only approximately 150 HeLa cells can find space on a spot of a diameter of 400 μm ($\approx 1.3 \times 10^{-3}$ cm^{-2}; see Table 6.1), and therefore a high transfection efficiency is necessary. The main prerequisite for a cell line is the "transfectability", although unlike in LPT the ability to take up nucleic acid is not sufficient. The "mode of

Table 6.1 Interdependency of tip diameter and spot size.

Needle tip diameter/μm	Spot diameter/μm	Spot size/cm^2	Max. no. of HeLa cells per spot
100	100–125	$\approx 8 \times 10^{-5}$	10
300	220–250	$\approx 3 \times 10^{-4}$	40
600	430–450	$\approx 1.3 \times 10^{-3}$	150

attachment/settlement" is also critical, since in RT the "transfectant" is provided only locally on the bottom of a dish. Ideally, seeded cells should settle down evenly over the dish and not move away after transfection. In order to achieve this, the cell seeding must be optimized in several respects. An often-observed problem is the tendency for certain cell lines not to attach on spots but rather next to them. However, seeding the cells at a high density forces them to settle down on the spots and ensures a maximum number of cells on each spot. Unfortunately, not all cell lines or experiments allow this strategy – an example is that of long-term experiments such as RNAi, where the cells require space for growth. In other cases, tight cell–cell contacts might interfere with the cellular response behavior, for example in mitosis. A more physiological (but also more elaborate) alternative to enhance cell attachment could – in certain cases – be replacement of the cell adhesion protein fibronectin, which is usually contained in the RT-transfectant mixture, or its supplementation with other reagents (e.g., poly-L-lysine, laminin).

Cell migration is another crucial factor, and characteristically occurs in different degrees for various cell lines. In experiments where cells must be seeded at a low density, cell migration can result in undefined spots and also cross-contamination between spots. However, these effects are almost negligible if the cells are plated at high density, as the migration is then spatially restricted.

6.2.2.3 Sample Preparation

The transfectant generally contains nucleic acid (usually plasmid DNA or siRNA), a transfection reagent, transfection reagent buffer, gelatin (to stabilize the transfection mixture and protect the spot from resolving after cell seeding), and an adhesion protein to allow cell attachment. Optionally, a fluorescently labeled marker molecule (e.g., oligonucleotide) can be cotransfected to monitor the transfection efficiency or the localization of the transfected cell colonies. As in LPT, the composition of the mixture must be optimized for each cell line individually, where the transfection reagent quality and quantity are the key elements. When preparing an entire 384-well multi-titer plate (MTP) for spotting, the use of a liquid-handling robot is recommended to minimize pipetting errors. For small sample numbers, however, multi-channel pipettes can be used.

6.2.2.4 Choice of Transfection Reagent

In RT, the transfection reagent must meet additional requirements to that of LPT. The use of inappropriate reagents will result in spots of low integrity, causing a "spreading out" of transfectant after addition of the cells, and hence cross-contamination or undefined spots. Furthermore, RT generally disallows any early exchange of medium after cell seeding, as it causes any loosely or unattached cells to be washed off. As an early medium exchange is recommended for certain transfection reagents, possible toxicity effects should be accounted for. A number of different transfection reagents should always be tested for each cell line. It is also recommended that different batches of the selected reagent to are compared, as their composition (e.g., lipid concentration) can vary, which in turn will affect the transfection efficiency, the standardization of a transfection protocol, and reproducibility.

6.2.2.5 Nucleic Acid Preparation

Another critical factor is the quality of the nucleic acid, which should be homogeneous, free of debris, and correctly buffered. Experiments have shown, that the best results are obtained with plasmid DNAs purified by using ion-exchange resins. DNAs purified using technologies based on silica membranes, glass wool or magnetic beads often result in poor transfection efficiencies.

6.2.2.6 Sample Scale: How Many Duplicates are Required?

Undoubtedly, HTS-approaches enable the acquisition of huge amounts of data and accelerate the progression of research. However, the resources needed can easily exceed the technical and infrastructural possibilities and, even more importantly, the budget for an average project. A main cost factor is for consumables such as glass-bottomed dishes, transfection reagents or siRNAs. The total amount of transfection mixture needed for printing (e.g., glass slides) is dependent on the number of duplicates and the print pins used. In order to minimize expenses, a realistic number of required samples should be estimated, including "reserve samples" for failed experiments and repeats. If an insufficient number of slides are printed, then a new source plate must be prepared. Storage of the source plate for more than 1 day is not recommended. For projects below industrial scale, less than 50 duplicates are normally printed, requiring a total volume of reagents of less than 20 µL per well. This volume is ideal for use in 384 shallow-well dishes.

Spotted samples can be stored for months in a dry environment (e.g., in the presence of drying pearls) at room temperature. The time needed for spotting depends mainly on the number of samples, but also on the spotter type/specifications (e.g., linear motor versus servo motor, number of needles, size of needles).

Hint: Never use samples immediately after spotting; a minimum drying period of 12 h is necessary for a good transfection result.

6.2.2.7 Choice of Microarrayer/Microspotting System (Spotter)

Type of Spotter A variety of different printing technologies are available, each with characteristic features in terms of sample deposition, spatial resolution, precision, deposition volume, speed and range of possible printing solutions. The spotters used for microarraying fall mainly into two categories:

- A *non-contact piezo inkjet spotter* dispenses the spotting solution through a micrometric nozzle onto the predefined regions of the target surface.

- *Contact spotters* with *solid pins* print the spotting solution onto the substrate surface, after taking it up from the source plate. In order to duplicate the spot, this cycle must be repeated. The use of so-called *split pins*, which have a reservoir for sample uptake (e.g., Stealth, MicroQuill®) allow printing multiple times on multiple substrates with one volume loading of sample. Direct contact between the liquid and a solid surface results in drop deposition.

However, because of the normally high viscosity of the RT printing solution (which contains gelatin), usually only a contact spotter with solid pins can be used reliably. The size of the printed spots is determined by the diameter of the tips used (see Table 6.1), the viscosity of the transfectant, and by the constitution of the glass surface.

Choice of Spotter
If a spotter needs to be purchased, certain features should be considered:

- the possible number of source plates
- source plate cooling, to stabilize the printing solution and to reduce evaporation
- stacker-option for automated source plate loading
- capacity for targets (substrates)
- type of targets (e.g., glass-bottomed dishes, 96- and 384-well MTP)
- tip cleaning technique (e.g., sonication or washing)
- barcode reader for plate identification and tracking
- closed chamber:
  - environmental control to regulate temperature and humidity
  - high-efficiency particle air (HEPA) filters to reduce dust contamination
- software, allowing:
  - flexibility for substrate definition (different sizes and formats which implies programming of at least two independent off-sets
  - definition of spot density
  - duplications inside each substrate
  - free definition of the needle position in X and Y in the print head (needle holder)

6.2.2.8 Layout Design (Spotting Pattern)

Unlike in manual pipetting, spotting robots are following a certain pipetting pattern, which becomes more complex with every additional needle used (normally up to 12 in X and up to four in Y). The spotter normally defines a subarray, which is dependent on the distance of the pins (4,5 mm) and the programmed spot-distance (a fraction of 4,5 mm). Arraying with one needle and a given spot distance of 1.5 mm would result in a subarray of 3×3 spots. A typical result for one or two needles is shown below:

MTP (source plate)

A1 A2 A3 A4 A5 A6 A7 A8 A9 A10 A11 A12 A13 A14 A15 A16 A17 A18

Target plates

Spotted array with one needle						Spotted array with two needles					
A1	A2	A3	A10	A11	A12	A1	A3	A5	A2	A4	A6
A4	A5	A6	A13	A14	A15	A7	A9	A11	A8	A10	A12
A7	A8	A9	A16	A17	A18	A13	A15	A17	A14	A16	A18

Consequently, the order in the source plate does not correspond to the position on the target substrate. Since usually a certain sample pattern on the target plate is required, a "back-transformation" of the target position layout into the source plate matrix is necessary. Employing software (e.g., in Excel) and barcode-technology for this time-consuming and error-prone step can facilitate this process and also allow tracking of the samples throughout the whole course, inclusive of data acquisition and analysis.

6.2.3
Summary

Since the discovery of FPs during the early 1990s, the impact of light microscopy for proteomic research has increased significantly. Unlike classical biochemical approaches, microscopy-based techniques allow the investigation of cellular processes and molecular functions either *in vivo* in fixed or living cells, and even on a single-cell level. However, these techniques are generally lacking in speed and throughput. The recent development of an array-based reverse transfection technique in combination with an advanced FRET microscopy approach made it possible to overcome this handicap. The *in vivo* screen described herein fulfills the requirements of high throughput and high content, without compromising sensitivity and single cell information.

6.2.4
Perspectives

This FLIM-based screening approach is generally adaptable to many projects for the investigation of molecular interactions or modifications. Further automation will result in much higher throughputs, allowing the performance of genome-wide screens within a short time period. Furthermore, it will be possible to analyze the dynamics of whole networks on a single chip.

6.2.5
Sample Preparation: Short Protocol

6.2.5.1 Recommended Equipment and Consumables

- General equipment and consumables:
  - Spotter
  - Low-speed centrifuge for MTP
  - Pipette tips 2.5 µL, 10 µL and 20 µL; filter tips are usually not required
  - Multi-well plate, recommended: 384 shallow well
  - Glass-bottomed dishes
  - The chemicals used are dependent on the cell line and the type of experiment

- Sample preparation for up to 200 samples:
  - Equipment needed: 2 µL and 10 µL single-channel pipettes and 10 µL multi-channel pipettes (8 or 12 channels)

various proteins at different levels. With protein expression profiles generated from homogenates or whole tissue samples it is difficult to identify reliably the cancer cell-specific proteins. Therefore, laser microdissection, with its unique ability separately to analyze protein expression in captured tumor cells and compare it with those from the surrounding tissue, will surely become an indispensable technology also in proteomic research [41]. The amount of protein retrieved from the dissected tissue usually is limited, besides proteins cannot be amplified as can nucleic acids. Therefore, the efficient coupling of protein analysis methods to microdissection depends mainly on the specificity of the preparation protocols, on the sensitivity of the downstream analyzing technologies, and finally on the degree of automation of the laser capture process.

Recently published data have proven that laser microdissection does not affect downstream protein analysis, and is suitable for most of the commonly used protein-analyzing technologies. However, sample preparation and staining procedures can have significant effects on protein retrieval and quality [42]. For optimal protein recovery the use of fresh-frozen unstained tissue is recommended, with the disadvantage of displaying poor morphology, while staining for better visualization with the commonly applied dyes, such as hematoxylin and eosin (H&E) or methylene blue, results in poor protein spectra. Cresyl violet produces results comparable to native tissue and is, to date, the least interfering stain [43]. Some immunostaining and fluorescent detection methods allow reasonable protein recovery [42]. Nevertheless in proteomics, staining remains a critical issue as it generally interferes with the protein analysis procedures. To counter this problem, software tools have been developed that allow the regions of interest on a stained first slide to be identified, and for the coordinates to be marked and transposed onto consecutive unstained sections (see for example the PALM RoboSoftware - "SerialSection" module in the software manual) [42].

Most of the established protein-analyzing methods such as two-dimensional polyacrylamide gel electrophoresis (2D-PAGE), as well as technologies based on mass spectrometry (MS) or nanoscale liquid chromatography (nano-LC), have been applied to analyze proteins from microdissected tissues [44,45]. Laser-microdissected samples are processed for protein expression analysis with nano-LC, matrix-assisted laser desorption ionization time of flight (MALDI-TOF), surface-enhanced laser desorption ionization time-of-flight (SELDI-TOF), or state-of-the-art chip techniques coupled to tandem-mass spectroscopy (MS/MS) [46–49].

6.3.3
Examples of Combined LMPC and Proteomic Analyses

6.3.3.1 LMPC and Preeclampsia
Preeclampsia is a pregnancy-associated disorder and the leading cause of maternal and perinatal death. Research has shown that preeclampsia is associated with molecular changes in trophoblasts, impairing their development and migration. To analyze trophoblast protein profiles, as few as 125 trophoblasts were collected by laser microdissection from frozen tissue sections of human placenta and their

peptide pattern was analyzed with MALDI-TOF MS/MS. Specific discriminating peptide signatures for trophoblast cells and surrounding villous stroma cells could be obtained. The analysis of peptide profiles from a small number of cells is important for the development of clinical markers to diagnose trophoblast-related pregnancy complications [50].

6.3.3.2 LMPC and Renal Cell Carcinoma

2D-PAGE followed by MALDI-TOF MS/MS has been used to investigate the protein composition of tumor and non-tumor cell areas in renal cell carcinoma. Normal kidney cells and metastatic renal carcinoma cells were separated by LMPC from 10-µm cryosectioned slices of standard clinical biopsies. Subsequently, 29 carcinoma-specific proteins could be identified that can be developed into markers for renal cell carcinoma and will allow the development of new therapies for this type of tumor [51].

6.3.3.3 LMPC and Hepatocellular Carcinoma

In order to investigate changes in complex protein patterns and to identify cancer-related proteins, single cells from histological sections of matched normal liver and hepatocellular carcinoma (HCC) were isolated by tissue microdissection and 83 distinct cellular proteins were examined by protein microarrays. This approach revealed differential expression between HCC and normal liver for 32 of the 83 proteins analyzed. The differential expression of these proteins was confirmed using Western blot analysis and tissue microarrays. The correlation of differentially regulated proteins with clinicopathologic data showed that proteins such as cyclin D1 and SOCS1 were associated with tumor prognosis. The data indicated that LMPC combined with an array-based approach facilitates the identification of new proteins associated with carcinogenesis, and offers the possibility of elucidating clinical cancer biomarkers. Protein-microarray technology allows questions to be addressed at various levels of protein interaction, and the analysis of thousands of proteins, simultaneously, within a single experiment. The use of miniaturized and multiplexed assay systems enables the measurement of up to several dozens of different analytes from limiting amounts of sample material, and the detection of marker proteins directly from histological sections or biopsy material [47].

6.3.3.4 LMPC and Brain Disorders

In psychiatric disorders, distinct cell populations become altered in their gene and protein expression, and therefore in their function. Due to the anatomic complexity of the brain, laser microdissection is especially important for retrieving homogeneous tissue samples for subsequent protein analyses [52]. Post-mortem Parkinson cerebellum slices were used for laser microdissection and protein expression analysis via nanoLC MS/MS, and 26 neuron-specific proteins could be identified from one single sample [48].

6.3.3.5 LMPC and Plant Biology

Vascular bundles from *Arabidopsis thaliana* stem tissue were dissected and collected via LMPC. Proteins were extracted and subjected to analysis, either by classical

2D-PAGE, or by nanoLC MS/MS. From 100 collected vascular bundles (~5000 cells), 68 specific proteins could be identified. Because of the significantly lower sample amount required for nanoLC MS/MS than for 2-D gel electrophoresis (2-DGE), the combination of LMPC and nanoLC MS/MS has a higher potential to promote comprehensive proteome analysis of specific tissues [53].

6.3.4
LMPC Adapted for Proteomic Applications

Laser microdissection and catapulting with the PALM MicroBeam enables the non-contact preparation of distinct cell areas or single cells in a highly automated manner (Figure 6.13) [42]. The technology is based on a pulsed UV-A laser that is coupled into a motorized research microscope and focused through the objective lenses. The tissue is disrupted within the narrow laser focus to produce a clear gap between selected and unwanted material. Subsequently, the excised area is transferred with a single laser pulse against gravity into an appropriate vessel (e.g., a microfuge cap). Both steps – the laser cutting as well as the laser-induced transport process – are based on the formation of an ultra-short microplasma that is confined to the minute laser focus and ensures safe specimen capture without adverse effects on their biomolecules. Depending on the desired cutting size, objectives of different magnification and numerical apertures are used, resulting in laser spot sizes down to 0.5 µm in diameter.

A unique feature of the LMPC technology that is of particular interest for high-throughput proteomic research, is the possibility to collect samples directly from routine glass-mounted tissue sections without the need of particular preparation techniques. To this purpose, areas of interest are first manually outlined with computer graphic tools. In a second step, the marked tissue areas are automatically excised and catapulted into the capture vial with multiple laser pulses ("AutoLPC" software function), fragmenting them into small flakes. This optomechanical disintegration could avoid the grinding step required in certain protein preparation protocols [48]. With this procedure, multiple areas can also be pooled into the same collection cap. This combination of direct catapulting and disintegration with no mechanical interference saves time, prevents specimen loss during grinding or pipetting, and also minimizes the risk of contamination with unwanted material.

As an alternative preparation technique, samples may be mounted onto a thin, inert membrane (i.e., polyethylene naphthalate (PEN) membrane) that serves as a stabilizing backbone. The main advantage of LMPC from membrane-mount tissue sections is preservation of the specimen's morphology after catapulting, and this allows verification of the successful transfer into the collection vial.

In order to meet the need for enhanced-throughput sampling, automated image analysis software with preset cell recognition algorithms has been implemented into the PALM MicroBeam. The object slide is screened either entirely or in pre-marked fields via the navigation function, and regions of interest that match the rules of the algorithms are outlined automatically. Along these marks the laser will cut and harvest rapidly and precisely [54]. As modern detection methods are often based on fluorescence techniques, the PALM MicroBeam can optionally be equipped with

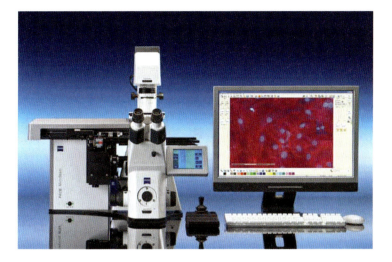

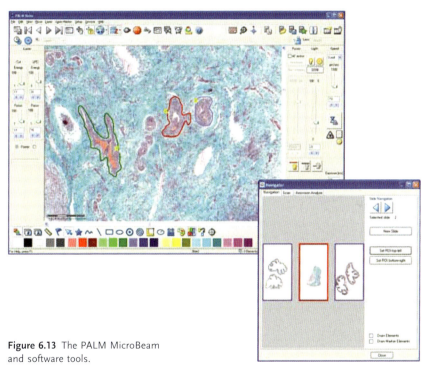

Figure 6.13 The PALM MicroBeam and software tools.

components for fluorescence microscopy, allowing LMPC under simultaneous fluorescence observation. Thus, the MicroBeam system is able to scan, detect, isolate and finally capture the specimen of interest, for example prespecified tissue areas, fluorescent-labeled rare cells, and metaphase or FISH-treated cells in fully automated manner. For the capture, a variety of collection vials is available which allow the distribution of different cell areas into distinct wells in the range of single caps up to 96-well collectors. All of these special modules and software tools tailor the MicroBeam system to an efficient technology to supplement functional proteomic research that enables the correlation of protein profiles with morphologically relevant features.

6.3.5
LMPC Combined with SELDI-TOF MS: A Promising Approach for Patient-Specific Analyses

Until now, 2D-PAGE has been considered the "workhorse" of proteomic research, often combined with MALDI-TOF analyzers. Although 2D-PAGE is able to resolve thousands of proteins, it is labor-intensive and requires large amounts of starting material. Moreover, the development of SELDI-TOF mass spectrometry has overcome many limitations of MALDI-TOF and 2D-PAGE. The SELDI MS ProteinChip technology, which was first described by Hutchens and Yip [55], utilizes chips with affinity surface coatings specifically to retain proteins based on their physicochemical characteristics prior to TOF MS analysis. The lysate of the catapulted specimen is applied onto the chip surface. The desired proteins are retained at the chromatographic surface, whilst the contaminants (such as buffer salts or detergents) are simply washed off. This easy purification step eliminates the need for pre-separation techniques and also allows quantitative evaluations. With SELDI MS, only a small number of cells is needed, and consequently this technology is ideal for small biopsies or microdissected tissue samples. In addition, its compatibility with laser microdissection and the catapulting procedure has been demonstrated in many studies [56–59,61–68].

A comprehensive study combining microdissected sample with the SELDI MS-based ProteinChip technology was performed with squamous cell carcinoma of the oral cavity, the pharynx and larynx (i.e., head and neck cancer; HNC) [59]. Worldwide, this common human malignancy affects approximately 500 000 patients each year, and the overall five-year survival rate among HNC patients is among the lowest of all major tumor types [60]. The search for molecular markers associated with the initiation and biological behavior of an individual tumor may help to improve diagnostics and provide prognostic information with regards to clinical course and outcome. However, in order to provide a more detailed insight into the complex genesis and progression of cancer entities, pure and exactly defined tissue – as well as highly parallel proteomic techniques – are required.

The separation of epithelial cells from the surrounding tissue constituents is especially challenging. In healthy tissue, the lining epithelium consists only of one or few cell layers, whereas in tumor areas the boundaries towards the normal pharyngeal tissue are rather irregular. Here, LMPC is the method of choice to capture homogeneous cell populations, thereby facilitating relevant conclusions during appropriate downstream analyses (Figure 6.14).

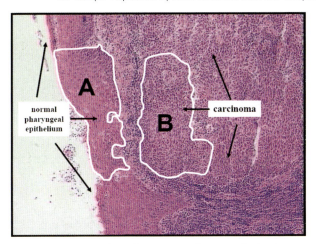

Figure 6.14 Hematoxylin and eosin-stained section of a head and neck tumor sample. The white line demonstrates the course of the laser cuts during microdissection of normal pharyngeal epithelium (A) and tumor cell complexes (B) [59].

In the above-mentioned study, the laser system captured pure populations of normal pharyngeal epithelium and homogeneous tumor squamous epithelial cells which were analyzed and compared on ProteinChip Arrays. (The detailed protocol is provided in Section 6.3.7.) Among a number of differentially expressed peaks, one signal showed a calculated mass (m/z value) of 35.90 kDa. Protein extracts of the same specimens were prepared and analyzed by 2D-PAGE. The differentially expressed spots showing an appropriate mass were cut from the 2-DGE gels, processed by in-gel digestion, examined by peptide mapping and tandem MS/MS, and subsequently identified as human annexin V, a calcium-binding protein that appears to be involved in the inhibition of blood coagulation. The assumption that annexin V is identical to the differentially expressed peak found earlier by ProteinChip analysis was confirmed with an immunodepletion assay. The localization of annexin V in tissue was subsequently verified on cryostat sections of the head and neck squamous-cell carcinoma (HNSCCs) by immunohistochemistry, using a monoclonal anti-annexin V antibody (Figure 6.15). Positive tissue areas were microdissected and reanalyzed on ProteinChip arrays to show that annexin V matched the differentially expressed peak found in the prior analysis (Figure 6.16). The comparative SELDI ProteinChip analysis of laser-captured pure tumor versus pure non-tumor tissue enables the highly precise, sensitive and specific discovery of tumor markers. A study based on this method confirmed that annexin V antibodies were present in the sera of patients with colorectal carcinoma [63]. Alternative approaches based on bulk tissue samples will not result in such unambiguous data, as here different cell types are always intermingled with the target cells.

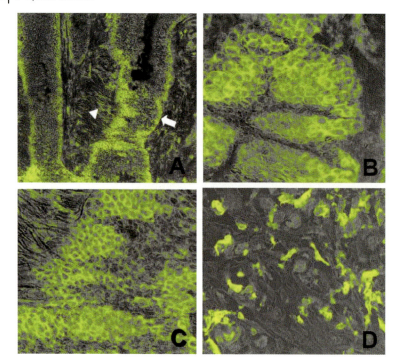

Figure 6.15 Immunohistochemistry of annexin V visualized by laser scanning microscopy. (A) Annexin V in squamous cell carcinoma (overview; original magnification, ×180). Positive staining (yellow-green) is visible especially in the periphery of the invasive growing tumor areas (arrow); slight deposition of annexin V could also been seen on collagenic fibers (arrowhead). (B) Detail of an invasive growing tumor area at higher magnification (original magnification, ×1500). (C) Detail of (A); intracytoplasmic localization and deposition on collagenic fibers (original magnification, ×1000). (D) Detection of annexin V in the cytoplasma of single tumor cells (original magnification, ×1700) [59].

In a further study, normal prostate stromal cells and cancerous epithelial cells were separated by LMPC, and the approximately 500 to 1000 freshly obtained prostate cells subsequently analyzed by SELDI-TOF MS. Although multiple specific protein patterns were reproducibly detected in both tumorous prostates and in controls, a specific 4.3-kDa peak was increased in the prostate tumor stroma compared to normal prostate [46].

In summary, the results from these studies corroborate that a combination of laser-assisted microdissection with SELDI-ProteinChip technology and immunohistochemistry can potentiate the identification and characterization of new specific biomarkers. This "Technical Triade" provides tremendous opportunities to identify tumor-specific proteins, which will significantly improve the detection and treatment of cancer (Figure 6.17) [61].

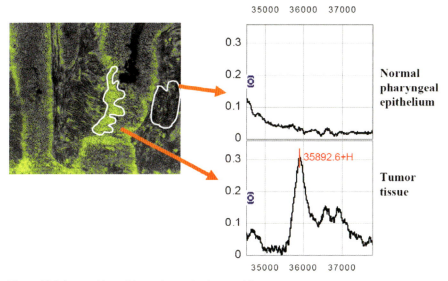

Figure 6.16 Areas with positive and negative immunohistochemical reaction were microdissected and analyzed on ProteinChip arrays. A signal with a molecular mass of 36.90 kDa (*) representing annexin V was detectable in protein lysates from positive areas, but was absent in the negative areas [59].

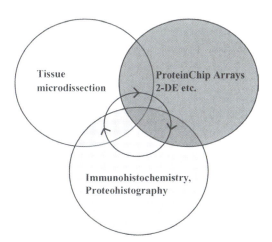

Figure 6.17 Technical triade for proteomic identification and characterization of cancer biomarkers. The starting point is tissue microdissection, where probes for ProteinChip arrays are gained. After profiling the biomarkers identified by two-dimensional electrophoresis (2-DE), immunodepletion, and other techniques, they were characterized by immunohistochemistry. Microdissection of immunohistologically positive areas and reanalyzes on ProteinChip arrays close this circle [61].

6.3.7.2 Laser Microdissection of Tissue Sections

Native air-dried cryostat tissue sections (8 μm thickness) are prepared on MembraneSlides (i.e., microscope slides spanned with a 1.35 μm PEN membrane) and processed for non-contact laser microdissection and catapulting (LMPC) using the PALM-MicroBeam system (P.A.L.M. Bernried, Germany).

- Tumor cell areas and normal epithelium are first localized and outlined on a H&E-stained section, but excised and catapulted from a non-stained subsequent serial section (see Figure 6.14).

- Multiple tissue areas containing several hundred cells each (in total about 3000–5000 cells) are pooled into the lid of a 500-μL reaction tube in less than 20 min. When tissue areas are too large to be catapulted, they may also be transferred manually into a tube by using a lancet needle [70].

- After centrifuging down the collected tissue areas (1000 ×g at room temperature), the proteins are extracted with a lysis buffer (100 mM Na-phosphate, pH 7.5, 5 mM EDTA, 2 mM $MgCl_2$, 3 mM 2-β-mercaptoethanol, 0.1% CHAPS, 500 μM Leupeptine, and 0.1 mM phenylmethylsulfonyl fluoride; PMSF) for 30 min on ice.

- Cell debris is removed by a 15-min centrifugation at 16 000 ×g at 4 °C, and the supernatant is either analyzed immediately or frozen in liquid nitrogen for a maximum of one day.

- A longer storage period would lead to a poorer protein spectrum.

6.3.7.3 ProteinChip Array Preparation and Analysis

The protein lysates from microdissected tissues are analyzed on a strong anionic exchanger array (SAX2; Ciphergen Biosystems Inc., Fremont, CA, USA).

- Sample targets on the arrays are equilibrated three times by applying 5 μL of binding/washing buffer (100 mM Tris-buffer, pH 8.5 containing 0.05% Triton X-100) for 20 min.

- After equilibration, the buffer is removed and 5 μL of fresh buffer added to each spot.

- Into this, 2 μL of sample extract is spiked and the ProteinChip Array is incubated in a humidity chamber for 90 min at 20 °C.

- The sample droplets are removed and each target is washed three times with 5 μL of binding/washing buffer.

- The targets are washed twice with 5 μL water to remove the buffer salts, and then air-dried.

- After the application of 2 × 0.5 μL saturated sinapinic acid [SPA; Sigma; dissolved in 50% acetonitrile (CAN; Sigma) containing 0.5% trifluoroacetic acid], mass analysis is performed in a ProteinChip Reader (PBS-II or PCS4000; Ciphergen

Biosystems Inc.), according to an automated data collection protocol. This includes an average of 195 laser shots to each spot, with a laser intensity of 200 and 270, respectively, depending on the measured region.

- Normalization of all spectra is performed using total ion current.

Cluster analysis of the detected signals and the determination of respective p-values is usually carried out with the Biomarker Wizard Program (Version 3.0; Ciphergen Biosystems Inc.). For p-value calculation, spectra with at least 10 signals in the range between 2 kDa and 20 kDa, and exhibiting a signal-to-noise (S/N) ratio of at least 5, should be selected and analyzed with the Mann-Whitney U-test for non-parametric data sets. The identification of differential expressed protein peaks is normally performed with one- or two-dimensional gel electrophoresis for separation, tandem MS and/or immunodepletion for identification, and immunohistochemistry for characterization [59,61].

6.3.8
Summary and Outlook

In summary, LMPC is a purification method which functions under high-resolution, microscopic control to isolate and accumulate cells of interest for genomic, transcriptomic, and proteomic analysis. LMPC enables the tracing and isolation of pure and rare cell populations, the correlation of cell development and subcellular processes with the cells' genomic and proteomic information, and the study of cellular function within the organism. Especially in tumor research this innovative technology has already provided new insights and better understanding of cellular malfunction and tumorigenesis. In conclusion, it is believed that most proteomic techniques can be combined with microdissection. However, based on the fact that detection systems become increasingly sensitive, laser microdissection and capture technologies should achieve indisputable standards in proteomics. To date, the combination of LMPC with proteomic techniques has allowed only the identification of rather abundant proteins. Now, however, it is necessary to develop proteomic analyses for high-content and high-throughput screening not only by improving MS techniques but also by developing computer algorithms for the rapid analysis of MS spectra [69]. The consequence of these developments should be better diagnosis and advanced treatment of disease.

Currently, important research efforts are under way to profile highly complex protein mixtures, the aim being to detect and identify disease-related biomarkers. Yet, if proteomics were to become practicable for the quantification of whole-cell lysates, it would have a distinct advantage over microarray studies in that proteins could be quantified in terms of their modification state and subcellular location [69].

In future, systems biology will rely increasingly on a combination of mRNA, proteome, and genetic data. Moreover, the comparison of data from functional genomic analyses with those from functional proteomic analyses of tissues of morphologically and functionally defined origin, should provide the ideal information in the support of clinical diagnostics. Such an approach is destined to become the basis of patient-tailored treatment.

Acknowledgments

These studies were partly supported by grants from the German Ministry of Education and Research (BMBF: Mikroso: 13N8257 and PhoNaChi: 13N8466). The authors wish to thank the PALM application laboratory for scientific assistance, and Sieglinde Hinteregger for project administration.

References

39 Schütze, K. and Lahr, G. (1998) Identification of expressed genes by laser-mediated manipulation of single cells. *Nat. Biotechnol.*, **16**, 737–742.

40 Micke, P., Östman, A., Lundeberg, J. and Ponten F. (2005) Laser-assisted cell microdissection using the PALM System. In: *Laser Capture Microdissection: Methods and Protocols.* (eds. G.I. Murray and S. Curran) Methods in Molecular Biology Series Humana Press, Inc., Totowa, NJ, USA, **293**, 151–166.

41 Ball, H.J. and Hunt, N.H. (2004) Needle in a haystack: microdissecting the proteome of a tissue. *Amino Acids*, **27**, 1–7.

42 Niyaz, Y., Stich, M., Sägmüller, B., Burgemeister, R., Friedemann, G., Sauer, U., Gangnus, R. and Schütze, K. (2005) Noncontact laser microdissection and pressure catapulting, in *Microarrays in Clinical Diagnostics* (eds. T. Joos and P. Fortina) Methods in Molecular Medicine Series, Humana Press, Inc., Totowa, NJ, USA, **144**, 1–24.

43 Fiedler, W., Hoppe, C., Schimmel, B., Koscielny, S., Dahse, R., Bereczki, Z., Claussen, U., Ernst, G. and von Eggeling, F. (2002) Molecular characterization of head and neck tumors by analysis of telomerase activity and a panel of microsatellite markers. *Int. J. Mol. Med.*, **9**, 417–423.

44 Shekouh, A.R., Thompson, C.C., Prime, W., Campbell, F., Hamlett, J., Herrington, C.S., Lemoine, N.R., Crnogorac-Jurcevic, T., Buechler, M.W., Friess, H., Neoptolemos, J.P., Pennington, S.R. and Costello, E. (2003) Application of laser capture microdissection combined with two-dimensional electrophoresis for the discovery of differentially regulated proteins in pancreatic ductal adenocarcinoma. *Proteomics*, **3**, 1988–2001.

45 An, H.J., Kim, D.S., Park, Y.K., Kim, S.K., Choi, Y.P., Kang, S., Ding, B. and Cho, N. H. (2006) Comparative proteomics of ovarian epithelial tumors. *J. Proteome Res.*, **5**, 1082–1090.

46 Wellmann, A., Wollscheid, V., Lu, H., Ma, Z.L., Albers, P., Schütze, K., Rohde, V., Behrens, P., Dreschers, S. and Ko, Y. and Wernert, N. (2002) Analysis of microdissected prostate tissue with ProteinChip arrays – a way to new insights into carcinogenesis and to diagnostic tools. *Int. J. Mol. Med.*, **9**, 341–347.

47 Tannapfel, A., Anhalt, K., Häusermann, P., Sommerer, F., Benicke, M., Uhlmann, D., Witzigmann, H., Hauss, J. and Wittekind, C. (2003) Identification of novel proteins associated with hepatocellular carcinomas using protein microarrays. *J. Pathol.*, **201**, 238–249.

48 Sauber, C., Sägmüller, B., Neumann, M. and Kretschmar, H.A. (2004) Identification of proteins in post-mortem human brain tissue by laser microdissection/pressure catapult and nano-LC/MS/MS. Agilent Application Note 5989-0895EN.

49 Schad, M., Lipton, M.S., Giavalisco, P., Smith, R.D. and Kehr, J. (2005) Evaluation of two-dimensional electrophoresis and liquid chromatography-tandem mass spectrometry for tissue-specific protein profiling of laser-microdissected plant samples. *Electrophoresis*, **26** (14), 2729–2738.

50 de Groot, C.J., Steegers-Theunissen, R.P., Guzel, C., Steegers, E.A. and Luider, T.M. (2005) Peptide patterns of laser dissected human trophoblasts analyzed by matrix-assisted laser desorption/ionisation-time of flight mass spectrometry. *Proteomics*, **5** (2), 597–607.

51 Poznanovic, S., Wozny, W., Schwall, G.P., Sastri, C., Hunzinger, C., Stegmann, W., Schrattenholz, A., Buchner, A., Gangnus, R., Burgemeister, R. and Cahill, M.A. (2005) Differential radioactive proteomic analysis of microdissected renal cell carcinoma tissue by 54 cm isoelectric focusing in serial immobilized pH gradient gels. *J. Proteome Res.*, **4** (6), 2117–2125.

52 Burnet, P.W.J., Eastwood, S.L. and Harrison, P.J. (2004) Laser-assisted microdissection: Methods for the molecular analysis of psychiatric disorders at a cellular resolution. *Biol. Psychiatry*, **55**, 107–111.

53 Schad, M., Mungur, R., Fiehn, O. and Kehr, J. (2005) Metabolic profiling of laser microdissected vascular bundles of *Arabidopsis thaliana*. *Plant Methods*, **1**, 2.

54 Niyaz, Y. and Sägmüller, B. (2005) Non-contact microdissection and pressure catapulting: Automation via object-oriented image processing. *Medical Laser Application*, **20**, 223–232.

55 Hutchens, T.W. and Yip, T.T. (1993) New desorption strategies for the mass spectrometric analysis of macromolecules. *Rapid Commun. Mass Spectrom.*, **7**, 576–580.

56 Wright, G.L., Cazares, L.H., Leung, S.M., Nasim, S., Adam, B.L., Yip, T.T., Schellhammer, P.F., Gong, L. and Vlahou, A. (1999) ProteinChip® surface enhanced laser desorption/ionization (SELDI) mass spectrometry: a novel protein biochip technology for detection of prostate cancer biomarkers in complex protein mixtures. *Prostate Cancer Protstatic Dis.*, **2**, 264–276.

57 von Eggeling, F., Davies, H., Lomas, L., Fiedler, W., Junker, K., Claussen, U. and Ernst, G. (2000) Tissue-specific microdissection coupled with ProteinChip® array technologies: Applications in cancer research. *Biotechniques*, **29**, 1066–1070.

58 von Eggeling, F., Junker, K., Fiedler, W., Wollscheid, V., Durst, M., Claussen, U. and Ernst, G. (2001) Mass spectrometry meets chip technology: A new proteomic tool in cancer research? *Electrophoresis*, **22**, 2898–2902.

59 Melle, C., Ernst, G., Schimmel, B., Bleul, A., Koscielny, S., Wiesner, A., Bogumil, R., Moller, U., Osterloh, D., Halbhuber, K.J. and von Eggeling, F. (2003) Biomarker discovery and identification in laser microdissected head and neck squamous cell carcinoma with ProteinChip® technology two-dimensional gel electrophoresis, tandem mass spectrometry, and immunohistochemistry. *Mol. Cell Proteomics*, **2**, 443–452.

60 Forastiere, A., Koch, W., Trotti, A. and Sidransky, D. (2001) Medical progress – Head and neck cancer. *N. Engl. J. Med.*, **345**, 1890–1900.

61 Melle, C., Ernst, G., Schimmel, B., Bleul, A., Koscielny, S., Wiesner, A., Bogumil, R., Moller, U., Osterloh, D., Halbhuber, K.J. and von Eggeling, F. (2004) A technical triade for proteomic identification and characterization of cancer biomarkers. *Cancer Res.*, **64**, 4099–4104.

62 Melle, C., Kaufmann, R., Hommann, M., Bleul, A., Driesch, D., Ernst, G. and von Eggeling, F. (2004) Proteomic profiling in microdissected hepatocellular carcinoma tissue using ProteinChip technology. *Int. J. Oncol.*, **24**, 885–891.

63 Melle, C., Ernst, G., Schimmel, B., Bleul, A., Thieme, H., Kaufmann, R., Mothes, H., Settmacher, U., Claussen, U., Halbhuber, K.J. and von Eggeling, F. (2005) Discovery and identification of alpha-defensins as low abundant tumor-derived serum markers in colorectal cancer. *Gastroenterology*, **129**, 66–73.

64 Melle, C., Osterloh, D., Ernst, G., Schimmel, B., Bleul, A. and von Eggeling, F. (2005) Identification of proteins from colorectal cancer tissue by two-dimensional gel electrophoresis and SELDI mass spectrometry. *Int. J. Mol. Med.*, **16**, 11–17.

65 Melle, C., Ernst, G., Schimmel, B., Bleul, A., Kaufmann, R., Hommann, M., Richter, K.K., Daffner, W., Settmacher, U., Claussen, U. and von Eggeling, F. (2005) Characterization of pepsinogen C as a potential biomarker for gastric cancer using a histo-proteomic approach. *J. Proteome Res.*, **4**, 1799–1804.

66 Melle, C., Ernst, G., Schimmel, B., Bleul, A., Mothes, H., Kaufmann, R., Settmacher, U. and von Eggeling, F. (2006) Different expression of calgizzarin (S100A11) in normal colonic epithelium adenoma and colorectal carcinoma. *Int. J. Oncol.*, **28**, 195–200.

67 Melle, C., Bogumil, R., Ernst, G., Schimmel, B., Bleul, A. and von Eggeling, F. (2006) Detection and identification of heat shock protein 10 as a biomarker in colorectal cancer by protein profiling. *Proteomics*, **6**, 2600–2608.

68 Cazares, L.H., Adam, B.L., Ward, M.D., Nasim, S., Schellhammer, P.F., Semmes, O.J. and Wright, G.L. (2002) Normal benign, preneoplastic, and malignant prostate cells have distinct protein expression profiles resolved by surface enhanced laser desorption/ionization mass spectrometry. *Clin. Cancer Res.*, **8**, 2541–2552.

69 Steen H. and Mann M. (2004) The ABCs (and XYZs) of peptide sequencing. *Nat. Rev.: Mol. Cell Biol.*, **5**, 699–711.

70 Böhm, M., Wieland, I., Schütze, K. and Rübben, H. (1997) MicroBeam MOMeNT non-contact laser microdissection of membrane-mounted native tissue. *Am. J. Pathol.*, **151**, 63–67.

6.4
Sample Preparation for Flow Cytometry

Derek C. Davies

6.4.1
Introduction

Proteomics enables cells to be profiled on the basis of their protein expression in terms of their population. However, there are times when although a heterogeneous cell population is under consideration, only a subset of these cells is of interest. This is where cell sorting – or separation of the population of interest – is vital. Two main approaches have been used to sort cells: (i) by using magnetic beads and a separation column; or (ii) by using flow cytometry.

Magnetic bead separation is useful when a large number ($>10^8$) of cells is required or when the starting population is large [71]. It is also less expensive and technically easier to perform than flow cytometry. However, cell sorting using a flow cytometer has several main advantages:

- more than one population (up to four populations is possible) may be sorted and retrieved;
- the cell purity should be in excess of 98%;
- it is possible to sort on the level of fluorescence; that is, bright versus dim populations;
- it is possible to sort on a range of fluorescent protein expression as well as on cellular functionality or nucleic acid content.

Flow cytometry may be defined as a means of measuring multiple parameters from cells or particles as they flow, one by one, through a laser beam and a sensing point. In many ways a flow cytometer is simply a large and very sophisticated fluorescence microscope. The way in which cells interact with light at the sensing point indicates many things, as the signal obtained is dependent on factors such as cell size, cell granularity, refractive index, and nuclear to cytoplasmic ratio. The excitation of fluorochromes as they pass through a focused laser beam (or beams) can provide information about the presence and relative amount of fluorescence. Hence, any portion of the cell that can be tagged with a fluorochrome may be measured. The portion may be an antigen at the surface of the cell, in the cytoplasm or in the nucleus; it may be the DNA or RNA content of the cell; or it may be a functional aspect of the cell, such as apoptotic status or calcium ion flux. Moreover, several fluorochromes per cell may be detected using optical filters to detect those wavelengths of light that are specific for certain fluorochromes. Modern flow cytometers are capable of measuring up to 17 fluorochromes per cell, and with modification more are possible [72]. The key point of flow cytometry – and perhaps its most powerful aspect – is that the measurements are made on a cell-by-cell basis. Hence, a clear prerequisite for flow cytometric analysis or cell sorting is that the sample must be in single-cell suspension.

Although flow cytometric analysis provides an excellent means of answering a wide variety of biological questions, there are occasions when it is necessary to physically isolate and purify the cells of interest for further investigation. The use of cells in proteomics analysis represents just such an occasion.

Flow sorting may be defined as the process of physically separating the particles of interest from other particles in the sample. Sorters may be either mechanical or electrostatic (stream in air) in nature:

- Mechanical cell sorters operate by physically deflecting the cells of interest in an enclosed fluid-based system. This may be useful when dealing with biohazardous material as no aerosols are produced. However, mechanical sorters are slow, they produce dilute samples after sorting, and they can sort only a single defined population at any one time.
- Electrostatic cell sorters function by deflecting droplets electrostatically, whereby the cells, which are enclosed in a column of sheath fluid, are ejected into the air. This fluid column is broken into droplets by the vibration of a piezo-electric

crystal, in a similar manner to the action of an ink-jet printer [73]. The frequency and amplitude of this vibration are controlled so that the break-off occurs in a predictable and adjustable way. Sorting then depends on ensuring that a cell of interest is at the break-off point at a given time after the analysis. At this point, the stream is charged such that drops containing wanted cells carry either a positive or negative charge and are deflected into a collection vessel after passing through a high-voltage electrical field. In this way a highly purified population of cells is rapidly recovered, and the unwanted cells are diverted to waste.

It is beyond the scope of this chapter to discuss the minutiae of the mechanisms of action of sorters, although many texts are available on the subject [74,75]. At this point it must be stressed that the success of a sort depends almost entirely on the quality of the input sample, and this may be optimized in several ways. Thus, the key to successful cell sorting is careful preparation of the sample.

6.4.2
Sample Preparation for Flow Cytometry

The preparation of samples for flow cytometry will vary depending on the cell source and the requirements for each cell type. In all cases, the ultimate aim at the end of processing is to obtain a preparation that has maximum yield and viability. Whilst the relationship between these states is complex, and dependent on cell type, no single isolation technique will be applicable to all cell types. Therefore, a certain degree of procedure optimization will always be required. The principal requirement – which is common to all cell types – is that, on completion of processing for flow sorting, the sample must be in monodisperse suspension. However, in order to achieve this a number of points must be considered:

- What is the source of the cells to be sorted?
- How are the relevant cells identified?
- How can the sample be best kept prior to, and during, sorting?
- How are the cells to be kept after sorting but prior to further analysis?

6.4.2.1 Preparation from Cells in Suspension
If the sample under investigation is a cell line that grows in suspension (e.g., HL60, Jurkat) or is found naturally in suspension (e.g., peripheral blood), the processing is relatively straightforward, although care must be taken not to damage or lose cells during the procedure. For suspension cells, the basic processing involves taking a sample of the cell culture, centrifuging it to remove the medium, and then resuspending the cells in phosphate-buffered saline (PBS; see Section 6.4.8.1) before staining to identify the cells of interest.

6.4.2.2 Preparation from Adherent Cells

For adherent cells lines and solid tissues, the sample preparation can be more involved and precise. In general, the usual methods used for subculturing adherent cell lines are sufficient to produce a suspension of cells for flow cytometry. Cells should be treated with either an enzyme such as trypsin (0.25%), a chelating agent such as EDTA (0.2%), or a combination of the two (0.25% trypsin + 0.2% EDTA in PBS). On occasion it is necessary to avoid trypsin as some antigens are stripped from the surface of the cell by the action of the enzyme. Alternative enzymes that may be used include collagenase (0.1%), dispase (0.1%), or commercial preparations such as Accutase (Innovative Cell Technologies, San Diego, CA, USA). It should be noted that some enzymes used for cell dissociation can undergo autolysis; hence, solutions should either be prepared immediately before use or stored at $-20\,°C$. If the antigen under investigation is sensitive to both enzymatic and chelating methods, then scraping the cells from the surface with a rubber policeman may be sufficient, although this can lead to an increased number of cell clumps. Such clumps may be broken by pipetting and/or syringing through a 21-gauge needle, or by passing the sample though a stainless steel or nylon mesh. Meshes of various pore size are available, the most suitable being 30 µm when small cells are used, and 70 µm when larger or adherent cells are used. Unfortunately, scraping and filtering can lead to high levels of cell loss, in addition to a possible loss of specific cells, and so should be avoided if at all possible.

6.4.2.3 Preparation from Solid Tissue

For tissues, gentle mechanical or enzymatic disaggregation will be needed, although no single technique will be applicable to all tissues. Spleen or bone marrow tissues can be minced, teased apart using a scalpel, pushed through a mesh (stainless steel or nylon), passed through a 25-gauge needle, or rubbed together between frosted slides. A final disaggregation can be achieved by vigorous pipetting, or by using EDTA with or without trypsin or collagenase. Solid tissues may also be treated with enzymatic agents such as trypsin, papain, collagenase, dispase, hyaluronidase or elastase.

It is also possible to retrieve nuclei from paraffin-embedded tissue. For this, thick (40–50 µm) sections are used, as thinner sections will lead to excessive cell debris. Following dewaxing, the tissue is digested with either pepsin [76] or collagenase/dispase [77] for a variable amount of time (this would be determined empirically, by microscopical observation), washed and stained. As this method will produce nuclei rather than whole cells, it only useful when DNA or nuclear proteins are to be investigated [78].

With all of these methods for adherent cells or solid tissue, it is important to control the medium in which the cells are suspended, the time of digestion, the temperature and pH of the dissociation solution, the enzyme concentration, and the way in which the enzyme is inhibited. However, for a particular tissue type these biological approaches are more easily standardized than for mechanical digestion.

6.4.2.4 General Considerations

With all methods of cell preparation it is important to avoid cell loss, and this is especially important when the starting cell numbers are low. The pipetting of supernatants rather than decanting, and ensuring that the centrifuge brakes are not active, will all help in this respect. Although clumps may be removed (as described above) by meshing, this can lead excessive cell loss. On completion of processing, the cells in all samples should be counted and, if possible, this should be performed objectively using an automated system, for example the Vi-Cell (Beckman Coulter, Miami, USA) rather than with a hemocytometer, which may be rather objective. It is also useful to check the preparation microscopically to ensure that it is free from excessive clumps and/or debris and, if unfixed, that the cell viability is good. At this stage it is possible to use Ficoll or a similar density separation reagent to clean up the preparations of debris or red cells if applicable. In addition – and most importantly – all of the above procedure should be carried out under aseptic conditions.

6.4.3
Identification of Relevant Cells

Staining for antigenic determinants, functional state or DNA content should only be performed after processing is complete and a single cell suspension has been produced. If antibodies are used to identify the relevant cell population, it is necessary to consider whether these might cause any downstream effects that would affect post-sorting processing; ideally, the antibodies should neither alter nor modulate cell function or behavior. Pilot experiments will be necessary to establish the best temperature and optimal time for staining, and the optimal antibody concentrations. It is important to realize that, when preparing large numbers of cells for sorting, the amount of antibody used must be increased. With today's sophisticated cell sorters it is also possible to use multiple fluorochromes: between six and eight is now common, and up to 17 is possible. However, in such large experiments the inclusion of controls become extremely important in order to be able specifically to define the sorted population. Digital signal processing, which has recently been introduced in many cytometers, has led to problems in data displays [79]. However, the wealth of information obtained in a multicolor experiment can offset these problems. The relevance of good staining is that it enables the cytometer operator to identify, specifically, the cell population that is to be sorted.

6.4.4
Cell Sorting

6.4.4.1 Cells and Samples
When a preparation of single cells has been obtained, it is important that they remain dispersed. *Cell clumping* during sorting can cause problems by clogging the sample line or the nozzle of the cytometer, and it may also artificially skew results by the selective sequestration of cells. Samples may be filtered at this stage, using nylon mesh as described previously. It is possible to reduce clumping by using a calcium- and magnesium-free buffer while the cells are being sorted. The cell concentration is also a consideration; the optimal concentration will depend on the cytometer itself and will

vary with the sheath pressure and nozzle size used. In general, small cells can be run at a higher concentration than larger cells; and cells may be sorted faster on a high-pressure cell sorter. Keeping both the sample and the collection tubes cool (4 °C) can also prevent clumping. Although the principle of hydrodynamic focusing within the cytometer will help in measuring cells individually, this is not perfect and some cell doublets will always be present, despite the precautions taken. This may have a deleterious effect on sorting purity if, for example, a positive cell of interest is measured at the same time as a negative; both would be sorted as a positive event, but there would be a concomitant reduction in sort purity. Many cell sorters have the ability to distinguish clumps and doublets based on the time taken to pass through the laser beam, and this facility should always be used during a cell sort, if available.

Dead cells present in the sample can increase sorting time, alter scatter profiles, take up antibodies non-specifically, and also show an increased autofluorescence. Although these cells may not be overly important if the cells are to be re-grown, it would be important to exclude them if specific numbers of cells were needed for functional studies, or if the cells were to be used to analyze protein expression or mRNA content.

A variety of dyes can be used specifically to exclude dead cells. As cells die, their membranes become permeable to DNA-binding dyes that are normally excluded due to the action of membrane pumps. The dye used will depend on the cytometer, the light sources available and any other fluorochromes being used, but the most widely used examples include DAPI (4′,6-diamidino-2-phenylindole), propidium iodide, 7-aminoactinomycin D, and TO-PRO-3.

Dead cells can also cause other problems, in that DNA from lysed cells is very "sticky" and may cause cell clumping. In this case, the addition of DNase I (20–100 mg mL^{-1} DNase with 5 mM $MgCl_2$) will aid the production of a single-cell suspension by digesting any free DNA.

It is important to maintain an *optimal pH* for the cells being sorted. This is particularly important in high-pressure sorters, as increased pressure can lead to a reduction in pH if the cells are not adequately buffered. In these cases, the addition of 25 mM HEPES will help to keep the cells in optimal condition.

6.4.4.2 Cytometer Considerations

Although sample preparation is important, there are also other considerations that lead to a successful sort. It is important to know at what percentage the cells of interest are present in the whole population, and how many cells are required for any subsequent experiment. These values will provide a good indication of the length of time that a sort will take, and the number of cells that need to be in the starting population.

There are several practical questions for which the sorter operator will need to know the answers:

- Is the sort aseptic?
- How many (and which) fluorochromes have been used?
- What is the size (diameter) of the cells to be sorted?

When these factors and known and the sample is running, it is a relatively simple matter to exclude dead cells (by DNA-binding dye uptake), to exclude debris by its light-scattering characteristics, to exclude cell doublets as far as possible by their time of flight through the laser beam, and finally to include the relevant cell population on the basis of its fluorescence characteristics. At this point it is worth mentioning that *thresholding* is a more important consideration that it is in analytical cytometry. Most cytometers use an electronic threshold – a hurdle above which an event must pass before it is analyzed. Cells that do not reach this threshold are in effect "invisible" to the machine and may reach the sorted tube. This may be especially problematic if the cells are being used for the PCR amplification of DNA or RNA. Hence, it is advisable to set the minimum threshold possible before commencing a sort.

The user will have to make a judgment as to the purity of the required sort, and this will influence the sort mode used. Most modern cell sorters provide the operator with the ability to vary the sort mode, depending on the wishes of the user. The default mode is to sort as pure a population as possible, but sometimes a precious population is needed and in these cases it is possible to override the coincidence abort circuit so that all the desired cells are sorted (enrichment sort). It is also possible to sort very stringently only a desired number of cells; this is useful in cell cloning, and also when given numbers of a specific population need to be compared. The cells should be collected into a medium appropriate to their fate; this may be a culture medium, a fixative, or an RNA preservative reagent such as Trizol. At the end of the sort, if possible, it is advisable to re-analyze a portion of the sorted cells to assess the purity (percentage of cells that fulfill the sort criteria) and recovery (number of cells in the sorted tube) of the sort, and how efficient it has been in terms of cell yield. Cell viability should also be checked pre- and post-sort and, if possible, functionality should also be assessed. This can be particularly important if a high-pressure sort has been performed. When a sample has been cell-sorted, downstream experiments will still require that the recovered cells are processed efficiently and effectively in order to avoid cell loss and cell death.

6.4.5
Application Example

Figure 6.19 illustrates an example of a two-color sort. A flow sorter is able to isolate defined cell populations with a very high degree of purity, with greater than 98% generally being possible. The yield of a sample will depend on the percentage of the cells of interest, and also to a large extent on the sample preparation. The example in Figure 6.19 shows a good preparation where there is minimal debris (panel A), few doublets (panel B) and very little cell death (panel C). The definition of two subsets of cells (panel D) allows each to be sorted; the results of the sort are shown in panels E and F. Very small ($<0.05\%$) populations can, in this way, be identified and isolated by flow cytometry.

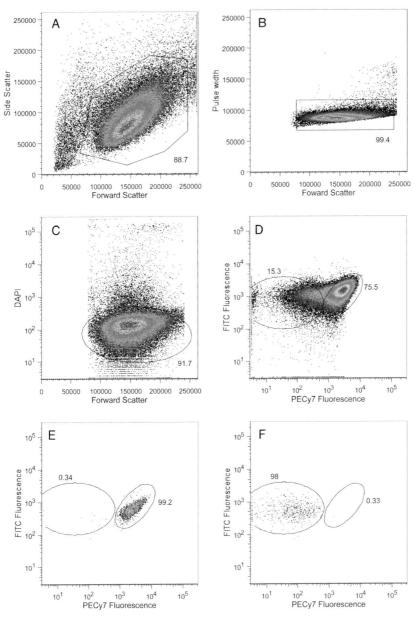

Figure 6.19 Murine bone marrow cells have been prepared as in Section 6.4.8.3 and stained with antigens against two surface antigens. One is labeled with fluorescein isothiocyanate (FITC) and one with PECy7 (phycoerythrin-cyanine 7 tandem conjugate). (A) The sample is initially gated on the scatter characteristics of the cells. (B) Cells that fall within this gate are then gated to include only single cells, excluding events that take longer to traverse the laser beam. (C) Cells are then gated on the basis of their DAPI-negativity. (D) Only cells that fulfill all three gating criteria are examined. Gates are set to sort the single FITC population and the double-positive population. (E, F) The results of the sort. The percentage of cells falling within the sort gates is indicated. The sort was performed at 10 000 cells per second on a FACS Aria (Becton Dickinson, San Jose, USA).

6.4.6
Summary

Flow cytometry is a particularly efficient method of collecting large numbers of cells of interest from a heterogeneous population. Successful cell sorting depends absolutely on the quality of the input sample, and care should be taken to optimize each stage of the preparative techniques. By following established protocols it is possible to produce single-cell suspensions of good viability and functionality, which is desirable in any post-sorting application.

6.4.7
Perspectives

What impact has proteomics had on the sorting field? The number of cells needed and the desire for these to be the relevant population has increased the need for high pressure, and therefore high-speed, sorters. Although cell sorting by flow cytometry is based on relatively simple operating procedures, the operation of such sorters is highly skilled and experience-based. In addition, cell sorters are relatively expensive pieces of equipment, but as they can be used for time-consuming sorts of large cell numbers as well as for more rapid high-throughput screening applications, the integration of this technology into core facilities or shared resources is now also more common and should be encouraged.

6.4.8
Recipes for Beginners

6.4.8.1 Cultured Suspension Cells

- Using cultured suspension cells or primary blood cells, place the cells directly from a culture flask into 50-mL conical tubes and centrifuge at $400 \times g$ for 5 min at $4\,°C$.
- Discard the supernatant and resuspend the pellet in medium; this may be culture medium or PBS supplemented with 1% bovine serum albumin. PBS is made as follows: 8 g NaCl; 0.5 g KCl; 1.43 g Na_2HPO_4; 0.25 g KH_2PO_4; dissolve in 1 L of distilled H_2O. Adjust pH to 7.2 and autoclave for 20 min. When calcium and magnesium are required in the buffer, add 133.4 mg $CaCl_2$ and 100 mg $MgCl_2$ per liter, prior to autoclaving.
- Re-centrifuge at $400 \times g$ ($4\,°C$) and discard the supernatant. Count cells and resuspend at an appropriate concentration (which will vary with sorter used).

6.4.8.2 Adherent Cells

- Remove culture medium by suction or by decanting. Harvest cells by using trypsin (0.25%; stock solution should be prepared by dissolving the enzyme in 0.001 N

HCl before diluting with medium or PBS; aliquots should be stored at $-20\,°C$) or EDTA (0.2%). A combination of trypsin and EDTA can also be used. Add 4 mL trypsin to 16 mL EDTA; this solution can be kept for 1 week at $4\,°C$, but should be warmed to $37\,°C$ before use.
- Monitor the monolayer microscopically at intervals, and shake the flask. This can be done at room temperature, but the flasks may be warmed to $37\,°C$ if dissociation is slow.
- When cells are detached, add medium with 10% serum to neutralize the enzyme.
- Check cell viability.
- Transfer cells to 50-mL conical tubes, centrifuge at $400\times g$ for 5 min at $4\,°C$.
- Discard supernatant, count cells, and resuspend in PBS. Once the cells are in suspension the antigen staining or addition of other fluorescent probes may be performed.

6.4.8.3 Solid Tissue

- Place tissue in a sterile Petri dish and add 20 mL trypsin-EDTA solution.
- Tease tissue apart using needle and scalpel, or alternatively use an automated cell preparation device such as a MediMachine (Dako, Ely, UK).
- Decant cells into tube, centrifuge at $400\times g$ for 5 min at $4\,°C$, as above.
- Leave in enzyme for 15 min, but check for disaggregation.
- Add medium containing 10% fetal calf serum to neutralize enzyme.
- Check cell viability.
- Transfer cells to 50-mL conical tubes, centrifuge at $400\times g$ for 5 min at $4\,°C$.
- Discard supernatant, count cells, and resuspend in PBS at an appropriate concentration.

References

71 Miltenyi, S., Muller, W., Weichel, W. and Radbruch, A. (1990) High gradient magnetic cell separation with MACS. *Cytometry*, **11**, 231–238.

72 Perfetto, S.P., Chattopadhyay, P.K. and Roederer, M. (1994) Seventeen-colour flow cytometry: unravelling the immune system. *Nat. Rev. Immunol.*, **4**, 648–655.

73 Herzenberg, L.A., Sweet, R.G. and Herzenberg, L.A. (1976) Fluorescence-activated cell sorting. *Sci. Am.*, **234**, 108–117.

74 Shapiro, H.M. (2003) *Practical Flow Cytometry*, Wiley Liss, New York.

75 Ormerod, M.G. (2000) *Flow Cytometry*, Oxford University Press, Oxford.

76 Hedley, D.W., Friedlander, M.L. and Taylor, I.W. (1985) Application of DNA flow cytometry to paraffin-embedded archival material for the study of aneuploidy and its clinical significance. *Cytometry*, **6**, 327–333.

77 Corver, W.E., ter Haar, N.T., Dreef, E.J., Miranda, N.F.C.C., Prins, F.A., Jordanova, E.S., Cornelisse, C.J. and Fleuren, G.J. (2005) High-resolution multi-parameter DNA flow cytometry enables detection of tumour and stromal cell subpopulations in paraffin-embedded tissues. *J. Pathol.*, **206**, 233–241.

78 Crockett, D.K., Lin, Z., Vaughn, C.P., Lim, M.S. and Elenitoba-Johnson, K.S. (2005) Identification of proteins from formalin-fixed paraffin-embedded cells by LC-MS/MS. *Lab. Invest.*, **85**, 1405–1415.

79 Herzenberg, L.A., Tung, J., Moore, W.A., Herzenberg, L.A. and Parks, D.R. (2006) Interpreting flow cytometry data: a guide for the perplexed. *Nat. Immunol.*, **7**, 681–865.

7
Chromatography

7.1
Sample Preparation for HPLC-Based Proteome Analysis

Egidijus Machtejevas and Klaus K. Unger

7.1.1
Introduction

In general, high-performance liquid chromatography (HPLC) analysis sample preparation is considered as the most essential step having a direct impact on the analytical results. Sample preparation, including direct injection techniques, serves the following purposes:

- it separates the analyte from other undesired components to enhance resolution,
- it removes interferences that affect the accuracy of quantitation,
- it concentrates the analyte of interest to improve sensitivity.

Numerous reviews have been produced on the state of the art in this field [1–3], and the general trends in the analysis of proteome samples have also been reviewed [4].

Sample preparation in proteomics applying multidimensional liquid chromatography/mass spectrometry (LC/MS) has changed its paradigm significantly as compared to classical single-column HPLC analysis. In proteomics, MS and bioinformatics are the core technologies which drive the field. Thus, LC – even when performed as a multidimensional tool – is often correctly coined as the sample clean-up technology for MS. In this sense, all separation techniques could be considered as sample preparation techniques for MS. The intermediate purification of peptide mixtures is achieved by using chromatographic techniques such as solid-phase extraction (SPE), multidimensional protein identification technology (MudPIT), multidimensional HPLC, and two-dimensional (2-D) electrophoresis.

Protein purification is a critical step in the process. For example, in human blood serum, some 90% of the protein content of serum is composed of 10 basic proteins.

Proteomics Sample Preparation. Edited by Jörg von Hagen
Copyright © 2008 WILEY-VCH Verlag GmbH & Co. KGaA, Weinheim
ISBN: 978-3-527-31796-7

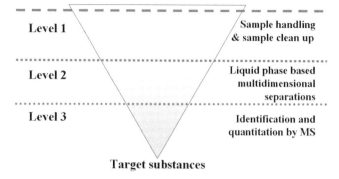

Figure 7.1 Selective filters in liquid chromatography. Each level could be reached employing different techniques. For example: Level 1, solid-phase extraction, size-exclusion chromatography (SEC), isoelectric focusing or isotachophoresis; Level 2, SEC, reverse-phase chromatography, hydrophobic interaction chromatography, hydrophilic interaction chromatography, immobilized metal-affinity chromatography, or affinity chromatography; Level 3, mass spectrometry (MS), tandem MS/MS.

The remaining 10% of serum consists of trace amounts of millions of different proteins. Thus, the partial purification of proteins is necessary so that proteins in trace amounts can be identified and their exact structural analysis performed. The situation is best demonstrated by the magic triangle shown in Figure 7.1. In order to analyze low-abundant proteins, it is necessary first to remove the highly abundant proteins effectively and then to apply a sequence of selective filters (LS separation modes) in order to resolve the analytes of interest down to a miniaturized platform to meet the MS boundary conditions. Today, it is generally accepted knowledge that the removal of highly abundant proteins such as albumin consecutively leads to a loss of low-abundant species. Furthermore, the application of several orthogonal LC separation systems also bears the danger of severe sample losses due to adsorption on the separation and capture column and sample transfer. Sample preparation for LC is a complex operation that often causes the greatest variability in analytical results. Preferably, sample manipulations should be kept at a minimum and as simple as possible. The logical consequence of this situation is to reduce significantly the number of dimensions at the multidimensional LC system, and to move the selective filters to the front – that is, to include them already in the sample clean-up process. The fact that a significant amount of information is lost as a result of analyte losses during sample clean-up has led to the development of several novel approaches, and these will be briefly described in this chapter.

7.1.2
Problems Related to Direct Sample Injection in HPLC

Although chromatographic separation techniques are well suited to the analysis of complex multi-component samples, the analysis of proteomes requires well-designed sample preparation procedures. During the early years of chromatographic studies, attempts were made to inject biological samples directly onto the HPLC column [5], but it was quickly realized that this let to a rapid loss in column performance, the chromatographic column selectivity was altered, and the column back-pressure increased because of the irreversible adsorption of matrix compounds. Thus, a useful means of sample preparation is required to reduce sample complexity and to remove components which would tend to bind irreversibly.

The direct injection of plasma or serum samples is especially detrimental to chromatographic columns (both, reversed- and normal-phase) which utilize mobile phases containing 15% or more organic solvent [6]. The problem here is that these samples contain large amounts of proteins, which are precipitated and or denatured and subsequently adsorbed onto the packing material; this in turn leads to back-pressure build-up, changes in retention time, decreased column efficiency, and reduced column capacity [7,8]. The decreased efficiency most likely results from the denatured protein inhibiting diffusion mass transport of the analyte to the packing material surface [9]. It should be noted that a significant precipitation of proteins leading to very large increases in back-pressure after only a few injections, requires at least 70% methanol content in the mobile phase. A study in which the precipitation of serum proteins was measured using light-scattering identified precipitation cut-offs of 25% for acetonitrile, 20% for 2-propanol, and 10% for tetrahydrofuran, with the remainder of the mobile phase being 0.1 M phosphate buffer, pH 6.8 [10]. It is important also to note that the temperature and/or time plays an important role in protein precipitation. Therefore, precision in sample handling time and temperatures should be monitored. Another concern is the mobile-phase pH, as precipitation will occur in at the protein's pI. Although all reversed-phase packing materials are susceptible to the adverse affects of proteins, in particular wide-pore supports (which have sufficiently large pores to allow the passage of protein molecules) show augmented adverse effects [11].

7.1.3
Trial and Error Selection of the Sample Preparation Method

When performing sample clean-up, the situation outlined by the paradigm "if you don't know where you are going, any road will do" should be avoided. It is most likely that proteomics will never offer a "cook-book" approach, for the reasons discussed below.

A proteomic analysis of a sample usually consists of four steps: (1) the extraction of proteins from the sample; (2) their separation; (3) their detection; and finally (4) an identification/analysis of the individual separated peptides. It is of major importance to pay particular attention to the sample extraction, as any error or losses that occur during this stage will have a major influence on the results.

Proteins are either located in different cell compartments (cytoplasma, a range of intracellular organelles) or they are secreted as extracellular proteins in various body is essential to remove nonproteinaceous solid materials and ions prior to the separation of urinary proteins in order to obtain high resolution. Consequently, several urine preconcentration methods have been reported, among the most common being dialysis and lyophilization [14], filtration [15], ultracentrifugation, and precipitation [16]. However, protein losses were much higher after ultrafiltration than after dialysis–lyophilization as compared to precipitation.

In the present authors' study, thawed serum samples were inspected visually and then centrifuged for 10 min at least at 2000 rpm at room temperature if particulate matter was observed. Centrifugation proved to be better than filtration as there was minimal sample loss and lower costs (as disposable supplies were not used). Filtration through a 0.22-μm membrane filter will eliminate bacteria, thus increasing sample stability and preventing accidental pollution by proteomes from microorganisms. However, the correct filter must be used as both the filter holder and membrane may have adsorption capabilities with certain capacity towards some analyte molecules. As adsorption may vary depending on the molecules present, the relative abundances of the analyte components may be shifted if filtration is conducted without due care. Although this effect cannot be fully eliminated, the most acceptable filters can be selected following a degree of trial and error. In general, borosilicate- and cellulose-containing filters should be avoided as they possess high protein-binding properties; polysulfonate and Durapore (Millipore, USA) membranes are useful for many sample clean-up applications. From a practical standpoint, it is advised that the first few drops (ca. 0.5 mL) of the filtered solution is directed to waste, as these will face the highest protein losses, even at minimal adsorption. The sample volume should match the filter size in order to maintain surface adsorption capacity at a minimum.

7.1.5
Specific Approaches Applied to Sample Clean-Up in Proteomics

7.1.5.1 Miniaturized Extraction Techniques
Miniaturized solid-phase extraction devices were developed recently by Millipore [17]. The microtips are packed with a small amount (0.1 mg) of C18 reversed-phase sorbent or of other sorbents. The volume of the packed bed is about 0.5 mL, which allows the extracted analyte to be eluted in a small volume of mobile phase (1–5 μL). This volume range is compatible with MALDI-TOF-MS sample volume requirements. The Millipore ZipTip is designed to be compatible with laboratory pipettes and auto pipettors. C18 ZipTips are used primarily for biopolymer desalting for MALDI-TOF-MS. The purified biopolymer can be eluted directly onto a MALDI target. In the near future, packed tips with affinity-like interactions may become a powerful microextraction tool for routine biomarker level monitoring.

7.1.5.2 Most Abundant Component Depletion
Most biofluids contain large amounts of well-known proteins such as albumin and immunoglobulins (IgGs), which can overwhelm the separation system and make the detection of low-abundant proteins and peptides very difficult. It is thus advantageous to remove these proteins prior to digestion and separation. In addition

to the previously described approaches, which are based on size-exclusion fractionation, a number of alternative approaches exist for reducing the overall protein load by the specific adsorption of albumin and IgG onto affinity matrices [18–22]. Whilst, normally, an affinity matrix is highly specific, in high-content samples the affinity ligand may show only limited specificity. Degrees of specificity exist between highly selective immunoaffinity matrices and less-selective, but more robust, affinity supports using synthetic ligands. In an effort to reduce the amount of albumin obtained from human serum, several affinity matrices have been evaluated based on antibodies or dye ligands. For example, antibody-mediated albumin removal was efficient and selective, while dye–ligand chromatography (a technique used extensively in protein chromatography) was surprisingly effective [23], especially with regards to high binding capacities and long column lifetime, though at the expense of selectivity. The main problem related to such depletion is the nature of the proteins and peptides, which often form complexes. As the most abundant species are adsorbed as complexes, many interesting and desired components may be lost, or their concentrations reduced with different ratios. This effect is also heavily dependent on the experimental conditions employed (buffer, pH, temperature, etc.). It should also be borne in mind that this step may alter the initial sample composition drastically, and not only towards the undesired molecules (E. Machtejevas, unpublished results).

7.1.5.3 Affinity-Enrichment Approaches

Often, certain classes of target molecules may be present in very small amounts and must be selectively isolated or enriched before identification. Affinity chromatography, which selectively retains proteins or peptides based on biospecific interactions, has been employed in several multidimensional separation methods for the selective trapping of proteins of interest (see Figure 7.2d). Although, in general an affinity matrix is highly specific, in high-content samples the affinity ligand may exhibit limited specificity; for example, phosphorylated, glycosylated or derivatized amino acids are often targeted by affinity chromatography in the first dimension. Immobilized metal-affinity chromatography (IMAC) has been reported as an online combination of IMAC-Fe^{3+} and reversed-phase chromatography for identifying phosphorylated peptides from 2-D peptide maps or synthetic peptide mixtures [24,25].

Recently, the group of Heck reported a novel analytical procedure for phosphopeptide enrichment employing a developed nanoflow 2D-LC-MS/MS set-up for the characterization of both non-phosphorylated and phosphorylated peptides in two separate measurements [26]. The procedure relies on the unique ion-exchange properties of Titanosphere, a new type of column material that consists of spherical particles of titanium oxide. In comparison with IMAC, this chromatographic approach has the advantage that fewer column handling steps are required, and therefore, it seems to be a more robust enrichment procedure for the selective enrichment and characterization of phosphopeptides from complex mixtures.

One of the most published multidimensional separation techniques that employ an affinity chromatography step is based on the use of isotope-coded affinity tags (ICAT) [27]. In this strategy, cysteinyl residues within proteins are modified with a thiol-reactive reagent that contains a biotin moiety. The proteins are enzymatically digested and the modified peptides recovered using immobilized avidin chromatography. The main purpose of affinity chromatography isolating only the cysteinyl-containing peptides is to reduce the complexity of the sample; however, as the ICAT-based strategy is designed for global proteomic studies, the post-affinity chromatography sample is still quite complex.

Surface capture technologies recently have been merged with MS and named surface-enhanced laser desorption/ionization mass spectrometry (SELDI-MS). While this strategy is not dependent upon, or constrained to, any particular technology platform, it has been pioneered and particularly advocated by Petricoin, Liotta and Chan at John Hopkins Hospital, together with scientists from Ciphergen Biosystems (Fremont, California), all using some variation of Ciphergen's SELDI-MS approach. SELDI is a variant of MALDI in which the metal target surface is coated and metal affinity are immobilized. In the Ciphergen instrument, these targets (or "chips" in their nomenclature) have eight or 16 discrete target spots. These surfaces are intended selectively to retain a subset of proteins from complex biomaterials such as serum or cell lysate [28,29]. Recently, Liotta et al. have developed an interesting approach to harvest endogenous peptides in blood. This group are applying porous polymeric nanoparticles with a porous shell and affinity ligands inside. The porous shell act as a size-exclusion barrier, and the peptides are captured inside the porous nanoparticles and enriched. After completion of harvest the captured peptides are then released either by a temperature switch or by electro-elution [30].

The protein equalizer technology was recently introduced by Boscheti and Righetti. This approach comprises a diverse library of combinatorial ligands bound to polymeric beads. When the functionalized beads are contacted with the sample of widely differing protein composition and abundances, they are able "to equalize" the protein population, by sharply reducing the concentration of the most abundant components while simultaneously enhancing the concentration of the most dilute components [31].

7.1.6
On-Line Sample Clean-Up Approaches

Most sample-preparation procedures are performed manually and are thus time-consuming and laborious. On-line sample clean-up and on-column concentrations avoid this disadvantage. There are a number of important features to be gained by having the liquid-phase separation system operated on-line. The overall yield in most cases is improved compared to off-line approaches and methodologies. Moreover, the exposed surfaces – which usually are the main cause of sample losses – are kept to a minimum. The overall precision can also be controlled by having yields above 50%. It is possible to handle yields that are lower, but it is generally a major analytical challenge to obtain operational stability within such analytical processes. The direct injection of samples onto HPLC columns is substantially advantageous in the clinical laboratory in terms of its time- and labor-saving capabilities, in addition to other advantages described below. General direct injection methods have been devised which deal with the problem of many different proteins being present in the sample. The available methods include the pre-column technique, restricted-access materials, and chromatography in mobile phases containing surfactant. High-performance affinity chromatography is also a direct injection technique that surely will demonstrate its power in the near feature. The characteristic and performance of each direct injection technique are comprehensively discussed below for the analysis of biological samples.

The *pre-column technique*, which currently is the most frequently reported direct injection method, utilizes two columns in series (pre-column and analytical) connected by a switching valve. The most common pre-column technique employs a reversed-phase pre-column and a reversed-phase analytical column. Most often, the sample is injected into an aqueous mobile phase flowing through a pre-column (1–4 cm in length, 3–4.6 mm i.d.) which retains lipophilic compounds but passes non-retained hydrophilic compounds to waste. The switching valve is then changed so that the components retained on the pre-column are eluted onto the analytical column by increasing the solvent strength of the mobile phase. This technique serves the dual function of concentrating the analyte and removing the hydrophilic substances. The pre-column injection technique has many advantages compared to traditional sample preparation techniques. Typically, it saves time in comparison to the labor-intensive liquid–liquid extraction and precipitation techniques. Other advantages over the SPE technique include a high degree of reproducibility and high recoveries such that the addition of an internal standard is not required. Another advantage of the pre-column technique over other direct injection techniques is its superior detection limit capabilities due to its ability to cope with the injection of large sample volumes.

Several types of column-switching design have been applied. Although the back-flush design is most often used because it minimizes band broadening [32], some groups prefer the forward-flush mode to protect the analytical column from possible impurities at the column head. Some of the designs also include a guard column placed before the analytical column. The use of an on-line 0.5-pm filter is also recommended, but this requires periodic replacement. A design that incorporates a second pre-column parallel to the first has been employed which increases sample throughput, by alternating injection on one pre-column and back-flushing of retained compounds on the other pre-column. Another design has two pre-columns in series, which provides extra versatility in the clean-up, and also allows for the injection of larger sample volumes because contamination of the analytical column is minimized.

Several design considerations have been introduced for pre-columns in order to minimize the adverse effects of the proteins. Large-diameter packing materials (25–50 pm) are most often used in pre-columns; typically, two to three times more injections can be made prior to pressure build-up problems on pre-columns packed with 50-pm material compared to 10-pm material [33]. A systematic study of pre-column particle size on performance was carried out to demonstrate the effect on pressure build-up and column efficiency, and clearly demonstrated the advantage of a larger-sized packing material. In most cases, a scheduled replacement of pre-columns, on-line filters, and/or guard columns was required in order to attain the maximum lifetime for the analytical column.

It is also important to note that, in a comparison of a particulate and a silica monolithic guard column, the RAM-SCX particulate column became clogged much

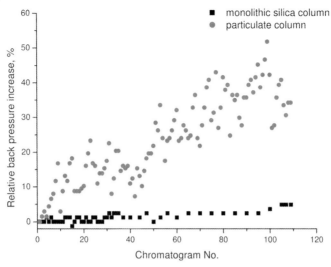

Figure 7.3 Changes in back-pressure of the particulate (25-μm particles) and the monolithic silica columns injected with filtered human plasma samples. Particulate column: 25 × 4 mm i.d., RAM-SCX. Monolithic silica column: 5 × 4.6 mm i.d., Chromolith guard column. Sample human plasma, 50 μL per injection.

more rapidly than the monolithic guard column (Figure 7.3). A total of 120 injections of plasma (50 μL each) led to an increase of approximately 6 bar in the particulate RAM-SCX column back-pressure, whilst the increase at the monolithic guard column was only approximately 1 bar. The lifetime of the short silica monolithic columns used as a trap column or as a guard column depends heavily on the type and volume of bio-fluid injected. The data shown graphically in Figure 7.4 demonstrate column efficiency diminution when applying filtered urine and plasma. Following plasma injection, the column performance fell drastically when the half-column volume was injected (Figure 7.4b and d). For urine, the column stability was at least 20-fold higher (Figure 7.4a and c), and clearly related to the sample complexity.

7.1.7
Restricted Access Technology

Direct injection techniques are generally preferable, as problems involved in off-line sample pretreatments, such as time-consuming procedures, errors and risk of low recoveries, can be readily avoided. However, reported methods in which direct injection techniques have been used, including a replaceable guard column inserted before the analytical column and a column-switching system, were only able to endure the injection of less than 100 μL of plasma (ca. 5 mL in total) because conventional reversed-phase material had been used as precolumns. This is a serious limitation of routine proteome screening in fully automated mode, as the column replacements shorten the system "work alone" time.

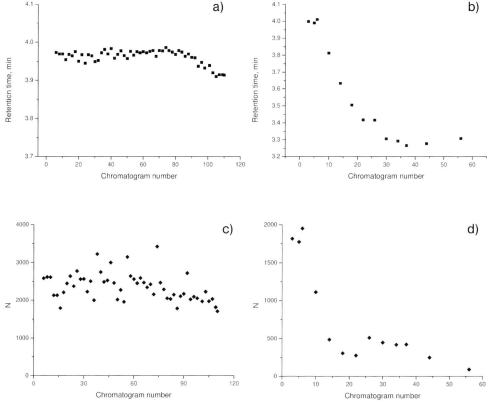

Figure 7.4 Changes in monolithic column performance after injecting filtered human urine (a, c) and plasma (b, d) samples. Monolithic silica column: 5×4.6 mm i.d., Chromolith guard column. Samples: human urine 100 µL per injection, human plasma 50 µL per injection. N = number of theoretical plates.

One sample clean-up concept which was examined with some success was originally developed by Hageston and Pinkerton [34], who designed a HPLC column the packing of which had a hydrophilic external surface and a hydrophobic internal surface, which acted as a reversed-phase material. In 1991, Desilets et al. [35] introduced the term of "restricted access", which designates a support family that allows the direct injection of biological fluids by limiting the accessibility of interaction sites within the pores to small molecules only. The term "restricted access material" (RAM) is a general term for a packing material having a hydrophobic interior covered by a hydrophilic barrier. The hydrophilic barrier allows the passage of small molecules to the hydrophobic part of the stationary phase, while sterically preventing large molecules, such as proteins, from interacting with this part of the stationary phase. Macromolecules are excluded and may interact only with the outer surface of the particle support coated with hydrophilic groups, which in turn minimizes the adsorption of matrix proteins. Boos and Rudolphi [36] suggested a classification of RAM sorbents with respect to their surface chemistry. Indeed, these authors discerned phases with different types of bonding in external and internal surfaces (bimodal phases) and phases with a unique bonding to both surfaces (unimodal phases). There are several types, including internal-surface reversed phase, semi-permeable surface, shielded hydrophobic phase, and mixed functional phase. A survey on the current state of the art of RAM-columns in sample pre-treatment is provided by Souverain *et al.* [37]. A new RAM support developed by Boos and Grimm [38] is based on strong cation exchange (SCX)-diol modification to improve performances in terms of efficiency, retention, and reproducibility (Figure 7.5). Those supports were able to withstand several hundred plasma or serum injections (total volume 6–7 mL), without loss in performance. The concept and methodology were successfully used for the sample clean-up of peptides and proteins from biofluids by extending the range of available materials employing cation- and anion-exchanger RAM [39]. Specific non-silica-based RAM were also developed for the investigation of food [40] and environmental matrices [41]. Vijayalakshmi and coworkers [42] have recently presented a new RAM referred to as a "bi-dimensional chromatographic support", which operates on a size-exclusion mode and an affinity or pseudoaffinity mode.

RAM columns possess a dual function: first, they operate as size-exclusion columns to remove high molecular-weight proteins and other undesired constituents. The size characteristics of proteins in pure size-exclusion chromatography (SEC) are known to be highly dependent on eluent composition such as pH, ionic strength (I)

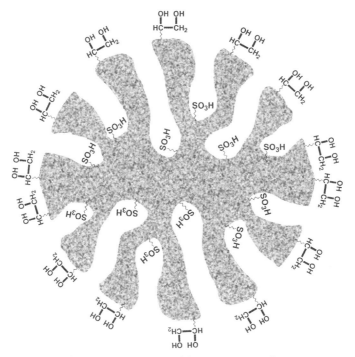

Figure 7.5 Schematic representation of the RAM-SCX particle.

of the buffer (which includes salt type and concentration) and on the flow rate [43]. Ionic strength and pH, however, can vary significantly among biofluids such as plasma and urine. The consequence will be that the sample clean-up procedures must be adjusted individually with respect to each type of biological sample, and standardized protocols must be elucidated. Second, the RAM column serves as trap or capture column to selectively enrich target compounds in a reversed-phase mode or in an ion-exchange mode.

The mass loadability of SPE and RAM columns plays a key role in executing the sample clean-up. It is advisable to work below the overload regime of the column; otherwise, displacement effects and other phenomena such as secondary interaction by adsorbed species might take place, which will lead to non-reproducible results [44]. The latter statement is especially important when the task is to monitor medium- to low-abundant proteins, when large sample volumes (in the milliliter range) are usually applied. As the column lifetime is known to be limited, a control measure must also be applied to check the condition of the RAM-SCX column and, if necessary, to replace it. In general, the column endured about 200 injections of urine.

RAM are currently used in many on-line SPE applications. Although the compatibility of these stationary phases with direct biological sample injection is high, it must still be borne in mind that the samples must be filtered or centrifuged prior to injection to remove the solid contaminants and precipitations. Even so, some components tend to agglomerate/precipitate with time as the samples are queued up in autosampler. Hence, an additional in-line filters is highly recommended.

Restricted access materials with strong cation-exchange surface functionality were chosen to extract positively charged peptides from a complex plasma or urine samples by utilizing their charge and charge distribution properties. The size exclusion was approximately 15 000 Da for proteins. As a result, a complex peptide mixture can be effectively resolved and concentrated prior to further separation in a second dimension. Strong cation-exchange chromatography is generally implemented as a primary separation technique due to its potential for increased mass load capacity, whilst reversed-phase chromatography is a perfect complement as a secondary separation technique because of its ability to remove salts and its direct compatibility with MS through electrospray ionization. The characteristic properties of RAM-SCX columns are described in detail elsewhere [45]. Following sample clean-up with the RAM-SCX column, it is important to transfer effectively the trapped peptides, desalt them, and direct them into the mass spectrometer. The overall system repeatability was tested using human hemofiltrate, and found to be acceptable [46], despite the fact that three subsequent columns are employed in the separation (Figure 7.6). When analyzing human urine, the mass spectra of selected peaks from two consecutive runs were found to be almost identical, there being only a slight shift in retention time and intensity. Such a set-up allows the investigation of differences between the patients at peptide level. A peptide map generated from a 150-µL urine injection containing about 1500 peptides is presented in Figure 7.7.

The particular advantage of RAM technology is that it can be integrated into a multidimensional LC system as a fully automated technique. An example of multidimensional separation platform with on-line sample clean-up for the analysis implementing a highly selective trapping step already in sample preparation, the method will significantly improve the overall performance of the analysis.

References

1 Gilar, M., Bouvier, E.S.P. and Compton, B.J. (2001) Advances in sample preparation in electromigration chromatographic and mass spectrometric separation methods. *J. Chromatogr. A*, **909**, 111–135.

2 Smith, R.M. (2003) Before the injection – modern methods of sample preparation for separation techniques. *J. Chromatogr. A*, **1000**, 3–27.

3 Saito, Y. and Jinno, K. (2003) Miniaturized sample preparation combined with liquid phase separations. *J. Chromatogr. A*, **1000**, 53–67.

4 Pandey, A. and Mann, M. (2000) Proteomics to study genes and genomes. *Nature*, **405**, 837–846.

5 Westerlund, D. (1987) Direct injection of plasma into column liquid chromatographic systems. *Chromatographia*, **24**, 155–164.

6 Adamovics, J.A. (1987) Determination of antibiotics and antimicrobial agents in human serum by direct injection onto silica liquid chromatographic columns. *J. Pharm. Biomed. Anal.*, **5**, 267–274.

7 Arvidsson, T., Wahlund, K.G. and Daoud, N. (1984) Procedures for direct injections of untreated blood plasma into liquid chromatographic columns with emphasis on a pre-column venting technique. *J. Chromatogr.*, **317**, 213–226.

8 Tice, P.A., Mazsaroff, I., Lin, N.T. and Regnier, F.E. (1987) Effects of large sample loads on column lifetime in preparative scale liquid chromatography. *J. Chromatogr.*, **410**, 43–51.

9 Hearn, M.T.W. (1982) High performance liquid chromatography and its application to protein chemistry. *Adv. Chromatogr.*, **20**, 1–82.

10 Pinkerton, T.C., Miller, T.D., Cook, S.E., Perry, J.A., Rateike, J.D. and Szczerba, T.J. (1986) The nature and use of internal surface reversed-phase columns: a new concept in high performance liquid chromatography. *J. BioChromatogr.*, **1**, 96–105.

11 Willams, D.E. and Kabra, P.M. (1990) Extended life for blood serum analysis columns using dual zone chromatographic materials. *Anal. Chem.*, **62**, 807–810.

12 Tammen, H., Schulte, I., Hess, R., Menzel, C., Kellmann, M., Mohring, T. and Schulz-Knappe, P. (2005) Peptidomic analysis of human blood specimens: Comparison between plasma specimens and serum by differential peptide display. *Proteomics*, **5**, 3414–3422.

13 John, H., Walden, M., Schafer, S., Genz, S. and Forssmann, W.-G. (2004) Analytical procedures for quantification of peptides in pharmaceutical research by liquid chromatography-mass spectrometry. *Anal. Bioanal. Chem.*, **378**, 883–897.

14 Rasmussen, H.H., Orntoft, T.F., Wolf, H. and Celis, J.E. (1996) Towards a comprehensive database of proteins from the urine of patients with bladder cancer. *J. Urol.*, **155**, 2113–2119.

15 Tantipaiboonwong, P., Sinchaikul, S., Sriyam, S., Phutrakul, S. and Chen, S.T. (2005) Different techniques for urinary protein analysis of normal and lung cancer patients. *Proteomics*, **5**, 1140–1149.

16 Thongboonkerd, V., McLeish, K.R., Arthur, J.M. and Klein, J.B. (2002) Proteomic analysis of normal human

urinary proteins isolated by acetone precipitation or ultracentrifugation. *Kidney Int.*, **62**, 1461–1469.
17 Pluskal, M.G. (2000) Microscale sample preparation. *Nat. Biotechnol.*, **18**, 104–105.
18 Georgiou, H.M., Rice, G.E. and Baker, M.S. (2001) Proteomic analysis of human plasma: failure of centrifugal ultrafiltration to remove albumin and other high molecular weight proteins. *Proteomics*, **1**, 1503–1506.
19 Kassab, A., Yavuz, H., Odabasi, M. and Denizli, A. (2000) Human serum albumin chromatography by Cibacron Blue F3GA-derived microporous polyamide hollow-fiber affinity membranes. *J. Chromatogr. B* **746**, 123–132.
20 Nakamura, K., Suzuki, T., Kamichika, T., Hasegawa, M., Kato, Y., Sasaki, H. and Inouye, K. (2002) Evaluation and applications of a new dye affinity adsorbent. *J. Chromatogr. A*, **972**, 21–25.
21 Wang, Y.Y., Cheng, P. and Chan, D.W. (2003) A simple affinity spin tube filter method for removing high-abundant common proteins or enriching low-abundant biomarkers for serum proteomic analysis. *Proteomics*, **3**, 243–248.
22 Govorukhina, N.I., Keizer-Gunnink, A., van der Zee, A.G.J., de Jong, S., de Bruijn, H.W.A. and Bischoff, R. (2003) Sample preparation of human serum for the analysis of tumor markers. Comparison of different approaches for albumin and γ-globulin depletion. *J. Chromatogr. A*, **1009**, 171–178.
23 Andrecht, S., Anders, J., Hendriks, R., Machtejevas, E. and Unger, K. (2004) Effiziente Probenvorbereitung für die Proteomanalyse von Körperflüssigkeiten. *Laborwelt*, **5** (2), 4–7.
24 Watts, J.D., Affolter, M., Krebs, D.L., Wange, R.L., Samelson, L.E. and Aebersold, R. (1994) Identification by electrospray ionization mass spectrometry of the sites of tyrosine phosphorylation induced in activated Jurkat T cells on the protein tyrosine kinase ZAP-70. *J. Biol. Chem.*, **269**, 29520–29529.
25 Li, S. and Dass, C. (1999) Iron(III)-immobilized metal ion affinity chromatography and mass spectrometry for the purification and characterization of synthetic phosphopeptides. *Anal. Biochem.*, **270**, 9–14.
26 Pinkse, M.W.H., Uitto, P.M., Hilhorst, M.J., Ooms, B. and Heck, A.J.R. (2004) Selective isolation at the femtomole level of phosphopeptides from proteolytic digests using 2D-nano LC-ESI-MS/MS and titanium oxide precolumns. *Anal. Chem.*, **76**, 3935–3943.
27 Gygi, S.P., Rist, B., Griffin, T.J., Eng, J. and Aebersold, R. (2002) Proteome analysis of low-abundance proteins using multidimensional chromatography and isotope-coded affinity tags. *J. Proteome Res.*, **1**, 47–54.
28 Petricoin, E.F., Zoon, K.C., Kohn, E.C., Barrett, J.C. and Liotta, L.A. (2002) Clinical proteomics: translating benchside promise into bedside reality. *Nature*, **1**, 683–695.
29 Petricoin, E.F., Belluco, C., Araujo, R.P. and Liotta, L.A. (2006) The blood peptidome: a higher dimension of information content for cancer biomarker discovery. *Nature*, **6**, 961–967.
30 Liotta, L.A. (2006) Plenary lecture: Spinning biological trash into diagnostic gold. Fifth Swedish Proteomics Society Symposium December 10–11, Uppsala, Sweden.
31 Righettia, P.G., Castagnaa, A., Antonuccia, F., Piubellib, C., Cecconia, D., Campostrinia, N., Rustichellia, C., Antoniolia, P., Zanussob, G., Monacob, S., Lomasc, L. and Boschett, E. (2005) Proteome analysis in the clinical chemistry laboratory: Myth or reality? *Clin. Chim. Acta*, **357**, 123–139.
32 Yamashita, K., Motohashi, M. and Yashiki, T. (1992) Automated high-performance liquid chromatographic method for the

simultaneous determination of cefotiam and delta 3-cefotiam in human plasma using column switching. *J. Chromatogr.*, **577**, 174–179.

33 Werkhoven-Goewie, C.E., De Ruiter, C., Brinkman, U.A.Th., Frei, R.W., De Jong, G.J., Little, C.J. and Stahel, O. (1983) Automated determination of drugs in blood samples after enzymatic hydrolysis using precolumn switching and post-column reaction detection. *J. Chromatogr.*, **255**, 79–90.

34 Hagestam, I.H. and Pinkerton, T.C. (1985) Internal surface reversed-phase silica supports for liquid chromatography. *Anal. Chem.*, **57**, 1757–1763.

35 Desilets, C.P., Rounds, M.A. and Regnier, F.E. (1991) Semipermeable-surface reversed-phase media for high-performance liquid chromatography. *J. Chromatogr.*, **544**, 25–39.

36 Boos, K.S. and Rudolphi, A. (1997) The use of restricted-access media in HPLC Part I – Classification and review. *LC-GC Int.*, **15**, 602–611.

37 Souverain, S., Rudaz, S. and Veuthey, J.-L. (2004) Restricted access materials and large particle supports for on-line sample preparation: an attractive approach for biological fluids analysis. *J. Chromatogr. B*, **801**, 141–156.

38 Boos, K.S. and Grimm, C.H. (1999) High-performance liquid chromatography integrated solid-phase extraction in bioanalysis using restricted access precolumn packings. *Trends Anal. Chem.*, **18**, 175–180.

39 Wagner, K., Miliotis, T., Marko-Varga, G., Bischoff, R. and Unger, K.K. (2002) An automated on-line multidimensional HPLC system for protein and peptide mapping with integrated sample preparation. *Anal. Chem.*, **74**, 809–820.

40 Bovanova, L. and Brandsteterova, E. (2000) Direct analysis of food samples by high-performance liquid chromatography. *J. Chromatogr. A*, **880**, 149–168.

41 Hogendoorn, E.A., Dijkman, E., Baumann, B., Hidalgo, C., Sancho, J.-V. and Hernandez, F. (1999) Strategies in using analytical restricted access media columns for the removal of humic acid interferences in the trace analysis of acidic herbicides in water samples by coupled column liquid chromatography with UV detection. *Anal. Chem.*, **71**, 1111–1118.

42 Pitiot, O., Poraht, J., Guzmann, R. and Vijayalakshmi, M.A. (2004) Ninth International Symposium on Biochromatography 5–7 May, Bordeaux, France, Programme & Abstracts, 34.

43 Quaglia, M., Machtejevas, E., Hennessy, T. and Unger, K.K. (2006) Gel Filtration – size-exclusion chromatography (SEC) of biopolymers: Optimization strategies and trouble shooting in *HPLC Made to Measure – A Practical Handbook for Optimization* (ed. S. Kromidas), Wiley-VCH, Weinheim, Germany, pp. 383–403.

44 Willemsen, O., Machtejevas, E. and Unger, K.K. (2004) Enrichment of proteinaceous materials on a strong cation-exchange diol silica restricted access material (RAM): protein-protein displacement and interaction effects. *J. Chromatogr. A*, **1025**, 209–216.

45 Machtejevas, E., Denoyel, R., Meneses, J.M., Kudirkaitė, V., Grimes, B.A., Lubda, D. and Unger, K.K. (2006) Sulphonic acid strong cation-exchange restricted access columns in sample cleanup for profiling of endogenous peptides in multidimensional liquid chromatography. Structure and function of strong cation-exchange restricted access materials. *J. Chromatogr. A*, **1123**, 38–46.

46 Machtejevas, E., Andrecht, S., Lubda, D. and Unger, K.K. (2007) Monolithic silica columns of various format in automated sample clean-up/multidimensional liquid chromatography/mass spectrometry for peptidomics. *J. Chromatogr. A*, **1144**, 97–101.

47 Majors, R.E. (1995) Sample preparation perspectives: trends in sample preparation. *LC-GC*, **13**, 742–746.

7.2
Sample Preparation for Two-Dimensional Phosphopeptide Mapping and Phosphoamino Acid Analysis

Anamarija Kruljac-Letunic and Andree Blaukat

7.2.1
Introduction

Protein phosphorylation is a key post-translational modification that is involved in the regulation of variety of cellular processes spanning from signal transduction, cell cycle control, and cell differentiation to metabolism and development [48,49]. It acts as a switch that turns protein activity "on" and "off" in a rapid and reversible manner. Therefore, identification of protein phosphorylation sites is often the first and crucial step towards an understanding of protein function. Despite a growing knowledge of many phosphorylation consensus sequences, phosphorylation sites can rarely be accurately predicted from the protein sequence alone, and in most cases the experimental determination of phosphorylation sites is compulsory.

Two-dimensional (2-D) phosphopeptide mapping is a powerful analytical method that originally was developed in the laboratory of Hunter [50], but since then has been widely used for the identification of protein phosphorylation sites [51,52]. The technique is based on the enzymatic digestion of ^{32}P-labeled phosphoproteins and separation of the resulting peptides in two dimensions by high-voltage electrophoresis and liquid chromatography on thin-layer cellulose (TLC) plates. Following elution from the TLC plate, separated phosphopeptides are subjected to Edman degradation and phosphoamino acid analysis that reveal the position of phosphoamino acid within the phosphopeptide and the type of phosphorylation (on serine, threonine or tyrosine residue), respectively. Reproducible phosphopeptide maps can be obtained from samples containing only a few hundred ^{32}P-disintegrations per minute, often corresponding to sub-femtomolar levels of phosphoproteins. Hence, the 2-D phosphopeptide mapping technique is very sensitive and, in cases when very low levels of phosphorylated proteins are available, even superior to mass spectrometry (MS) approaches.

7.2.2
Important Aspects in Sample Preparation Procedures

The transient nature of many phosphorylation events, the often low stoichiometry of phosphorylation at a given site, and/or the low copy number of phosphoproteins per cell makes direct analysis of sites of phosphorylation often difficult, regardless of the method of analysis. Therefore, initial considerations during the sample preparation procedure for 2-D phosphopeptide mapping are maintenance of protein phosphorylation states and, if necessary, enrichment of the phosphoproteins of interest. The phosphorylation state is usually preserved by the addition of protein phosphatase inhibitors (added as a mixture of different inhibitors) during lysis and immunoprecipitation steps, as well as by manipulation of samples at 4°C.

Due to the high sensitivity of the method, and given that sufficient radioactivity can be incorporated into the phosphoprotein of interest, success should normally be achieved in obtaining satisfying phosphopeptide maps even without phosphoprotein enrichment. Nevertheless, enrichment procedure might improve the appearance of the map and make weaker phosphopeptides more visible. How phosphoproteins of interest and the whole phosphoproteome can be enriched is discussed elsewhere [53].

After separation by SDS–PAGE, the immunoprecipitated proteins should be transferred electrophoretically onto nitrocellulose membrane, because the recovery of proteins from gels is laborious, time-consuming, and may also result in significant losses of radioactive material [54]. On the contrary, the recovery of proteins from nitrocellulose is much more efficient and takes place during enzymatic digestion of the protein to peptides. It is important therefore that, before proceeding with a proteolytic digest, the excised pieces of nitrocellulose membrane are soaked in 0.5% polyvinylpyrrolidone, which blocks the protein binding sites on the nitrocellulose that may otherwise capture the proteolytic enzyme. A protein-free blocking procedure is crucial because proteins would compete with the radiolabeled phosphoprotein for the active protease.

The proteolytic digest is usually performed using trypsin, and repeated twice in order to obtain complete digestion of the protein. However, sometimes (especially after trypsin digestion of high molecular-weight proteins) data obtained from Edman degradation suggest the presence of too many candidate sites. In that case, more information about particular phosphopeptide can be gained by secondary digestion with another enzyme, such as V8 protease. Additionally, the whole protein can be digested with another enzyme and the resulting phosphopeptide map compared with the one obtained by trypsin digestion. The detailed list of proteolytic enzymes that can be utilized, together with their sequence specificity and reaction conditions, are described by van der Geer and Hunter [54].

Cysteine and methionine residues within the peptide can exist in several oxidation states and show different mobilities during chromatography. In order to avoid the separation of these different oxidation states in the chromatography dimension, and thereby an increased complexity of phosphopeptide maps, the samples are quantitatively oxidized using performic acid. During oxidation the samples should be kept at 4°C, because incubation with performic acid at a higher temperature may result in unwanted side reactions, such as fragmentation of the peptide backbone.

Finally, oxidized phosphopeptides are dissolved in electrophoresis buffer. Currently, three different electrophoresis buffers are commonly used, each differing in their composition and providing different pH values and hydrophobicity. As it is usually impossible to predict which buffer will provide the best separation for the particular peptide mixtures, all three buffers should be tested. Nevertheless, most peptides dissolve well in pH 1.9 buffer, and streaking of phosphopeptides is generally less often visible using this buffer.

7.2.3
Application Example

The identification of tyrosines 308 and 782 in the Arf GTPase-activating protein ASAP1 as the major sites phosphorylated by the non-receptor protein tyrosine kinase Pyk2 represents an example of the successful phosphorylation sites identification by the 2-D phosphopeptide mapping technique [55]. The experiment was performed as follows. HEK293T cells cotransfected with ASAP1 and Pyk2 or ASAP1 alone were lysed and subjected to an anti-Flag immunoprecipitation to isolate Flag-ASAP1 and associated proteins. The complexes were incubated with [γ-^{32}P]ATP in kinase buffer for 30 min at 30°C. Reactions were then stopped and samples subjected to 8% SDS–PAGE, transferred onto nitrocellulose membranes, and analyzed by autoradiography (Figure 7.9A). Phosphorylated ASAP1 was excised from membranes and digested *in situ* with trypsin. A fraction of the tryptic radiolabeled ASAP1 peptides was separated on TLC plates by high-voltage electrophoresis and ascending chromatography, and localization of phosphopeptides was detected by PhosphorImager analysis. In Figure 7.9B, a circled cross illustrates where the samples were applied, and "+" and "−" indicate the polarity during electrophoresis. A second peptide fraction was hydrolyzed with hydrochloric acid at 110°C, subjected to a phosphoamino acid analysis and visualized using a PhosphorImager. Phosphorylated amino acids were identified by comigrations with ninhydrin-stained standards that are indicated by dashed circles (Figure 7.9C).

Phosphopeptides 1 and 2 were extracted from the cellulose matrix, subjected to 20 cycles of Edman degradation, and the cleaved amino acids collected and analyzed using a PhosphorImager. The content of ^{32}P radioactivity in each degradation cycle was quantified and expressed in arbitrary units (AU). In peptide 1, a radioactivity peak was found in sequencing cycle 3, which suggested a phosphorylated tyrosine in the corresponding position of the peptide. A similar analysis of peptide 2 proposed a phosphorylated tyrosine in position 6 (Figure 7.9D). A list of tryptic peptides with a tyrosine in position 3 (tyrosines 82, 782, and 1094) or 6 (tyrosines 273 and 308) obtained by tryptic digestion of ASAP1 was generated using the PeptideMass tool available at http://us.expasy.org/tools/peptide-mass.html (Figure 7.9E). To confirm that the predicted Pyk2 phosphorylation sites are targeted by Pyk2 phosphorylation, ASAP1-Y308F and ASAP1-Y308/782F mutants were phosphorylated by Pyk2 *in vitro*, digested *in situ* with trypsin, and subjected to the same 2-D phosphopeptide analysis as the phosphorylated wild-type ASAP1. Phosphopeptides were localized by PhosphorImager analysis. A circled cross illustrates where samples were applied, while "+" and "−" indicate the polarity during electrophoresis (Figure 7.9F).

7.2.4
Summary

Two-dimensional phosphopeptide mapping is a powerful analytical method that is widely used for the identification of protein phosphorylation sites. It is based on the digestion of ^{32}P-labeled phosphoprotein with site-specific protease and separation of

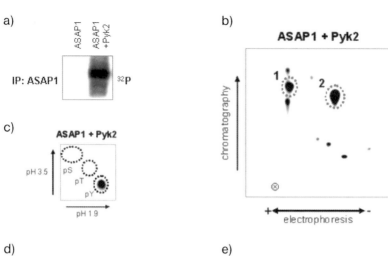
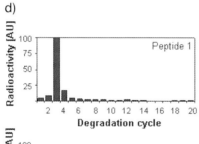
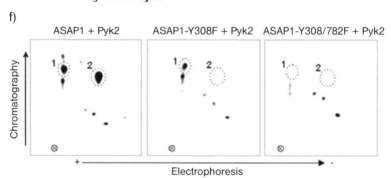

Figure 7.9 Example of identification of protein phosphorylation sites by 2-D phosphopeptide mapping and phosphoamino acid analysis techniques. See text for details.

the obtained phosphopeptides in two dimensions on TLC plates. Thereafter, the separated peptides are recovered from the TLC plate and subjected to Edman sequencing and phosphoamino acid analysis in order to reveal the position of phosphoamino acid within the phosphopeptide and the type of phosphorylation, respectively. This chapter includes a basic protocol describing the sample preparation steps for the 2-D phosphopeptide mapping and phosphoamino acid analysis procedures, and also discusses the most important issues that should be considered during preparation.

7.2.5
Perspective

Due to recent advances in MS techniques for phosphoprotein analysis, 2-D phosphopeptide mapping today has a limited space of application [56]. However, the direct analysis of sites of phosphorylation – particularly those phosphorylated *in vivo* – has often proved difficult, even when sensitive mass spectrometric instrumentation are used. Therefore, when the ionization of phosphopeptides and/or low abundance of the phosphoprotein of interest is an issue and causes phosphopeptides not to be visualized by MS, the 2-D phosphopeptide mapping technique is the preferred method of choice and will more likely result in successful determination of phosphorylation sites in particular phosphoproteins. Furthermore, whilst MS is very reliable for delivering positive results (e.g., the detection of specific phosphorylated amino acid residues), a failure to identify phosphorylated peptides by MS does not rule out the presence of phosphorylation sites in the protein of interest.

Unfortunately, the inconvenience associated with the use of radioactivity, together with the multiple steps during the rather time-consuming procedure, makes this traditional phosphorylation analysis method less attractive for the high-throughput pipelines that are required for phosphoproteome analysis.

7.2.6
Recipe for Beginners

7.2.6.1 2-D Phosphopeptide Mapping Procedure

- Isolate the ^{32}P-labeled phosphoprotein of interest from *in-vitro* kinase reactions or from cell lysates obtained from *in-vivo* ^{32}P-labeling using standard immunoprecipitation procedures. Perform immunoprecipitation in the presence of phosphatase inhibitor cocktails containing for example 1–10 mM sodium fluoride, 0.1–2 µM sodium orthovanadate, up to 50 mM β-glycerophosphate, and 0.5–2 mM EGTA.

- Separate the immunoprecipitated proteins by SDS–PAGE and transfer them onto a nitrocellulose membrane.

- Wrap the nitrocellulose membrane in a thin plastic foil and expose it on X-ray film or PhosphorImager screen. The nitrocellulose membrane should not be allowed to dry, as this may affect the tryptic digest. An overnight film exposure or 2–4 h on a PhosphorImager should be sufficient to obtain satisfactory results.

- Overlay the autoradiogram or a 1 : 1 print of the PhosphorImager analysis with the nitrocellulose membrane, mark the band of interest by a needle, and cut it out. For the perfect overlay, the use of a commercial phosphorescent ruler or radioactive ink (a few drops of a ^{14}C source in regular ink) is suggested. The precision of the cuts should be verified by re-exposing the nitrocellulose membrane on X-ray film or PhosphorImager screen.

- Transfer the membrane pieces to 200 µL of 0.5% polyvinylpyrollidone (PVP) in 0.6% acetic acid, and incubate at 37°C for 30 min to block protein-binding sites on the nitrocellulose that may otherwise capture trypsin. Measure the membrane-bound radioactivity in a β-counter using a Cherenkov program.

- Aspirate the blocking solution and wash the nitrocellulose pieces three times with water.

- Add 200 µL of freshly prepared 50 mM ammonium bicarbonate solution containing 1 µg of modified sequencing grade trypsin and incubate overnight at 37°C.

- Transfer the supernatant to a fresh tube, wash the nitrocellulose pieces twice with water for 15–30 min at 37°C, and pool the supernatants. Measure the free and residual membrane-bound radioactivity in a β-counter. About 75–90% of ^{32}P should be released from the nitrocellulose. If this is not the case, membranes can be incubated with trypsin a second time. The amount of radioactivity in phosphopeptides that is needed to obtain high-quality 2-D maps depends on the individual protein, but ca. 1000 cpm can be considered as a good starting point.

- Dry the samples in a Speedvac concentrator (2–6 h).

- Prepare a fresh performic acid solution by mixing 9 parts formic acid with 1 part 30% hydrogen peroxide solution, and incubate for 1 h at room temperature. Add 50 µL of performic acid to chilled peptide samples and incubate for 1 h on ice. This treatment will oxidize cysteine and methionine residues and improve yields of a second trypsin cleavage step.

- Stop the reaction by diluting samples with 500 µL of water and freezing at −80°C.

- Dry the samples in a Speedvac concentrator (6–12 h). At this stage, a high and very stable vacuum is necessary to avoid irreversible oxidative damage of samples. When using new equipment, a test run with 500 µL of frozen water is suggested: the water should stay frozen during the whole vacuum drying process, otherwise the vacuum must be improved.

- Add 50 µL of freshly prepared 50 mM ammonium bicarbonate solution and sonicate samples in a water bath for 5 min to dissolve peptides. Add 1 µg of modified sequencing grade trypsin, mix, and incubate overnight at 37°C.

- Add 140 μL of pH 1.9 electrophoresis buffer (formic acid:acetic acid:water, 44:156:1800, v/v/v), mix and centrifuge (13 000 rpm, room temperature) for 5 min. Transfer 180 μL of the supernatant to a fresh tube, carefully avoiding insoluble material, and dry in a Speedvac concentrator. The remaining 20 μL and eventually insoluble material may be used for a phosphoamino acid analysis to determine the overall phosphoamino acid composition of the protein (see first step in Section 7.2.6.1).

- Dissolve phosphopeptides in 5–10 μL of electrophoresis buffer, mix intensively, and centrifuge (13 000 rpm, room temperature) for 1 min.

- Apply supernatants in minute portions onto cellulose TLC plates.

- For detailed instructions of where to spot samples on TLC plates, and how to perform the 2-D phosphopeptide analysis, consult the manual of the Hunter thin-layer peptide mapping electrophoresis system [57] or Boyle et al. [50].

7.2.6.2 Phosphoamino Acid Analysis Procedure

- Add 200 μL of 6 M HCl to 20 μL of left-over material from the indicated step in Section 7.2.6.1, and incubate at 110°C for 1 h in a carefully sealed tube (screw cap tubes recommended) to hydrolyze the proteins/peptides.

- Add 300 μL of water, mix, and dry in a Speedvac concentrator for 4–6 h.

- Add 500 μL of water, mix, and centrifuge (13 000 rpm, room temperature) for 5 min. Take 450 μL of the supernatant, avoiding any insoluble material, transfer to a fresh tube, and dry in a Speedvac concentrator. Repeat this stage.

- Dissolve the samples in 5 μL of pH 1.9 electrophoresis buffer and add to each 1.5 μg of non-radioactive phosphoserine, phosphothreonine, and phosphotyrosine standards. Apply supernatants in minute portions onto cellulose TLC plates.

- For detailed instructions of where to spot samples on TLC plate and how to perform 2-D phosphopeptide analysis, consult the manual of the Hunter thin-layer peptide mapping electrophoresis system [57] or Boyle et al. [50].

References

48 Hunter, T. (1995) Protein kinases and phosphatases: the yin and yang of protein phosphorylation and signaling. Cell, **80**, 225–236.

49 Hunter, T. (2000) Signaling – 2000 and beyond. Cell, **100**, 113–127.

50 Boyle, W.J., van der Geer, P. and Hunter, T. (1991) Phosphopeptide mapping and phosphoamino acid analysis by two-dimensional separation on thin-layer cellulose plates. Methods Enzymol., **201**, 110–149.

51 Ronnstrand, L., Mori, S., Arridsson, A.K., Eriksson, A., Wernstedt, C., Hellman, U., Claesson-Welsh, L. and Heldin, C.H. (1992) Identification of two C-terminal autophosphorylation sites in the PDGF beta-receptor: involvement in the interaction with phospholipase C-gamma. EMBO J., **11**, 3911–3919.

52 Blume-Jensen, P., Wernstedt, C., Heldin, C.H. and Ronnstrand, L. (1995) Identification of the major phosphorylation sites for protein kinase C in kit/stem cell factor receptor in vitro and in intact cells. *J. Biol. Chem.*, **270**, 14192–14200.

53 Delom, F. and Chevet, E. (2006) Phosphoprotein analysis: from proteins to proteomes. *Proteome Sci.*, **4**, 15.

54 van der Geer, P. and Hunter, T. (1994) Phosphopeptide mapping and phospho-amino acid analysis by electrophoresis and chromatography on thin-layer cellulose plates. *Electrophoresis*, **15**, 544–554.

55 Kruljac-Letunic, A., Moelleken, J., Kallin, A., Wieland, F. and Blaukat, A. (2003) The tyrosine kinase Pyk2 regulates Arf1 activity by phosphorylation and inhibition of the Arf-GTPase-activating protein ASAP1. *J. Biol. Chem.*, **278**, 29560–29570.

56 Areces, L.B., Matafora, V. and Bachi, A. (2004) Analysis of protein phosphorylation by mass spectrometry. *Eur. J. Mass Spectrom. (Chichester Eng)*, **10**, 383–392.

57 Boyle, W.J. and Hunter, T. *Hunter Thin Layer Peptide Mapping Electrophoresis System* C.B.S. Scientific Company, Inc., Del Mar, CA, USA.

8
Structural Proteomics

8.1
Exploring Protein–Ligand Interactions by Solution NMR

Rudolf Hartmann, Thomas Stangler, Bernd W. König, and Dieter Willbold

8.1.1
Introduction

Complex biological processes rely on specific interactions of proteins with other polypeptides, such as receptors in signal transduction, chaperons in protein folding, or antigens in immune defense, with nucleic acids during gene transcription and translation, with complex carbohydrates in cell recognition, but also with small molecules such as hormones, substrates, or drugs. A detailed understanding of biological function at the molecular level clearly requires an in-depth interaction analysis not only from a structural point of view but also with regard to the specificity, strength and mechanism of the interaction. Dynamic aspects certainly play a crucial role in molecular recognition.

The nuclear magnetic resonance (NMR) spectroscopic analysis of complex formation goes far beyond a static "picture" of the three-dimensional (3-D) structure of tightly bound complexes. Solution NMR is in the unique position to probe both, protein structure and dynamics, at atomic resolution. NMR is sensitive to the dynamic processes that occur over a wide range of time scales, from picoseconds to seconds. Taken together, this makes NMR a powerful and versatile tool for the study of molecular processes. Interactions, characterized with dissociation constants (K_D) in the nano- to millimolar range, are accessible using a diverse collection of NMR techniques. NMR parameters utilized for interaction analysis include chemical shift, spin relaxation rates, translational diffusion rates, cross-relaxation rates, and saturation transfer phenomena [1,2]. A prominent strength of NMR is the ability to closely monitor the binding of a ligand to a receptor. In contrast to most other techniques, NMR is also sensitive to the transient and dynamic aspects of interactions, and also to low-affinity binding.

Proteomics Sample Preparation. Edited by Jörg von Hagen
Copyright © 2008 WILEY-VCH Verlag GmbH & Co. KGaA, Weinheim
ISBN: 978-3-527-31796-7

8.1.2
Localization of Interaction Sites by Chemical Shift Perturbation (CSP) Mapping

The major aim of many studies on protein interactions with large or small molecules is delineation of the interaction site on one or both components. The most frequently used NMR method for mapping protein–ligand interactions to the surface of a target protein is based on chemical shift changes resulting from complex formation [3]. The ligand may be of any type – for example, an ion, a small molecule, or a large protein – as long as the NMR signals of the target protein are easily detectable. Heteronuclear single quantum correlation (HSQC)-based experiments allow the detection of ligand binding over a wide range of K_D values. Binding is detected by comparing two-dimensional (2-D) ^{1}H-^{15}N HSQC spectra of the ^{15}N-enriched target protein in the absence and presence of ligand to elucidate ligand-induced changes in chemical shift. Each non-proline amino acid of the protein gives rise to a correlation signal of the backbone amide group in the ^{1}H-^{15}N HSQC experiment. These correlation signals can be assigned by other 2-D and 3-D NMR experiments to the corresponding amino acid. Typically, perturbations are considered significant if $\Delta\delta > 0.05$ ppm for at least two peaks in the spectrum [24,27]. A series of ^{1}H-^{15}N-HSQC spectra of ^{15}N-labeled target protein is recorded with increasing amounts of non-labeled ligand. Chemical shifts of nuclei close to the binding interface are very likely perturbed by the presence of the ligand. The CSP data map the binding site to the amino acid sequence, or onto the 3-D structure of the target, if known. Chemical shift mapping is not restricted to the contact site on the target protein. Instead, nuclei of both interaction partners directly involved in binding can be identified by the CSP methodology, provided that appropriate isotope-labeling schemes and/or NMR experiment combinations are chosen. CSP-based delineation of contact surfaces might be challenging in the presence of allosteric effects [5], or even impossible if extended conformational changes of the target occur upon binding. In both cases, chemical shifts of nuclei far away from the binding site may also change.

HSQC-based NMR titrations do not only unravel the contact sites but may also allow estimates to be made on the strength, stoichiometry, specificity, and kinetic aspects of the interaction [3]. The kinetics of binding can be classified based on the overall spectral changes observed during titration. If complex dissociation is fast on the chemical shift and spin–spin relaxation time scales, only a single correlation is observed for each pair of directly bound ^{1}H-X spins in the HSQC. This situation is referred to as "fast chemical exchange", and is typically observed for low-affinity binding. The corresponding chemical shifts are residence time-weighted averages of the shifts in the free and bound states, respectively. Peaks in the HSQC spectrum that belong to interface residues move gradually from free to bound state positions upon titration with ligand (see Figure 8.1). Binding constants for identified ligands can be determined by monitoring chemical shift changes as a function of ligand concentration [25]. Data are then fitted using a least-squares grid search varying the values of K_D and the chemical shift of ligand-saturated protein according to

$$K_D = \{([P]_0 - x)([L]_0 - x)\}/x$$

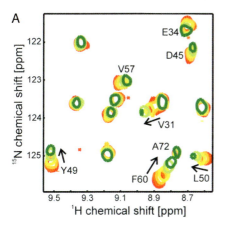 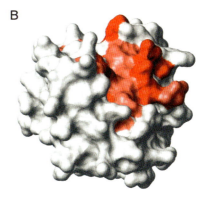

Figure 8.1 Analysis of GABARAP–peptide interaction by HSQC titration. (A) Superposition of a selected region of HSQC spectra obtained upon titration of GABARAP with increasing amounts of a peptide ligand (from red over yellow to green). Only $^1H-^{15}N$ correlations of the ^{15}N GABARAP are visible. HSQC peaks that show substantial chemical shift changes are annotated. The chemical environment of the corresponding nuclei changes due to peptide binding. The gradual shift of resonances indicates fast exchange of the ligand on and off the binding site. (B) Surface representation of the NMR structure of GABARAP [28]. Residues with pronounced chemical shift changes due to peptide binding are colored in red. A continuous patch is obtained that aligns along a groove on the surface of GABARAP and most likely represents the peptide-binding site. (Reproduced from [30]).

where $[P]_0$ and $[L]_0$ are the total molar concentrations of target and ligand, respectively, and x is the molar concentration of the target/ligand complex given by

$$x = (\delta_{obs} - \delta_{free})/\Delta$$

where δ_{obs} and δ_{free} are the chemical shifts for the target molecule at each ligand concentration and in the absence of ligand, respectively, and Δ is the chemical shift difference at saturating levels of ligand [26].

In contrast, in case of slow chemical exchange, which is typically observed for strong binding with small dissociation rate constants, separate HSQC peaks are observed for free and bound interface residues (Figure 8.2). Titration with ligand causes decreasing and increasing intensities of free and bound state resonances, respectively. The intensity ratio of peak pairs that belong to the same site reflects the relative occupation of the two states. Detailed analysis of the exchange kinetics may require separate assignment of free and bound state resonances. Chemical exchange that is intermediate on the chemical shift time scale typically results in extensive line broadening or complete disappearance of NMR signals of nuclei that are located in the interface.

Ring currents have a pronounced effect on the chemical shift of nearby nuclei. If the ligand contains aromatic ring systems, the surface analysis of CSP data may allow very precise mapping of the binding site on the surface of the target, as

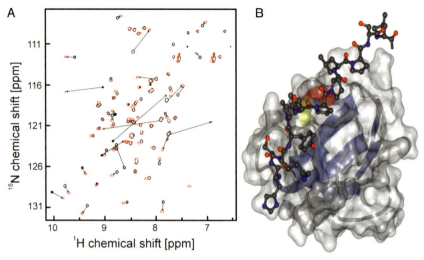

Figure 8.2 Chemical shift perturbation (CSP) mapping of human Hck SH3 domain residues involved in binding to a 12-residue peptide. (A) Superposition of ^{1}H, ^{15}N-HSQC spectra of ^{15}N isotope-labeled Hck SH3 recorded in the absence (black) and with an equimolar amount (red) of peptide. Titration of Hck SH3 with peptide (not shown) results in a stepwise disappearance of some of the free state signals in parallel with the appearance of new resonances. The new signals progressively increase in intensity and arise from the bound state protein conformation. This behavior indicates an exchange of the peptide between free and bound state that is slow compared to the observed difference in chemical shift. Arrows indicate the most pronounced resonance displacements. Residues with substantial chemical shift changes very likely belong to the peptide-binding site of the Hck SH3 domain. (B) Ribbon and surface diagram of the high-resolution structure of human Hck SH3 (PDB entry 2A4Y). The bound peptide is shown in a ball-and-stick representation. The NMR structure of the complex is based on 354 inter- and 2146 intramolecular NOE-derived ^{1}H–^{1}H distances [29]. (Reproduced from [30]).

demonstrated for ligand binding to hepatitis C virus (HCV) NS3 helicase and protease [6].

To prepare the sample for a HSQC-titration, the ^{15}N-labeled protein must be solved with a concentration of 50 μM to 200 μM in H$_2$O/phosphate or phosphate-buffered saline (PBS) buffer to monitor the exchangeable amide protons. The concentration of the ligand depends on the K_D value of the protein–ligand complex, and must cover a range from zero to approximately 10-fold excess of ligand over protein to saturate the binding sites of the protein. The ligand and ligand–protein complex must be soluble in the buffer system of the protein. In some cases, the addition of 10 to 20% dimethyl sulfoxide (DMSO) may help to increase the solubility of the ligands.

8.1.3
Saturation Transfer Difference Spectroscopy

The basis of the saturation transfer difference (STD) NMR technique is an efficient transfer of magnetization over the entire protein via dipolar coupling [7,8]. The selective saturation of a protein ^{1}H NMR signal will rapidly spread the magnetization to all protons of the protein by intramolecular spin diffusion. In addition, intermolecular magnetization transfer will cause saturation of protons of those ligands that transiently bind to the receptor. The STD method does not require isotope labeling, and is best suited for receptor concentrations in the low micromolar range, a large excess of potential binders (typically 50- to 100-fold), and a large difference in the size and rotational diffusion characteristics of receptor and ligand. There is no limitation to the size of the protein. The addition of 10–20% DMSO may increase the solubility of hydrophobic ligands. Spectral artifacts resulting from suppression of the strong ^{1}H signal of the water can be reduced by recording STD experiments in D$_2$O-based phosphate or PBS buffer. A regular ^{1}H NMR spectrum of a receptor and a mixture of potential binders recorded without saturation will be dominated by the signals of the highly concentrated small molecules. In a second spectrum recorded after the selective saturation of a receptor resonance, the NMR signals of the receptor and of receptor-binding ligand molecules will have diminished intensity. The difference spectrum calculated from these two one-dimensional (1-D) NMR spectra contains signals of positive binders only (cf. Figure 8.3). Irradiation

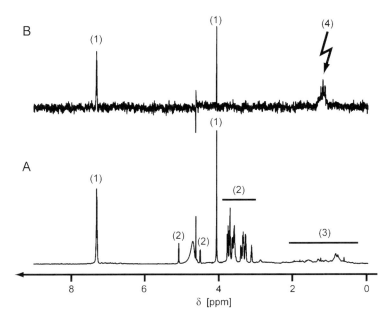

Figure 8.3 Saturation transfer difference (STD) spectroscopy proves interaction of trypsin with benzylamine. (A) ^{1}H NMR spectrum of a sample containing 2 mM benzylamine (1), 2 mM glucose (2), and 20 μM trypsin (3) in ^{2}H$_2$O. Weak interaction of trypsin with benzylamine is known [12], glucose was added as negative control. (B) Difference of ^{1}H NMR spectra recorded without and after selective saturation of protein resonances at ~1 ppm (4). Only benzylamine signals, but not the glucose resonances, are attenuated following saturation. An accumulation of 512 difference spectra clearly shows benzylamine signals (1). The spectral region around the saturation frequency (4) is slightly disturbed. (Reproduced from [30]).

for the selective saturation of protein resonances must be confined to a spectral region that is devoid of ligand signals. During application of the selective pulse train, typically for several seconds, there is a rapid turnover of weakly binding ligand molecules on each receptor; that is, the sample volume is constantly "pumped" with saturated ligand molecules because the relaxation time of the small ligand molecule is much longer than the relaxation time of the protein. This pumping effect is responsible for the extraordinary high sensitivity of the STD approach which makes it very attractive for compound screening [7]. However, very tight binding will strongly reduce the turnover rate and may cause complete relaxation of ligand magnetization in the bound state – that is, strong binders might go undetected. The STD approach has also been successfully applied to study the binding of small molecules to an integral membrane protein [9], virus particles [10], and living cells [11]. The STD technique is easily adapted for use in 2-D NMR pulse schemes [8] such as homonuclear ^{1}H–^{1}H or heteronuclear ^{1}H–^{13}C correlation experiments. An unambiguous identification of binding ligand molecules directly from mixtures of potential ligands is possible by recording 2-D STD spectra. For quantification of the STD effect, the STD amplification factor was introduced

$$STD_{af} = (I_0 - I_{sat}) \cdot (\text{Ligand excess})/I_0$$

where I_0 is the integral intensity of one unsaturated proton and I_{sat} is the integral intensity of the proton after saturation. The size of STD_{af} is not directly correlated to the affinity of the ligand, but an analysis of the different STD_{af} for all of the proton resonances of a ligand will provide important structural information.

The binding epitope of the ligand can be identified by the STD_{af}. Those protons in immediate contact with the receptor experience the strongest saturation, while protons further away from the contact site are less affected [8,13,14]. Reliable interface mapping requires a relatively high off-rate of the ligand and a large excess of ligand over receptor. Widely different longitudinal relaxation rates of ligand protons [15] or non-uniform saturation of receptor resonances [16] may complicate the quantitative interpretation of differential saturation in terms of binding epitopes.

STD NMR benefits from a high proton density in the receptor which supports efficient saturation transfer. Target molecules with few protons may profit from a variant of the STD method named WaterLOGSY [17,18]. Within this scheme, the proton resonance of bulk water is either selectively saturated or inverted. Soluble proteins are surrounded by a layer of reversibly bound water. Efficient magnetization transfer between small water and ligand molecules is feasible only via binding to the large receptor. The observation of negative nuclear Overhauser effects (NOEs) between the ligand and water protons clearly indicates ligand binding to the receptor.

8.1.4
Ligand Screening by NMR

NMR screening of compound libraries is now one of the established methods for inhibitor design and drug lead generation in pharmaceutical research [19,20]. NMR is in the unique position to identify weakly binding ligands and to provide structural details that may aid in the rational design of potent drug-leads based on the initial hits. A diverse array of NMR-based methods has been developed which allow the selection of ligands from a large compound library that bind to a given receptor. One of the first implementations of NMR screening is based on CSP in heteronuclear 2-D NMR spectra of isotope-labeled receptor upon the addition of test compound mixtures. Low-affinity ligands identified in this way can be optimized and linked together to produce a high-affinity ligand; this approach is known as "SAR by NMR" [21].

Many of the established screening methods rely on observation of the small ligand molecules which yield relatively simple NMR spectra [7]. Ligand-based methods are restricted to transient binding, and usually function at relatively low receptor concentrations and do not require isotope labeling. Screening methods monitor NMR observables that change drastically upon binding of the ligand to a large receptor. Small molecules exhibit slow relaxation rates (R_1, R_2), vanishing or weakly negative 2-D-NOESY cross-peaks, and large translational diffusion rates. In contrast, large receptor molecules and receptor-bound ligands show fast relaxation rates, positive NOEs, and smaller translational diffusion rates. Binding can also be verified by the observation of ^{1}H magnetization transfer from receptor to ligand, which only occurs for molecules intimately bound to the receptor and requires some minimum life time of the bound complex. A disadvantage of ligand-based methods is the lack of information on the location of the ligand-binding site on the receptor, which complicates the design of potent drug leads by covalently linking ligands that weakly bind to proximal sites on the target. Second-site screening techniques partially address this issue [22]. The need for high ligand concentrations might pose a solubility problem, and may result in false hits due to unspecific binding. Finally, techniques that are based on transient binding may fail to detect strongly binding ligands due to complete relaxation of the ligand magnetization in the bound state. Binding assays that exploit competition with previously identified ligands to the same site can avoid this problem [23].

Beside the methods for exploring protein–ligand interaction described here so far, NMR is also suitable to determine the high-resolution 3-D structure of a given complex. This is by far the most laborious approach, and is limited in respect to the molecular weights of the components. A more comprehensive overview of NMR applications for investigation of structure and function of proteins has been published recently [30].

References

1 Takeuchi, K. and Wagner, G. (2006) NMR studies of protein interactions. *Curr. Opin. Struct. Biol.*, **16**, 109–117.

2 Otting, G. (1993) Experimental NMR techniques for studies of protein-ligand interactions. *Curr. Opin. Struct. Biol.*, **3**, 760–768.

3 Zuiderweg, E.R. (2002) Mapping protein-protein interactions in solution by NMR spectroscopy. *Biochemistry*, **41**, 1–7.

4 Pellecchia, M., Montgomery, D.L., Stevens, S.Y., van der Kooi, C.W., Feng, H.P., Gierasch, L.M. and Zuiderweg, E.R. (2000) Structural insights into substrate binding by the molecular chaperone DnaK. *Nat. Struct. Biol.*, **7**, 298–303.

5 Stevens, S.Y., Sanker, S., Kent, C. and Zuiderweg, E.R. (2001) Delineation of the allosteric mechanism of a cytidylyltransferase exhibiting negative cooperativity. *Nat. Struct. Biol.*, **8**, 947–952.

6 McCoy, M.A. and Wyss, D.F.. (2002) Spatial localization of ligand binding sites from electron current density surfaces calculated from NMR chemical shift perturbations. *J. Am. Chem. Soc.*, **124**, 11758–11763.

7 Meyer, B. and Peters, T. (2003) NMR spectroscopy techniques for screening and identifying ligand binding to protein receptors. *Angew. Chem. Int. Ed. Engl.*, **42**, 864–890.

8 Mayer, M. and Meyer, B. (1999) Characterization of ligand binding by saturation transfer difference NMR spectroscopy. *Angew. Chem. Int. Ed. Engl.*, **38**, 1784–1788.

9 Meinecke, R. and Meyer, B. (2001) Determination of the binding specificity of an integral membrane protein by saturation transfer difference NMR: RGD peptide ligands binding to integrin alphaIIbbeta3. *J. Med. Chem.*, **44**, 3059–3065.

10 Benie, A.J., Moser, R., Bauml, E., Blaas, D. and Peters, T. (2003) Virus-ligand interactions: identification and characterization of ligand binding by NMR spectroscopy. *J. Am. Chem. Soc.*, **125**, 14–15.

11 Claasen, B., Axmann, M., Meinecke, R. and Meyer, B. (2005) Direct observation of ligand binding to membrane proteins in living cells by a saturation transfer double difference (STDD) NMR spectroscopy method shows a significantly higher affinity of integrin alpha(IIb)beta3 in native platelets than in liposomes. *J. Am. Chem. Soc.*, **127**, 916–919.

12 Markwardt, F., Landmann, H. and Walsmann, P. (1968) Comparative studies on the inhibition of trypsin plasmin, and thrombin by derivatives of benzylamine and benzamidine. *Eur. J. Biochem.*, **6**, 502–506.

13 Mayer, M. and Meyer, B. (2001) Group epitope mapping by saturation transfer difference NMR to identify segments of a ligand in direct contact with a protein receptor. *J. Am. Chem. Soc.*, **123**, 6108–6117.

14 Herfurth, L., Ernst, B., Wagner, B., Ricklin, D., Strasser, D.S., Magnani, J.L., Benie, A.J. and Peters, T. (2005) Comparative epitope mapping with saturation transfer difference NMR of sialyl Lewis(a) compounds and derivatives bound to a monoclonal antibody. *J. Med. Chem.*, **48**, 6879–6886.

15 Yan, J., Kline, A.D., Mo, H., Shapiro, M.J. and Zartler, E.R. (2003) The effect of relaxation on the epitope mapping by saturation transfer difference NMR. *J. Magn. Reson.*, **163**, 270–276.

16 Jayalakshmi, V. and Krishna, N.R. (2002) Complete relaxation and conformational exchange matrix (CORCEMA) analysis of intermolecular saturation transfer effects in reversibly forming ligand-receptor complexes. *J. Magn. Reson.*, **155**, 106–118.

17 Dalvit, C., Pevarello, P., Tato, M., Veronesi, M., Vulpetti, A. and Sundstrom, M. (2000) Identification of compounds with binding affinity to proteins via magnetization transfer from bulk water. *J. Biomol. NMR*, **18**, 65–68.

18 Dalvit, C., Fogliatto, G., Stewart, A., Veronesi, M. and Stockman, B. (2001) WaterLOGSY as a method for primary NMR screening: practical aspects and range of applicability. *J. Biomol. NMR*, **21**, 349–359.

19 Pellecchia, M., Sem, D.S. and Wuthrich, K. (2002) NMR in drug discovery. *Nat. Rev. Drug Discov.*, **1**, 211–219.

20 Lepre, C.A., Moore, J.M. and Peng, J.W. (2004) Theory and applications of NMR-based screening in pharmaceutical research. *Chem. Rev.*, **104**, 3641–3676.

21 Shuker, S.B., Hajduk, P.J., Meadows, R.P. and Fesik, S.W. (1996) Discovering

high-affinity ligands for proteins: SAR by NMR. *Science*, **274**, 1531–1534.
22 Jahnke, W., Florsheimer, A., Blommers, M.J., Paris, C.G., Heim, J., Nalin, C.M. and Perez, L.B. (2003) Second-site NMR screening and linker design. *Curr. Top. Med. Chem.*, **3**, 69–80.
23 Dalvit, C., Flocco, M., Knapp, S., Mostardini, M., Perego, R., Stockman, B.J., Veronesi, M. and Varasi, M. (2002) High-throughput NMR-based screening with competition binding experiments. *J. Am. Chem. Soc.*, **124**, 7702–7709.
24 Hajduk, P.J., Boyd, S., Nettesheim, D., Nienaber, V., Severin, J., Smith, R., Davidson, D., Rockway, T. and Fesik, S.W. (2000) Identification of novel inhibitors of urokinase via NMR-based screening. *J. Med. Chem.*, **43**, 3862–3866.
25 Hajduk, P.J., Sheppard, G., Nettesheim, D., Olejniczak, E.T., Shuker, S.B., Meadows, R.P. and Fesik, S.W. *et al.* (1997) Discovery of potent nonpeptide inhibitors of stromelysin using SAR by NMR. *J. Am. Chem. Soc.*, **119**, 5818–5827.
26 Fesik, S.W. and Hajduk, P.J. (1997) US Patent 5698401.
27 Hajduk, P.J., Gerfin, T., Boehlen, J.M., Haberli, M., Marek, D. and Fesik, S.W. (1999) High-throughput nuclear magnetic resonance-based screening. *J. Med. Chem.*, **42**, 2315–2317.
28 Stangler, T., Mayr, L.M. and Willbold, D. (2002) Solution structure of the GABAA receptor associated protein GABARAP: Implications for biological function and its regulation. *J. Biol. Chem.*, **277**, 13363–13366.
29 Schmidt, H., Hoffmann, S., Tran, T., Stoldt, M., Stangler, T., Wiesehan, K. and Willbold, D. (2007) Solution structure of a Hck SH3 domain ligand complex reveals novel interaction modes. *J. Mol. Biol.*, **365**, 1517–1532.
30 Stangler, T., Hartmann, R., Willbold, D. and König, B.W. (2006) Modern high resolution NMR for the study of structure dynamics and interactions of biological macromolecules. *Z. Phys. Chem.*, **220**, 567–613.

8.2
Sample Preparation for Crystallography

Djordje Musil

8.2.1
Introduction

With the development of genomics and proteomics, and their application in basic biological research and the biotechnologies, there is an increasing need for three-dimensional (3-D) structural knowledge of proteins, nucleic acids and multi-macromolecular assemblies by X-ray crystallography. The 3-D structures provide a basis for understanding catalytic mechanisms, ligand-binding and interaction in multi-macromolecular assemblies. Furthermore, redundancies in structural elements and motifs emerging from the increasing number of structurally known proteins suggest that the number of protein structural frameworks that can be found in Nature is not only finite, but also manageable. The goal of classifying and cataloguing all protein structures may be reached in the foreseeable future. Of course, to achieve this goal, crystals diffracting at high resolution are needed [31].

The crystallographic structural determination of proteins is a demanding multidisciplinary task that requires protein construct design, optimized expression and purification procedures, crystallization trials, solution of the crystallographic phase problem, and model building. The progress made in recombinant technology during the 1980s and 1990s has not only warranted the expression of proteins in large quantities, but also allowed for the target protein to be modified in a form that facilitates protein crystallization. This, together with technology-driven progress in protein structure determination (protein crystallization robotics capable of dispensing volumes in the nanoliter range, third-generation synchrotron X-ray sources, more sensitive X-ray detectors with a faster readout and ever greater computational power with equivalent software development), has resulted in an enormous increase in the number of solved 3-D protein structures. The Protein Database (PDB) snapshot from 2nd January 2007 includes a total of 40 933 experimentally determined coordinate files, whereby 34 672 structures have been determined by X-ray crystallography [32].

In spite of these remarkable results, macromolecular structure determination is not a simple task. The growth of crystals of sufficient diffraction quality can be a major obstacle, and is in most cases the key limiting step in macromolecular X-ray structure determination. As this bottleneck often emanates from the poor sample quality or inadequate protein preparation used for crystallization trials, it is essential to retain high-quality standards when preparing protein for crystallization.

8.2.2
Use of Recombinant Proteins in Crystallization

There is no particular single strategy for producing protein for crystallization that is universally successful. Since, at each experimental step, there are usually several options at the experimentalist's disposal, these must be tested and decisions made on case-by-case basis.

During recent decades, much experimental knowledge has been accumulated through protein crystallization studies, these having shown that some proteins are more sensitive to variations in their physical, chemical and biochemical environment than are others. In order to reduce this variation, and therewith to facilitate protein crystallization, high-quality protein samples must be used. Generally, the protein target for crystallization must be extremely pure, highly homogeneous, soluble, and stable throughout the crystallization experiments. However, the ultimate test of the value of a purified and formulated protein sample is determined by concrete crystallization experiments.

In order to begin crystallization screening experiments, milligram quantities of the target protein are needed. Although new technologies such as microfluidic devices [35] or microarrays [36] have significantly reduced the amount of protein necessary for crystallization screening, they are not yet suitable for high-throughput crystallization experiments. In the past, a 5000-fold purification was not unusual when protein material was isolated from its natural source, but the degree of purification required in the case of recombinant proteins often lies in the range that is two orders of magnitude lower [37]. This, together with the possibility of tailoring protein constructs according to the experimentalists' needs, is a clear advantage in favor of the recombinant methods. Indeed, a survey of crystallographic studies published by Derewenda and coworkers in 2003 [38] indicated that over 90% of crystal structures are based on recombinant material.

Careful consideration of the protein constructs and the choice of the expression system can notably facilitate the folding, yield and purification of the target protein. Simultaneously, the construct can be designed to obtain a more homogeneous and stable protein. Some factors that may cause difficulties in protein crystallization are listed in Table 8.1. Further details and examples of the optimization of protein

Table 8.1 Some factors that may impair protein crystal growth.

Factors impairing protein crystal growth	Examples	Some remedies
Sample purity	Chemical and/or macromolecular impurities	Further purification
Inhomogeneity due to post-translational modifications	Phosphorylation, acetylation, amidination, methylation, glycosylation	Mutation of phosphorylation and glucosylation sites, change of the expression system
Inhomogeneity due to incomplete folding or partial denaturation		Heat treatment [40], coexpression with fusion proteins [41]
Protein flexibility: Inter-domain movements Flexible loops Surface side chains (Lys, Glu) disfavoring crystallization	Flexible inter-domain linkers	Expressing only domain of interest, complexing with other macromolecules and/or small molecule ligands, limited proteolysis, redesigning loops (mutations, deletions), mutating Lys and Glu at the protein surface
Glycosylation	Large and flexible carbohydrate chains, protein destabilization due to deglycosylation	Deglycosylation, change of the expression system
Undesired proteolytic degradation	N- and C-terminal heterogeneity, internal cleavage, incorrectly processed recombinant products	Protease inhibitors throughout purification
Chemically damaged residues	Oxidized Cys and Trp, deamidinated Asn and Gln	Reducing agents
Unspecific aggregates	Hydrophobic patches at the surface that lead to protein aggregation	Adding detergents, limited concentration of chaotropes, complexing with other macro-molecules and/or small molecule ligands, coexpression with fusion proteins

constructs for crystallization are available in Refs. [38,43–45]. In addition, Giegé and McPherson have summarized some general concerns regarding macromolecular samples aimed at crystallization studies [42].

The first – and perhaps single most important – step in the crystallization of a macromolecule is its purification and characterization [33]. Currently, there is a vast literature available on protein purification [57–61].

Samples must be pure in terms of lacking contaminants, but also be relatively monodisperse [62–64], homogeneous, and "conformationally pure", lacking denatured species and other structural microheterogeneities that might adversely affect crystal growth. Microheterogeneities, such as variations in the primary, secondary, tertiary and quaternary structure, as well as the presence of various aggregation states, may also affect a crystal's purification and crystallization properties [33]. An overview of the steps involved in protein production for structural studies is provided in Figure 8.4.

8.2.3
Protein Solubility and Crystallization

The solubility pattern of the target protein, and accordingly its tendency to crystallize under given conditions, is highly protein-specific. The solubility and crystallizability of a protein are defined by individual protein properties, such as its hydrophobic content, charge distribution at the surface, and its structural flexibility.

The solubility of a substance at a given temperature is defined as the concentration of the dissolved solute, which is in equilibrium with its solid form. The equilibrium constant K is related to the free energy change ΔG by:

$$\Delta G = -RT \ln K \tag{1}$$

where R is the universal gas constant and T is the thermodynamic temperature. The free energy change is related to the enthalpy and entropy changes by the equation:

$$\Delta G = \Delta H - T \Delta S \tag{2}$$

Thus, the solute will be more soluble if the decrease in the free energy of the system is larger.

The dissolving process for simple ionic compounds in water is enthalpy-driven – that is, the enthalpy change is dominant. Although the solubility of proteins and other biological macromolecules can be described in a similar way as the solubility of inorganic salts, in considering protein solubility account must be made for a relatively large entropic contribution to the total free energy change. Water is a highly polar solvent, and is characterized by its hydrogen bonding network. When a protein dissolves in a water solution, it will disturb the hydrogen bond network, leading to an energetically unfavorable decrease in entropy. Of course, the hydrogen bond rearrangement is only one of the factors that contribute to the free energy

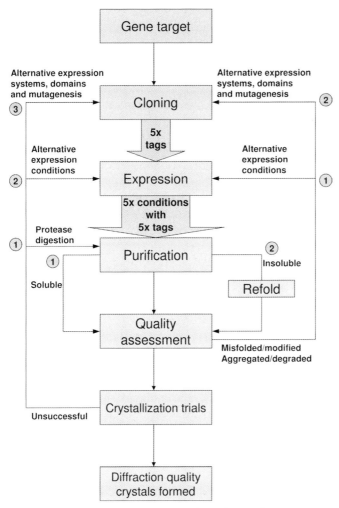

Figure 8.4 An overview of the steps involved in high-throughput protein production for structural studies. With the correct positive and negative controls it should be possible to pass through such a flowchart only once in order to successfully obtain protein samples suitable for structural analysis. The numbers indicate preferences in the flowchart diagram. (From [34]).

change. The sample solubility is related to the total change in both enthalpy and entropy. An inability to exactly calculate the total free energy change of the system means that prediction of the protein solubility is difficult.

The properties of the solvent, such as pH, dielectric constant and ionic strength, define the net charge of the protein and its folding. In general, proteins are dissolved in a water solution, with or without addition of buffers and other chemicals. This

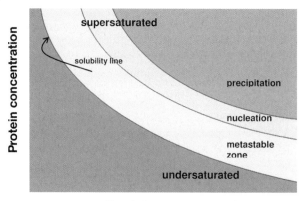

Figure 8.5 A protein solubility diagram.

solvent composition defines the specific physico-chemical conditions of the protein–solvent mixture, which in turn determines the protein solubility.

The change in protein solubility which depends on the concentration of the precipitating agent in solution is shown, schematically, in Figure 8.5. A system with a solute concentration below the solubility limit is undersaturated, whereas in a system where the solute concentration exceeds the solubility limit the solution is supersaturated. At the solubility limit – that is at equilibrium – there is no shift toward either solid or the solution state. Accordingly, the crystals cannot grow from a solution that is barely saturated – in order to achieve crystal growth, a solution must be supersaturated.

A supersaturated solution can be further subdivided into three zones: metastable; nucleation; and precipitation (Figure 8.5). The characteristic of the metastable zone is that nucleation and crystal growth do not occur spontaneously; crystals can only be grown in the metastable zone if nucleation has been induced by seeding with pre-existing crystals from other solutions, or by introducing a disturbance into the system that will promote heterogeneous nucleation. Once a system has reached a sufficient supersaturation (nucleation zone), then nucleation and crystal growth will occur spontaneously. At very high supersaturation the protein will precipitate from solution in an amorphous state – that is, it has reached the precipitation zone.

8.2.4
Protein Crystallization

Protein crystallization can often be seen as a difficult task, and is usually the major bottleneck in 3-D structure determination by protein crystallography. Crystallization is a multiparametric process that is influenced by number of physical, chemical, and biochemical factors (see Table 8.2). These factors are not independent of each other, and their influence on the crystallizability of a particular protein may vary. A systematic and complete variation of even most commonly optimized crystallization parameters such as pH, temperature, protein concentration, concentration of

Table 8.2 Factors affecting crystallization (from Ref. [33]).

Physical	Chemical	Biochemical
Temperature/temperature variation	pH	Purity of the macromolecule/impurities
Surfaces	Precipitant type	Ligands, inhibitors, effectors
Methodology/approach to equilibrium	Precipitant concentration	Aggregation state of the macromolecule
Gravity	Ionic strength	Post-translational modifications
Pressure	Specific ions	Source of macromolecule
Time	Degree of supersaturation	Proteolysis/hydrolysis
Vibrations/sound/mechanical perturbations	Reductive/oxidative environment	Chemical modifications
Electrostatic/magnetic fields	Concentration of the macromolecule	Genetic modifications
Dielectric properties of the medium	Metal ions	Inherent symmetry of the macromolecule
Viscosity of the medium	Cross-linkers/polyions	Stability of the macromolecule
Rate of equilibration	Detergents/surfactants/amphophiles	Isoelectric point
Homogeneous or heterogeneous nucleants	Non-macromolecular impurities	History of the sample

various precipitating agents and additives, would lead to a combinatorial explosion in a number of crystallization experiments. Clearly, this is not a feasible approach, although several commonly applied screening strategies are available to overcome this obstacle:

- *Incomplete factorials*: Carter and Carter [46] applied the incomplete factorial approach in experiment design, based on the simultaneous variation of several parameters. Instead of testing all possible combinations of experimental conditions (in their case 4032), it was sufficient to use a set of 35 initial search conditions to provide an even sampling of the entire experimental volume. The results were ranked in nine quality categories, and the effects of each of the 25 factors were evaluated statistically.
- *Sparse matrix sampling*: Jancarik and Kim designed a sparse matrix of trial conditions based on known or published crystallization conditions [47]. These authors grouped the crystallization parameters into three categories: pH and buffer materials; additives; and precipitating agents. This approach became a basis for the first commercially available crystallization screens.
- *Grid screens*: In this systematic approach, 2-D matrices were made and a single factor was sampled on each of the two axes. A description and the history of

the method can be found in Refs. [48–50,55]. In grid screens, the precipitant concentration is usually varied against pH [65]. A 2-D array has very limited coverage of a multi-dimensional space, so that an exhaustive search by screening many pair of variables must be performed and this, in turn, requires large amounts of protein.

The sparse matrix approach was extended by including further precipitating agents and additives, as well as by combining it with systematic screening [51,52], resulting in new screening protocols. Today, many crystallization screens are available commercially [53], and most are adapted to the microtiter 96-well format and therefore convenient to use on automated liquid dispenser systems (crystallization robots).

With few exceptions, the crystallization screening will rarely result in crystals of desired size and diffraction quality. After identifying conditions under which crystalline protein material is obtained, the optimization experiments must usually be started. Optimization is normally performed on a grid around the original conditions from the screen. The sampling around the original conditions can be either uniform or, like in Hardin–Sloane designs, non-uniform [54].

- *Crystallization techniques:* The most commonly used crystallization techniques are vapor diffusion, batch (micro-batch), and dialysis [57]. Vapor diffusion and micro-batch techniques are suitable for the application of automated crystallization systems (crystallization robots), and are therefore the methods of choice for crystallization screening. The principles of these techniques are shown schematically in Figure 8.6.

  Because it is very simple to set up, the sitting drop vapor diffusion technique is the most widely used in crystallization screening experiments. The solution to be crystallized against is dispensed at the bottom of the well, and the protein solution is dispensed into the small depression above the reservoir solution (Figure 8.6a). The two solutions are spatially separated. The crystallization trial is started by adding a certain amount of the reservoir solution to the protein droplet and sealing the system. As the system is sealed, the only transport of water or any other volatile component is between the droplet and the reservoir solution. The system reaches equilibrium slowly through vapor diffusion. In an ideal case, the protein in the droplet reaches supersaturation and crystal growth takes place. Some examples of 24- and 96-well microtiter plates used for protein crystallization are shown in Figure 8.7.

Micro-batch is another method suitable for the both manual and automated crystallization experiments (see Figure 8.6b). An automated system for micro-batch protein crystallization under an oil film was initially described by Chayen et al. [56]. The micro-batch crystallization technique allows the precise conditions of crystal growth to be known because they are reached instantly upon mixing the protein and the precipitant solution. Before mixing, the protein solution is undersaturated. After mixing, the precipitant solution influences the protein solubility to yield a mixture that is supersaturated with respect to protein. In contrast to the vapor diffusion technique, micro-batch does not suffer from concentration gradients

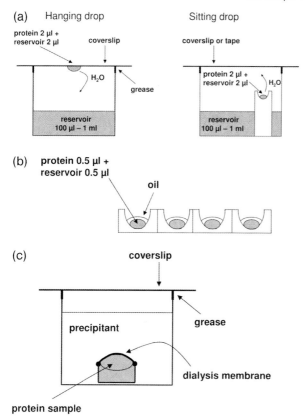

Figure 8.6 Most common crystallization techniques: (a) vapor diffusion, (b) microbatch, and (c) dialysis.

within the sample during the diffusion process. Different effects can be achieved, depending on which type of oil is used to cover the crystallization droplets. While paraffin oil does not allow water diffusion and the crystallization conditions are fixed, silicon oil will have water diffusing through it and the droplet will dehydrate in the course of time. In a modification of the method, where various mixtures of silicon and paraffin oil are used, the dehydration rate of the droplet can be fine-tuned.

A set-up for crystallization by dialysis is shown schematically in Figure 8.6c. Although it is not suitable for high-throughput crystallization, dialysis is the method that allows an extreme flexibility in changing conditions in a crystallization experiment. The very same protein sample can be dialyzed and eventually crystallized after testing and changing several precipitant solutions. Of course, this is possible only if the protein does not suffer irreversible damage during the experiments.

The corresponding paths in the solubility diagram for each of the three crystallization techniques are shown in Figure 8.8.

290 8 Structural Proteomics

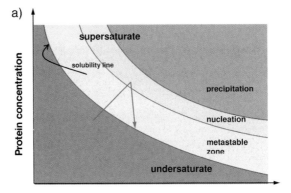

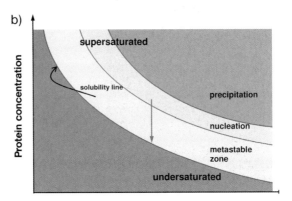

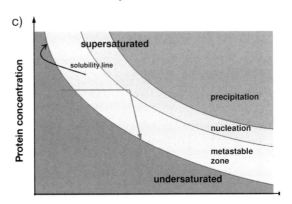

Figure 8.7 Solubility diagram in crystallization experiments using (a) vapor diffusion, (b) batch, (c) dialysis methods.

24-well plate

VDX   Corning

96-well plate

Corning

Figure 8.8 Microtiter plates for protein crystallization.

8.2.5
Practical Examples

Crystallization procedures for several proteins are described at http://www.rigaku.com/protein/crystallization.html (courtesy of J. Pflugrath). Additional crystallization recipes can be found at the homepage of Hampton Research (http://www.hamptonresearch.com). The description of the crystallization of human α-thrombin is described below.

- Materials
  - Human α-thrombin (Enzyme Research Laboratories, 1801 Commerce Drive, South Bend, IN 46628, USA)
  - Hirugen (aa 54–65 of hirudin Tyr^{63} sulfated) (Sigma-Aldrich; GDFEEIPEEY*LQ)

- Prepare the following thrombin solution
  - 4–6 mg mL^{-1} human α-thrombin
  - ∼2 mg mL^{-1} hirugen
  - 50 mM sodium phosphate buffer, pH 7.3
  - 1 mM NaN$_3$
  - 375 mM NaCl

- Crystallize using vapor diffusion technique against
  - 100 mM sodium phosphate buffer, pH 7.3
  - 1 mM NaN$_3$
  - 20–30% PEG 8000 (PEG 4000 or PEG6000 should also be effective)

References

31 Ducruix, A. and Giegé, R. (eds) (1999) *Crystallization of Nucleic Acids and Proteins*, Oxford University Press.

32 http://www.pdb.org.

33 McPherson, A. (1998) *Crystallization of Biological Macromolecules*, CSHL Press, pp. 67–126.

34 Stevens, A.R.C. (2000) *Structure*, **8**, R177–R185.

35 Juárez-Martínez, G., Steinmann, P., Roszak, A.W., Isaacs, N.W. and Cooper, J.M. (2002) *Anal. Chem.* **74**, 3505–3510.

36 Hansen, C.L., Skordalakes, E., Berger, J.M. and Quake, S.R. (2002) *Proc. Natl. Acad. Sci. USA*, **99**, 16531–16536.

37 Hughes, S.H. and Stock, A.M. (2001) in *International Tables for Crystallography F.* (eds M.G. Rossmann and E. Arnold), Kluwer Academic Publishers, Dordrecht, pp. 65–80.

38 Derewenda, Z.S. (2004) *Methods*, **34**, 354–363.

39 Stevens, A.R.C. (2000) *Structure*, **8**, R177–R185.

40 Pusey, M.L., Liu, Z.-J., Tempel, W., Praissman, J., Lin, D., Wang, B.-C., Gavira, J.A. II. and Ng, J.D. (2005) *Prog. Biophys. Mol. Biol.* **88**, 359–386.

41 Smyth, D.R., Mrozkiewicz, M.K., McGrath, W.J., Listwan, P. and Kobe, B. (2003) *Protein Sci.*, **12**, 1313–1322.

42 Giegé, R. and McPherson, A. (2001) in *International Tables for Crystallography F.* (eds M.G. Rossmann and E. Arnold), Kluwer Academic Publishers, Dordrecht, pp. 89.

43 Dale, G.E., Oefner, Ch. and D'Arcy, A. (2003) *J. Struct. Biol.* **142**, 88–97.

44 Chance, M.R., Bresnick, A.R., Burley, S.K., Jiang, J.-S., Lima, C.D., Sali, A., Almo, S.C., Bonanno, J.B., Buglino, J.A., Boulton, S., Chen, H., Eswar, N., He, G., Huang, R., Ilyin, V., McMahan, L., Pieper, U., Ray, S., Vidal, M. and Wang, L.K. (2002) *Protein Sci.*, **11**, 723–738.

45 Kwong, P.D., Wyatt, R., Desjardins, E., Robinson, J., Culp, J.S., Hellmig, B.D., Sweet, R.W., Sodroski, J. and Hendrickson, W.A. (1999) *J. Biol. Chem.* **274**, 4115–4123.

46 Carter, C.W., Jr. and Carter, C.W. (1979) *J. Biol. Chem.* **254**, 12219–12223.

47 Jancarik, J. and Kim, S.-H. (1991) *J. Appl. Crystallogr.* **24**, 409–411.

48 McPherson, A. (1985) *Methods Enzymol.* **114**, 112–120.

49 McPherson, A. (1985) *Methods Enzymol.* **114**, 120–125.

50 McPherson, A. (1985) *Methods Enzymol.* **114**, 125–127.

51 Cudney, B., Patel, S., Weisgraber, K., Newhouse, I. and McPherson, A. (1994) *Acta Crystallogr.*, **D50**, 414–423.

52. Newman, J., Egan, D., Walter, T.S., Meged, R., Berry, I., Ben Jelloul, M., Sussman, J.L., Stuart, D.I. and Perrakis, A. (2005) *Acta Crystallogr.*, **D61**, 1426–1431
53. Emerald Biosystems: http://www.emeraldbiosystems.com; Fluka: http://www.sigmaaldrich.com; Hampton Research: http://www.hamptonresearch.com; Jena Bioscience: http://www.jenabioscience.com; Molecular Dimensions: http://www.moleculardimensions.com; Nextal/Qiagen: http://www.nextalbiotech.com.
54. Carter, C.W. (1999) in *Crystallization of Nucleic Acids and Proteins* (eds A. Ducruix and R. Giegé), Oxford University Press, pp. 75–120.
55. McPherson, A. (1998) *Crystallization of Biological Macromolecules*, Cold Spring Harbor Laboratory Press, pp. 271–329.
56. Chayen, N.E., Shaw Stewart, P.D., Maeder, D.L. and Blow, D.M. (1990) *J. Appl. Crystallogr.*, **23**, 297–302.
57. McPherson, A. (1998) *Crystallization of Biological Macromolecules*, Cold Spring Harbor Laboratory Press, pp. 159–214.
58. Scopes, R.K. (1993) *Protein Purification: Principles and Practice*, 3rd edn. Springer-Verlag, New York.
59. Roe, S. (ed) (2001) *Protein Purification Techniques: A Practical Approach*, 2nd edn. Oxford University Press, pp. 27–38.
60. Coligan, J.E., Dunn, B.M., Ploegh, H.L., Speicher, D.W. and Wingfield, P.T. (eds) (1995–2007) *Current Protocols in Protein Science*, John Wiley & Sons, Inc.
61. Hatti-Kaul, R. and Mattiasson, B. (eds) (2003) *Isolation and Purification of Proteins*, Marcel Dekker, Inc.
62. Zulauf, M. and D'Arcy, A. (1992) *J. Cryst. Growth* **122**, 102–106.
63. D'Arcy, A. (1994) *Acta Crystallogr.*, **D50**, 469–471.
64. Bergfors, T.M. (1999) *Protein Crystallization: Techniques, Strategies and Tips*, International University Line, La Jolla, pp. 29–38.
65. Weber, P.C. (1990) *Methods*, **1**, 31–37.

9
Interaction Analysis

9.1
Sample Preparation for Protein Complex Analysis by the Tandem Affinity Purification (TAP) Method

Bertrand Séraphin and Andrzej Dziembowski

9.1.1
Introduction

Due to the development of high-speed DNA sequencing and computing tools for handling large quantities of data, the last few years of the 20th century experienced an extraordinary accumulation of genetic information, including the determination of the first full genomic sequences. These analyses revealed the existence of numerous open reading frames (ORFs) encoding thousand of proteins. If the function of some of these proteins could be predicted from biocomputing analyses, the sequences of protein of unknown function were identified at an astonishing rate. Henceforth, protein databases – which originally contained exclusively the sequences of functionally characterized proteins – are now mostly filled with predicted translation products. Given the central role of proteins in mediating biological activities, the development of new technologies that would allow the parallel analysis of numerous protein samples became necessary, and these were regrouped under the name "proteomic analyses". Many parameters must be determined in order to establish the protein profile of a given cell: this involves not only knowing which proteins are expressed but also their states of modification (phosphorylation, methylation, acetylation, N-terminal modification, processing sites, etc.), their localizations, and their association with partner proteins. Moreover, each of these parameters must be determined quantitatively, including protein levels that vary over several orders of magnitudes, and the relationships between the various parameters must also be determined as one modification may affect another type of modification and/or the localization of a given factor. Thus, unlike DNA analyses – where sequence information is most often sufficient (given that in most cases one or two gene copies are present per cells)– proteomic analyses necessitate the determination of a wide variety of both qualitative and quantitative data. Thus, many different technologies are required for proteomic analyses; these include protein sequence determination and the identification of post-translational modifications which, today, are most often be performed by using mass spectrometry (see the other chapters in this book). The assessment of absolute or relative protein levels can be made by using a variety of techniques, including by dimensional gel electrophoresis, Western blotting, or mass spectrometry. However, given that proteins in cells are present over a wide range of concentrations, it is often difficult to obtain a comprehensive analysis of all protein levels for any one given cell. A similar limitation applies to post-translational modifications, which often are not only difficult to detect but also highly labile. The analysis of protein interaction is also difficult, both because a given protein may interact with many different partners, and because such an association may be transient in nature. Again, it was necessary to develop a variety of methodologies in order to allow such analyses at the proteomic scale. In this chapter, the details are described of the technique used to prepare samples for analysis by "tandem affinity purification" (TAP), a method developed to allow the identification of interacting protein partners, at the proteomic scale.

9.1.2
The Problem with Regards to Sample Preparation: The Pros and the Cons

Various methodologies have been developed to allow the identification of protein interaction patterns in high-throughput proteomic experiments. Some are based exclusively on molecular biology techniques, and this clearly applies to the two-hybrid approach [1]. Although such strategies benefit from the standard methodologies used for handling nucleic acids, they suffer from the drawback that these interactions are tested in an artificial environment, such as different protein concentrations, an absence of possible protein modifications, different cellular compartments, and an absence of some partners.

Alternatively, protein interaction may be identified in biochemical experiments. When using such an approach, the protein of interest (the "target") is purified and specifically associated partners are identified, most often by mass spectrometry analyses. The various strategies used for biochemical identification of interacting protein partners can be classified based on the nature of the method used for purification (e.g., classical biochemical approach, use of antibodies, use of tagged protein) and whether the partners were presupposed to interact *in vivo* or whether the interaction was reconstituted *in vitro*. For the latter strategy, the target protein, which often is produced recombinantly, is incubated in an extract containing putative partners. After an incubation which allows the partner to associate, the target protein is selectively recovered (most often using tag-based purification or immunoprecipitation with specific antibodies) together with interacting extract factors which are then identified. One frequently used strategy of this type is the so-called "GST pull-down method" [2]. This type of approach has two main advantages:

- Large quantities of the target protein can be used. By the law of mass action, this allows the association of the target protein even with low-affinity partners or those present at low concentration.

- The recombinant protein does not need to be expressed in the target host cells; thus, it is applicable to every protein from which biological material is naturally available.

Unfortunately, however, these methods suffer from several drawbacks. Indeed, the procedure may only succeed if the interacting partners are not already tightly bound to the endogenous target protein, and if the reassembly process can take place in the extract. This may not always be the case, as complex formation may occur in several steps, some of which require catalytically active factors and thus may not occur efficiently in extracts. Often, endogenous proteins are also present in stable assemblies, and consequently these factors will not be available to interact with the target protein, thereby reducing the power of this approach.

Because of this limitation, the option was to purify the protein complexes preassembled in cells. As proteomic-scale analyses require a high throughput and wide coverage of the protein expressed in cells under study, classical biochemical purification was deemed unsuitable. Indeed, this strategy requires the empirical development of an adapted purification procedure for each of the target proteins, as these all differ in their biochemical properties. Strategies based on coimmunoprecipitation could be envisaged, but these require the generation of many antisera – a step which is itself difficult, rather time-consuming, and expensive. The major limitation of antibody-based strategies is that antibody quality is highly variable. Thus, cross-reactions and/or the presence of non-specific antibodies in whole sera may often lead to the coprecipitation of non-specific contaminants. Antibody affinities are also highly variable; low-affinity sera will result in low yields, thus preventing the purification of low-abundance target and their partners.

A strategy based on the purification of tagged protein expressed in cells of interest was therefore selected. Indeed, this allows the purification of the target protein under relatively standard conditions, and so permits the (parallel) processing of numerous samples. The selection of high-affinity tags ascertains that, even the lowest abundance target protein and their partners will be recovered. Because some levels of contaminants constituted mostly by abundant cellular proteins are always found in any purification step, the new our strategy was based on two successive affinity purification steps [3]. This feature is essential if low-abundance targets and interacting factors (which often are present in substoichiometric amounts) are to be recovered and must be discriminated specifically from background contaminants.

Despite these advantages, this strategy has also certain limitations. First, the need to express in the cells of interest a recombinant gene copy encoding a tagged fusion of the target protein limits its application to cells and/or organisms that can be genetically manipulated. When methodologies for recombinant protein expression are available, transformed cells will in most cases express both the target protein fused to the purification tag and an untagged version encoded by the natural copy of the gene. Thus, the tagged protein will be in competition with an untagged factor for complex assembly, and this may limit the recovery yields. It is also desirable to maintain the expression of the target protein close to its natural level, in order to avoid the formation of an unspecific interaction. This feature is sometimes difficult to achieve, especially if sequences controlling expression of the target factor have not been characterized. These various problems (construction of recombinant organisms, limited ability to inhibit expression of the endogenous target copies, inaccurate control of expression) stem from the limits of the present technology for gene manipulation. However, they will differ from organism to organism, and many parameters may be improved for a given target if sufficient efforts are invested.

Other limitations result directly from the use of a recombinant tagged factor for purification: the mere presence of a tag on a given protein may affect its function, leading in some cases to incorrect cellular localization, an inability to interact with partners, and/or instability. The only remedy in such cases seem to be to change the fusion protein itself, either by fusing the purification tag at another location or by selecting another purification tag. Because of the possible steric hindrance of the tag, some authors have suggested that smaller tags should be preferred. However, if smaller tags are to be tolerated as they fit into crowded environments, they may be inaccessible in such contexts and therefore useless for purification purposes. Moreover, smaller tags are unlikely to adopt a stable fold and thus may be more prone to interaction with "non-specific" partners (e.g., chaperone) than larger tags that will adopt an easily soluble globular structure. Interestingly, large-scale tagging experiments have revealed that proteins are usually highly tolerant to the addition of a tag at their C-terminus; for example, over 93% of the essential yeast proteins remained functional when fused to a ~20 kDa tag [4]. It is likely that steric problems explain the 7% of cases where such a large addition was not tolerated. The addition of a longer linker between the target protein and the purification tag, or fusion of the tag at the protein N-terminus, might have been envisaged to solve such problems.

Overall, affinity purification appears to be the method of choice to identify protein complexes and to analyze protein interaction. Although the method has some clear limitations and will not be successful for all target proteins, it does offer the possibility of rapidly analyzing complex formation and protein interaction, thereby allowing for a change in the scale of analysis of protein samples. Moreover, because analyses are performed at the protein level from assemblies recovered from the natural host, direct information on interaction is gained. This contrasts with other strategies, such as the two-hybrid analysis, where partners may be missing and for which interaction must be inferred from indirect data. The success of affinity purification for complexes relies on the use of high-affinity tags that will allow the recovery of low-abundance complex with high yield, and under purification conditions that should be sufficiently gentle to maintain even weak biologically relevant interactions.

Comparative testing led the group of the present authors to choose a calmodulin-binding domain (CBD) and IgG binding domains of *Staphylococcus aureus* protein A (ProtA) for efficient purification under native conditions, and to combine them separated with a tobacco etch virus (TEV) protease cleavage site to create the original TAP tag (Figure 9.1) [3,5]. This tag has proved extremely powerful for large-scale studies (e.g., Ref. [6]), and an example of its use is described in the following section.

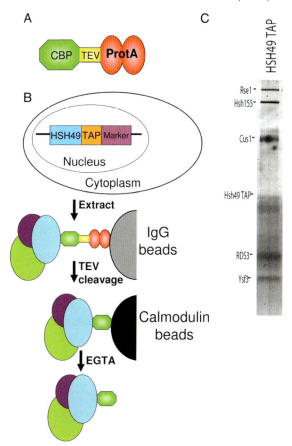

Figure 9.1 Principle and example of tandem affinity purification (TAP) purification. (A) Structure of the TAP tag. For the C-terminal tag, the calmodulin binding peptide (CBP, green) is followed by a cleavage site for the TEV protease (TEV, yellow) and two IgG binding modules of the *Staphylococcus aureus* protein A (ProtA, the two modules are depicted in red). These modules are in the opposite order for the N-terminal TAP tag. (B) Principle of a TAP purification. The TAP purification of the Hsh49 protein and associated yeast SF3b splicing factor [7] is shown as an example. The C-terminal TAP tag coding sequence and an associated marker are inserted in the yeast genome downstream of the HSH49 coding sequence in a haploid strain. The construction is performed by transforming a PCR fragment containing the TAP-Marker cassette flanked by genomic sequences in such a way that homologous recombination will fuse the entire HSH49 coding sequence in frame with the TAP coding sequence. After verification, the strain is used to prepare extracts. These are used for a first affinity purification on IgG agarose (or Sepharose) beads. Retained complexes are selectively eluted by TEV protease cleavage before a second affinity purification step on calmodulin agarose beads. Purified complexes can then be eluted by addition of EGTA that chelates calcium, analyzed using gel electrophoresis. (C) Purified proteins interacting with the tagged target may be identified by mass spectrometry. Proteins copurifying with Hsh49 were fractionated using gel electrophoresis and stained with silver. The position of migration of the yeast SF3b subunits are indicated [7].

9.1.3
Application Example

As an example of the application of the TAP tag, purification of the SF3b complex from yeast will be described at this point. The SF3b complex is an hexameric protein assembly involved in pre-mRNA splicing [7]. To purify this particle, the TAP tag was fused to a protein, Hsh49, which is a specific subunit of this factor. For this purpose, a homologous recombination strategy was used to insert the TAP tag encoding cassette, and a selection marker, downstream of the Hsh49 coding sequence (Figure 9.1b). This strategy allows for the in-frame fusion of the TAP tag coding sequence and the target encoding ORF. Thus, in the recombinant strain, expression of the fusion protein remains under the control of the natural promoter regulatory sequences. It should be noted, however, that other sequences which naturally control expression of the target gene may have been disrupted (e.g., if sequences located downstream of the ORF regulated translation, this control would have been lost). This powerful strategy also allows the elimination of all untagged protein from the cell, as the single endogenous gene of the haploid yeast strain is converted into the tagged one. This ensures that only TAP tagged version of the factor is produced, and demonstrates further that the tagged version of the gene is functional, because the Hsh49 protein is essential for yeast viability and the tagged strain grows essentially like the wild-type. Unfortunately, in many other organisms, this powerful strategy for TAP tag fusion cannot be applied and the construction of recombinant cells or organisms expressing the desired tagged target must be performed in accordance with the specific strategies suitable for the generation of recombinant cells in the target species. Strains generated by homologous recombination were validated by polymerase chain reaction (PCR) to ensure that insertion of the tag occurred correctly and by Western blotting using reagents detecting the protein A moiety of the tag to ascertain that the tagged protein was produced.

For TAP purification, the tagged strain is routinely grown to the late logarithmic phase in 2 L of rich medial this allows the recovery of approximately 10 g of cells. The cells are broken using a French press and the cell debris eliminated by centrifugation at $90\,000 \times g$ for 85 min at 4°C (Ti 45 Beckman rotor, or equivalent rotors; for protocol details, see: http://www.embl-heidelberg.de/ExternalInfo/seraphin/TAP.html). The resultant extract is then dialyzed against a glycerol-containing buffer before freezing. Again, although this extract preparation procedure was shown empirically to work well for many yeast complexes, it is certainly neither optimal for every cellular assembly nor adapted for other organisms. The TAP purification is initiated by mixing the defrosted extract with beads carrying IgG molecules (usually of rabbit or human origin, as these bind tightly to protein A). After a first binding step (Figure 9.1b), the unbound material was eluted and the proteins interacting non-specifically were removed by extensive washing. In order to elute the bound complex, TEV protease is incubated with the beads in an appropriate buffer. The eluate is then incubated with beads carrying calmodulin molecules in a calcium-containing buffer; this allows interaction of the CBD peptide with calmodulin. The unbound contaminants are then eluted and residual traces eliminated by additional washing steps. The target complex is then eluted by the addition of EGTA that chelates calcium, thereby disrupting the CBD–calmodulin interaction, which is calcium-dependent. Eluted fractions are then concentrated and their content analyzed by gel electrophoresis (Figure 9.1c). The various protein bands are then identified, revealing Hsh49 partners. This final is currently most often performed using mass spectrometry analysis.

9.1.4
Summary

Protein purification is a method of choice to identify partners of a given factor. Although a standard biochemical purification may be used for this purpose, this often requires a large amount of work to develop a specific purification strategy for a protein of interest. In order to facilitate and "standardize" protein complex purification, a method was developed based on two successive affinity purification steps; this was the TAP strategy [3,5]. This method has proven useful for purifying and characterizing the composition of numerous protein complexes from a wide variety of organisms. Given the ever-growing number of putative protein sequences present in databases, it is likely that the TAP method will remain a valuable tool for the identification of protein interaction during the forthcoming years. Moreover because TAP-purified complexes remain active, the TAP method will certainly become a useful tool for the functional characterization of the numerous proteins identified by genome-sequencing projects.

9.1.5
Perspective

Since the first description of the TAP tag, additional tags allowing the recovery of protein complexes by two (or more) successive affinity purification steps have been described (e.g., Ref. [8]). Although, in most cases advantages for the specific tag combination selected have been described, at present it is difficult to draw conclusions regarding the real advantages of one tag over another because comparative data are often unavailable, or are limited to a few examples. Future large-scale studies will most likely reveal which tag combination is best suited for complex purification, although it is possible that the optimal tag may vary from one target protein to the next. While awaiting the outcome of these optimization studies, the TAP tag may be safely used for complex purification, as evidenced by the numerous analyses reporting its use in many different biological systems.

9.1.6
Recipe for Beginners

Detailed protocols for the use of the TAP method are available, including buffer compositions ([3,5]; http://www.embl-heidelberg.de/ExternalInfo/seraphin/TAP.html). A basic description of the various steps is as follows:

- The construction of cells or organisms expressing the desired TAP tagged target protein should be performed following strategies specifically available for the system under study. Validation of expression of the target protein is achieved with Western blotting.

- Extracts are prepared using, again, a strategy that is adapted both to the organism of interest and to the complex analyzed. Conditions should be kept as gentle as possible in order to maintain complex integrity. Presence of the complex can at all stages be assessed by detection of the tagged protein with Western blotting.

- The extract is mixed with washed IgG beads (200 μL of suspension, for 10 mL of extracts or ca. 200 mg of protein) for 2 h. Unbound material is discarded and beads are extensively washed.

- TEV protease is added and incubated with beads for 2 h at 16°C.

- The eluate is recovered and calcium is added; this fraction is then incubated with calmodulin beads for 1 h.

- Unbound material is discarded and beads are extensively washed.

- A buffer containing EGTA is added to elute the bound complex.

- The purified proteins are recovered, concentrated and analyzed by gel electrophoresis before identification by mass spectrometry.

References

1 Serebriiskii, I. (2005) Yeast two-hybrid system for studying protein-protein interactions, in *Protein-Protein Interactions: A Molecular Cloning Manual* (Eds. E.A. Golemis and P.D. Adams), Cold Spring Harbor Laboratory Press, Cold Spring Harbor, pp. 167–194.

2 Einarson, M.B., Pugacheva, E.N. and Orlinick, J.R. (2005) Identification of protein-protein interactions with glutathione-S-transferase fusion proteins: in *Protein-Protein Interactions: A Molecular Cloning Manual* (Eds E.A. Golemis and P.D. Adams), Cold Spring Harbor Laboratory Press, Cold Spring Harbor, pp. 81–104.

3 Rigaut, G., Shevchenko, A., Rutz, B., Wilm, M., Mann, M. and Séraphin, B. (1999) A generic protein purification method for protein complex characterization and proteome exploration. *Nat. Biotechnol.*, **17** (10), 1030–1032.

4 Ghaemmaghami, S., Huh, W.K., Bower, K., Howson, R.W., Belle, A., Dephoure, N., O'Shea, E.K. and Weissman, J.S. (2003) Global analysis of protein expression in yeast. *Nature*, **425** (6959), 737–741.

5 Puig, O., Caspary, F., Rigaut, G., Rutz, B., Bouveret, E., Bragado-Nilsson, E., Wilm, M. and Séraphin, B. (2001) The tandem affinity purification (TAP) method: a general procedure of protein complex purification. *Methods*, **24** (3), 218–229.

6 Gavin, A.C., Bosche, M., Krause, R., Grandi, P., Marzioch, M., Bauer, A., Schultz, J., Rick, J.M., Michon, A.M., Cruciat, C.M., Remor, M., Hofert, C., Schelder, M., Brajenovic, M., Ruffner, H., Merino, A., Klein, K., Hudak, M., Dickson, D., Rudi, T., Gnau, V., Bauch, A., Bastuck, S., Huhse, B., Leutwein, C., Heurtier, M.A., Copley, R.R., Edelmann, A., Querfurth, E., Rybin, V., Drewes, G., Raida, M.,

Bouwmeester, T., Bork, P., Séraphin, B., Kuster, B., Neubauer, G. and Superti-Furga, G. (2002) Functional organization of the yeast proteome by systematic analysis of protein complexes. *Nature*, **415** (6868), 141–147.

7 Dziembowski, A., Ventura, A.P., Rutz, B., Caspary, F., Faux, C., Halgand, F., Laprevote, O. and Séraphin, B. (2004) Proteomic analysis identifies a new complex required for nuclear pre-mRNA retention and splicing. *EMBO J.*, **23** (24), 4847–4856.

8 Honey, S., Schneider, B.L., Schieltz, D.M., Yates, J.R. and Futcher, B. (2001) A novel multiple affinity purification tag and its use in identification of proteins associated with a cyclin-CDK complex. *Nucleic Acids Res.*, **29** (4), E24.

9.2
Exploring Membrane Proteomes

Filippa Stenberg and Daniel O. Daley

9.2.1
Introduction

Around one quarter of all proteins in a cell are membrane-embedded [9,10]. Considering that many soluble proteins are also tethered to cellular membranes through lipid moieties, hydrophobic patches, charge-interactions, or in membrane–protein complexes, it is not unreasonable to suggest that 30 to 40% of a cellular proteome is membrane-associated. Membrane-embedded proteins carry out many pivotal roles in a cell: they control the uptake of nutrients and ions, they sense the external environment, and generate energy during respiration and photosynthesis. As a result of these pivotal roles, these proteins constitute an important target for the pharmaceutical industry, where more than 70% of current drug targets are membrane proteins [11].

Unfortunately, membrane proteins (as a whole) have been poorly studied compared to their soluble counterparts. This difference stems from the fact that they are generally incompatible with commonly used biochemical methodologies, and therefore inaccessible to high-throughput proteomic and structural genomic approaches. Although no specific physical property of a membrane protein can account for this incompatibility, membrane proteins can be investigated in proteomic studies if considerations are made and sample preparation/biochemical assays are adapted accordingly.

In this chapter, some of the pitfalls of working with membrane proteins will be highlighted, and considerations that must be made so that they do not become casualties in proteomic studies identified. In particular, the chapter will cover the basic problems of defining, separating, and identifying membrane proteomes. In addition, problems associated with obtaining an interaction network and obtaining structural information will be discussed.

9.2.2
Defining Membrane Proteomes

In comparison to many other classes of proteins, the identification of membrane proteins from whole genome sequences is not difficult. Sequence-based prediction programs (i.e., TMHMM [9]) can, with ≤99% certainty, distinguish multi-spanning α-helical membrane proteins from soluble proteins. This means that the main component of a membrane proteome can be easily identified in minutes from the genome sequence alone. This exercise has been carried out for almost all sequenced genomes, and indicates that 20 to 30% of proteins in all cells are membrane- embedded [9,10].

Prediction programs still struggle with the identification of two subclasses of the membrane proteome; single-spanning α-helical membrane proteins and β-barrel membrane proteins. Single-spanning α-helical membrane proteins cannot easily be distinguished from soluble proteins with a secretion peptide, and their exact number therefore remains unclear. β-barrel membrane proteins are only present in the outer membranes of Gram-negative bacteria, mitochondria and chloroplasts, and are thought to represent 2 to 3% of a cellular proteome [12–14]. Again, this number is difficult to assess as β-barrel-type membrane proteins are not particularly hydrophobic, which makes them difficult to distinguish from soluble proteins that contain β-sheets. Despite these difficulties, some progress has been made in the identification of both single-spanning α-helical [15], and β-barrel membrane proteins [13,16].

Predicted membrane proteomes can also be cross-referenced with protein localization algorithms, to obtain a predicted organellar proteome [17]. However, it should be noted that protein localization predictors are less accurate for membrane proteins than for soluble proteins [18]. All in all, an accurate assessment of the membrane proteome of a cell can be made from genome sequence using sequence-based predictors.

9.2.3
Separation of Membrane Proteomes

A key problem in the characterization of a membrane proteome arises when trying to separate the constituent proteins in an experimental setting. In recent years the most popular approach for the experimental separation of proteomes has been two-dimensional electrophoresis (2-DE) – that is, isoelectric focusing (IEF) followed by a second dimension of SDS–PAGE. Although this approach has been widely used on almost all cellular/subcellular proteomes, it is now well established that it is discriminatory to membrane proteins as their alkaline pH makes them incompatible with commonly used pI ranges during IEF [19,20]. More significantly, their hydrophobic nature makes them poorly soluble in the aqueous media used for IEF, where they irreversibly precipitate and do not transfer efficiently into the SDS–PAGE dimension [19–21]. This fact is supported by the observation that there is an apparent limit in the GRAVY (Grand Average of Hydrophobicity) score for proteins entering into 2-DE [22]. Although various solubilization "tricks" have been trialed to overcome these difficulties (for a review, see Ref. [20]), the prognosis for membrane proteins in 2-DE is still poor, and they are generally considered refractory to the approach.

Alternative gel-based proteomic methods have therefore been developed specifically for membrane proteins (see Table 9.1). To date, the majority of these methods have not been accepted by the proteomics community because of the limited number of proteins which can be separated by their use. In contrast, the Blue Native (BN)–PAGE method of

Table 9.1 Gel-based methods for the separation of membrane proteins.

Method	Principle	Reference
BN–/SDS–PAGE	Membrane proteins are solubilized as complexes, then mixed with Coomassie dye to give them a net negative charge and separated according to their molecular mass in a native gel. The BN-gel can be analyzed in a second dimension by SDS–PAGE, where individual proteins are resolved in ``vertical channels'' according to their molecular mass.	[23]
CN–/SDS–PAGE	Same basic principle as for BN–PAGE, but Coomassie dye is omitted from sample and cathode buffer. Protein complexes are therefore resolved by their own net charge. Resolution is not as good as BN–PAGE, but the approach is milder on complexes than BN–PAGE.	[78]
16-BAC/SDS–PAGE	Membrane proteins are solubilized by the cationic detergent benzyldimethyl-*n*-hexadecylammonium chloride (16-BAC) and separated in an acidic buffer and gel system. To identify the individual proteins in each complex, the BAC gel is analyzed in a second dimension by SDS–PAGE.	[79]
dSDS–PAGE	Membrane proteins are separated by successive rounds of SDS–PAGE. Protein spots in the second dimension are dispersed around a diagonal. As hydrophobic proteins travel faster in SDS–PAGE, they move off the diagonal, and are thus separated from soluble proteins.	[80]
Bicine–dSDS–PAGE	Same principle as for dSDS–PAGE, but using a bicine buffer system, which claims to resolve more membrane proteins.	[81]
TFE–dSDS–PAGE	Membrane protein samples are first separated by SDS–PAGE where very stable protein interactions are maintained. In the second dimension SDS-gels are exposed to small membrane-active alcohols (i.e., TFE) and further separated by a second dimension of SDS–PAGE. Protein complexes not affected by the alcohols will migrate on a diagonal, whereas protein complexes that are dissociated by the alcohols will migrate off the diagonal.	[82]

CN, Clear native; TFE, 2,2,2-triflouroethanol.

Schägger and von Jagow [23] has been widely used for the separation of membrane proteins from whole-cell lysates [24], mitochondria [25–29], chloroplasts [30], peroxisomes [31], and bacteria [32–36]. In the BN–PAGE method, membrane proteins are solubilized as complexes by mild detergents and mixed with Coomassie dye to give them a net negative charge. They are then separated according to their molecular mass in a native gel (for a review, see Ref. [37]). Individual proteins in each complex can be

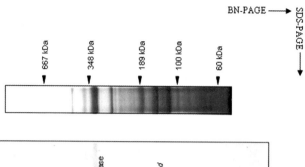

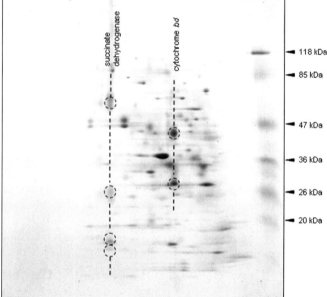

Figure 9.2 BN–/SDS–PAGE separation of the inner membrane proteome of E. coli. Protein complexes were separated according to their molecular mass in the BN–PAGE (top strip), and constituent proteins resolved into vertical channels in the SDS–PAGE. The gel has been stained with Coomassie. Two "channels" which contain membrane protein complexes (i.e., succinate dehydrogenase and cytochrome bd) are indicated. (Adapted from [34]).

further analyzed by a second dimension of SDS–PAGE (Figure 9.2). In SDS–PAGE, the proteins are consequently resolved into vertical "channels", enabling visualization of the individual constituents. BN–PAGE is therefore unique from other gel-based methods in that it enables protein interactions to be explored. It is also an attractive proteomic solution for basic expression profiling of membrane proteins, and can be used to complement traditional 2-DE [17] as it does not discriminate against membrane proteins. A BN–PAGE protocol for beginners is included later in this chapter.

Finally, traditional SDS–PAGE continues to provide a useful platform to separate membrane proteomes. Although resolution is far inferior to that of 2-D approaches, it can be effective if coupled to sensitive mass spectrometric techniques such as LC/MS/MS (see below).

9.2.4
Experimental Identification of Membrane Proteins

The difficulties in handling membrane proteins do not end after gel-based separation. Membrane proteins are also notoriously difficult to identify by conventional mass spectrometry – that is, when peptide fragments are used to obtain a "mass fingerprint". In this approach, trypsin is most commonly used to digest protein samples, because it reliably cleaves proteins at predictable sites (i.e., arginine (R) and lysine (K) residues), that generate peptides in the preferred mass range for mass spectrometric analysis.

However, biased R and K distributions in membrane proteins mean that tryptic digestion yields fragments that are often incompatible with mass spectrometric identification. The uneven distribution of R and K in membrane proteins stems from the fact that they are energetically unfavorable to be in a membrane environment [38]. Further, as stated by the "positive-inside rule" [39], they are preferentially positioned in cytoplasmic loops [40,41]. Tryptic digestion of membrane proteins therefore yields peptides that are often too long and hydrophobic for mass spectrometry [21,42]. As a consequence, membrane proteins are usually identified from peptides in their soluble domains. Proteins that do not have sufficiently large soluble domains will not yield enough tryptic peptides to obtain satisfactory sequence coverage of the protein, and therefore a statistically significant match in database searches.

An easy solution to the problem of identifying membrane proteins has unfortunately not yet surfaced. One obvious suggestion has been to use proteases that do not discriminate against membrane proteins. For example, cyanogen bromide (in conjunction with trypsin) cleaves proteins to a length and hydrophobicity which make them amenable to mass spectrometric analysis [42–44]. Alternatively, addition of the detergent octyl-b-glucopyranoside during the extraction of peptides from gel pieces can prevent losses of the more hydrophobic peptides [44]. A technically more challenging solution is the "shaving" of soluble domains from membrane proteins by digestion with non-specific proteases, such as proteinase K in a basic pH [45]. Although this approach improves sequence coverage by generating overlapping peptides, it is not an end-point phenomenon like trypsin digestion, and must therefore be carefully monitored [46].

To alleviate the need for high sequence coverage, it is possible to use more sensitive mass spectrometric techniques such as MS/MS, where proteins can be unambiguously identified by as little as two tryptic peptides [47]. Further, MS/MS can be coupled to non-gel separation techniques such as liquid chromatography (LC) (the "shotgun approach") to obtain high-throughput identifications of membrane proteins [45,48]. This approach is probably the most effective method currently available for the identification of membrane proteomes.

9.2.5
Mapping Membrane Interactomes

A major goal in the post-genomics era is to delineate the myriad of protein interactions within a cell. A number of high-throughput studies have attempted to

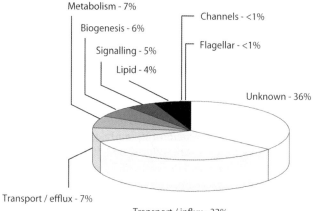
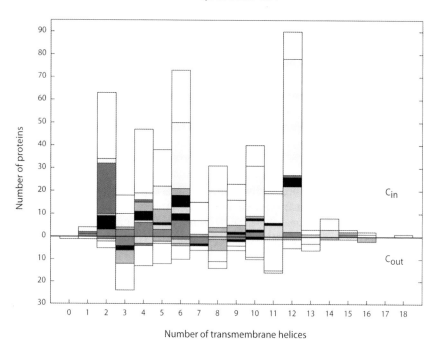

Figure 9.3 Functional and topological characterization of the *E. coli* inner membrane proteome. Top panel: Fraction of the proteome (737 proteins) assigned to different functional categories. Bottom panel: Numbers of proteins with different topologies (601 proteins in total). C_{in} topologies are plotted upwards, C_{out} downwards. Again, each functional category is shown. (Reproduced from [67]).

proteomes remain ill-defined. Unfortunately, no specific physical property can explain why membrane proteins are recalcitrant, but growing awareness of their importance has fuelled innovative approaches to solve this problem. Indeed, recently conducted studies have begun to shed light on these "forgotten" proteomes.

9.2.8
Recipe for Beginners

BN–/SDS–PAGE can be used to separate membrane proteomes for expression profiling, or to explore membrane interactomes. In this basic protocol, sample preparation usually begins with the preparation of membrane fractions.

9.2.8.1 Sample Preparation

- Resuspend approximately 100 μg of the prepared membrane fraction in 85 μL of ACA_{750} buffer (750 mM n-amino-caproic acid, 50 mM Bis-Tris, 0.5 mM Na_2EDTA, pH 7.0).

- Solubilize the membrane–protein complexes by addition of 0.5–1.5% (w/v) n-dodecyl β-D-maltoside and mix gently but thoroughly.

- Incubate the sample on ice for 20 min.

Note: The choice of detergent and detergent concentration should be empirically determined for each sample.

- Clear the solubilized membrane protein complexes by centrifugation at 264 000 × g for 30 min at 4 °C.

- Add the supernatant to 15 μL of G250 solution [5% (w/v) Coomassie G250 in ACA_{750} buffer].

9.2.8.2 BN–PAGE

Gel dimensions must be carefully considering when pouring BN-gels, so that they are compatible in size and thickness for the second-dimension SDS–PAGE. Typically, a Hoefer SE-600 system is used, with gel dimensions of 14 cm × 16 cm × 1.5 mm.

- Prepare a gradient separating gel [5–15% acrylamide, 250 mM n-amino-caproic acid, 25 mM Bis-Tris pH 7.0, 10% (w/v) glycerol].

- Cover with a stacking gel [4% acrylamide 250 mM n-amino-caproic acid, 25 mM Bis-Tris, pH 7.0].

- Prepare cathode buffer [50 mM Tricine, 15 mM Bis-Tris, 0.02% (w/v) Coomassie G250, pH 7.0 and 0.03% (w/v) n-dodecyl β-D-maltoside] and anode buffer [50 mM Bis-Tris, pH 7.0].

- Add sample to the gel and begin electrophoresis at 75 V for 1 h (in cold room).

- Change electrophoresis conditions to 10 mA/gel with a current cap of 380 V.

- Destain the gel [10% (v/v) acetic acid, 30% (v/v) methanol, 2% (w/v) glycerol] for visualization of the protein complexes.

9.2.8.3 SDS–PAGE

Lanes can be cut it into strips for analysis in a second dimension.

- Soak the BN-gel strip for 20 min in equilibration buffer [2% (w/v) SDS, 250 mM Tris–HCl, pH 6.8].

- Place on top of a standard SDS–PAGE (set without a comb in the stacking gel) and electrophorese as per standard protocols.

Note: Placing the BN strip on top of the SDS-gel is most easily done by plying the glass plates apart.

Acknowledgments

These studies were supported by a grant from the Swedish Research Council to D.O. D. The authors thank Erik Granseth for help with the figures.

References

9 Krogh, A., Larsson, B., von Heijne, G. and Sonnhammer, E.L. (2001) *J. Mol. Biol.*, **305**, 567–580.

10 Wallin, E. and von Heijne, G. (1998) *Protein Sci.*, **7**, 1029–1038.

11 Byrne, B. and Iwata, S. (2002) *Curr. Opin. Struct. Biol.*, **12**, 239–243.

12 Wimley, W.C. (2002) *Protein Sci.*, **11**, 301–312.

13 Zhai, Y. and Saier, M.H., Jr (2002) *Protein Sci.*, **11**, 2196–2207.

14 Casadio, R., Fariselli, P., Finocchiaro, G. and Martelli, P.L. (2003) *Protein Sci.*, **12**, 1158–1168.

15 Käll, L., Krogh, A. and Sonnhammer, E.L. (2004) *J. Mol. Biol.*, **338**, 1027–1036.

16 Martelli, P.L., Fariselli, P., Krogh, A. and Casadio, R. (2002) *Bioinformatics*, **18** (Suppl 1), S46–S53.

17 Heazlewood, J.L., Tonti-Filippini, J., Verboom, R.E. and Millar, A.H. (2005) *Plant Physiol.*, **139**, 598–609.

18 Millar, A.H., Whelan, J. and Small, I. (2006) *Curr. Opin. Plant Biol.*, **9**, 610–615.

19 Santoni, V., Molloy, M. and Rabilloud, T. (2000) *Electrophoresis*, **21**, 1054–1070.

20 Molloy, M.P. (2000) *Anal. Biochem.*, **280**, 1–10.

21 Klein, C., Garcia-Rizo, C., Bisle, B., Scheffer, B., Zischka, H., Pfeiffer, F., Siedler, F. and Oesterhelt, D. (2005) *Proteomics*, **5**, 180–197.

22 Wilkins, M.R., Gasteiger, E., Sanchez, J.C., Bairoch, A. and Hochstrasser, D.F. (1998) *Electrophoresis*, **19**, 1501–1505.

23 Schägger, H. and von Jagow, G. (1991) *Anal. Biochem.*, **199**, 223–231.

24 Camacho-Carvajal, M.M., Wollscheid, B., Aebersold, R., Steimle, V. and Schamel, W.W. (2004) *Mol. Cell. Proteomics*, **3**, 176–182.

25 Jänsch, L., Kruft, V., Schmitz, U.K. and Braun, H.P. (1996) *Plant J.*, **9**, 357–368.

26 Cruciat, C.M., Brunner, S., Baumann, F., Neupert, W. and Stuart, R.A. (2000) *J. Biol. Chem.*, **275**, 18093–18098.

27 van Lis, R., Atteia, A., Mendoza-Hernandez, G. and Gonzalez-Halphen, D. (2003) *Plant Physiol.*, **132**, 318–330.

28 Krause, F., Reifschneider, N.H., Vocke, D., Seelert, H., Rexroth, S. and Dencher, N.A. (2004) *J. Biol. Chem.*, **279**, 48369–48375.

29 Cardol, P., Gonzalez-Halphen, D., Reyes-Prieto, A., Baurain, D., Matagne, R.F. and Remacle, C. (2005) *Plant Physiol.*, **137**, 447–459.

30 Ciambella, C., Roepstorff, P., Aro, E.M. and Zolla, L. (2005) *Proteomics*, **5**, 746–757.

31 Agne, B., Meindl, N.M., Niederhoff, K., Einwachter, H., Rehling, P., Sickmann, A., Meyer, H.E., Girzalsky, W. and Kunau, W.H. (2003) *Mol. Cell*, **11**, 635–646.

32 Stroh, A., Anderka, O., Pfeiffer, K., Yagi, T., Finel, M., Ludwig, B. and Schagger, H. (2004) *J. Biol. Chem.*, **279**, 5000–5007.

33 Herranen, M., Battchikova, N., Zhang, P., Graf, A., Sirpio, S., Paakkarinen, V. and Aro, E.M. (2004) *Plant Physiol.*, **134**, 470–481.

34 Stenberg, F., Chovanec, P., Maslen, S.L., Robinson, C.V., Ilag, L.L., von Heijne, G. and Daley, D.O. (2005) *J. Biol. Chem.*, **280**, 34409–34419.

35 Farhoud, M.H., Wessels, H.J., Steenbakkers, P.J., Mattijssen, S., Wevers, R.A., van Engelen, B.G., Jetten, M.S., Smeitink, J.A., van den Heuvel, L.P. and Keltjens, J.T. (2005) *Mol. Cell. Proteomics*, **4**, 1653–1663.

36 Lasserre, J.P., Beyne, E., Pyndiah, S., Lapaillerie, D., Claverol, S. and Bonneu, M. (2006) *Electrophoresis*, **27**, 3306–3321.

37 Eubel, H., Braun, H.P. and Millar, A.H. (2005) *Plant Methods*, **1**, 11.

38 Hessa, T., Kim, H., Bihlmaier, K., Lundin, C., Boekel, J., Andersson, H., Nilsson, I., White, S.H. and von Heijne, G. (2005) *Nature*, **433**, 377–381.

39 von Heijne, G. (1986) *J. Mol. Biol.*, **192**, 287–290.

40 Ulmschneider, M.B., Sansom, M.S. and Di Nola, A. (2005) *Proteins*, **59**, 252–265.

41 Nilsson, J., Persson, B. and von Heijne, G. (2005) *Proteins*, **60**, 606–616.

42 Eichacker, L.A., Granvogl, B., Mirus, O., Muller, B.C., Miess, C. and Schleiff, E. (2004) *J. Biol. Chem.*, **279**, 50915–50922.

43 van Montfort, B.A., Doeven, M.K., Canas, B., Veenhoff, L.M., Poolman, B. and Robillard, G.T. (2002) *Biochim. Biophys. Acta*, **1555**, 111–115.

44 van Montfort, B.A., Canas, B., Duurkens, R., Godovac-Zimmermann, J. and Robillard, G.T. (2002) *J. Mass Spectrom.*, **37**, 322–330.

45 Wu, C.C. and Yates, J.R. III (2003) *Nat. Biotechnol.*, **21**, 262–267.

46 Rabilloud, T. (2003) *Nat. Biotechnol.*, **21**, 508–510.

47 Steen, H. and Mann, M. (2004) *Nat. Rev. Mol. Cell. Biol.*, **5**, 699–711.

48 Washburn, M.P., Wolters, D. and Yates, J.R., III (2001) *Nat. Biotechnol.*, **19**, 242–247.

49 Uetz, P., Giot, L., Cagney, G., Mansfield, T.A., Judson, R.S., Knight, J.R., Lockshon, D., Narayan, V., Srinivasan, M., Pochart, P., Qureshi-Emili, A., Li, Y., Godwin, B., Conover, D., Kalbfleisch, T., Vijayadamodar, G., Yang, M., Johnston, M., Fields, S. and Rothberg, J.M. (2000) *Nature*, **403**, 623–627.

50 Zhu, H., Bilgin, M., Bangham, R., Hall, D., Casamayor, A., Bertone, P., Lan, N., Jansen, R., Bidlingmaier, S., Houfek, T., Mitchell, T., Miller, P., Dean, R.A., Gerstein, M. and Snyder, M. (2001) *Science*, **293**, 2101–2105.

51 Rain, J.C., Selig, L., De Reuse, H., Battaglia, V., Reverdy, C., Simon, S., Lenzen, G., Petel, F., Wojcik, J., Schachter, V., Chemama, Y., Labigne, A. and Legrain, P. (2001) *Nature*, **409**, 211–215.

52 Ito, T., Chiba, T., Ozawa, R., Yoshida, M., Hattori, M. and Sakaki, Y. (2001) *Proc. Natl. Acad. Sci. USA*, **98**, 4569–4574.

53 Gavin, A.C., Bosche, M., Krause, R., Grandi, P., Marzioch, M., Bauer, A., Schultz, J., Rick, J.M., Michon, A.M.,

Cruciat, C.M., Remor, M., Hofert, C., Schelder, M., Brajenovic, M., Ruffner, H., Merino, A., Klein, K., Hudak, M., Dickson, D., Rudi, T., Gnau, V., Bauch, A., Bastuck, S., Huhse, B., Leutwein, C., Heurtier, M.A., Copley, R.R., Edelmann, A., Querfurth, E., Rybin, V., Drewes, G., Raida, M., Bouwmeester, T., Bork, P., Seraphin, B., Kuster, B., Neubauer, G. and Superti-Furga, G. (2002) *Nature*, **415**, 141–147.

54 Ho, Y., Gruhler, A., Heilbut, A., Bader, G.D., Moore, L., Adams, S.L., Millar, A., Taylor, P., Bennett, K., Boutilier, K., Yang, L., Wolting, C., Donaldson, I., Schandorff, S., Shewnarane, J., Vo, M., Taggart, J., Goudreault, M., Muskat, B., Alfarano, C., Dewar, D., Lin, Z., Michalickova, K., Willems, A.R., Sassi, H., Nielsen, P.A., Rasmussen, K.J., Andersen, J.R., Johansen, L.E., Hansen, L.H., Jespersen, H., Podtelejnikov, A., Nielsen, E., Crawford, J., Poulsen, V., Sorensen, B.D., Matthiesen, J., Hendrickson, R.C., Gleeson, F., Pawson, T., Moran, M.F., Durocher, D., Mann, M., Hogue, C.W., Figeys, D. and Tyers, M. (2002) *Nature*, **415**, 180–183.

55 Giot, L., Bader, J.S., Brouwer, C., Chaudhuri, A., Kuang, B., Li, Y., Hao, Y.L., Ooi, C.E., Godwin, B., Vitols, E., Vijayadamodar, G., Pochart, P., Machineni, H., Welsh, M., Kong, Y., Zerhusen, B., Malcolm, R., Varrone, Z., Collis, A., Minto, M., Burgess, S., McDaniel, L., Stimpson, E., Spriggs, F., Williams, J., Neurath, K., Ioime, N., Agee, M., Voss, E., Furtak, K., Renzulli, R., Aanensen, N., Carrolla, S., Bickelhaupt, E., Lazovatsky, Y., DaSilva, A., Zhong, J., Stanyon, C.A., Finley, R.L., Jr, White, K.P., Braverman, M., Jarvie, T., Gold, S., Leach, M., Knight, J., Shimkets, R.A., McKenna, M.P., Chant, J. and Rothberg, J.M. (2003) *Science*, **302**, 1727–1736.

56 Butland, G., Peregrin-Alvarez, J.M., Li, J., Yang, W., Yang, X., Canadien, V., Starostine, A., Richards, D., Beattie, B., Krogan, N., Davey, M., Parkinson, J., Greenblatt, J. and Emili, A. (2005) *Nature*, **433**, 531–537.

57 Gavin, A.C., Aloy, P., Grandi, P., Krause, R., Boesche, M., Marzioch, M., Rau, C., Jensen, L.J., Bastuck, S., Dumpelfeld, B., Edelmann, A., Heurtier, M.A., Hoffman, V., Hoefert, C., Klein, K., Hudak, M., Michon, A.M., Schelder, M., Schirle, M., Remor, M., Rudi, T., Hooper, S., Bauer, A., Bouwmeester, T., Casari, G., Drewes, G., Neubauer, G., Rick, J.M., Kuster, B., Bork, P., Russell, R.B. and Superti-Furga, G. (2006) *Nature*, **440**, 631–636.

58 Krogan, N.J., Cagney, G., Yu, H., Zhong, G., Guo, X., Ignatchenko, A., Li, J., Pu, S., Datta, N., Tikuisis, A.P., Punna, T., Peregrin-Alvarez, J.M., Shales, M., Zhang, X., Davey, M., Robinson, M.D., Paccanaro, A., Bray, J.E., Sheung, A., Beattie, B., Richards, D.P., Canadien, V., Lalev, A., Mena, F., Wong, P., Starostine, A., Canete, M.M., Vlasblom, J., Wu, S., Orsi, C., Collins, S.R., Chandran, S., Haw, R., Rilstone, J.J., Gandi, K., Thompson, N.J., Musso, G., St. Onge, P., Ghanny, S., Lam, M.H., Butland, G., Altaf-Ul, A.M., Kanaya, S., Shilatifard, A., O'Shea, E., Weissman, J.S., Ingles, C.J., Hughes, T.R., Parkinson, J., Gerstein, M., Wodak, S.J., Emili, A. and Greenblatt, J.F. (2006) *Nature*, **440**, 637–643.

59 Fields, S. and Song, O. (1989) *Nature*, **340**, 245–246.

60 Stagljar, I. and Fields, S. (2002) *Trends Biochem. Sci.*, **27**, 559–563.

61 Piehler, J. (2005) *Curr. Opin. Struct. Biol.*, **15**, 4–14.

62 Johnsson, N. and Varshavsky, A. (1994) *Proc. Natl. Acad. Sci. USA*, **91**, 10340–10344.

63 Johnsson, N. and Varshavsky, A. (1994) *EMBO J.*, **13**, 2686–2698.

64 Miller, J.P., Lo, R.S., Ben-Hur, A., Desmarais, C., Stagljar, I., Noble, W.S. and Fields, S. (2005) *Proc. Natl. Acad. Sci. USA*, **102**, 12123–12128.

65 Hu, C.D., Chinenov, Y. and Kerppola, T.K. (2002) *Mol. Cell*, **9**, 789–798.

66 Remy, I. and Michnick, S.W. (2004) *Methods*, **32**, 381–388.

67 Daley, D.O., Rapp, M., Granseth, E., Melen, K., Drew, D. and von Heijne, G. (2005) *Science*, **308**, 1321–1323.

68 Kim, H., Melen, K., Osterberg, M. and von Heijne, G. (2006) *Proc. Natl. Acad. Sci. USA*, **103**, 11142–11147.

69 Rigaut, G., Shevchenko, A., Rutz, B., Wilm, M., Mann, M. and Seraphin, B. (1999) *Nat. Biotechnol.*, **17**, 1030–1032.

70 Wagner, S., Bader, M.L., Drew, D. and de Gier, J.W. (2006) *Trends Biotechnol.*, **24**, 364–371.

71 White, S.H. (2004) *Protein Sci.*, **13**, 1948–1949.

72 Oberai, A., Ihm, Y., Kim, S. and Bowie, J.U. (2006) *Protein Sci.*, **15**, 1723–1734.

73 Melén, K., Krogh, A. and von Heijne, G. (2003) *J. Mol. Biol.*, **327**, 735–744.

74 Rapp, M., Drew, D., Daley, D.O., Nilsson, J., Carvalho, T., Melen, K., De Gier, J.W. and Von Heijne, G. (2004) *Protein Sci.*, **13**, 937–945.

75 Arai, M., Fukushi, T., Satake, M. and Shimizu, T. (2005) *Comput. Biol. Chem.*, **29**, 379–387.

76 Bernsel, A. and Von Heijne, G. (2005) *Protein Sci.*, **14**, 1723–1728.

77 Granseth, E., Daley, D.O., Rapp, M., Melen, K. and von Heijne, G. (2005) *J. Mol. Biol.*, **352**, 489–494.

78 Schägger, H., Cramer, W.A. and von Jagow, G. (1994) *Anal. Biochem.*, **217**, 220–230.

79 Macfarlane, D.E. (1989) *Anal. Biochem.*, **176**, 457–463.

80 Rais, I., Karas, M. and Schagger, H. (2004) *Proteomics*, **4**, 2567–2571.

81 Williams, T.I., Combs, J.C., Thakur, A.P., Strobel, H.J. and Lynn, B.C. (2006) *Electrophoresis*, **27**, 2984–2995.

82 Spelbrink, R.E., Kolkman, A., Slijper, M., Killian, J.A. and de Kruijff, B. (2005) *J. Biol. Chem.*, **280**, 28742–28748.

10
Post-Translational Modifications

10.1
Sample Preparation for Phosphoproteome Analysis
René P. Zahedi and Albert Sickmann

10.1.1
Introduction

Phosphorylation is among the most abundant post-translational modifications (PTMs) in proteins. Due to the interplay of phosphatases and kinases, phosphorylation and dephosphorylation events are highly dynamic and regulate a broad variety of cellular processes, including protein folding, function, activity, interaction, targeting and degradation. Since phosphorylation disorders may have a dramatic impact on cell vitality, much effort has been expended to extend knowledge of the signaling transduction pathways. In this respect, proteomic approaches are the method of choice, as PTMs cannot be readily predicted from the genome.

Although 30% of all human proteins are assumed to be at least transiently phosphorylated [1], the enormous complexity and vast dynamic range of biological samples render proteomic shotgun approaches inappropriate. Moreover, the low stoichiometry of phosphoproteins (ratio phosphorylated:non-phosphorylated counterparts) poses a major analytical challenge.

Indeed, these issues can be addressed by specific sample preparation for phosphoproteomics, usually aiming at the reduction of sample complexity as well as dynamic range by the selective enrichment of phosphopeptides from complex samples.

Although, currently four major types of phosphorylation are known – namely O-phosphates, N-phosphates, acylphosphates, and S-phosphates in this chapter attention will be focused exclusively on O-phosphates, as these are by far the best studied.

10.1.2
General Sample Preparation

In general, sample preparation for phosphoproteomics must fulfill certain major prerequisites, and in this context special attention must be paid to the lability of the

Proteomics Sample Preparation. Edited by Jörg von Hagen
Copyright © 2008 WILEY-VCH Verlag GmbH & Co. KGaA, Weinheim
ISBN: 978-3-527-31796-7

O-phosphates phosphoserine (pSer) and phosphothreonine (pThr) in alkaline milieu. Furthermore, the amenability to all types of Ser-, Thr-, and Tyr-kinases and phosphatases must be borne in mind, especially with regard to differential studies.

Hence, all sample preparation steps – beginning from cell lysis – should be accomplished at low temperatures and in the presence of respective inhibitors in order to avoid the introduction of artificial modifications *in vitro*. Moreover, the high affinity of phosphopeptides towards metal ions must also be taken into account, as this might lead to irreversible losses on etched steel surfaces, for example in non-bioinert high-performance liquid chromatography (HPLC) systems [2].

For the analysis of highly complex samples, the introduction of additional prefractionation steps prior to actual phosphopeptide enrichment might be considered, for example ion-exchange chromatography and isoelectric focusing (IEF) on the peptide level, or gel electrophoresis on the protein level. Moreover, the analysis of (sub)cellular compartments enables a more sensitive analysis and additional localization data.

10.1.3
Reduction of Sample Complexity

Several approaches for the reduction of sample complexity have been combined with a subsequent enrichment of phosphopeptides, including gel electrophoresis [3] and IEF [4]. However, the compatibility of utilized buffers with downstream methods must be ascertained in order to maximize the efficiency and selectivity of phosphopeptide enrichment and, consequently, the yield.

10.1.3.1 Gel Electrophoresis

A reduction of sample complexity can easily be achieved by separating a protein mixture via one-dimensional or two-dimensional gel electrophoresis (1-DE or 2-DE). Afterwards, specific bands or spots can be excised, washed, subjected to proteolytic digestion and subsequently to phosphopeptide enrichment or to specific mass spectrometric analysis, for instance by neutral loss scanning [5], precursor ion scanning [6], or multiple reaction monitoring [7].

Phosphopeptides are often difficult to detect after a simple gel electrophoretic separation due to their low stoichiometry, and therefore many studies include a precedent immunoprecipitation based on phosphospecific antibodies.

Generally, non-gel-based methods are preferred in phosphoproteomics, as all gel electrophoretic methods are limited with regard to protein molecular weight. Besides, substantial loss of material can occur due to incomplete transfer into the gel matrix as well as incomplete in-gel digestion or subsequent extraction of peptides. Furthermore, the common 2-DE set-up is not compatible with hydrophobic proteins and therefore is inappropriate for the analysis of membrane receptors.

10.1.3.2 Isoelectric Focusing

Microscale solution- [8] or free-flow IEF [9] can be utilized for a reasonable reduction of sample complexity on the protein as well as on the peptide level. However, it must be borne in mind that the phosphorylation of proteins and peptides might induce p*I*-shifts of up to several units [10]. Nevertheless, despite the still limited resolvable pH area, IEF in liquid media does not imply as strong limitations with regards to size or hydrophobicity of proteins as do gel electrophoretic procedures.

10.1.4
Methods for Phosphopeptide/Protein Enrichment

A variety of more or less selective enrichment procedures have been established in phosphoproteomic research, of which the most common approaches will be discussed in the following paragraphs.

Mostly, phosphospecific enrichment is accomplished on the peptide level. In this context, the compatibility of upstream sample buffers with following enrichment procedures must be assured, especially with regards to interfering salts or detergents. Therefore, often additional prepurification steps using, for example, reversed-phase (RP) cartridges are introduced prior to phosphopeptide enrichment.

The chromatographic approaches discussed here can either be accomplished using HPLC instruments or as one-pot reactions in appropriate sample vials and spin columns. In the latter case, special attention must be paid to the complete removal of washing buffers, as residual liquid may dramatically increase the unspecific non-phosphopeptide background.

10.1.4.1 Immunoprecipitation
While antibodies raised against pSer and pThr are phospho-motif-specific, only recognizing pSer and pThr within respective consensus sequences [11], pTyr antibodies, such as PY-20, PY-100 and 4G10, are phospho-amino acid-specific and have been utilized successfully in the characterization of signaling pathways [12,13]. Generally, in immunoprecipitation-based approaches the eluates are separated via 1-DE or 2-DE and subsequently analyzed by mass spectrometry. Alternatively, immunoprecipitation can be accomplished on the peptide level after digestion [14]. Thereby, the co-purification of non-tyrosine-phosphorylated interaction partners (e.g., SH2 domain-containing proteins) is bypassed. Additionally, non-phosphopeptides from phosphoproteins are also depleted. Consequently, the total number of identified phosphopeptides can be improved.

10.1.4.2 Immobilized Metal Ion Affinity Chromatography (IMAC)
With regards to the fractionation of protein samples, IMAC has a longstanding record [15]. In 1986, Andersson *et al.* introduced it as a technique for the specific isolation of phosphoproteins [16], and more than 10 years later IMAC was established as a powerful tool in proteome research [17]. Although, in recent years several alternative methods for the enrichment of phosphorylated species have been devised, IMAC still represents the most widely used approach in phosphoproteomic research today.

The principle of IMAC is based on the binding of negatively charged phosphate groups to positively charged metal ions due to Coulomb interactions. Here, special attention must be paid to the chosen buffer conditions [18]. While binding and washing are accomplished under acidic conditions, enriched phosphopeptides are eluted using pH- or salt gradients steps.

Nevertheless, acidic peptides are often co-purified as they also interact with metal ions. Ficarro et al. overcame this drawback by esterifying all carboxylate groups with HCl-saturated, dried methanol prior to IMAC, resulting in a substantial improvement of phosphopeptide enrichment [19]. Yet, the reaction conditions are critical and must guarantee a quantitative reaction without byproducts; otherwise the sensitivity and identification rate in the subsequent mass spectrometric analyses are significantly reduced.

Nowadays, a variety of different resins is available, comprising Fe^{3+}, Ga^{3+}, Al^{3+} and Zr^{2+}, all of which demonstrate different efficiencies and specificities.

Recently, Ndassa et al. reported an improved IMAC protocol for large-scale phosphoproteomic applications, yielding a five- to 10-fold improved recovery of singly and multiply phosphorylated peptides on an HPLC set-up by modifying buffer compositions during the resin reconstitution and washing steps [20].

10.1.4.3 Metal Oxides

In recent years, the use of metal oxides for the enrichment of phosphopeptides has emerged as a promising alternative to the well-established protocols. Thus, it has been shown that compounds with phosphate groups build self-assembling monolayers on the surfaces of metal oxides [21,22]. Consequently, a variety of approaches for phosphopeptide enrichment based on different metal oxides have been introduced, mainly TiO_2, ZrO_2, and $Al(OH)_3$.

Titanium Dioxide Originally utilized for the enrichment of phosphates and phospholipids [23], Sano et al. extended this new approach towards phosphopeptides [24,25]. Pinkse et al. then introduced a TiO_2-based online 2-D-nano-LC-MS/MS set-up which enabled the identification of phosphorylation sites in the low femtomole range [26]. In contrast, in an offline MALDI-MS-based solution, Larsen et al. added 2,5-dihydroxybenzoic acid (DHB) to the sample loading buffer, and this led to an increased specificity and consequently improved phosphopeptide yield, as DHB has a higher affinity towards TiO_2 bindings sites than most non-phosphorylated peptides [27]. However, DHB interferes with RP liquid chromatography (RP-LC), thereby impeding the use of neat online solutions. Only recently, Mazanek et al. developed an improved protocol based on the addition of 1-octanesulfonic acid (OSA) and DHB to the loading buffer; this resulted in an improved selectivity towards phosphopeptides as well as an enhanced trapping on RP precolumns [28]. As a cationic ion-pair agent, OSA shows high affinity to acidic carboxyl groups, thereby blocking acidic peptide residues and preventing unspecific binding to the TiO_2 phase.

Recently, TiO_2-coated magnetic beads (Fe_3O_4/TiO_2 core/shell) have been introduced [29] which facilitate the application of one-pot enrichment without commercially available, but cost-intensive microtips.

Zirconium Dioxide In a recent study, Kweon et al. evaluated the ability of ZrO_2 microtips to enrich for phosphopeptides, and compared this metal oxide to TiO_2 [30].

In the opinion of these authors, while both materials were seen to be highly selective, they noticed a slight, but very useful, difference: whereas ZrO_2 preferably enriches monophosphorylated peptides, TiO_2 tends to enrich particularly multiply phosphorylated species. Hence, the complementary use of these two approaches should promote optimal yields. In this context, microtips comprising a combination of ZrO_2 and TiO_2 materials are already commercially available.

Aluminum Hydroxide Wolschin *et al.* reported the use of metal oxide/hydroxide affinity chromatography (MOAC) based on $Al(OH)_3$-coated beads for the enrichment of phosphorylated species on the protein as well as the peptide level [31]. While the complexity of the peptide sample was quite low, the enrichment of phosphoproteins was monitored by Pro-Q-Diamond staining [32] after 1-DE followed by LC-MS/MS analyses of respective protein bands. However, the mass spectrometry data only provided mere protein identifications; although, from other reports, most of the identified proteins are known to be phosphorylated in related species, a precise verification of phosphorylation – or even phosphorylation sites based on mass spectrometric data – is missing.

Conclusion Although much effort has been expended to improve and optimize the metal oxide-based enrichment of phosphorylated species, these approaches still lack convincing data from the analysis of complex biological samples. From the current state, metal oxide approaches can be mainly recommended as a last step of enrichment, for example after gel electrophoresis or preceding enrichment procedures.

As metal oxides have already been successfully used for the enrichment of phospholipids [23], the presence of residual membrane fragments and lipids derived from sample preparation may pose a major obstacle to the analysis of cell lysates.

10.1.4.4 Cation-Exchange Chromatography

In 2004, Beausoleil *et al.* described the application of strong cation exchange chromatography (SCX) for the enrichment of phosphopeptides from complex mixtures [3]. Presuming that, at pH 2.7, most tryptic peptides carry a net charge of 2^+ or 3^+, phosphates retain an additional negative charge leading to a net charge of 1^+ for most mono-phosphorylated peptides. Thereby, a specific enrichment of phosphopeptides can be achieved by applying appropriate SCX gradients.

In the respective study, HeLa cell nuclei were separated by 1-DE and the gel was cut into 10 fractions, which were subjected to in-gel digestion [33], albeit with large volumes. The separation of the respective eluates by SCX and subsequent mass spectrometric analyses by MS^3 on an LCQ Deca XP ion trap led to the identification of 2002 phosphorylation sites from 967 proteins.

Although it is argued, that SCX enrichment lacks selectivity towards phosphopeptides and leads to the co-purification of a whole variety of non-phosphorylated peptides (e.g., amino-terminal acetylated and carboxy-terminal peptides [34]), this method is highly compatible with complex mixtures, but should be applied in combination with additional enrichment procedures or with phospho-specific scanning techniques on the mass spectrometry level (e.g., MS^3, neutral loss scanning, precursor ion scanning) in order to maximize phosphopeptide yield.

10.1.4.5 Derivatization Approaches

Many different derivatization approaches for phosphopeptides have been developed to address three major issues: (i) the suppression of phosphopeptide ionization; (ii) the lability of the phosphoester group; and (iii) the specific enrichment of phosphopeptides from complex mixtures.

β-Elimination Most derivatization approaches are based on β-elimination of pSer and pThr to the unsaturated amino acids dehydroalanine and 2-aminodehydrobutyric acid by alkali treatment. By subsequent Michael-type addition of a nucleophile [35], functional groups allowing for specific enrichment or separation of phosphopeptides from non-phosphorylated peptides, for instance by diagonal chromatography, can be introduced. However, it must be borne in mind, that β-elimination might result in racemization of α-carbon atoms in the peptide backbone, thereby generating diastereomers which may lead to peak broadening during RP-LC [36].

The introduction of biotin moieties after β-elimination enables highly selective enrichment based on avidin affinity chromatography [37,38].

Another approach combined phosphopeptide enrichment by IMAC with on-resin β-elimination and Michael-type addition of 2-aminoethanethiol. Thereby, the recovery of multiply phosphorylated peptides could be improved [39].

In 2005, Brittain et al. described fluorous solid-phase extraction (FSPE) as a novel enrichment strategy in phosphoproteomics [40]. After β-elimination of pSer and pThr, 1H,1H,2H,2H-perfluorooctane-1-thiol was used as the Michael-type nucleophile. Based on the strong nature of fluorine–fluorine interactions, a selective enrichment of phosphopeptides led to the identification of more than 60 phosphopeptides form Jurkat T-cells.

However, all β-elimination-based methods suffer from the same inherent drawbacks. First, arylphosphates (pTyr) do not undergo β-elimination and therefore are not accessible to any approaches based on this strategy. Second, there is a potential risk of false-positive phosphorylation site localization due to partial elimination of non-phosphorylated Ser and Thr residues as well as O-glycans [41,42]. Third, the reaction conditions must be monitored carefully in order to guarantee quantitative reactions without interfering byproducts, for example deamidation [43].

Nevertheless, many different strategies for phosphopeptide enrichment based on β-elimination provide a high sensitivity down to the femtomole range.

Phosphoamidates A derivatization approach, which is not biased against pTyr was introduced by Zhou et al. [44]. After generating phosphoamidate bonds by a carbodiimide-catalyzed condensation between peptides and excess amine, derivatized phosphopeptides are captured by solid-phase extraction due to the binding of introduced free sulfhydryl groups to immobilized iodoacetyl groups. After acidic hydrolysis of the phosphoamidate bonds, intact phosphopeptides can be recovered. However, due to a high number of derivatization and HPLC purification steps, this approach is not only time-consuming but also accompanied by loss of material. Therefore, Tao *et al.* enhanced this approach by coupling phosphopeptides to a polyamine dendrimer via phosphoamidate chemistry [45], leading to a reduction in the number of sample processing steps. Thus, after an initial anti-phosphotyrosine immunoprecipitation, 97 phosphotyrosine-containing proteins and interaction partners, corresponding to a total of 75 pTyr sites and 80 pSer/pThr sites, could be identified and quantified from Jurkat T cells treated with pervanadate for 2 and 10 min, respectively.

Combined Fractional Diagonal Chromatography Due to their different physicochemical properties, phosphopeptides and their corresponding non-phosphorylated counterparts show differences in their chromatographic behavior in RP-LC. Thus, Gevaert *et al.* utilized combined fractional diagonal chromatography (COFRADIC) to enrich for phosphopeptides after phosphatase treatment [46]. However, the extent of the induced retention time shift depends on both: the type of O-phosphorylation (pTyr shifts larger towards more hydrophobic retention times) as well as the number of phosphorylation sites within the peptide. Although COFRADIC approaches pose an elegant strategy to deal with highly complex mixtures, by using this approach an unambiguous localization of the specific site of phosphorylation is not possible for peptides containing more than one potential O-phosphorylated amino acid (Ser, Thr, Tyr) within the respective sequence. Moreover, artificial retention time shifts due to modifications unintentionally introduced between primary and secondary COFRADIC separations (e.g., deamidation) might lead to false-positive results. Therefore, before the primary LC run, the sample is split into two aliquots. One aliquot is treated with phosphatase and subsequently incubated with trypsin in the presence of $H_2^{18}O$, leading to the incorporation of ^{18}O at the C-termini of the peptides; the other aliquot serves as an untreated control. After pooling the two samples the primary run is accomplished. Next, the primary fractions are treated with phosphatase and separated in the secondary run. In this way non-phosphorylated peptides which show a retention time shift can be distinguished from phosphorylated peptides by subsequent mass spectrometry, as the latter emerge as ^{16}O singlet and not as $^{16}O/^{18}O$ doublet.

10.1.5
Summary

Currently, a whole variety of different methods for phosphopeptide enrichment is available. Although modern mass spectrometers provide the opportunity to detect phosphopeptides by specific scanning techniques in the low femtomole range, a special focus should be set to appropriate sample processing in order to diminish the background of non-phosphorylated peptides. Preferably, the enrichment of phosphopeptides should be combined with specific scanning techniques, as all introduced methods to a certain degree also enrich for non-phosphorylated peptides which might hamper the identification of phosphopeptides in complex mixtures.

As all of these described methods tend to enrich different subsets of phosphopeptides with different specificity, their use should depend on the goals of the respective study. In principle, for large-scale analyses a combination of different approaches is recommended. Yet, the number of processing steps should be kept as low as possible in order to reduce sample loss and unintentional introduction of peptide modifications *in vitro*, which is of particular importance for quantitative approaches.

10.1.6
Perspective

Since the general focus is set increasingly on quantitative proteomics, there is a clear tendency to move from the mere identification of phosphorylation sites towards determining the relative ratios between differential samples. Thus, the dynamics of signal transduction processes can be described precisely by monitoring hundreds of phosphorylation sites in a time-dependent manner. For this purpose, the introduction of stable isotope labels to phosphopeptides is necessary. As the phosphorylation of amino acids is a highly dynamic process which depends on the complex interplay between multiple kinases and phosphatases, the reaction conditions as well as the sample processing steps must be considered very carefully in order to circumvent the induction of artificial phosphorylation/dephosphorylation events that may lead to an incorrect description of complex cellular processes. In general, quantitative phosphoproteomic approaches should always include the usage of kinase- and phosphase inhibitor cocktails.

10.1.7
Recipe for Beginners: IMAC

Generally, parameters such as buffer conditions or incubation times must be determined empirically with regard to the respective sample. Therefore, this recipe must be considered to be a suggestion rather than a stringent guideline:

- Digest the sample with trypsin in 50 mM ammonium bicarbonate (extra pure) pH 7.8 overnight at 37 °C.
- Dilute the sample solution in order to reduce salt concentration.
- Lyophilize the sample, as residual water reduces esterification yield.
- Prepare a solution of 2 M methanolic HCl by careful dropwise addition of 160 µL acetyl chloride (reagent grade) to 1 mL anhydrous methanol (LC grade) under stirring and cooling.
- Redissolve the lyophilized sample in 100 µL of methanolic HCl and incubate for 2 h at room temperature.
- Lyophilize the sample once more. Purify with RP cartridge.

- Wash the IMAC resins (PHOS Select™ iron affinity gel; Sigma Aldrich) in loading buffer consisting of 30% acetonitrile (LC grade), 250 mM acetic acid (*pro analysi* (p.a.); high purity), pH 2.7 using a spin column (SigmaPrep™; Sigma Aldrich).
- Repeat the previous step twice.
- Reconstitute the sample in loading buffer.
- Add the sample to the washed IMAC beads.
- Incubate for 30 min, with mixing.
- Discard the supernatant and wash the beads twice with loading buffer.
- Wash the beads once with deionized water to remove residual loading buffer.
- Add elution buffer consisting of 200 mM sodium phosphate (p.a.), pH 8.4. Subsequent MALDI-TOF-MS requires an additional desalting step. Alternatively, elute with 400 mM ammonium hydroxide (p.a.).
- Acidify the eluates to pH 2.4 to prevent β-elimination.
- Analyze the eluates by mass spectrometry, generally LC-MS/MS.

Acknowledgments

The authors thank the Deutsche Forschungsgemeinschaft (DFG) for continuous support within the SFB 688 and the FZT 82.

References

1 Cohen, P. (2002) The origins of protein phosphorylation. *Nat. Cell Biol.*, **4** (5), E127–E130.
2 Tuytten, R., Lemiere, F., Witters, E., Van Dongen, W., Slegers, H., Newton, R.P., Van Onckelen, H. and Esmans, E.L. (2006) Stainless steel electrospray probe: a dead end for phosphorylated organic compounds? *J. Chromatogr. A*, **1104** (1–2), 209–221.
3 Beausoleil, S.A., Jedrychowski, M., Schwartz, D., Elias, J.E., Villen, J., Li, J., Cohn, M.A., Cantley, L.C. and Gygi, S.P. (2004) Large-scale characterization of HeLa cell nuclear phosphoproteins. *Proc. Natl. Acad. Sci. USA*, **101** (33), 12130–12135.
4 Cantin, G.T., Venable, J.D., Cociorva, D. and Yates, J.R. III (2006) Quantitative phosphoproteomic analysis of the tumor necrosis factor pathway. *J. Proteome Res.*, **5** (1), 127–134.
5 Resing, K.A., Johnson, R.S. and Walsh, K.A. (1995) Mass spectrometric analysis of 21 phosphorylation sites in the internal repeat of rat profilaggrin, precursor of an intermediate filament associated protein. *Biochemistry*, **34** (29), 9477–9487.
6 Huddleston, M.J., Annan, R.S. and Carr, S.A. (1993) Selective detection of

phosphopeptides in complex mixtures by electrospray liquid chromatography/mass spectrometry. *J. Am. Soc. Mass Spectrom.*, **4** (9), 710–717.

7 Unwin, R.D., Griffiths, J.R., Leverentz, M.K., Grallert, A., Hagan, I.M. and Whetton, A.D. (2005) Multiple reaction monitoring to identify sites of protein phosphorylation with high sensitivity. *Mol. Cell. Proteomics*, **4** (8), 1134–1144.

8 Tang, H.Y. and Speicher, D.W. (2005) Complex proteome prefractionation using microscale solution isoelectrofocusing. *Expert Rev. Proteomics*, **2** (3), 295–306.

9 Burggraf, D., Weber, G. and Lottspeich, F. (1995) Free flow-isoelectric focusing of human cellular lysates as sample preparation for protein analysis. *Electrophoresis*, **16** (6), 1010–1015.

10 Zhu, K., Zhao, J., Lubman, D.M., Miller, F.R. and Barder, T.J. (2005) Protein pI shifts due to posttranslational modifications in the separation and characterization of proteins. *Anal. Chem.*, **77** (9), 2745–2755.

11 Gronborg, M., Kristiansen, T.Z., Stensballe, A., Andersen, J.S., Ohara, O., Mann, M., Jensen, O.N. and Pandey, A. (2002) A mass spectrometry-based proteomic approach for identification of serine/threonine-phosphorylated proteins by enrichment with phospho-specific antibodies: identification of a novel protein, Frigg, as a protein kinase A substrate. *Mol. Cell. Proteomics*, **1** (7), 517–527.

12 Pandey, A., Podtelejnikov, A.V., Blagoev, B., Bustelo, X.R., Mann, M. and Lodish, H.F. (2000) Analysis of receptor signaling pathways by mass spectrometry: identification of vav-2 as a substrate of the epidermal and platelet-derived growth factor receptors. *Proc. Natl. Acad. Sci. USA*, **97** (1), 179–184.

13 Zhang, Y., Wolf-Yadlin, A., Ross, P.L., Pappin, D.J., Rush, J. and Lauffenburger, D.A. and White, F.M. (2005) Time-resolved mass spectrometry of tyrosine phosphorylation sites in the epidermal growth factor receptor signaling network reveals dynamic modules. *Mol. Cell. Proteomics*, **4** (9), 1240–1250.

14 Rush, J., Moritz, A., Lee, K.A., Guo, A., Goss, V.L., Spek, E.J., Zhang, H., Zha, X.M., Polakiewicz, R.D. and Comb, M.J. (2005) Immunoaffinity profiling of tyrosine phosphorylation in cancer cells. *Nat. Biotechnol.*, **23** (1), 94–101.

15 Porath, J., Carlsson, J., Olsson, I. and Belfrage, G. (1975) Metal chelate affinity chromatography, a new approach to protein fractionation. *Nature*, **258** (5536), 598–599.

16 Andersson, L. and Porath, J. (1986) Isolation of phosphoproteins by immo-bilized metal (Fe3+) affinity chromato-graphy. *Anal. Biochem.*, **154** (1), 250–254.

17 Neville, D.C., Rozanas, C.R., Price, E.M., Gruis, D.B., Verkman, A.S. and Townsend, R.R. (1997) Evidence for phosphorylation of serine 753 in CFTR using a novel metal-ion affinity resin and matrix-assisted laser desorption mass spectrometry. *Protein Sci.*, **6** (11), 2436–2445.

18 Feuerstein, I., Morandell, S., Stecher, G., Huck, C.W., Stasyk, T., Huang, H.L., Huber, L.A. and Bonn, G.K. (2005) Phosphoproteomic analysis using immobilized metal ion affinity chromatography on the basis of cellulose powder. *Proteomics*, **5** (1), 46–54.

19 Ficarro, S.B., McCleland, M.L., Stukenberg, P.T., Burke, D.J., Ross, M. M., Shabanowitz, J., Hunt, D.F. and White, F.M. (2002) Phosphoproteome analysis by mass spectrometry and its application to *Saccharomyces cerevisiae*. *Nat. Biotechnol.*, **20** (3), 301–305.

20 Ndassa, Y.M., Orsi, C., Marto, J.A., Chen, S. and Ross, M.M. (2006) Improved immobilized metal affinity chromato-graphy for large-scale phospho-proteomics applications. *J. Proteome Res.*, **5** (10), 2789–2999.

21 Folkers, J.P., Gorman, C.B., Laibinis, P. E., Buchholz, S. Whitesides, G.M. (1995)

Self-assembled monolayers of long chain hydroxamic acids on the native oxides of metals. *Langmuir*, **11** (3), 813–824.

22 Hofer, R., Textor, M. and Spencer, N.D. (2001) Alkyl phosphate monolayers, self assembled from aqueous solution onto metal oxide surfaces. *Langmuir*, **17** (13), 4014–4020.

23 Ikeguchi, Y. and Nakamura, H. (2000) Selective enrichment of phospholipids by titania. *Anal. Sci.*, **16** (5), 541–543.

24 Sano, A. and Nakamura, H. (2004) Chemo-affinity of titania for the column-switching HPLC analysis of phosphopeptides. *Anal. Sci.*, **20** (3), 565–566.

25 Sano, A. and Nakamura, H. (2004) Titania as a chemo-affinity support for the column-switching HPLC analysis of phosphopeptides: application to the characterization of phosphorylation sites in proteins by combination with protease digestion and electrospray ionization mass spectrometry. *Anal. Sci.*, **20** (5), 861–864.

26 Pinkse, M.W., Uitto, P.M., Hilhorst, M.J., Ooms, B. and Heck, A.J. (2004) Selective isolation at the femtomole level of phosphopeptides from proteolytic digests using 2D-NanoLC-ESI-MS/MS and titanium oxide precolumns. *Anal. Chem.*, **76** (14), 3935–3943.

27 Larsen, M.R., Thingholm, T.E., Jensen, O.N., Roepstorff, P. and Jorgensen, T.J. (2005) Highly selective enrichment of phosphorylated peptides from peptide mixtures using titanium dioxide microcolumns. *Mol. Cell. Proteomics*, **4** (7), 873–886.

28 Mazanek, M., Mitulovic, G., Herzog, F., Stingl, C., Hutchins, J.R.A., Peters, J.M. and Mechtler, K. (2006) Titanium dioxide as a chemo-affinity solid phase in offline phosphopeptide chromatography prior to HPLC-MS/MS analysis. *Nature Protocols*, **1**, 1977–1987.

29 Chen, C.T. and Chen, Y.C. (2005) Fe_3O_4/TiO_2 core/shell nanoparticles as affinity probes for the analysis of phosphopeptides using TiO_2 surface-assisted laser desorption/ionization mass spectrometry. *Anal. Chem.*, **77** (18), 5912–5919.

30 Kweon, H.K. and Hakansson, K. (2006) Selective zirconium dioxide-based enrichment of phosphorylated peptides for mass spectrometric analysis. *Anal. Chem.*, **78** (6), 1743–1749.

31 Wolschin, F., Wienkoop, S. and Weckwerth, W. (2005) Enrichment of phosphorylated proteins and peptides from complex mixtures using metal oxide/hydroxide affinity chromatography (MOAC). *Proteomics*, **5** (17), 4389–4397.

32 Steinberg, T.H., Agnew, B.J., Gee, K.R., Leung, W.Y., Goodman, T., Schulenberg, B., Hendrickson, J., Beechem, J.M., Haugland, R.P. and Patton, W.F. (2003) Global quantitative phosphoprotein analysis using Multiplexed Proteomics technology. *Proteomics*, **3** (7), 1128–1144.

33 Shevchenko, A., Wilm, M., Vorm, O. and Mann, M. (1996) Mass spectrometric sequencing of proteins silver-stained polyacrylamide gels. *Anal. Chem.*, **68** (5), 850–858.

34 Gruhler, A., Olsen, J.V., Mohammed, S., Mortensen, P., Faergeman, N.J., Mann, M. and Jensen, O.N. (2005) Quantitative phosphoproteomics applied to the yeast pheromone signaling pathway. *Mol. Cell. Proteomics*, **4** (3), 310–327.

35 Meyer, H.E., Hoffmann-Posorske, E., Korte, H., Heilmeyer, J. and Ludwig, M.G. (1986) Sequence analysis of phosphoserine-containing peptides: Modification for picomolar sensitivity. *FEBS Lett.*, **204** (1), 61–66.

36 Heilmeyer, L.M., Jr., Serwe, M., Weber, C., Metzger, J., Hoffmann-Posorske, E. and Meyer, H.E. (1992) Farnesylcysteine, a constituent of the alpha and beta subunits of rabbit skeletal muscle phosphorylase kinase: localization by conversion to S-ethylcysteine and by tandem mass spectrometry. *Proc. Natl. Acad. Sci. USA*, **89** (20), 9554–9558.

37 Oda, Y., Nagasu, T., and Chait, B.T. (2001) Enrichment analysis of phosphorylated proteins as a tool for probing the

phosphoproteome. *Nat. Biotechnol.*, **19** (4), 379–382.

38 Goshe, M.B., Conrads, T.P., Panisko, E.A., Angell, N.H., Veenstra, T.D. and Smith, R.D. (2001) Phosphoprotein isotope-coded affinity tag approach for isolating and quantitating phosphopeptides in proteome-wide analyses. *Anal. Chem.*, **73** (11), 2578–2586.

39 Thompson, A.J., Hart, S.R., Franz, C., Barnouin, K., Ridley, A. and Cramer, R. (2003) Characterization of protein phosphorylation by mass spectrometry using immobilized metal ion affinity chromatography with on-resin beta-elimination and Michael addition. *Anal. Chem.*, **75** (13), 3232–3243.

40 Brittain, S.M., Ficarro, S.B., Brock, A. and Peters, E.C. (2005) Enrichment and analysis of peptide subsets using fluorous affinity tags and mass spectrometry. *Nat. Biotechnol.*, **23** (4), 463–468.

41 Rusnak, F., Zhou, J. and Hathaway, G.M. (2004) Reaction of phosphorylated and O-glycosylated peptides by chemically targeted identification at ambient temperature. *J. Biomol. Tech.*, **15** (4), 296–304.

42 McLachlin, D.T. and Chait, B.T. (2003) Improved beta-elimination-based affinity purification strategy for enrichment of phosphopeptides. *Anal. Chem.*, **75** (24), 6826–6836.

43 Karty, J.A. and Reilly, J.P. (2005) Deamidation as a consequence of beta-elimination of phosphopeptides. *Anal. Chem.*, **77** (14), 4673–4676.

44 Zhou, H., Watts, J.D. and Aebersold, R. (2001) A systematic approach to the analysis of protein phosphorylation. *Nat. Biotechnol.*, **19** (4), 375–378.

45 Tao, W.A., Wollscheid, B., O'Brien, R., Eng, J.K., Li, X.J., Bodenmiller, B., Watts, J.D., Hood, L. and Aebersold, R. (2005) Quantitative phosphoproteome analysis using a dendrimer conjugation chemistry and tandem mass spectrometry. *Nat. Methods*, **2** (8), 591–598.

46 Gevaert, K., Staes, A., Van Damme, J., De Groot, S., Hugelier, K., Demol, H., Martens, L., Goethals, M. and Vandekerckhove, J. (2005) Global phosphoproteome analysis on human HepG2 hepatocytes using reversed-phase diagonal LC. *Proteomics*, **5** (14), 3589–3599.

10.2
Sample Preparation for Analysis of Post-Translational Modifications: Glycosylation

David S. Selby, Martin R. Larsen, Miren J. Omaetxebarria, and Peter Roepstorff

10.2.1
Introduction

Proteins are involved in most of the molecular processes of life. These activities are predominantly controlled by a combination of the rates of protein synthesis, degradation and post-translation modifications (PTMs). While synthesis can be investigated at the DNA or RNA level, the characterization of PTMs – including those involved in degradation – requires direct analysis at the protein level. Previously, the details have been reported of hundreds of different types of PTM, ranging from relatively straightforward modifications such as enzymatic processing and methylation, to more complex modifications such as glycosylation. Each of the many

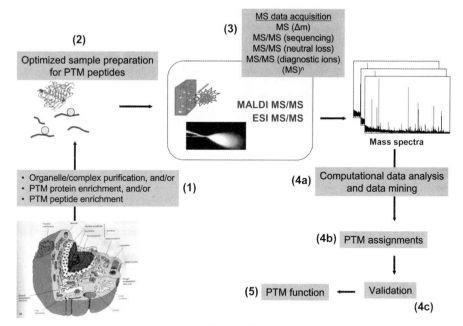

Figure 10.1 Diagrammatic representation of the modification specific proteomics approach, for the characterization of post-translation modifications (PTMs).

possible modifications affects the physico-chemical characteristics of the proteins and peptides, with the large range of characteristics making it very difficult to determine all of the differently modified forms of proteins present in a cell with any one specific sample preparation or analysis procedure. This is particularly the case when examining complex mixtures of proteins, such as those found in proteomic samples.

One general strategy that can be used to probe PTMs in complex protein mixtures is the modification specific proteomics approach [47]. This approach (see Figure 10.1) typically involves combining some form of subproteome and/or modification specific enrichment technique with PTM-specific sample preparation, MS data acquisition, and data analysis. As glycosylation is the most common and complex PTM [48], in this chapter examples will be used of the analysis of protein glycosylation to illustrate the application of a modification-specific proteomics approach.

Glycosylation is known to have many different biological roles [49,50]. These roles vary from those which relate primarily to general effects of the size and shape of the glycan, such as protein folding and assembly of protein complexes [51], to those which depend upon the specific configuration of branched glycan structures, such as cell recognition, cell–cell interaction and immune responses [52]. A simple form of glycosylation, O-GlcNAc, is even thought to be involved in signal transduction within the cell [53].

Mammalian and plant glycoproteins commonly contain three types of glycan, with N- and O- linkage being most frequent [54]:

- N-linked glycans, where the linkage is via an amide bond to asparagine in an Asn-Xxx-(Ser, Thr or Cys) motif, with Xxx being any amino acid.
- O-linked glycans, where the glycan structure is attached to serine or threonine.
- The carbohydrate portion of glycosylphosphatidylinositol lipid anchors.

The primary structure of the more common N- and O-linked glycans, consists of branching carbohydrate structures (see Figure 10.2 for some examples), which can be contrasted with the linear primary structure of DNA, RNA, and proteins. There are also other types of glycosylation, such as the simple O-GlcNAcylation [53], and C-glycosylation of tryptophan [55]; however, samples for these specialized cases are prepared and analyzed in a different manner and will not be considered further in this chapter.

Full characterization of protein glycosylation requires determination of:

- the protein sites that are glycosylated,
- the level of occupancy at each site,
- the glycan structures,
- the total number of different glycoforms.

Given the heterogeneity of glycan structures, no single technique is capable of providing a full structural analysis, although nuclear magnetic resonance (NMR) is perhaps the most powerful structural tool. Unfortunately, NMR is very insensitive when compared to mass spectrometry (MS) and chromatography; thus, MS-based approaches are often used in proteomic or other studies, where only a small amount of sample is available.

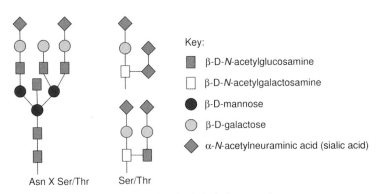

Figure 10.2 Examples of N-linked and O-linked glycopeptide structures.

It is often difficult to determine all of the glycoforms present on a given protein, as: (i) only a limited proportion of a given protein is likely to be modified; (ii) several sites may be glycosylated on an individual protein; and (iii) the modifying group may be heterogeneous. Glycosylation, in particular, is known to be highly heterogeneous and a given site may have over 20 glycan structures [56]. The biological relevance of having so many different structures attached to a single site is not well understood; however, it may be speculated that the heterogeneity of the glycan structures provides a mechanism for fine-tuning interactions to fit the dynamic environment of a living cell.

The limitations of mass spectrometric analysis (particularly ionization difficulties and the inability to resolve isobaric species, such as the different hexose sugars) mean that it is difficult to characterize glycoproteins. (Note: It is sometimes possible to use high-energy collision-induced dissociation to resolve some isobaric glycan species by generating cross-ring cleavages, but the spectra generated in this manner are complex and difficult to interpret.) Characterization has required the use of separation techniques or highly selective detection methods. These approaches typically commence with gel separation, followed by enzymatic and/or chemical reactions, with liquid chromatography and fluorescence detection, or matrix-assisted laser desorption/ionization (MALDI) or electrospray ionization (ESI) mass spectrometry detection [57], or previously MS with fast-atom bombardment [58,59].

The main methods that have been used for preparation of samples for analysis of protein glycosylation can be divided into two main categories:

- sample preparation for detailed characterization of a relatively pure protein,
- sample preparation for analysis of low level, complex proteomic samples.

This chapter deals primarily with the application of the modification-specific proteomics approach to the second category.

10.2.2
Advantages and Disadvantages of Different Sample Preparation Methods

Powerful methods have been developed to enable the thorough characterization of protein glycosylation. The traditional methods involve the chemical or enzymatic release of the carbohydrates, followed by various techniques such as sequential use of specific exoglycosidases (see, e.g., QA-Bio http://glycotools.qa-bio.com/ for a list) and high-performance liquid chromatography (HPLC) purification or MS to characterize the glycan structures [60]. Typically, a combination of separation techniques and analytical methods is required to complete the analysis. The major advantage of this type of approach is that it can often provide full characterization of the glycan structures. The disadvantages are that information on the site of attachment is lost, these methods are very time-consuming and – even more importantly – they generally require microgram quantities of purified protein. This requirement for significant quantities of purified protein means that it is not directly compatible with glycoproteomic studies, where the aim is to obtain

10.2.5
Perspective

A number of protocols that allow the site-specific assignment of glycans in glycoproteins on a proteomics sensitivity level are now available. The relative quantification of site occupancy, as well as of different types of glycan structure on a given site, is also possible – though this represents a more challenging investigation. There is still room for improvement, however, and future applications of the techniques will, without doubt, result in the optimization of existing protocols and the development of new techniques.

10.2.6
Recipe for Beginners: Enrichment of Glycopeptides with a HILIC Microcolumn

This procedure can be used to enrich N-linked and O-linked glycopeptides from a digest prepared either in solution from a protein extract or from a gel band. It is recommended that, when first attempted, the procedure be used with samples at the picomole level. It is also recommended that each microcolumn is used only once, in order to avoid the cross-contamination of samples.

10.2.6.1 Materials

- ZIC-HILIC chromatographic media, (ZIC-HILIC, silica 10 µm, 200 Å, Sequant AB, Umeå, Sweden), suspended in HPLC-grade acetonitrile or methanol. This material contains zwitterionic sulfobetain groups, which provides superior enrichment when compared to bare silica HILIC materials.

- Poros R2 and OLIGO R3 chromatographic media, 20 µm (Applied Biosystems, CA, USA).

- C-8 StageTips made from GELoader Tips, available from Proxeon Biosystems A/S (Odense, Denmark).

- Disposable 1-mL syringe, fitted to the GELoader with a cut-down 200-µL tip (see Figure 10.3).

- Washing solution: 80% acetonitrile, 19.5% 18 MΩ water, 0.5% formic acid.

- Elution solution: 0.5% aqueous formic acid.

- N-glycosidase F (PNGase F) (Roche Diagnostics, Mannheim, Germany).

- 50 mM ammonium bicarbonate in 18 MΩ, pH 7.8.

Optional: make up the 50 mM ammonium bicarbonate in 50% ^{18}O water.

10.2.6.2 Procedure: Purification of Glycopeptides

- Prepare washing solution and elution solution freshly on the day of use.

- Make up the digested sample in washing solution.

- GELoader tip microcolumns packed with HILIC beads are created by depositing a few microliters of HILIC chromatographic media slurry into 10 μL of washing solution. The HILIC beads are packed on top of the C8 plug. The length of the column is dependent on the amount of peptides to be analyzed; a 3- to 5-mm column is sufficient for up to 20 pmol. Gentle air pressure is used (plastic syringe) to pack the column.

- Clean the microcolumn with 15 μL of elution solution, flushing the solution through with gentle pressure from the syringe.

- Condition the column by flushing with 30 μL of washing solution.

- Load the digest sample containing glycopeptides onto the column using the syringe. Ensure that the volume loaded is at least 10 μL.

- Wash unbound material from the column with 20–40 μL of washing solution.

- Elute the glycosylated peptides from the HILIC material with 7–15 μL of elution solution and *retain the eluate*.

- Glycopeptides with relatively hydrophobic peptides may still be bound to the C8 plug. Elute these glycopeptides from the plug with 3 μL of wash solution and *pool with eluate from the previous step*.

- Dry the glycopeptides down and store until required for MS analysis. When the solution volume is reduced to 10 μL or less, it is recommended to analyze an aliquot by MALDI-TOF MS, using a glycopeptide-compatible matrix, such as 2,5-dihydroxybenzoic acid. Signals separated by masses such as 146 (fucoses), 162 (hexoses), 203 (N-acetyl hexosamines) and 291 (sialic acid) indicate successful enrichment of glycopeptides.

10.2.6.3 Procedure: Deglycosylation of N-Linked Glycopeptides

- Redissolve the enriched glycopeptides in the 50 mM ammonium bicarbonate solution and add 0.2 units of PNGase F.

- Incubate at 37 °C from 3 h to overnight; this should remove all N-linked glycans from the peptides, other than those containing β(1-3) core fucosylation.

- Store in the freezer until required for MS analysis.

The asparagines where glycans were attached will be converted (deamidation) into aspartate (+1 Da mass) and the intact glycans will be released by this procedure, allowing further analysis of both the glycan and peptide pools, as required. The optional use of 50% ^{18}O water for the bicarbonate buffer provides for doublet (+1 and +3 Da) deamidation peaks for the formerly glycosylated peptides, ensuring that they will not be confused with other deamidated peptides.

References

47 Jensen, O.N. (2004) Modification-specific proteomics: characterization of post-translational modifications by mass spectrometry. *Curr. Opin. Chem. Biol.*, **8**, 33–41.

48 Sharon, N. and Lis, H. (1997) Glycoproteins: structure and function in *Glycosciences: status and perspectives*, H-.J. Gabius and S. Gabius (eds.) Chapman & Hall, Weinheim, Germany, pp. 133–162.

49 Varki, A. (1993) Biological roles of oligosaccharides – all of the theories are correct. *Glycobiology*, **3**, 97–130.

50 Varki, A., Cummings, R., Esko, J., Freeze, H., Hart, G. and Marth, J. (1999) *Essentials of Glycobiology*, Cold Spring Harbor Press, Cold Spring Harbor, New York.

51 Helenius, A. and Aebi, M. (2001) Intracellular functions of N-linked glycans. *Science*, **291**, 2364–2369.

52 Rudd, P.M., Elliott, T., Cresswell, P., Wilson, I.A. and Dwek, R.A. (2001) Glycosylation and the immune system. *Science*, **291**, 2370–2376.

53 Wells, L., Vosseller, K. and Hart, G.W. (2001) Glycosylation of nucleocyto-plasmic proteins: Signal transduction and O-GlcNAc. *Science*, **291**, 2376–2378.

54 Harvey, D.J. (1999) Matrix-assisted laser desorption/ionization mass spectrometry of carbohydrates. *Mass Spectrom. Rev.*, **18**, 349–450.

55 Hofsteenge, J., Muller, D.R., Debeer, T., Loffler, A., Richter, W.J. and Vliegenthart, J.F.G. (1994) New-type of linkage between a carbohydrate and a protein: C-glycosylation of a specific tryptophan residue in human Rnase U-S. *Biochemistry*, **33**,13524–13530.

56 Mortz, E., Sareneva, T., Julkunen, I. and Roepstorff, P. (1996) Does matrix-assisted laser desorption/ionization mass spectrometry allow analysis of carbohydrate heterogeneity in glyco-proteins? A study of natural human interferon-gamma. *J. Mass Spectrom.*, **31**, 1109–1118.

57 Kuster, B., Krogh, T.N., Mortz, E. and Harvey, D.J. (2001) Glycosylation analysis of gel-separated proteins. *Proteomics*, **1**, 350–361.

58 Tsai, P.K., Dell, A., Ballou, C.E. (1986) Characterization of acetylated and acetolyzed glycoprotein high-mannose core oligosaccharides by fast-atom-bombardment mass spectrometry. *Proc. Natl. Acad. Sci. USA*, **83**, 4119–4123.

59 Thomas-Oates, J.E. and Dell, A. (1989) Fast atom bombardment-mass spectrometry strategies for analysing glycoprotein glycans. *Biochem. Soc. Trans.*, **17**, 243–245.

60 Rudd, P.M., Guile, G.R., Kuster, B., Harvey, D.J., Opdenakker, G. and Dwek, R.A. (1997) Oligosaccharide sequencing technology. *Nature*, **388**, 205–207.

61 Hart, C. and Schulenberg, B., Steinberg, T.H., Leung, W.Y. and Patton, W.F. (2003) Detection of glycoproteins in polyacryl-amide gels and on electroblots using Pro-Q Emerald 488 dye, a fluorescent periodate Schiff-base stain. *Electrophoresis*, **24**, 588–598.

62 Cummings, R.D. (1994) Use of lectins in analysis of glycoconjugates, in *Guide to Techniques in Glycobiology; Methods in Enzymology Series*, Vol. 230, (eds W.J. Lennarz and G.W. Hart), Academic Press, San Diego, CA. pp. 66–86.

63 Yang, Z.P. and Hancock, W.S. (2004) Approach to the comprehensive analysis of glycoproteins isolated from human serum using a multi-lectin affinity column. *J. Chromatogr. A*, **1053**, 79–88.

64 Takegawa, Y., Deguchi, K., Keira, T., Ito, H., Nakagawa, H. and Nishimura, S. (2006) Separation of isomeric 2-aminopyridine derivatized N-glycans and N-glycopeptides of human serum immunoglobulin G by using a zwitterionic type of hydrophilic-interaction chromatography. *J. Chromatogr. A*, **1113**, 177–181.

65 Hagglund, P., Bunkenborg, J., Elortza, F., Jensen, O.N. and Roepstorff, P. (2004) A new strategy for identification of N-glycosylated proteins and unambiguous assignment of their glycosylation sites using HILIC enrichment and partial deglycosylation. *J. Proteome Res.*, **3**, 556–566.

66 Jebanathirajah, J.N., Steen, H. and Stensballe, A., Jensen, O.N. and Roepstorff, P. (2002) in Modification specific Proteomics: An integrated strategy for glyco- and phosphospecific proteomics. Proceedings of the 50th ASMS Conference on Mass Spectrometry and Allied Topics, Orlando, FL, June 2–6.

67 Erdjument-Bromage, H., Lui, M., Lacomis, L., Grewal, A., Annan, R.S., McNulty, D.E. and Carr, S.A., Tempst, P. (1998) Examination of micro-tip reversed-phase liquid chromatographic extraction of peptide pools for mass spectrometric analysis. *J. Chromatogr. A*, **826**, 167–181.

68 Gobom, J., Nordhoff, E., Mirgorodskaya, E. and Ekman, R. and Roepstorff, P. (1999) Sample purification and preparation technique based on nanoscale reversed-phase columns for the sensitive analysis of complex peptide mixtures by matrix-assisted laser desorption/ionization mass spectrometry. *J. Mass Spectrom.*, **34**, 105–116.

69 Wilm, M. and Mann, M. (1996) Analytical properties of the nanoelectrospray ion source. *Anal. Chem.*, **68**, 1–8.

70 Larsen, M.R. and Cordwell, S.J. and Roepstorff, P. (2002) Graphite powder as an alternative or supplement to reversed-phase material for desalting and concentration of peptide mixtures prior to matrix-assisted laser desorption/ionization-mass spectrometry. *Proteomics*, **2**, 1277–1287.

71 Larsen, M.R. and Hojrup, P. and Roepstorff, P. (2005) Characterization of gel-separated glycoproteins using two-step proteolytic digestion combined with sequential microcolumns and mass spectrometry. *Mol. Cell. Proteomics*, **4**, 107–119.

72 Jensen, S.S., Heegaard, N.H.H. and Jakobsen, L. and Larsen, M.R. (2006) in: Exploring the Sialiome using titanium dioxide and mass spectrometry. Proceedings of the 54th ASMS Conference on Mass Spectrometry and Allied Topics, Seattle, WA, May 28–June 1, abstract A060707.

73 Selby, D.S., Larsen, M.R., Evans, G. and Roepstorff, P. (2006) in Glycosylation of porcine seminal plasma proteins: selective enrichment and characterisation, Proceedings of the 54th ASMS Conference on Mass Spectrometry and Allied Topics, Seattle, WA, May 28–June 1, Abstract A060302.

74 Pinkse, M.W.H. and Uitto, P.M., Hilhorst, M.J., Ooms, B. and Heck, A.J.R. (2004) Selective isolation at the femtomole level of phosphopeptides from proteolytic digests using 2D-nanoLC-ESI-MS/MS and titanium oxide precolumns. *Anal. Chem.*, **76**, 3935–3943.

75 Larsen, M.R., Thingholm, T.E., Jensen, O.N., Roepstorff, P. and Jorgensen, T.J.D. (2005) Highly selective enrichment of phosphorylated peptides from peptide mixtures using titanium dioxide microcolumns. *Mol. Cell. Proteomics*, **4**, 873–886.

76 Larsen, M.R., Jensen, S.S., Jacobsen, L.A. and Heegaard, N.H.H. (2006) Exploring the sialiome using titanium dioxide

chromatography and mass spectrometry. *Mol. Cell Proteomics* (in print).
77 Omaetxebarria, M., Larsen, M., Hooper, N., Arizmendi, J., Jensen, O. (2005) in Titanium dioxide, a promising approach for GPI-anchored peptide enrichment, HUPO 4th Annual World Congress, August 29–September 1, Munich, Germany.
78 Gabius, H.J., Siebert, H.C., Andre, S., Jimenez-Barbero, J. and Rudiger, H. (2004) Chemical biology of the sugar code. *Chembiochem*, **5**, 741–764.
79 Cummings, R.D. (1997) Lectins as tools for glycoconjugate purification and characterization, in *Glycosciences: status and perspectives*, (eds Gabius H-.J., and Gabius S.), Chapman & Hall, Weinheim, Germany. pp. 191–199.

11
Species-Dependent Proteomics

11.1
Sample Preparation and Data Processing in Plant Proteomics

Katja Baerenfaller, Wilhelm Gruissem, and Sacha Baginsky

11.1.1
Introduction

Mass spectrometry is an increasingly important tool for molecular and cellular biology, and also for the emerging field of systems biology. With the availability of gene and genome sequence databases and recent technical and conceptual advancements, mass spectrometry (MS)-based proteomics is now also widely applied for the study of many different aspects in plant sciences. An earlier review on plant proteome analyses [1] reported, that between February 2004 and February 2006, about 200 original articles focusing on plant proteomics have been published, which represents a clear increase in activity in this field.

The interest in plant proteomics is fundamental, because plants are an essential food source for the human and animal diet, and because a thorough understanding of plant biology is necessary to secure food supplies in a changing environment. Studies at the protein level are important to reveal the molecular mechanisms underlying plant growth, development, and interactions with the environment, as proteins contain several dimensions of information, including the abundance, state of modification, subcellular localization, three-dimensional (3-D) structure, and their association with each other and/or with biomolecules of different types. Because proteomics systematically identifies and characterizes proteins, it is suitable to provide information on protein abundance and protein regulation. As such it also integrates information on translation efficiency and enzyme activity with genome-structure and transcription. Thus, proteome analysis has become an important source of information about protein expression, splice variants, and the erroneous or incomplete prediction of gene structures in databases. The analysis and quantification of both proteins and mRNA is complementary, and is necessary for a complete understanding of how a cell functions. As yet, the correlation between levels of mRNA and protein in *Arabidopsis* is poor [2], and ultimately it is most likely the concentration of proteins and their interactions that are the real effectors in the cell.

Plant proteome studies, whilst undeniably important, still present a major challenge, as the analysis of plant tissues poses a number of practical difficulties, and as only three plant genomes have been sequenced and published to date (May 2007). In this chapter, an overview is provided of plant-specific aspects of proteome analyses, the aim being to provide suitable protocols for sample preparation and data analysis in plant proteomics experiments.

11.1.2
Plant-Specific Considerations in Proteomics

11.1.2.1 Cell Walls

Plant cells are typically surrounded by a cell wall, which is indispensable for the intactness of the plant cell, as it is able to withstand the pressure created by the difference in osmolarity between the inside of the cell and the surrounding medium. Cell walls maintain cell shape, the direction of cell growth, the overall architecture of the plant, and also protect the plant from pathogen attack. The principal components of cells walls are three types of polysaccharide, namely cellulose, hemicellulose, and pectins. Additionally, cell wall proteins represent a minor fraction of the cell wall mass.

Cell walls pertain to proteomics analyses in two different ways. First, they necessitate special procedures to render inner-cellular proteins accessible; second, the cell wall proteins themselves are the focus of proteomics studies, because they are essential constituents of cell walls, being involved in the modification of cell-wall components, wall structure, signaling and interactions with plasma membrane proteins at the cell surface [3].

The two main strategies used to disrupt cell walls are to: (i) apply mechanical force; or (ii) degrade them enzymatically with cellulases, hemicellulases, or pectinases [4]. Mechanical force is usually applied by subjecting the plant material to grinding with a mortar and pestle, or by disrupting it in a Waring blender.

The isolation of cell wall proteins presents a challenge because the proteins associated with the cell wall may be lost during the isolation procedure. In addition, the polysaccharides may interact with intracellular proteins after cell disruption, resulting in a high degree of contaminant proteins [5]. A protocol for the purification of cell wall proteins is provided in Section 11.1.3.1.

11.1.2.2 Plastids

Plants contain an additional organelle, which is characteristic of photoautotroph eukaryotes, the plastid. Plastids have essential biosynthetic and metabolic activities, which include photosynthetic carbon fixation, the storage of products such as starch, and the synthesis of amino acids, fatty acids, and secondary metabolites. Plastids are derived from undifferentiated proplastids, from which they differentiate into chromoplasts (in fruits and petals), etioplasts (in dark-grown leaf tissue), chloroplasts (in photosynthetically active leaf tissue), amyloplasts (in roots), and elaioplasts (in seed endosperm). The small plastid genome codes for only about 90 proteins, as most of the genes coding for plastid proteins were lost to the nucleus during the course of evolution. Information about the plastid proteome therefore may provide insights into protein sorting and targeting, and about proteome dynamics during plastid differentiation (for reviews, see Refs. [6,7]). In fact, recent studies on the chloroplast proteome revealed the presence of nucleus-encoded chloroplast proteins without any recognizable plastid localization signal, which illustrates that the assessment of the plastid proteome exclusively by targeting prediction tools will be only partial [8,9]. Accounting for this complexity are the sensitivity and specificity constraints of protein targeting prediction software, potentially unreliable gene predictions, and the fact that targeting may apparently also occur through alternative routes [10]. Proteomics studies on plastid proteins will undoubtedly help to further improve our understanding of intracellular protein targeting and basic cellular functions; consequently, detailed protocols for the extraction of proteins from *Arabidopsis* chloroplasts and rice etioplasts are provided in Section 11.1.3.2.

11.1.2.3 Protein Extraction from Plant Tissue

The extraction of proteins from plant material is often more challenging than from the tissues of other organisms, mainly because plant cells are often rich in compounds which interfere with protein stability, separation, and analysis. Furthermore, plant cells often contain relatively low protein concentrations.

The ideal extraction protocol would remove, reproducibly, the full complement of proteins in a given plant sample, with low contamination by other molecules. However, due to the different properties of the proteins in terms of their molecular weight, charge, hydrophobicity, post-translational processing, and complexation with other molecules, as well as the extremely diverse properties of different plant tissues, no single extraction protocol can cope with all of these materials. The appropriate protocol will therefore depend on the properties of the plant tissue and the proteins of interest, as well as the downstream analysis that is to be performed. Three standard protein extraction procedures are provided in Section 11.1.3.

11.1.2.4 Extraction from Recalcitrant and Resistant Tissue

Those features of plant tissues which render them of special interest as a food source or for commercial benefit, are also those features which cause problems in their proteomic analyses. These materials include pigments, carbohydrates, polyphenolics, polysaccharides (particularly starch), pectins, waxes, lipids, organic acids, terpenes and secondary metabolites in fruits and vegetables, lignins and phenols in woody tissue, etc. Plant tissues containing low protein contents and high levels of interfering compounds are considered recalcitrant for proteomic analysis.

In one recent study, different extraction protocols (e.g., phenol extraction, trichloroacetic acid (TCA)/acetone extraction), as well as a variation of the latter in which the plant tissue was first resuspended in aqueous buffer before precipitation with TCA/acetone, were compared for tissues from tomato, banana, avocado, and orange [11]. The results were assessed with regard to the protein yield and the quality of two-dimensional (polyacrylamide) gel electrophoresis (2-DE) that could be achieved after employing these protocols. The conclusion was that phenol extraction performed best for these tissues. Another method for extracting proteins from recalcitrant tissue was employed for olive leaves, and comprised a phenol extraction (in the presence of sodium dodecylsulfate; SDS) from tissue which had been finely ground and thoroughly washed with TCA and acetone [12]. A refinement of the method included an additional washing of the ground material with methanol, and was shown to be effective for the extraction of proteins from a variety of plant tissues from different species such as bamboo, pine, orange, sugar cane, redwood, tobacco, and grape [13]. This refined method, which is termed phenol/SDS extraction, is described in detail in Section 11.1.3.6.

In yet another recent study, three different protocols for the extraction of proteins from recalcitrant tissue, this time represented by banana and apple, were compared [14]. The protein yield, the number of well-resolved polypeptides after 2-DE, and the suitability of the protocol for MS analysis were each evaluated. The best-performing method comprised a protein extraction step with hot SDS, followed by TCA/acetone precipitation; this is referred to as SDS–TCA/acetone, and is described in Section 11.1.3.6.

A further challenge involves protein extraction from mechanically resistant tissue such as wood in woody perennials, because wood contains large amounts of phenols and lignins. Nonetheless, Vander Mijinsbrugge *et al.* managed to extract proteins from the bark of poplar, basically by applying the phenol extraction procedure described in Section 3.D [15].

11.1.2.5 Dynamic Range Limitations

As with many other proteomics studies, plant proteomics also faces the problem of a huge dynamic range of protein concentrations in the plant cell. Rubisco, the most abundant protein on earth, accounts for up to 50% of the soluble protein in green plant tissue, and such severely hampers the identification of low-abundance proteins. While commercially supported solutions to circumvent this limitation by affinity depletion of Rubisco have become available recently, the only way to alleviate this problem so far has been to fractionate the proteins in a way that as much Rubisco as possible is contained in one fraction, with as many other proteins as possible in the other fractions. An example of such a fractionation technique is provided in Section 11.1.3.7.

11.1.2.6 Proteomics in As-Yet Unsequenced Organisms

Until now (May 2007), only the sequences of *Arabidopsis thaliana* [16], rice (*Oryza sativa*) [16] and black cottonwood (*Populus trichocarpa*) [17] have been completed and published (for an overview on genome sequencing projects of plants view the Genome Projects Database on the NCBI website [18]). In contrast, proteome studies have been performed on 35 different plant species between February 2004 and

February 2006 [1]. From this it is clear that the database sequence resources for plants are severely limited. As protein identification is generally based on the *in-silico* match between experimentally determined and theoretically derived peptide masses from the database, this constitutes a significant bottleneck in proteomics studies of species with unsequenced genomes. Currently, approaches are being developed to circumvent these limitations, including combinations of standard database searches with database-independent approaches and/or sequence-similarity database searches. A detailed description is provided in Section 11.1.4.

11.1.3
Sample Preparation Protocols

The following section includes example protocols which cover the above-described challenges specific to plant materials. A more general description of how to proceed with the extracted proteins is also provided.

The choice of an appropriate protocol depends largely on the plant tissue under investigation, as well as on the sample requirements of the planned downstream analysis. The scheme in Figure 11.1 provides an overview of the different methods described here, in which the arrows depict their combination in the described examples.

The first step in sample preparation consists of isolating the appropriate plant material. This will in most cases be the collection of a specific plant tissue, and, if needed, the isolation of particular subcellular compartments. Protocols for the

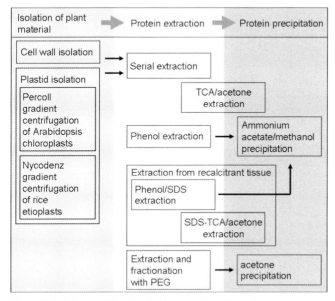

Figure 11.1 Scheme of the described protocols and how they are connected in the example methods.

isolation of cell walls and plastids, both of which are typical for plant cells, are provided in Section 11.1.3.1.

The next step is to extract proteins from the isolated plant material. The standard protein extraction protocols involve precipitation of the proteins in TCA/acetone solution (see Section 11.1.3.3 for a detailed protocol), followed by solubilization of the precipitate in a detergent-containing buffer. This protocol is used mainly to overcome the problem of low concentrations of proteins per weight of plant tissue, and also helps to avoid interference with metabolites of the plants secondary metabolism, although some polymeric contaminants are often co-extracted. Protein precipitation is also thought to avoid the risk of protein degradation caused by the high activity of proteases in plants. On the other hand, the incomplete precipitation or resolubilization of proteins can lead to selective losses of proteins. Furthermore, TCA will precipitate nucleic acids longer than about 20 nucleotides, and in some cases it may hydrolyze proteins.

One method which is complementary to TCA/acetone precipitation is that of phenol extraction (see Section 11.1.3.4), as both methods combine ease of use with effective removal of contaminant material. These two methods provide the most generically useful extraction methods for plant proteins, especially if the samples are to be subjected to 2-DE, because after precipitation the protein sample can be resuspended in any buffer compatible with 2-DE.

For studies which do not include 2-DE, and for which the buffer requirements for subsequent analyses are therefore less stringent, it might be preferable to avoid the possible bias imposed by the precipitation step. In the serial extraction protocol, the proteins are both extracted and fractionated according to their differential solubility by using a multiple buffer system with increasing protein solubilization capacity (see Section 3.E.). When the plant material contains high levels of Rubisco that interfere with downstream analyses, an extraction and fractionation with polyethylene glycol might be advantageous (see Section 11.1.3.7).

Another issue to be considered here is the extraction of proteins from recalcitrant tissue, because special conditions are required to deal with the interfering compounds. Two protocols developed to meet these specific requirements are provided in Section 11.1.3.6.

11.1.3.1 Cell Wall Protein Extraction

In the protocol described by Feiz et al. [5], 11-day-old etiolated hypocotyls are homogenized in a Waring blender in 1 mL low ionic strength buffer (0.4 M sucrose, 5 mM acetate buffer, pH 4.6, protease inhibitors) per 30 g hypocotyl fresh weight. Polyvinylpyrrolidone (PVP; 100 mg g^{-1} fresh tissue) is added, the homogenate is incubated while stirring for 30 min at 4 °C, and then centrifuged at 1000×g for 15 min at 4 °C. The pellet is resuspended in 0.6 M sucrose, 5 mM acetate buffer, pH 4.6, and re-centrifuged. The resulting pellet is resuspended in 1 M sucrose, 5 mM acetate buffer, pH 4.6, and re-centrifuged. The pelleted material is washed on a nylon net with 5 mM acetate buffer, pH 4.6 (3 L per 16 g fresh tissue) to remove contaminating intracellular proteins. The cell wall fraction is then further ground in liquid nitrogen with a mortar and pestle.

Cell wall proteins are serially extracted from the resulting lyophilized cell walls (0.65 g) by two sequential extractions with 6 mL 0.2 M $CaCl_2$, 5 mM acetate buffer, pH 4.6, protease inhibitors, followed by two extractions with 6 mL 2 M LiCl, 5 mM acetate buffer, pH 4.6, protease inhibitors. The final extraction step is carried out by boiling the residual cell walls in 12 mL of 62.5 mM Tris–HCl, pH 6.8, 4% SDS, 50 mM DTT for 5 min. At each step, the mixture is centrifuged at $4000 \times g$ for 15 min at 4 °C; the supernatant is then collected and the sedimented cell walls resuspended by vortexing at room temperature in the subsequent extraction buffer. The protein fractions obtained after each extraction are desalted and stored until further analysis.

11.1.3.2 Plastid Isolation

Plastid isolation is exemplified by two protocols, one using *Arabidopsis* chloroplasts, and the other rice etioplasts. Basically, the same protocol as described recently for chloroplast isolation was applied to isolate chromoplasts from bell pepper [19]. However, this protocol is not applicable for the isolation of etioplasts as they contain a very dense paracrystalline prolamellar body, which leads to their sedimentation in a Percoll gradient. Thus, a Nycodenz gradient was applied for the isolation of rice etioplasts.

- Isolation of *Arabidopsis* Chloroplasts
  This protocol is based on that described by Kleffmann *et al.* [9]. Note: All steps of the chloroplast isolation are performed at 4 °C:
  - Leaves from *Arabidopsis thaliana* plants are homogenized with a Waring blender or with a mortar and pestle in homogenization buffer containing 0.45 M sorbitol, 2.5 mM $MgCl_2$, 50 mM HEPES/KOH, pH 7.8, and passed through two layers of Miracloth 1R (Calbiochem).
  - The filtrate is centrifuged at $1000 \times g$ for 5 min to remove cellular debris.
  - The supernatant is removed and re-centrifuged at $4000 \times g$ for 20 min.
  - The pellet containing the chloroplasts is resuspended in homogenization buffer, and 4 mL of suspension loaded onto a linear Percoll gradient developed from 10% to 60% Percoll in isolation buffer (0.33 M sorbitol, 2.5 mM $MgCl_2$, 50 mM HEPES/KOH, pH 7.8).
  - The gradient is centrifuged at $8000 \times g$ for 20 min.
  - Intact chloroplasts are collected, washed twice in isolation buffer, and loaded onto a second Percoll gradient. In total, the chloroplasts are purified by three consecutive Percoll gradient centrifugations.

  The purity and intactness of the chloroplasts can be assessed by fluorescence microscopy, enzyme assays, and Western blotting. Contamination of the plastids (e.g., with mitochondria or peroxisomes) can be assessed with fumarase and catalase enzyme assays, or by antibody detection of compartment-specific proteins.

- Isolation and Purification of Rice Etioplasts
  This protocol is based on the method published by von Zychlinski *et al.* [20]. *Note*: All steps of the etioplast isolation are performed at 4 °C:

- Rice shoots are harvested, cut into small pieces, and homogenized in etioplast isolation solution [EIS: 10 mM HEPES/KOH, pH 7.8, 2 mM EDTA, 2 mM $MgCl_2$, 1 mM tetrasodium pyrophosphate, 600 mM sorbitol, 0.2% (w/v) bovine serum albumin (BSA)] using a Waring blender.
- The homogenate is filtered first through two layers and then through four layers of Miracloth.
- The gradient is centrifuged at $8000 \times g$ for 20 min to remove cellular debris.
- The supernatant is re-centrifuged at $8000 \times g$ for 10 min to pellet the etioplast material.
- The pellet containing the plastids is resuspended in EIS, and the suspension adjusted with 50% Nycodenz stock solution [50% (w/v) Nycodenz in EIS without sorbitol and BSA, but with 5 mM dithiothreitol (DTT)] to a final Nycodenz concentration of 30%.
- The Nycodenz step gradient consisting of 6 mL 25%, 8 mL 20%, 6 mL 15% and 3 mL 10% Nycodenz (the 50% Nycodenz stock solution is diluted with EIS plus 5 mM DTT to the required Nycodenz concentrations) is then cast on top of 5 mL of the organelle suspension.
- The gradient is centrifuged at $8000 \times g$ for 45 min.
- The two yellowish bands at the interface of 20–15% and 25–20% Nycodenz contain the highest amount of intact plastids.
- The bands are pooled, diluted threefold (v/v) with EIS plus 5 mM DTT, and centrifuged at $8000 \times g$ for 5 min to remove residual Nycodenz.
- The resulting pellet is resuspended in 20 mL EIS and centrifuged at $500 \times g$ for 10 min.
- The pellet is resuspended as before and re-centrifuged at $500 \times g$ for 10 min to produce the final plastid pellet.

11.1.3.3 Protein Extraction with TCA/Acetone

The method described here was as applied by Isaacson *et al.* [21]:

- Frozen tissue is ground in a mortar with liquid nitrogen.

- The powder is suspended in 10% (v/v) TCA in acetone (prepared by adding 1 vol. of 100% w/v TCA in water to 9 vols. of acetone) with 2% (v/v) β-mercaptoethanol added just before use (5–15 mL TCA/acetone solution per gram tissue).

- The proteins are precipitated overnight at $-20\,°C$ and then sedimented by centrifugation at $5000 \times g$ for 30 min at $4\,°C$.

- The supernatant is discarded and the pellet washed by adding an equal volume of ice-cold acetone and re-centrifuging at $5000 \times g$ for 10 min at $4\,°C$.

- The washing step is repeated twice.

- The protein pellet is then dried and later resuspended in an appropriate buffer. Heating to $40\,°C$ and/or sonication can improve the solubilization of the proteins [15].

11.1.3.4 Phenol Extraction

For phenol extraction, the method of Carpentier et al. [22] may be used; this is based on a method reported by Hurkman and Tanaka [23]:

- The plant material is ground in a mortar in the presence of liquid nitrogen.

- The homogenate (100–150 mg) is resuspended in 500 µL ice-cold extraction buffer [50 mM Tris–HCl, pH 8.5, 5 mM EDTA, 100 mM KCl, 1% (w/v) DTT, 30% (w/v) sucrose, 1×complete protease inhibitor cocktail (Roche)], and vortexed for 30 s.

- To this is added 500 µL ice-cold Tris-buffered phenol, pH 8.0, and the sample is vortexed for 15 min at 4 °C.

- After centrifugation at 6000×g for 3 min at 4 °C, the phenolic phase is collected and re-extracted with an equal volume of extraction buffer (500 µL).

- After vortexing and centrifugation as described above, the proteins in the phenolic phase are precipitated overnight with 5 volumes 0.1 M ammonium acetate in methanol at −20 °C.

11.1.3.5 Serial Extraction

This method, as described by Kleffmann et al. [9], was applied to isolated chloroplasts. If the serial extraction is to be used to extract proteins from whole plant cells, the cell walls must first be disrupted, either by grinding with a mortar and pestle or by mixing with a Waring blender.

The serial extraction protocol comprises sequences of steps for tissue disintegration, protein solubilization, and removal of insoluble material by ultracentrifugation:

- Step 1: The proteins are solubilized in 40 mM Tris–HCl, pH 8.0, 5 mM $MgCl_2$, 1 mM DTT, 2× protease inhibitor cocktail (Roche). The supernatant fraction obtained after ultracentrifugation at 100 000×g for 45 min at 4 °C contains only buffer-soluble, mainly cytosolic proteins.

  Note: All of the following steps are carried out at room temperature, because urea bnd SDS tend to precipitate at low temperatures.

- Step 2: Chaotropes (e.g., urea) are added to aid extraction of the hydrophobic, structure-associated proteins. For this, the pellet from the first extraction is solubilized in 8 M urea, 20 mM Tris-base, 5 mM $MgCl_2$, 20 mM DTT and 2× protease inhibitor cocktail, followed by ultracentrifugation (as detailed above). This second extraction supernatant contains peripheral membrane proteins.

- Step 3: The integral membrane proteins are then retrieved in two sequential steps with the aid of detergents with increasing solubilization capacity. In the first step, the pellet from extraction 2 is solubilized in 7 M urea, 2 M thiourea, 20 mM Tris-base, 40 mM DTT, 2% CHAPS, 1% Brij 35, 2× protease inhibitor cocktail.

- Step 4: The pellet from extraction 3 is then solubilized in 40 mM Tris base, 4% SDS, 40 mM DTT, 2× protease inhibitor cocktail, which should solubilize all remaining highly hydrophobic membrane proteins.

11.1.3.6 Extraction from Recalcitrant and Resistant Tissue

- Phenol/SDS Extraction
  This method was applied by Wang et al. [13] for the extraction of proteins from bamboo, orange, pine, sugar cane, redwood, tobacco and grape, and is based on a refinement of the protocol published earlier [12]:
  - The plant tissue sample is ground to a very fine powder with a mortar and pestle in liquid nitrogen.
  - The powder (100–300 mg) is sequentially washed with 2 mL cold 10% TCA in acetone, 2 mL cold 0.1 M ammonium acetate in 80% methanol, and 2 mL cold 80% acetone, and then left to dry at room temperature.
  - At each washing step the pellet is resuspended completely by vortexing, and the suspension is centrifuged at $10\,000 \times g$ for 3 min at 4 °C. Each wash solution is pre-chilled at $-20\,°C$ for at least 1 h.
  - The dry powder is subsequently resuspended in 0.8 mL Tris-buffered phenol, pH 8.0, and 0.8 mL dense SDS buffer [30% (w/v) sucrose, 2% SDS, 0.1 M Tris–HCl, pH 8.0, 5% β-mercaptoethanol], mixed thoroughly, incubated for 5 min, and centrifuged at $16\,000 \times g$ for 3 min at 4 °C.
  - The upper phenol phase is transferred to a new 2-mL tube, which is subsequently filled with at least four volumes of cold 0.1 M ammonium acetate in methanol.
  - The proteins are left to precipitate at $-20\,°C$ for at least 10 min, and then pelleted by centrifugation at $16\,000 \times g$ for 5 min at 4 °C.
  - On discarding the supernatant a white pellet should be visible; this is washed once with 2 mL 100% methanol, once with 2 mL 80% acetone, and then left to dry.

- SDS–TCA/Acetone Extraction
  The following protocol is essentially as described by Song et al. [14] for the extraction of proteins from banana and apple:
  - Tissue is ground to a fine powder in liquid nitrogen with mortar and pestle.
  - The powdered tissue (2 g) is suspended in 10–12 mL of 2% SDS extraction buffer (2% SDS, 60 mM DTT, 20% glycerol, 40 mM Tris–HCl, pH 8.5).
  - The sample is heated to 95 °C for 8 min and, after cooling, is centrifuged at $8000 \times g$ for 15 min at 4 °C.
  - The proteins in the supernatant are precipitated by adding 3 volumes of 10% TCA in acetone containing 20 mM DTT, incubating at $-20\,°C$ for 45 min, and then centrifuging at $18\,000 \times g$ for 10 min at $-4\,°C$ (see Note below).
  - The pellet is washed with ice-cold acetone containing 20 mM DTT, precipitated for 1 h at $-20\,°C$ and re-centrifuged at $20\,000 \times g$ for 10 min at $-4\,°C$ (see Note below).
  - The supernatant is discarded and the pellet left to dry.

  Note: Either a Jouan KR 4i or a Sorvall RC-6 centrifuge (both Thermo Scientific) is allowed for such low temperatures and can be used for these centrifugation steps (personal communication).

11.1.3.7 Extraction and Fractionation with Polyethylene Glycol (PEG)

This protocol was described by Kim et al. [24] for the extraction of proteins from rice leaf blades:

- The plant tissue (2 g) is ground with a mortar and pestle in liquid nitrogen to a fine powder.

- The powder is suspended in 12 mL of ice-cold Mg/NP-40 extraction buffer [0.5 M Tris–HCl, pH 8.3, 2% (v/v) NP-40, 20 mM $MgCl_2$, 2% (v/v) β-mercaptoethanol, 1 mM phenylmethylsulfonyl fluoride (PMSF) and 1% (w/v) PVP)].

- The solution is centrifuged at $12\,000 \times g$ for 15 min at 4 °C.

- The pellet is discarded and to the supernatant is added a 50% (w/v) PEG stock solution to a final concentration of 10% PEG.

- The solution is incubated on ice for 15 min and then centrifuged at $1500 \times g$ for 15 min at 4 °C. (At this point the protein pellet constitutes the 10% PEG fraction.)

- The supernatant is adjusted to 20% PEG by adding 50% (w/v) PEG stock solution and then incubated and centrifuged as described above. (The resulting pellet constitutes the 10–20% PEG protein fraction.)

- The final supernatant is precipitated with acetone by adding 4 volumes of cold acetone, incubation at $-20\,°C$ for 30 min, and centrifugation at $3000 \times g$ for 10 min at 4 °C.

11.1.3.8 Stages Following Protein Extraction

The first stage after protein extraction usually consists of gel electrophoresis. In this respect, 2-DE constitutes a very powerful technique for the separation, visualization and quantification of proteins from tissue extracts, and is widely used in plant proteomics. Recently, there has also been a rapid increase in the design of systems capable of resolving, identifying and quantifying proteins without 2-DE. For example, one-dimensional (1-D) SDS–PAGE is commonly used for the fractionation and purification of extracts before MS analysis. The rise of these techniques has mainly been due to the availability of ever-faster and more sensitive mass spectrometers, which make high-throughput proteomics experiments with complex samples possible, and feasible.

Following electrophoresis, the proteins are usually digested in-gel and extracted from the gel pieces. The resultant peptides are often directly subjected to reverse-phase separation on HPLC systems coupled on-line to a mass spectrometer, or are spotted onto a MALDI plate. Alternatively, the peptides can be purified just before MS analysis with C18-based reverse-phase systems such as Sep-Pak cartridges (Waters) or ZipTip tips (Millipore).

For complex peptide mixtures, or if the interest lies in a specific subset of peptides such as post-translationally modified peptides, these can be prefractionated or enriched before MS analysis. A variety of methods are available that utilize multi-dimensional chromatography coupled with MS. One such approach is termed "multidimensional protein identification technology", abbreviated MudPIT [25]. In

MudPIT, a strong cation-exchange phase in the first dimension is followed by reverse-phase chromatography in the second dimension by packing a microcapillary tip column consecutively with two different chromatography phases and coupling this on-line to a mass spectrometer.

11.1.4
MS/MS Data Processing for Unsequenced Organisms

A generalized workflow showing how data processing for unsequenced organisms can be optimized to obtain as many positive protein identifications as possible is shown in Figure 11.2. The first step in such an analysis usually consists of a standard database search against an as-ample-as-possible plant protein database. Especially if the organism of interest does not have many proteins in the selected protein

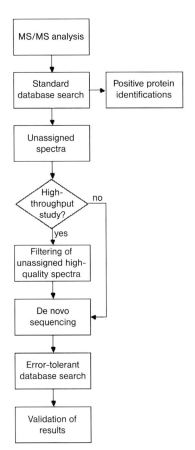

Figure 11.2 General workflow of data processing for unsequenced organisms. The validation of the results step will rely heavily on manual interpretation of spectra in case the filtering of high-quality spectra is omitted.

database, a standard database search might provide only very few protein identifications, as any discrepancy between the sequence of the measured peptide and the sequence of the database entry will impede its identification.

One way to alleviate this limitation may consist of a *de-novo* interpretation of unassigned tandem mass spectra, followed by an error-tolerant database search with MS-BLAST, as described for the characterization of the larvae venom of moths [26]. As MS-BLAST searches tolerate multiple mismatches and amino acid substitutions between queried sequences and database sequences, it can tolerate a remarkable polymorphism of protein sequences and enable the identification of as-yet missed proteins. The method described here was used on samples of limited complexity, namely peptides from isolated spots in 2-DE gels, and relied heavily on the manual interpretation of spectra. Thus, it is not applicable to high-throughput proteomics approaches.

Another approach, which also is applicable to high-throughput proteomics experiments, includes a quality filtering step for the identification of high-quality spectra, which had not been assigned in a standard database search. Only these spectra are then subjected to further analyses, as they might originate from true peptides not contained in the database [19,27]. This strategy requires a database-independent spectrum scoring scheme to distinguish putative peptide-derived spectra from low-quality noise and contaminant spectra. One tool available for performing such quality scoring of spectra is known as QUALSCORE [28]. The amino acid sequences can then be extracted directly from the high-quality MS/MS spectra in database-independent fashion by *de-novo* sequencing with one of the best performing tools available for *de-novo* sequencing, namely PepNovo [29] or NovoHMM [30]. The *de-novo* sequencing results are then filtered on the basis of a reliability score before being submitted to error-tolerant MS blast searches. The resulting MS-BLAST results are again validated, thereby increasing the identification of proteins from unsequenced organisms. This strategy has been successfully employed to increase the detection rate of proteins from bell pepper and spinach in proteome analyses [19,27].

11.1.5
Concluding Remarks

Plant proteomics is an increasingly important discipline in the field of plant science. To date, unanticipated dependencies within the plant's regulatory and metabolic networks, and the inherent robustness of biological systems pose a major obstacle in many studies aimed at biotechnological improvements of plants. Thus, a better understanding of plant biology is necessary to enable successful biotechnological advancements. As systems biology relates to the study of whole biological systems, of the interactions of their components, and how these interactions influence the function of the whole system, it promises to provide solutions to questions connected to protein network structures and dependencies. Furthermore, many questions such as the allergenicity of genetically modified und unmodified plants are of growing public interest. Recently, Hjerno *et al.* showed that proteomics is able to reveal information about allergens and allergen isoforms, allowing the analysis of allergens in different plant tissues [31]. Furthermore, proteomics is one strategy to investigate the mechanisms of plant pathogen interactions at the protein level. This knowledge is key to understanding the mechanisms of pathogen susceptibility and/or resistance, and also helps to define biotechnological targets in order to protect plants from pathogen attacks [32].

The methods and protocols described in this chapter cope with the many challenges that are specific to sample preparation and data processing in plant proteomics. An overview over the methods is provided, and the different protocols described, so that a rapid selection and implementation of the appropriate method is possible.

References

1 Rossignol, M., Peltier, J.B., Mock, H.P., Matros, A., Maldonado, A.M. and Jorrin, J.V. (2006) Plant proteome analysis: A 2004–2006 update. *Proteomics*, **6**, 5529–5548.

2 Greenbaum, D., Colangelo, C., Williams, K. and Gerstein, M. (2003) Comparing protein abundance and mRNA expression levels on a genomic scale. *Genome Biol.*, **4**, 117.

3 Jamet, E., Canut, H., Boudart, G. and Pont-Lezica, R.F. (2006) Cell wall proteins: a new insight through proteomics. *Trends Plant Sci.*, **11**, 33–39.

4 Capek, P., Renard, C.M. and Thibault, J.F. (1995) Enzymatic degradation of cell walls of apples and characterization of solubilized products. *Int. J. Biol. Macromol.*, **17**, 337–340.

5 Feiz, L., Irshad, M., Pont-Lezica, R.F., Canut, H. and Jamet, E. (2006) Evaluation of cell wall preparations for proteomics: a new procedure for purifying cell walls from *Arabidopsis* hypocotyls. *Plant Methods*, **2**, 10.

6 Baginsky, S. and Gruissem, W. (2004) Chloroplast proteomics: potentials and challenges. *J. Exp. Botany*, **55**, 1213–1220.

7 Baginsky, S. and Gruissem, W. (2006) *Arabidopsis thaliana* proteomics: from proteome to genome. *J. Exp. Botany*, **57**, 1485–1491.

8 Friso, G., Giacomelli, L., Ytterberg, A.J., Peltier, J.B., Rudella, A., Sun, Q. and Wijk, K.J. (2004) In-depth analysis of the thylakoid membrane proteome of *Arabidopsis thaliana* chloroplasts: new proteins, new functions, and a plastid proteome database. *Plant Cell*, **16**, 478–499.

9 Kleffmann, T., Russenberger, D., von Zychlinski, A., Christopher, W., Sjolander, K., Gruissem, W. and Baginsky, S. (2004) The *Arabidopsis thaliana* chloroplast proteome reveals pathway abundance and novel protein functions. *Curr. Biol.*, **14**, 354–362.

10 Villarejo, A., Buren, S., Larsson, S., Dejardin, A., Monne, M., Rudhe, C., Karlsson, J. and Jansson, S., et al. (2005) Evidence for a protein transported through the secretory pathway en route to the higher plant chloroplast. *Nat. Cell Biol.*, **7**, 1224–1231.

11 Saravanan, R.S. and Rose, J.K. (2004) A critical evaluation of sample extraction techniques for enhanced proteomic analysis of recalcitrant plant tissues. *Proteomics*, **4**, 2522–2532.

12 Wang, W., Scali, M., Vignani, R., Spadafora, A., Sensi, E., Mazzuca, S. and Cresti, M. (2003) Protein extraction for two-dimensional electrophoresis from olive leaf, a plant tissue containing high levels of interfering compounds. *Electrophoresis*, **24**, 2369–2375.

13 Wang, W., Vignani, R., Scali, M. and Cresti, M. (2006) A universal and rapid protocol for protein extraction from recalcitrant plant tissues for proteomic analysis. *Electrophoresis*, **27**, 2782–2786.

14 Song, J., Braun, G., Bevis, E. and Doncaster, K. (2006) A simple protocol for protein extraction of recalcitrant fruit tissues suitable for 2-DE and MS analysis. *Electrophoresis*, **27**, 3144–3151.

15 Vander Mijnsbrugge, K., Meyermans, H., Van Montagu, M., Bauw, G. and Boerjan, W. (2000) Wood formation in poplar: identification, characterization, and seasonal variation of xylem proteins. *Planta*, **210**, 589–598.

16 www.ncbi.nlm.nih.gov/Genomes/.

17 Initiative, T.A.G. (2000) Analysis of the genome sequence of the flowering plant *Arabidopsis thaliana*. *Nature*, **408**, 796–815.

18 Tuskan, G.A., Difazio, S., Jansson, S., Bohlmann, J., Grigoriev, I., Hellsten, U., Putnam, N. and Ralph, S., et al. (2006) The genome of black cottonwood, *Populus trichocarpa* (Torr. & Gray). *Science*, **313**, 1596–1604.

19 Siddique, M.A., Grossmann, J., Gruissem, W. and Baginsky, S. (2006) Proteome analysis of bell pepper (*Capsicum annuum* L.) chromoplasts. *Plant Cell Physiol.*, **47**, 1663–1673.

20 von Zychlinski, A., Kleffmann, T., Krishnamurthy, N., Sjolander, K., Baginsky, S. and Gruissem, W. (2005) Proteome analysis of the rice etioplast: metabolic and regulatory networks and novel protein functions. *Mol. Cell. Proteomics*, **4**, 1072–1084.

21 Isaacson, T., Damasceno, C.M., Saravanan, R.S., He, Y., Catala, C., Saladie, M. and Rose, J.K. (2006) Sample extraction techniques for enhanced proteomic analysis of plant tissues. *Nature Protocols*, **1**, 769–774.

22 Carpentier, S.C., Witters, E., Laukens, K., Deckers, P., Swennen, R. and Panis, B. (2005) Preparation of protein extracts from recalcitrant plant tissues: an evaluation of different methods for two-dimensional gel electrophoresis analysis. *Proteomics*, **5**, 2497–2507.

23 Hurkman, W.J. and Tanaka, C.K. (1986) Solubilization of plant membrane proteins for analysis by two-dimensional gel electrophoresis. *Plant Physiol.*, **81**, 802–806.

24 Kim, S.T., Cho, K.S., Jang, Y.S. and Kang, K.Y. (2001) Two-dimensional electrophoretic analysis of rice proteins by polyethylene glycol fractionation for protein arrays. *Electrophoresis*, **22**, 2103–2109.

25 Washburn, M.P., Wolters, D. and Yates, J.R., III (2001) Large-scale analysis of the yeast proteome by multidimensional protein identification technology. *Nat. Biotechnol.*, **19**, 242–247.

26 Shevchenko, A., de Sousa, M.M., Waridel, P., Bittencourt, S.T., de Sousa, M.V. and Shevchenko, A. (2005) Sequence similarity-based proteomics in insects: characterization of the larvae venom of the Brazilian moth *Cerodirphia speciosa*. *J. Proteome Res.*, **4**, 862–869.

27 Baginsky, S., Grossmann, J. and Gruissem, W. (2007) Proteome analysis of chloroplast mRNA processing and degradation. *J. Proteome Res.*, **6**, 809–820.

28 Nesvizhskii, A.I., Roos, F.F., Grossmann, J., Vogelzang, M., Eddes, J.S., Gruissem, W., Baginsky, S. and Aebersold, R. (2006) Dynamic spectrum quality assessment and iterative computational analysis of shotgun proteomic data: toward more efficient identification of post-translational modifications, sequence polymorphisms, and novel peptides. *Mol. Cell. Proteomics*, **5**, 652–670.

29 Frank, A. and Pevzner, P. (2005) PepNovo: de novo peptide sequencing via probabilistic network modeling. *Anal. Chem.*, **77**, 964–973.

30 Fischer, B., Roth, V., Roos, F., Grossmann, J., Baginsky, S., Widmayer, P., Gruissem, W. and Buhmann, J.M. (2005) NovoHMM: a hidden Markov model for de novo peptide sequencing. *Anal. Chem.*, **77**, 7265–7273.

31 Hjerno, K., Alm, R., Canback, B., Matthiesen, R., Trajkovski, K., Bjork, L., Roepstorff, P. and Emanuelsson, C. (2006) Down-regulation of the strawberry

Bet v1-homologous allergen in concert with the flavonoid biosynthesis pathway in colorless strawberry mutant. *Proteomics*, **6**, 1574–1587.

32 Padliya, N.D. and Cooper, B. (2006) Mass spectrometry-based proteomics for the detection of plant pathogens. *Proteomics*, **6**, 4069–4075.

11.2
Sample Preparation for MudPIT with Bacterial Protein Samples

Ansgar Poetsch and Dirk Wolters

11.2.1
Introduction

As bacteria have central importance in both, biotechnology and medicine, an understanding of the pathophysiology and cellular physiology of pathogenic organisms such as *Staphylococcus*, *Listeria*, *Mycobacterium*, and *Mycoplasma* is of high demand. On the other hand, bacteria such as *Corynebacterium glutamicum* or *Bacillus subtilis* have broad economic relevance in white biotechnology, mainly in the production of amino acids, antibiotics, vitamins, and proteins. The elucidation of regulatory pathways will lead to a better understanding of transport and energy processes in these bacteria, which is mandatory for example in strain improvement by metabolic engineering and the optimization of fermentation conditions [33]. Moreover, bacteria such as *B. subtilis* and *Escherichia coli* serve as versatile and extensively investigated model organisms for functional genomics, transcriptomics and proteomics under numerous conditions of stress and starvation [34]. Due to a genome-wide collection of single knockout strains, these bacteria serve as an ideal tool for physiologists and biochemists to correlate protein function with phenotypes, depending on a large number of different environmental growth conditions to achieve a comprehensive array of the adaptational network. Today, in the so-called "post-genomic era", a large number of organisms are completely sequenced and databases serve as the backbone of integrated systems biology approaches. Mass spectrometry (MS) is the analytical method of choice for such high-throughput protein identification and quantification technology.

Since the growth conditions are easily reproducible and the proteome of bacteria is simple compared to that of higher eukaryotic organisms, almost every analytical MS strategy was applied or developed on bacteria [35,36].

Two-dimensional polyacrylamide gel electrophoresis (2-D PAGE), which today is used routinely to separate complex protein mixtures, was pioneered 1975 by O'Farrell and Klose, and subsequently adapted to microbiology by Neidhardt for *E. coli* [37–39]. Whilst several thousand protein spots can be resolved using this powerful technology, only a few hundred different proteins can be routinely identified by mass spectrometry from these gels. 2-D PAGE comprises several steps – which are in total rather labor-intensive – to process a protein sample: these comprise the first dimension of isoelectric focusing, the second dimension of sodium dodecylsulfate

(SDS) gel separation, staining with suitable dyes, spot detection and picking, in-gel digestion, and subsequent mass spectrometric analysis, preferably using matrix-assisted laser desorption ionization (MALDI) time of flight (TOF) instruments. Recently developed robot systems may automate this workflow, although a complete automation has not yet been achieved. 2-D PAGE is a highly visible tool, and differences between two cell states can quickly be identified on the protein level. Therefore, this technology is widely used especially for hydrophilic, non-membrane proteins, usually referred to as "soluble" proteins. Despite these benefits, 2-D PAGE does have some intrinsic drawbacks, which are difficult to overcome. Despite the labor-intensive procedure, 2-D gels have a limited dynamic range, which makes detection of low-abundant proteins somewhat of a challenge [40]. In addition, 2-D gels have limitations regarding the pI (3–10); whereas, the detection of basic proteins is especially difficult, low- and high-molecular-weight proteins and hydrophobic membrane proteins are not compatible with 2-D gels, as the latter precipitate in the first dimension by losing their charge during the isoelectric focusing step. Recently, several excellent reviews have described the powerful application of 2-D PAGE for the profiling of bacterial proteomes under different growth conditions in order to study cellular physiology and pathways in a systems biology-driven approach [33,35].

In this chapter, the relatively new technology of multidimensional protein identification technology – otherwise known as "MudPIT" – is introduced, which overcomes some of the drawbacks of 2-D gels.

11.2.2
The MudPIT Technology

MudPIT is based on a gel-free, 2-D, high-performance liquid chromatography (HPLC) approach initially developed by Yates and colleagues (Figure 11.3) [41–43]. A complex protein mixture is denatured and digested with proteases. The core of the MudPIT technology is a fritless, fused silica capillary (50–100 µm i.d.), where at one end of the capillary a fine tip is pulled using a laser puller [44]. The capillary is packed with a strong cation-exchange (SCX) resin back-to-back with a reversed-phase (RP) material. The column is interfaced with a quaternary HPLC system coupled directly to a tandem mass spectrometer. Consecutive salt steps applying increasing salt concentrations followed by reversed-phase gradients are

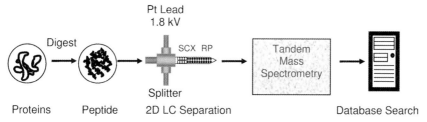

Figure 11.3 An outline of multidimensional protein identification technology (MudPIT). SCX, strong cation exchange; RP, reversed-phase.

alternated, until the peptides are completely depleted from the SCX resin. Depending on the complexity of the mixture, 10 to 15 salt steps are applied. Peptides are subjected online to the mass spectrometer where collision-induced dissociation (CID) fragments ions in a tandem mass spectrum. The SEQUEST algorithm matches spectra to database peptide sequences. Typically, a large quantity of ~30 000 to 50 000 tandem mass spectra are generated within a 24-h period. The MudPIT technology has been proven to be a powerful high-throughput tool with which to shotgun entire small proteomes.

The MudPIT technology was pioneered by Washburn, Wolters and Yates in 2001 for use with *Saccharomyces cerevisiae*, wherein a total of 1484 proteins was identified [42,43], among which were 131 integral membrane proteins. Subsequently, Koller et al. designed a proteomic survey of metabolic pathways in rice by using MudPIT and a 2-D gel-based approach [45]. A total of 2528 proteins was detected from the roots, leaves and seeds. Interestingly, only 15% of the root proteins, 18% of seed proteins and 40% of leaf proteins were identified by 2-D gels compared to MudPIT. Later, Thompson's group investigated the dynamics of the *Shewanella oneidensis* response to chromate stress at different exposure time intervals by using online MudPIT, and compared the data with their genomic microarray analysis [46]. As bioreduction of the highly toxic Cr(VI) to relative innocuous Cr(III) leads to detoxification of soil and groundwater, these bacteria appear to be a promising candidate for environmental biology purposes. To date, a total of 2370 proteins has been elucidated in different regulatory networks involving proteins in oxidative stress protection, detoxification, iron and sulfur acquisition, and SOS-controlled DNA repair mechanisms. For example, Yates et al. applied MudPIT to characterize a selected antigenic protein of *Bacillus anthracis* after affinity purification using monoclonal and polyclonal antibodies [47]. The surface protein EA1 was identified and proved to be a specific biomarker for different strains of *B. anthracis*. These authors suggested that this approach, using a basic 2-h MudPIT run for the screening of unknown samples for robust identification of the human pathogen, would be a crucial prerequisite to the consequent treatment of contaminated individuals or areas, respectively. Skipp and colleagues combined a two-technologies approach, namely MudPIT and GeLC-MS/MS, to shotgun the proteome of the bacterial pathogen *Chlamydia trachomatis* [48]. For the latter technology, 25 lanes were cut out and the proteins digested from a 1-D SDS gel. In this way, a total of 298 proteins was identified with GeLC-MS/MS compared to 117 with the two-column MudPIT approach, leading to an overall total of 328 proteins. The relatively low number of reported proteins from the MudPIT experiment suggests that the newly developed MudPIT system generally demands a degree of expert experience compared to the more error-tolerant and well-established 1-D SDS gel approach used by biologists and biochemists (see Section 11.2.6).

A combined multi-technology approach consisting of MudPIT, 2-D PAGE and SDS–PAGE combined with multidimensional microcapillary LC-MALDI-MS/MS and transcriptomics was applied to analyze the *Staphylococcus aureus* proteome [49]. Gene chip data revealed that 97% of the 2596 open reading frames (ORFs) were detected during the post-exponential phase, while 591 proteins were identified with the proteomics approach.

Basically, there are two ways in which a MudPIT or MudPIT-like experiment (e.g., online or offline MudPIT) can be performed. The most-often described technology is the typical online MudPIT set-up, which can easily be automated, as a pure peptide fraction is loaded onto the SCX column. For the offline set-up, instead of using salt step gradients to deplete peptides from the SCX and alternating the RP gradients, a regular gradient is utilized first to elute and fractionate the peptides. In the offline mode, two separate columns are used and the system is usually not automated; however, for increased column performance non-volatile salts and an organic modifier can be used, which may be incompatible with online MudPIT. After buffer exchange, each fraction is consecutively loaded and separated on the RP gradient.

Coldham and Woodward subsequently applied this offline MudPIT approach to study the *Salmonella typhimurium* proteome. A total 816 proteins was identified from 20 fractions collected from the SCX phase, including 34 outer membrane proteins [50]. Likewise, Vollmer *et al.* collected up to 45 peptide fractions from a linear 120-min SCX gradient, identifying a total of 339 proteins from *E. coli* [51]. In general, the application of online and offline MudPIT experiments is less strict than described here. The largest protein profile of *E. coli* was generated using an offline MudPIT approach, which uses nine salt steps to elute from the SCX first, after which the fractions were consecutively processed by RP-LC/MSMS. Thus, a total of 1147 different proteins was identified, including 99 membrane proteins. Some 97% of these proteins were confirmed by detection of their corresponding mRNA using Affymetrix GeneChips [52]. One other example of an offline MudPIT experiment using 10 ammonium acetate salt steps in the first dimension investigated the human pathogen *Staphylococcus aureus*. Here, the proteomes of growing and non-growing cells were analyzed with 2-D gel-based and non-gel-based approaches, whereby 473 proteins were identified using the former technology and 650 with the latter [53].

The discussions of online or offline MudPIT achieves better results remain ongoing, and the advantages or disadvantages of either approach have not yet been clearly defined. Nonetheless, this chapter will focus on the online MudPIT approach, which has proved to be highly successful in many biological experiments, as described here.

11.2.3
Membrane Proteins and MudPIT

As mentioned above, MudPIT is especially suited to the identification of membrane proteins, although this class of proteins was largely overlooked by investigators using 2-D PAGE. Overall, integral membrane proteins are structurally and functionally much less characterized than other proteins, this being reflected by the fact that less than 1% of proteins in the Protein Data Bank are integral membrane proteins [54]. Nevertheless, the extreme importance of this class of proteins is clear, as 20 to 30% of all genes in an organism code for integral membrane proteins [55]. These proteins are involved in central cellular processes and form the major protein class for many drug targets [56]. Due to their location in the lipid bilayer, membrane proteins are amphipathic, and the targets for tryptic cleavage – lysine and arginine – are mainly absent from transmembrane (TM) helices, being found only in the hydrophilic part of the protein. The size of exposed hydrophilic domains varies among integral membrane proteins from large (e.g., epidermal growth factor receptor) to small (e.g., rhodopsin). As most integral membrane proteins are of low abundance, all proteomics protocols usually involve prefractionation steps for the enrichment of the membrane fraction. Examples include organelle separation by free-flow electrophoresis [57], and the removal of membrane-associated proteins by chaotropes [58] or alkaline pH washes [59]. In the next step, the membranes are either treated directly with protease or solubilized with detergents, such that the intact proteins are further separated before digestion. Even after enrichment, however, a high percentage of the identified proteins are usually not membrane proteins, regardless of the proteomics approach applied. In conclusion, despite their great importance – and consequent considerable effort in method development [60] – the current approaches aimed at obtaining a comprehensive coverage of the membrane proteome are far from satisfactory. Generally, 2-D PAGE is unsuitable for the separation of integral membrane proteins [61], and even if membrane proteins are reported as being identified via 2-D PAGE they are usually not very hydrophobic and contain not more than one predicted TM domain [62]. By comparison, the combination of SDS–PAGE and LC-ESI MS/MS has been applied with greater success, although problems such as protein insolubility and the loss of hydrophobic peptides, which prevent protein identification, remain [63]. These can be overcome by a variety of protein-digestion strategies using isolated intact membranes, followed by peptide separation using MudPIT (Figure 11.4). Membranes can be solubilized in 90% formic acid and digested with cyanogen bromide (CNBr) [43] which, as a chemical reagent, cleaves embedded membrane proteins under these harsh conditions. Consequently, the pH is increased to pH 8.5 and the sample is digested with endoproteinase Lys C in 8 M urea; after a fourfold dilution with 100 mM ammonium bicarbonate, trypsin generates smaller peptides that are more amenable to analysis by tandem MS. The MudPIT approach revealed 131 integral membrane proteins which, at the time of writing is the largest number of membrane protein identifications for a single organism. The same group developed an additional strategy, which uses proteinase K under basic conditions. At pH 11, Wu et al. reported the cleavage of proteins to reasonably sized 6- to 20-mers, whereas lower pH conditions would result in close-to-complete digestion down to dipeptides. Whilst cleavage with CNBr/trypsin [43] in formic acid or proteinase K in aqueous buffer [64] yields almost exclusively peptides of exposed hydrophilic domains from membrane proteins (e.g., loops), digestion in 60% methanol also permits the analysis of peptides from transmembrane helices [65].

Recently, the present authors developed for *Corynebacterium glutamicum* a modified and almost close-to-complete integral membrane protein enrichment and cleavage procedure, and subsequently used this for the analysis with MudPIT [43]. The membranes were first incubated with trypsin in aqueous buffer to remove hydrophilic domains and membrane-associated proteins, followed by chymotrypsin/trypsin digestion in 60% methanol to release peptides originating from the TM helices. Alternatively, intact membranes were treated with CNBr/trypsin to identify membrane proteins by exposed domains [43]. Besides a distinct number of cytosolic

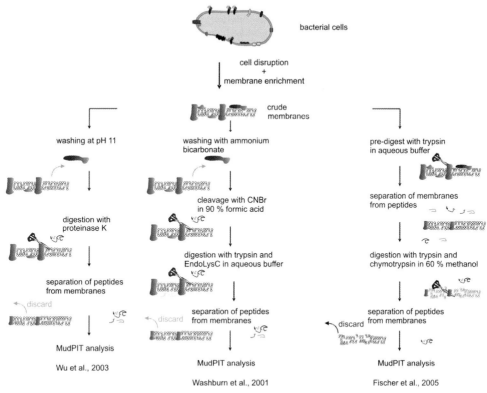

Figure 11.4 Strategies for membrane proteomics using MudPIT.

or membrane-associated proteins, the combined data analysis from both digests yielded 326 integral membrane proteins (~50% of all predicted) covering membrane proteins both with small and large numbers of TM helices. In addition, membrane proteins with a high GRAVY (grand average hydrophobicity) score [66] were identified, and basic and acidic membrane proteins were seen to be evenly represented.

11.2.4
Quantitative MudPIT

Non-gel-based labeling strategies are favorable for differential membrane proteomics. The incorporation of stable isotopes into bacteria (especially ^{13}C and ^{15}N) is relatively robust, easily manageable, and suitable in combination with MudPIT [67]. Relative ratios are calculated through mass spectrometric peak heights between labeled and unlabeled samples. In 2003, Yates *et al.* suggested a strategy whereby they used the isotopically labeled sample simply as an internal standard [68], such that differences due to growth on minimal media or incomplete labeling would be canceled out. The *E. coli* degradosome composition was analyzed by a combination of affinity purification with FLAG epitope-tagged RNase E coupled with a label-free quantitative MudPIT approach [69]. The normalized SEQUEST score values are given as a percentage of RNase E, and thus may be taken as a quantitative estimate of the relative protein abundance. Similar findings were obtained from the Yates laboratory, where it was shown that the spectral count (the number of successful peptide sequencing events) directly correlated to protein abundance [70]. In this chapter, due to reasons of space limitation, it is impossible to provide a comprehensive overview of all quantification strategies compatible with MudPIT (e.g., ICAT, iTRAQ, or SILAC). For this reason, the reader is referred to reviews [71,72] and to Chapter 4.4 in this book.

11.2.5
Limitations of MudPIT

Although many research groups have begun to shotgun entire proteomes with MudPIT, complex mixtures of biological origin still overwhelm the resolution capability of many analytical technologies, including MudPIT. In practice, MudPIT tackles protein identification on the peptide level, referred to as a "bottom-up" approach. Therefore, a complex mixture consisting of a few hundred proteins can easily become a separational problem involving a few thousand peptides. Taking into account that more than 200 different post-translational modifications are known [73], and that some 30% of all proteins in a proteome may be phosphorylated in higher organisms, usually more than one peptide elute at the same time [74]. Moreover, the dynamic range between high- and low-abundant proteins may be quite diverse. The human plasma proteome comprises a dynamic range of at least 10 decades [75]; by comparison, the dynamic range determined for MudPIT is approximately four decades [42]. As a consequence, peptides compete in the ionization process, and ion suppression becomes significant – at least for the less-abundant peptides. As mass spectrometers have a finite duty cycle, the actual sample rate is always under-represented. Especially for low-abundant proteins it might be a random event, in which one of the co-eluting peptides is fragmented during CID. Therefore, the reproducibility of single peptide hits between two consecutive runs might be less than 50%. Durr *et al.* reported a saturation level of identified proteins from the same sample after approximately five to seven MudPIT experiments, analyzing tryptic digests of soluble mouse lung cytosolic proteins [76]. With the recent development of faster-scanning and more sensitive mass spectrometers (e.g., the linear ion trap mass spectrometer), this intrinsic problem of such profiling approaches by MudPIT might, in time, become less serious. Other approaches such as peak parking are being introduced to increase proteome coverage [77]. Here, the flow rate of the pump is decreased significantly or even stopped, so that the mass spectrometer is allowed an increased time to sample a higher number of chromatographic peaks, which show peak width of only 10 to 30 s at regular flow rates.

The prefractionation of organelles and membranes before running a MudPIT experiment simplifies the sample and significantly improves the number of detected proteins [78]. Froehlich *et al.* performed a sucrose step gradient to isolate mixed-envelope membrane from *Arabidopsis thaliana* prior to off-line MudPIT [79]. A 2-D LC system itself can be used to fractionate on the protein level using SCX-RPLC [80] or SEC-RPLC [81]. Consequently, choosing one of these strategies easily multiplies the number of samples which readily overwhelm the number of experiments that one MS analyst can process.

Nevertheless, a complete proteome coverage of any organism is unlikely to become feasible in the near future.

11.2.6
Pitfalls of MudPIT

The purity of the sample for MudPIT is crucial in the extreme. As high sensitivity is related to small column diameters, clogging of the columns and subsequent complete loss of the sample represents a severe problem. In particular, membrane samples must be free of particles and contaminants before being loaded onto a column. Sucrose, glycerol or high salt concentrations must also be removed, because these samples regularly fail to load correctly. Protein precipitation with for example trichloroacetic acid (TCA), desalting with solid-phase extraction cartridges, or the use of cut-off filters to eliminate non-digested proteins might be useful for generating a suitable sample. Less high salt concentrations (<500 mM) can be desalted on an online triphasic column, where 3 to 4 cm of RP material is packed onto the SCX resin.

Due to the high sensitivity of the mass spectrometer in combination with MudPIT, the solvents used should be of the highest grade, and contaminants such as keratin (from skin, hair, etc.) should be strictly eliminated.

As the column dimensions are relatively small, the dead volume in nano-LC – and especially in MudPIT experiments – becomes a crucial point. The user should be especially careful about fittings and column packing, as irregular packing or gaps will ruin any nano-LC separation.

11.2.7
Summary

Today, non-gel-based MudPIT or MudPIT-like approaches are rapidly emerging and proving to represent a complementary technology to the 30-year-old "gold standard" of 2-D PAGE. Sample preparation including protease-amenable conditions and well-designed prefractionation, in combination with a deeper understanding of MudPIT, is a crucial step for the proteome-wide analysis of biological samples, where bacteria serve as handy model organisms.

11.2.8
Perspective

Nowadays, systems biology is one of the worldwide key areas in which research teams are working, and the elucidation and understanding of global pathways have proved to be especially fruitful when studying model organisms such as bacteria. Tools such as MudPIT have contributed – and will continue to contribute – to this ambitious project involving the large-scale analysis of proteomes and subproteomes. Although correlation between protein expression and transcription level is not restricted to single loci but to entire pathways, it appears that protein expression in protein complexes is regulated to large extent by post-translational modifications [82]. Both needs may be addressed by MudPIT: due to its excellent resolving power, MudPIT is ideally suited to the identification of post translational modifications (PTMs) [83], and protein complexes can be reliable elucidated with tandem affinity purification (TAP)-MudPIT [84]. Improved fractionation methods, reliable quantification strategies, robust data-mining software, and faster and more-sensitive mass spectrometers which are currently under development, will undoubtedly provide research groups with the opportunity to examine much more closely these complex biological processes. Surely, MudPIT will in the future become as versatile a tool as gene chips are today?

11.2.9
Recipe for Beginners: MudPIT: Soluble and Membrane Proteins

- Lyse bacterial cells (~5 g wet weight) with a French press or another procedure.
- Sediment the unbroken cells at $6000 \times g$ for 15 min at 4 °C, collect supernatant and repeat centrifugation.
- Pellet the membranes by ultracentrifugation at $100\,000 \times g$ for 30 min at 4 °C.
- Digest membranes and soluble proteins separately (~600 µg proteins according to Lowry protein assay). Try to keep solvent volume as low as possible.
- Digest soluble proteins directly in 60% MeOH, 25 mM ammonium bicarbonate (Ambic) at pH 8.5 with trypsin (protein/protease weight ratio = 50 : 1) overnight or at the least for 4 h at 37 °C.
- Reduce the pH to 3 with formic acid.
- Lyophilize the solution to eliminate organic solvent to a final volume of 20–50 µL. Retain the soluble fraction for loading onto the MudPIT column.
- Optional soluble proteins can be reduced in 10 mM dithiothreitol in 8 M urea at 37 °C for 30 min and alkylated in 50 mM iodoacetamide in the dark for another 30 min. For subsequent digestion with trypsin, the urea concentration must be less than 2 M.
- Digest membranes with trypsin (protein/protease weight ratio = 20 : 1) in 25 mM Ambic (pH 8.5) overnight or at least for 4 h at 37 °C.
- Pellet the membranes by ultracentrifugation at $100\,000 \times g$ for 30 min at 4 °C.
- Wash membranes twice with 1 mL ice-cold deionized water, and re-centrifuge ($100\,000 \times g$ for 30 min at 4 °C).

11.2 Sample Preparation for MudPIT with Bacterial Protein Samples

- Resuspend pellet in 600 µL 100% methanol, sonicate 2 × 10 min in ultrasonic bath.
- Add 25 mM Ambic to a final methanol concentration of 60% and adjust pH to 8.5 with ammonium hydroxide.
- Add sequencing-grade trypsin (Promega) and chymotrypsin (Roche) to sample [each 1 : 50 protease/sample (w/w)].
- Perform proteolysis in organic solvent overnight at 37 °C.
- Remove undigested proteins and membrane fragments with 10-kDa cut-off filters (Microcon YM-3, Millipore).
- Pull a fine fritless spray tip (1–3 µm) at the end of a 100-µm internal diameter (i.d.) × 365 µm outer diameter (o.d.) fused silica capillary with a P2000 laser puller [85].
  Settings such as filament, delay time, pull, heat and velocity, can be optimized as described. Alternatively, ready-to-pack fused silica columns or even-packed RP-SCX columns can be purchased from New Objective [86]. First, pack 8–12 cm of 3–5 µm RP resin into the capillary by means of a pressurized stainless steal vessel (Brechbühler Inc. [87]). A pressure of 80–120 bar of helium or argon forces a methanol slurry of the packing material through the column, followed by 3 cm of 3–5 µm SCX resin. If the sample must be desalted, a third phase consisting of 3–4 cm RP material can be packed on top of the SCX column.
- Equilibrate the column with Buffer A (2% acetonitrile and 0.1% formic acid) and adjust HPLC net flow rate to between 200 and 300 nL min^{-1}.
- Load up to 400 µg of sample with the same pressurized vessel or an autosampler.
- Connect column to a PEEK microtee or microcross (Upchurch [88]), depending on the split system.
  At this point a Pt-lead assures the connection for the high voltage (1.8 kV) by a liquid junction needed for ESI. Usually, quaternary analytical HPLC pumps require flow rates of 200–300 µL min^{-1} to deliver reproducible gradients, so that the final split ratio is ca 1 : 1000.
- Prepare four solvents: Buffer A (2% acetonitrile, 0.1% formic acid); Buffer B (80% acetonitrile, 0.1% formic acid); Buffer C (250 mM ammonium acetate, 2% acetonitrile, 0.1% formic acid); Buffer D (1.5 M ammonium acetate, 2% acetonitrile, 0.1% formic acid).
  Typical MudPIT chromatograms start with an equilibration step with Buffer A which can run up to 30 min. The next cycle incorporates a salt step gradient with 5% Buffer C, 95% Buffer A for 2 min. This elutes some of the peptides from the SCX phase.
- After a short re-equilibration with Buffer A for 7 min, a shallow 90-min gradient is generated up to 100% Buffer B. The following cycle differs from the one before only in the salt step concentration of Buffer C. Depending on the complexity of the

sample, cycles with 5%, 10%, 20%, 30%, 40%, 50%, 60%, 70%, 80%, 90%, 100% Buffer C can be applied. Finally, two cycles at the end apply Buffer D for 20 min and deplete all peptides from the SCX packing material.

- Set MS to acquire a full MS scan between 350 and 2000 m/z, followed by full MS/MS scans of the three most intensive ions from the preceding MS scan.

- Set relative collision energy for collision-induced dissociation to 35%, and enable dynamic exclusion of masses with a repeat count of 1 and a 3-min exclusion duration window.

References

33 Van Bogelen, R.A., Schiller, E.E., Thomas, J.D. and Neidhardt, F.C. (1999) Diagnosis of cellular states of microbial organisms using proteomics. *Electrophoresis*, **20**, 2149–2159.

34 Volker, U. and Hecker, M. (2005) From genomics via proteomics to cellular physiology of the Gram-positive model organism *Bacillus subtilis*. *Cell Microbiol.*, **7**, 1077–1085.

35 VerBerkmoes, N.C., Connelly, H.M., Pan, C. and Hettich, R.L. (2004) Mass spectrometric approaches for characterizing bacterial proteomes. *Expert Rev. Proteomics*, **1**, 433–447.

36 Han, M.J. and Lee, S.Y. (2006) The *Escherichia coli* proteome: past present, and future prospects. *Microbiol. Mol. Biol. Rev.*, **70**, 362–439.

37 Klose, J. (1975) Protein mapping by combined isoelectric focusing and electrophoresis of mouse tissues. A novel approach to testing for induced point mutations in mammals. *Humangenetik*, **26**, 231–243.

38 O'Farrell, P. (1975) High resolution two-dimensional electrophoresis of proteins. *J. Biol. Chem.*, **250**, 4007–4021.

39 Pedersen, S., Bloch, P.L., Reeh, S. and Neidhardt, F.C. (1978) Patterns of protein synthesis in *E. coli*: a catalog of the amount of 140 individual proteins at different growth rates. *Cell*, **14**, 179–190.

40 Gygi, S.P., Corthals, G.L., Zhang, Y., Rochon, Y. and Aebersold, R. (2000) Evaluation of two-dimensional gel electrophoresis-based proteome analysis technology. *Proc. Natl. Acad. Sci. USA*, **97**, 9390–9395.

41 Link, A.J., Eng, J., Schieltz, D., Carmack, E., Mize, G.J., Morris, D.R., Garvik, B.B. and Yates, J.R., III (1999) Direct analysis of protein complexes using mass spectrometry. *Nat. Biotechnol.*, **17**, 676–682.

42 Wolters, D.A., Washburn, M.P. and Yates, J.R., III (2001) An automated multidimensional protein identification technology for shotgun proteomics. *Anal. Chem.*, **73**, 5683–5690.

43 Washburn, M.P., Wolters, D. and Yates, J.R., III (2001) Large-scale analysis of the yeast proteome by multi-dimensional protein identification technology. *Nat. Biotechnol.*, **19**, 242–247.

44 Gatlin, C.L., Kleemann, G.R., Hays, L.G., Link, A.J. and Yates, J.R., III (1998) Protein identification at the low femtomole level from silver-stained gels using a new fritless electrospray interface for liquid chromatography-microspray and nanospray mass spectrometry. *Anal. Biochem.*, **263**, 93–101.

45 Koller, A., Washburn, M.P., Lange, B.M., Andon, N.L., Deciu, C., Haynes, P.A., Hays, L., Schieltz, D., Ulaszek, R., Wei, J., Wolters, D. and Yates, J.R., III (2002) Proteomic survey of metabolic pathways in rice. *Proc. Natl. Acad. Sci. USA*, **99**, 11969–11974.

46 Brown, S.D., Thompson, M.R., Verberkmoes, N.C., Chourey, K., Shah, M., Zhou, J., Hettich, R.L. and Thompson, D.K. (2006) Molecular dynamics of the *Shewanella oneidensis* response to chromate stress. *Mol. Cell. Proteomics*, **5**, 1054–1071.

47 Krishnamurthy, T., Hewel, J., Bonzagni, N.J., Dabbs, J., Bull, R.L. and Yates, J.R., III (2006) Simultaneous identification and verification of *Bacillus anthracis*. *Rapid Commun. Mass Spectrom.*, **20**, 2053–2056.

48 Skipp, P., Robinson, J., O'Connor, C.D. and Clarke, I.N. (2005) Shotgun proteomic analysis of *Chlamydia trachomatis*. *Proteomics*, **5**, 1558–1573.

49 Scherl, A., Francois, P., Bento, M., Deshusses, J.M., Charbonnier, Y., Converset, V., Huyghe, A., Walter, N., Hoogland, C., Appel, R.D., Sanchez, J.C., Zimmermann-Ivol, C.G., Corthals, G.L., Hochstrasser, D.F. and Schrenzel, J. (2005) Correlation of proteomic and transcriptomic profiles of *Staphylococcus aureus* during the post-exponential phase of growth. *J. Microbiol. Methods*, **60**, 247–257.

50 Coldham, N.G. and Woodward, M.J. (2004) Characterization of the *Salmonella typhimurium* proteome by semi-automated two dimensional HPLC-mass spectrometry: detection of proteins implicated in multiple antibiotic resistance. *J. Proteome Res.*, **3**, 595–603.

51 Vollmer, M., Horth, P. and Nagele, E. (2004) Optimization of two-dimensional off-line LC/MS separations to improve resolution of complex proteomic samples. *Anal. Chem.*, **76**, 5180–5185.

52 Corbin, R.W., Paliy, O., Yang, F., Shabanowitz, J., Platt, M., Lyons, C.E., Jr. Root, K., McAuliffe, J., Jordan, M.I., Kustu, S., Soupene, E. and Hunt, D.F. (2003) Toward a protein profile of *Escherichia coli*: comparison to its transcription profile. *Proc. Natl. Acad. Sci. USA*, **100**, 9232–9237.

53 Kohler, C., Wolff, S., Albrecht, D., Fuchs, S., Becher, D., Buttner, K., Engelmann, S. and Hecker, M. (2005) Proteome analyses of *Staphylococcus aureus* in growing and non-growing cells: a physiological approach. *Int. J. Med. Microbiol.*, **295**, 547–565.

54 Gao, F.P. and Cross, T.A. (2005) Recent developments in membrane-protein structural genomics. *Genome Biol.*, **6**, 244.

55 Wallin, E. and von Heijne, G. (1998) Genome-wide analysis of integral membrane proteins from eubacterial archaean, and eukaryotic organisms. *Protein Sci.*, **7**, 1029–1038.

56 Hopkins, A.L. and Groom, C.R. (2003) The druggable genome. *Nat. Rev. Drug Discov.*, **1**, 727–730.

57 Zischka, H., Weber, G., Weber, P.J., Posch, A., Braun, R.J., Buhringer, D., Schneider, U., Nissum, M., Meitinger, T., Ueffing, M. and Eckerskorn, C. (2003) Improved proteome analysis of *Saccharomyces cerevisiae* mitochondria by free-flow electrophoresis. *Proteomics*, **3**, 906–916.

58 Xiong, Y., Chalmers, M.J., Gao, F.P., Cross, T.A. and Marshall, A.G. (2005) Identification of *Mycobacterium tuberculosis* H37Rv integral membrane proteins by one-dimensional gel electrophoresis and liquid chromatography electrospray ionization tandem mass spectrometry. *J. Proteome Res.*, **4**, 855–861.

59 Fujiki, Y., Hubbard, A.L., Fowler, S. and Lazarow, P.B. (1982) Isolation of intracellular membranes by means of sodium carbonate treatment: application to endoplasmic reticulum. *J. Cell Biol.*, **93**, 97–102.

60 Wu, C.C. and Yates, J.R. (2003) The application of mass spectrometry to membrane proteomics. *Nat. Biotechnol.*, **21**, 262–267.

61 Santoni, V., Molloy, M. and Rabilloud, T. (2000) Membrane proteins and proteomics: un amour impossible? *Electrophoresis*, **21**, 1054–1070.

62 Molloy, M.P., Phadke, N.D., Chen, H., Tyldesley, R., Garfin, D.E., Maddock, J.R. and Andrews, P.C. (2002) Profiling the alkaline membrane proteome of *Caulobacter crescentus* with two-dimensional electrophoresis and mass spectrometry. *Proteomics*, **2**, 899–910.

63 Klein, C., Garcia-Rizo, C., Bisle, B., Scheffer, B., Zischka, H., Pfeiffer, F., Siedler, F. and Oesterhelt, D. (2005) The membrane proteome of *Halobacterium salinarum*. *Proteomics*, **5**, 180–197.

64 Wu, C.C., MacCoss, M.J., Howell, K.E. and Yates, J.R. (2003) A method for the comprehensive proteomic analysis of membrane proteins. *Nat. Biotechnol.*, **21**, 532–538.

65 Blonder, J., Conrads, T.P., Yu, L.R., Terunuma, A., Janini, G.M., Issaq, H.J., Vogel, J.C. and Veenstra, T.D. (2004) A detergent- and cyanogen bromide-free method for integral membrane proteomics: Application to *Halobacterium* purple membranes and the human epidermal membrane proteome. *Proteomics*, **4**, 31–45.

66 Kyte, J. and Doolittle, R.F. (1982) A simple method for displaying the hydropathic character of a protein. *J. Mol. Biol.*, **157**, 105–132.

67 Venters, R.A., Huang, C.C., Farmer, B.T., II, Trolard, R., Spicer, L.D. and Fierke, C.A. (1995) High-level 2H/13C/15N labeling of proteins for NMR studies. *J. Biomol. NMR*, **5**, 339–344.

68 MacCoss, M.J., Wu, C.C., Liu, H., Sadygov, R. and Yates, J.R., III (2003) A correlation algorithm for the automated quantitative analysis of shotgun proteomics data. *Anal. Chem.*, **75**, 6912–6921.

69 Regonesi, M.E., Del Favero, M., Basilico, F., Briani, F., Benazzi, L., Tortora, P., Mauri, P. and Deho, G. (2006) Analysis of the *Escherichia coli* RNA degradosome composition by a proteomic approach. *Biochimie*, **88**, 151–161.

70 Liu, H., Sadygov, R.G. and Yates, J.R. III (2004) A model for random sampling and estimation of relative protein abundance in shotgun proteomics. *Anal. Chem.*, **76**, 4193–4201.

71 Lill, J. (2003) Proteomic tools for quantitation by mass spectrometry. *Mass Spectrom. Rev.*, **22**, 182–194.

72 Blonder, J., Conrads, T.P. and Veenstra, T.D. (2004) Characterization and quantitation of membrane proteomes using multidimensional MS-based proteomic technologies. *Expert Rev. Proteomics*, **1**, 153–163.

73 Krishna, R.G. and Wold, F. (1993) Post-translational modification of proteins. *Adv. Enzymol. Relat. Areas Mol. Biol.*, **67**, 265–298.

74 Hubbard, M.J. and Cohen, P. (1993) On target with a new mechanism for the regulation of protein phosphorylation. *Trends Biochem. Sci.*, **18**, 172–177.

75 Anderson, N.L. and Anderson, N.G. (2002) The human plasma proteome: history character, and diagnostic prospects. *Mol. Cell. Proteomics*, **1**, 845–867.

76 Durr, E., Yu, J., Krasinska, K.M., Carver, L.A., Yates, J.R., Testa, J.E., Oh, P. and Schnitzer, J.E. (2004) Direct proteomic mapping of the lung microvascular endothelial cell surface in vivo and in cell culture. *Nat. Biotechnol.*, **22**, 985–992.

77 Davis, M.T. and Lee, T.D. (1998) Rapid protein identification using a microscale electrospray LC/MS system on an ion trap mass spectrometer. *J. Am. Soc. Mass Spectrom.*, **9**, 194–201.

78 Brunet, S., Thibault, P., Gagnon, E., Kearney, P., Bergeron, J.J. and Desjardins, M. (2003) Organelle proteomics: looking at less to see more. *Trends Cell Biol.*, **13**, 629–638.

79 Froehlich, J.E., Wilkerson, C.G., Ray, W. K., McAndrew, R.S., Osteryoung, K.W., Gage, D.A. and Phinney, B.S. (2003) Proteomic study of the *Arabidopsis thaliana* chloroplastic envelope membrane utilizing alternatives to traditional two-dimensional electrophoresis. *J. Proteome Res.*, **2**, 413–425.

80 Opiteck, G.J., Lewis, K.C., Jorgenson, J.W. and Anderegg, R.J. (1997)

Comprehensive on-line LC/LC/MS of proteins. *Anal. Chem.*, **69**, 1518–1524.

81 Opiteck, G.J., Ramirez, S.M., Jorgenson, J.W. and Moseley, M.A., III (1998) Comprehensive two-dimensional high-performance liquid chromatography for the isolation of overexpressed proteins and proteome mapping. *Anal. Biochem.*, **258**, 349–361.

82 Washburn, M.P., Koller, A., Oshiro, G., Ulaszek, R.R., Plouffe, D., Deciu, C., Winzeler, E. and Yates, J.R., III (2003) Protein pathway and complex clustering of correlated mRNA and protein expression analyses in *Saccharomyces cerevisiae*. *Proc. Natl. Acad. Sci. USA*, **100**, 3107–3112.

83 MacCoss, M.J., McDonald, W.H., Saraf, A., Sadygov, R., Clark, J.M., Tasto, J.J., Gould, K.L., Wolters, D., Washburn, M., Weiss, A., Clark, J.I. and Yates, J.R., III (2002) Shotgun identification of protein modifications from protein complexes and lens tissue. *Proc. Natl. Acad. Sci. USA*, **99**, 7900–7905.

84 Graumann, J., Dunipace, L.A., Seol, J.H., McDonald, W.H., Yates, J.R., III Wold, B.J. and Deshaies, R.J. (2004) Applicability of tandem affinity purification MudPIT to pathway proteomics in yeast. *Mol. Cell. Proteomics*, **3**, 226–237.

85 Sutter Instrument Company, www.sutter.com.

86 New Objective, www.newobjective.com.

87 Brechbühler, Inc., www.brechbuehler.com.

88 Upchurch Scientific, www.upchurch.com.

11.3
Sample Preparation for the Cell-Wall Proteome Analysis of Yeast and Fungi

Kai Sohn, Ekkehard Hiller, and Steffen Rupp

11.3.1
Introduction

The quality and completeness of proteome analyses correlate strictly with the quality of the preparation for the respective protein samples. Due to the extraordinary heterogeneity of the physical–chemical properties that are characteristic for the individual proteins, protein samples more often represent enrichments of subpopulations of proteins rather than a true representation for a given proteome. In addition to the individual amino acid composition, post-translational modifications, as well as covalent linkages of proteins, directly determine the behavior of the respective proteins. With respect to solubility, many hydrophobic proteins, for example, tend to precipitate in aqueous solutions and are consequentially depleted from complex protein samples. Moreover, extensive post-translational modifications (PTMs) such as glycosylations in eukaryotes substantially minimize the analyzable resolution of downstream applications, including high-resolution two-dimensional (2-D) polyacrylamide gel electrophoresis (2-D PAGE), for example. The situation becomes even worse for proteins that are covalently linked to subcellular structures. Many cell-wall proteins from fungi, for example, are directly attached via a so-called glycosylphosphatidylinositol- (GPI-) anchor to a complex glucan network of the outer cell wall, and in most cases such proteins are also heavily glycosylated. Both, covalent linkage to the cell wall, as well as massive PTMs confer structural properties to these proteins in order to meet the specific requirements of cell-wall proteins under physiological conditions. Many of these proteins are often involved in the stable attachment of the whole cell to different surface structures, or function as receptors in the perception of signal molecules from the outside. Strikingly, in pathogenic fungi such as *Candida albicans*, covalently linked cell-wall proteins are also virulence factors critical for pathogenesis in humans. Comprehensive analyses of the cell-wall proteomes of pathogenic fungi therefore represent a promising approach that will reveal a more detailed understanding about the molecular mechanisms involved during infection, and may also provide a collection of putative targets for the development of specific antimycotics.

11.3.2
Description of the Problem with Regards to Sample Preparation

The cell wall of fungi represents the major interface between the fungal cell and its respective environment. It is composed of a complex network of polysaccharides comprising β-1,3- and β1,6-linked glucans, as well as chitin. This polymeric structure provides a robust casing for the cell, and is equipped with specific collection of cell-wall proteins on the surface of this structure. According to the type of attachment to the cell wall, different types of cell-wall proteins can be discriminated in fungal cells. First, there are proteins that are attached to the surface by non-covalent interactions. These proteins are more or less loosely attached to the surface, and can be released by extraction applying relatively mild conditions. Among these proteins are secreted proteins such as proteases, as well as proteins originally identified as having a function in the cytosol, including glycolytic enzymes or heat-shock proteins, for example. The identification of those cytosolic proteins in the cell wall led to the proposition that such proteins represent so-called "moonlighting" proteins, with functions both in the cytosol and in the cell wall. However, in order to achieve an unambiguous identification of such cell-wall proteins, methods must be established which allow for the discrimination of both subcellular populations of proteins. For this purpose, surface proteins are first labeled using a biotin-tag. This tag contains an activated *N*-hydroxysuccinimide (NHS)-ester that is able to react specifically with amino groups of proteins, but is not membrane-permeable. For this reason, only extracellular proteins are labeled with biotin, in contrast to intracellular proteins. Biotin-labeled proteins can then be enriched by affinity purification using a streptavidin-matrix, and are subsequently identified. In addition to the proteins that are just loosely attached to the cell wall there are also proteins which are covalently linked via intermolecular bonds to the glucan and chitin network of this subcellular structure. Mostly, this type of cell wall linkage is mediated by so-called GPI-anchors that are transferred towards the cell-wall proteins during passage through the secretory pathway. Frequently, covalently linked cell-wall proteins are also heavily modified by post-translational glycosylations. According to the so-called "lollipop on a stick" model, these glycosylations (mainly O-glycosylation) stiffen the peptide backbone of the proteins to span the thickness of the cell-wall structure and to expose a functional domain to the extracellular space. In this context, the sample preparation for a proteomic analysis of this type of cell-wall protein not only requires disintegration of the covalent bonds linking these proteins to the cell wall, but also should aim at reducing the extensive PTMs to facilitate the unambiguous determination of their respective expression patterns. To this end, cell walls of fungal cells are purified and washed under high-stringency conditions using ionic detergents such as sodium dodecyl-sulfate (SDS) to remove non-covalently attached proteins or contaminants. The covalently-linked cell wall proteins are subsequently fragmented by the addition of highly purified proteases such as trypsin or Lys-C. Those domains of cell-wall proteins with PTMs such as O-glycosylations are protected against a proteolytic attack, and are therefore omitted from further analyses. Only the proteolysis of non-modified domains will result in the release of a respective set of peptides that can subsequently be analyzed using 2-D gel electrophoresis or mass spectrometry (MS), for example. As a result, only non-modified molecules – and therefore samples that are easier to analyze – are prepared using such an approach. However, due to a lower sequence-coverage for these cell-wall proteins, higher standards for the identification of proteins must be applied to meet the demands for an unambiguous identification using, for example, MS.

11.3.3
Application Example

Candida albicans as a model for the investigation of proteins localized to the cell wall, was analyzed, because it represents the most commonly isolated pathogenic fungus responsible for invasive mycosis in humans that is within the focus of clinical mycology [89]. As a commensal organism, present in the gastrointestinal or urogenital tract of 30 to 60% of the human population, this opportunistic pathogen is widely spread and a constant risk for immunocompromised patients. Mortality in patients suffering from systemic candidiasis is up to 42%, and thus is more prevalent than mortality linked to Gram-negative bacteria [90]. In recent years, some key mechanisms and factors critical for virulence in *C. albicans* were characterized, including polymorphic switches, secreted proteases and lipases, as well as proteins localized to the cell wall [91–93]. In particular, the cell wall as the first contact site between the pathogen and its host represents a promising target for therapeutic intervention to treat mycoses [94–96]. Some of the proteins integrated into the cell wall were shown to be targets of the immune system of patients with systemic candidiasis [97]. Because of the complex structure of the cell wall (polymeric glucan and chitin structures, post-translational modified and covalently-linked proteins) the analysis and identification of the cell-wall proteome requires sophisticated approaches [98]. Non-covalently proteins integrated into the cell wall can be labeled with sulfo-NHS-LC-biotin, which cannot pass through the cell membrane and can then be extracted and enriched by affinity chromatography (Figure 11.5). The isolated proteins are then separated by gel electrophoresis and identified using MS.

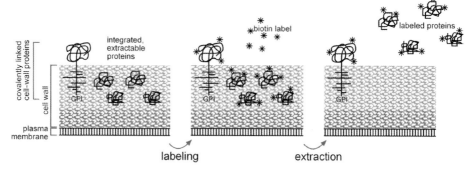

Figure 11.5 Preparation of non-covalently linked cell-wall proteins. Proteins that are non-covalently linked to the fungal cell wall are specifically labeled using a biotin tag which cannot pass through the plasma membrane. Subsequent extraction and affinity-enrichment of the labeled proteins allows for an analysis and identification.

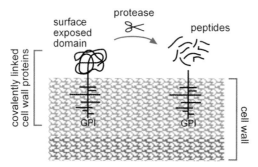

Figure 11.6 Preparation of covalently linked cell-wall proteins. The fungal cell walls are purified and non-covalently attached proteins, as well as contaminants, are removed by high-stringency washing steps. Proteolytic digestion of the accessible domains of the covalently linked cell-wall proteins leads to the release of soluble peptides which can be used to identify the respective proteins.

In order to identify covalently linked proteins, the cell walls are isolated and non-covalently bound proteins are removed by stringent washing steps. Thereafter, a proteolytic digestion is performed, to release peptides derived from unglycosylated surface-exposed domains of the covalently linked proteins (Figure 11.6). Separation by reversed phase-HPLC followed by analysis using MS, will then reveal the identification of the proteins.

11.3.4
Summary

Analyzing the cell wall proteome of yeast and fungi is a challenging task, especially with respect to covalently linked and glycosylated proteins. However, such cell-wall proteome analyses offer the opportunity to better understand at the molecular level for example host–fungus interactions that are critical for virulence in human pathogens. Techniques of sample preparation for cell-wall proteome analyses must overcome problems related to the release of covalently anchored proteins, as well as to the heterogeneity due to heavy glycosylation. Moreover, such techniques should also provide the basis to analyze soluble proteins more loosely attached to the cell surface. By using the biotin-label or the protease-release approach for the preparation of samples described in this chapter, in combination with techniques for a high-resolution separation and for the identification of peptides including 2-D gel electrophoresis and MS (MALDI-TOF, ESI-MS/MS), it became possible to analyze both soluble as well as covalently linked proteins of the cell wall of *C. albicans* [99,100]. Depending on the structure and the accessibility of the protein domains of the covalently linked cell-wall proteins, the protease approach generates peptides with molecular masses in the range of approximately 0.5 to 20 kDa. The bigger fragments (ca. 5 to 20 kDa) can be reproducibly separated and analyzed by 2-D gel electrophoresis, as they are readily soluble and only rarely modified. Spots can be visualized using standard staining methods including silver staining or fluorescent dyes. By applying this approach to the cell wall of *C. albicans*, it was possible to identify Als3p, an adhesin containing a GPI-anchor, that is expressed on the hyphal cell wall but not on the surface of blastospores [99]. In contrast, it was also found that Cht2, coding for a chitinase, is expressed on both, the cell wall of hyphae as well as of blastospores. In contrast to fragments between 5 and 20 kDa, that are generated by the protease approach, smaller peptides (0.5–5 kDa) can be analyzed more reliably using liquid chromatography in combination with tandem MS (LC-MS/MS).

When using the biotin-label approach, surface proteins are enriched that can be separated using standard separation techniques including one-dimensional (1-D)-PAGE and 2-D PAGE, as well as Western blotting. Unambiguous identification of proteins is then accomplished using peptide mass fingerprinting if the respective sequence databases are available, or by *de-novo*-sequencing. An analysis of the soluble cell-wall proteome of *C. albicans* revealed a complex composition of surface proteins, some of which surprisingly had cytosolic counterparts [100]. Among these is Tsa1p, commonly known as a thiol-specific antioxidant-like protein from the cytosol. It was possible to identify Tsa1p as a cell-wall component that is specifically expressed on hyphal walls, but not on the surface of blastospores. The expression of Tsa1p on the cell wall in addition to a cytosolic localization demonstrates that this technique is able to identify proteins with moonlighting functions [100,101]. In the meantime, the identification of such proteins on the surface of yeast and fungal cells has also been confirmed by independent approaches including immunohistochemical methods, indicating that the proposed techniques will provide an approved and reliable starting point for the preparation of protein samples from yeast and fungal cell walls.

11.3.5
Perspective

Proteome analysis of those cellular compartments which are refractory to most solubilization methods, such as the cell wall of fungi or plants, remain a major challenge for the proteome scientist. However, these methods are fundamentally required in order to learn more about the biosynthesis of these structures and their functionalities, including mechanisms of nutrient uptake to host–pathogen interaction. Especially the identification of proteins in the cell wall of fungi, as well as other organisms, with known functions in other parts of the cell, indicates that the cell wall as a compartment might be more complex than expected. The identification of novel unpredicted proteins involved in the biogenesis of the cell wall, also termed "moonlighting proteins", requires the use of proteomics in order to determine the players in this process, based on their actual localization. Genome-based approaches and homology-based predictions currently fail to point out these novel components. Both, the biotin and the protease approach, showed successfully how protein samples may be made available for global proteomic analyses. Although these methods still need to be refined in order to complete the picture of the cell-wall proteins in fungi, their application is not restricted to the fungal cell wall. Rather, these approaches are universal for all organisms which have a cell wall, including plants. Moreover, analyses of organisms which have no cell walls but rather have an extensive extracellular matrix might also become possible.

11.3.6
Recipe for Beginners

11.3.6.1 Cultures

- Cultivate fungal cultures at 30 °C in 10 mL YPD (1% yeast extract, 2% bacto peptone, 2% glucose) overnight.

- Inoculate fresh cultures with overnight culture to an optical density OD_{600nm} of 0.1.

- Cultivate to an $OD_{600\,nm}$ of 3 in 1000 mL of YPD medium at 30 °C.

- Harvest by centrifugation at 3500 ×g for 10 min at 4 °C.

11.3.6.2 Preparation of Soluble Cell-Surface Proteins

- Wash the cells three times with ice-cold phosphate-buffered saline (PBS) pH 7.4.

- Incubate for 1 h at 4 °C with 10 mg mL^{-1} Sulfo-NHS-LC-Biotin (Pierce, Rockford, IL, USA) in binding buffer (50 mM NaHCO$_3$, pH 8.5).

- Block the remaining reactive Sulfo-NHS-LC-Biotin by adding 2 volumes of 100 mM Tris–HCl, pH 8.0, and incubate for a further 30 min.

- Harvest the labeled cells by centrifugation at $3500 \times g$ for 10 min at 4 °C and wash them three times in cold PBS, pH 7.4.

- Disrupt the cell walls by applying three cycles in a French press (SLM Instruments, Champaign, IL, USA) at a pressure of 2×10^6 h Pa in disruption buffer (PBS, pH 7.4, 1% Nonidet P-40, 0.1% SDS) containing protease inhibitors.

- Centrifuge extracts at $3500 \times g$ for 5 min at 4 °C.

- Collect the supernatant and centrifuge at $15\,000 \times g$ for 10 min at 4 °C.

- Purify biotinylated proteins in this supernatant by affinity chromatography using immobilized NeutrAvidin (Pierce, Rockford, IL, USA).

- Equilibrate the NeutrAvidin-Agarose in PBS containing 1% NP-40 and 0.1% SDS prior to use.

- Incubate 1 mL of the extract from biotinylated cells with 100 μL immobilized NeutrAvidin at 4 °C overnight.

- Wash agarose beads after binding four times in cold PBS containing 1% NP-40 and 0.1% SDS and eluted with sample buffer (50 mM Tris–HCl pH 8, 2% SDS, 5% β-mercaptoethanol, 10% glycerol, bromophenol blue) for 5 min at 55 °C.

11.3.6.3 Preparation of Peptides from Covalently Linked Cell-Wall Proteins

- Wash fungal cells with ice-cold H_2O and with 10 mM Tris–HCl, pH 7.5.

- Resuspend cells in 10 mM Tris–HCl, pH 7.5 ($\sim 10^7$ cells mL^{-1}) and disintegrate by applying three cycles in a French Press at a pressure of 2×10^6 h Pa in the presence of a protease inhibitor cocktail.

- To remove any non-covalently linked proteins and intracellular contaminants, wash the isolated cell walls once with 1 M NaCl, three times with H_2O, and extract three times for 15 min at 100 °C with 50 mM Tris–HCl, pH 7.8, containing 2% SDS, 100 mM Na-EDTA, and 40 mM β-mercaptoethanol (5 mL buffer per gram cells wet weight).

- Wash the SDS-treated cell walls five times with water, and freeze-dry.

- Resuspend 5 mg of the dry cell wall in 100 μL 100 mM Tris–HCl, pH 8, 7.5 mM DTT, and incubate at 60 °C for 1 h to reduce the disulfide chains.

- Afterwards, alkylate by adding iodoacetamide to a final concentration of 15 mM and incubate for 30 min at room temperature in the dark.

- Add 10 μL acetonitrile and dilute with 100 mM Tris–HCl, pH 8, to a final volume of 200 μL. For digestion of the surface-exposed protein domains, incubate overnight at 37 °C with 5 μg trypsin (sequencing grade).

- Centrifuge at 16 000 × g for 10 min at room temperature to remove the remaining cell wall fragments.

- The supernatant contains the soluble peptide mixture derived from the SDS-resistant cell-wall proteins that can be applied to 2-D gel electrophoresis or tandem MS.

References

89 Calderone, R.A. (2002) *Candida and Candidiasis*, ASM Press, Washington, DC.

90 Kullberg, B.J. and Oude Lashof, A.M. (2002) Epidemiology of opportunistic invasive mycoses. *Eur. J. Med. Res.*, **7**, 183–191.

91 Hoyer, L.L. (2001) The ALS gene family of *Candida albicans*. *Trends Microbiol.*, **9**, 176–180.

92 Hube, B., Monod, M., Schofield, D.A., Brown, A.J. and Gow, N.A. (1994) Expression of seven members of the gene family encoding secretory aspartyl proteinases in *Candida albicans*. *Mol. Microbiol.*, **14**, 87–99.

93 Lo, H.J., Kohler, J.R., Di Domenico, B., Loebenberg, D., Cacciapuoti, A. and Fink, G.R. (1997) Nonfilamentous *C. albicans* mutants are avirulent. *Cell*, **90**, 939–949.

94 Klis, F.M., Boorsma, A. and De Groot, P.W. (2006) Cell wall construction in *Saccharomyces cerevisiae*. *Yeast*, **23**, 185–202.

95 Klis, F.M., de Groot, P. and Hellingwerf, K. (2001) Molecular organization of the cell wall of *Candida albicans*. *Med. Mycol.*, **39**, 1–8.

96 Monteoliva, L., Matas, M.L., Gil, C., Nombela, C. and Pla, J. (2002) Large-scale identification of putative exported proteins in *Candida albicans* by genetic selection. *Eukaryot. Cell*, **1**, 514–525.

97 Pitarch, A., Jimenez, A., Nombela, C. and Gil, C. (2006) Decoding serological response to *Candida* cell wall immunome into novel diagnostic, prognostic, and therapeutic candidates for systemic candidiasis by proteomic and bioinformatic analyses. *Mol. Cell. Proteomics*, **5**, 79–96.

98 Rupp, S. (2004) Proteomics on its way to study host-pathogen interaction in *Candida albicans*. *Curr. Opin. Microbiol.*, **7**, 330–335.

99 Sohn, K., Schwenk, J., Urban, C., Lechner, J., Schweikert, M. and Rupp, S. (2006) Getting in touch with Candida albicans: the cell wall of a fungal pathogen. *Curr. Drug Targets*, **7**, 505–512.

100 Urban, C., Sohn, K., Lottspeich, F., Brunner, H. and Rupp, S. (2003) Identification of cell surface determinants in *Candida albicans* reveals Tsa1p, a protein differentially localized in the cell. *FEBS Lett.*, **544**, 228–235.

101 Urban, C., Xiong, X., Sohn, K., Schroppel, K., Brunner, H. and Rupp, S. (2005) The moonlighting protein Tsa1p is implicated in oxidative stress response and in cell wall biogenesis in *Candida albicans*. *Mol. Microbiol.*, **57**, 1318–1341.

12
The Human Proteosome

12.1
Clinical Proteomics: Sample Preparation and Standardization

Gerd Schmitz and Carsten Gnewuch

12.1.1
Introduction

The proteome of an organism is the protein complement of its genome. The proteome is highly dynamic and varies according to cell type and functional state of the cell. The proteome may reflect immediate and characteristic changes in response to disease processes and external stimulation. As the actual proteome – that is, a set of all proteins present in a body fluid, cell, tissue, or organism at a certain point of time – represents only a subset of all possible gene products, the proteome cannot be directly predicted from genome information. Any protein may exist in multiple forms that respond within a particular cell or between different cells via post-translational modifications (PTMs) or degradation processes that affect protein structure, localization, function, and turnover. Like the proteome, the proteolytic degradation products of the proteome – the so called low-molecular-weight (LMW) range plasma proteome – may also have the potential to contain disease-specific information. The identified disease-specific peptides appear to be fragments of endogenous high-abundant proteins, such as transthyretin; or fragments of low-abundance cellular and tissue proteins, such as BRCA-2 [1,2]. Also, catabolic pathways lead to degradation of proteins, with the extent of this degradation being highly dependent on preanalytical variables such as temperature or time of storage of the clinical sample. Emerging novel nanotechnology strategies that make use of these LMW biomarkers either *in vivo* or *ex vivo* will greatly enhance the ability to discover and characterize molecules for early disease detection or stratification, and will expand the prognostic capability of current proteomic modalities. In addition, pathophysiological processes including proteolytical activities – for example, tumor proteases are detectable in the plasma peptidome [1]. The discovery of sensitive and specific panels of biomarkers rather than a single marker protein for early disease detection and risk stratification or prognostic evaluation holds the key to future

Proteomics Sample Preparation. Edited by Jörg von Hagen
Copyright © 2008 WILEY-VCH Verlag GmbH & Co. KGaA, Weinheim
ISBN: 978-3-527-31796-7

treatment of complex diseases such as metabolic and vascular disease, cancer, and neurodegenerative disorders.

Recent advances in proteomic analysis, both in the field of affinity binder technologies and in liquid-phase technologies such as tandem mass spectrometry (MS/MS), high-resolution liquid-phase separation methods and advances in chemometry and biometry for large-scale data analysis, form the basis for the high-throughput and high-content analyses required in clinical proteomics. They also offer the possibility of introducing these research tools into laboratory diagnostics to screen for risk factors, to identify new disease-specific or stage-specific biomarkers, and to find novel markers for therapeutic drug monitoring or new therapeutic targets. Thus, clinical proteomics, has the potential to complement genomics, metabolomics, lipidomics, glycomics, and transcriptomics including splice variant analysis in order to gain a better understanding of disease processes.

Whilst metabolomics had undergone previous advances over several years in laboratory medicine (e.g., newborn screening), clinical proteomics is at present on the verge of entering the hospital arena. However, certain essential criteria for the successful use of proteomics within a clinical environment must be fulfilled:

- High-throughput analysis platforms must provide reproducible protein patterns in a clinically acceptable turnaround time and also acquire the instrumental stability of laboratory analyzers. Platforms must also fit into the workflow of a clinical laboratory.

- Bioinformatic algorithms must include chemometry, data reduction and conversion into actionable health information; they must also be robust and capable of being integrated into standard laboratory information systems.

- Preanalytical conditions must be standardized and improved for the development of clinically applicable tests.

Before commencing any investigations to discover novel biomarkers, a defined patient subset that is matched in every possible epidemiologic and physiologic parameter, including age, gender, hormonal status, treatment, and hospitalization status, must be identified. For biomarker development, Standard Operating Procedure (SOP)-driven biobanks must be integrated into the diagnostic workflow with validated storage conditions for all proteomic applications. Indeed, the standardized and also SOP-driven preparation of patient samples is one of the most urgent challenges for reproducible and clinical useful results.

12.1.2
The Preanalytical Phase: Sample Preparation, Standardization, and Quality Management

Although advances in genomics and proteomics have led to high expectations of clinical biomarker discovery, the generation of validated biomarkers needs more attention towards the preanalytical stage, such as sample retrieval, preparation, and processing [3]. The preclinical discovery phase may lead to interesting biomarkers which must be validated before transformation into clinical proteomics. The type of sample needed, and their processing, may be quite different for the two phases. In general, there is a major difference between the requirements of a high-throughput proteomic profiling (clinical proteomics) approach and protein identification and the in-depth characterization (low-throughput) of protein samples in the preclinical phase. Another point is the preparation of the patient; for example, for metabolomic parameters it is necessary that the patient is fasted before sample collection. Diurnal rhythms for parameters such as hormones or cytokines must also be considered.

12.1.2.1 Standardization of the (Pre)-Analytical Process

Examples of systematic bias errors resulting in false-positive, as well as false-negative, results include:

- Preanalytical variables, such as systematic differences in study populations and/or sample collection, handling, and pre-processing procedures.

- Within-class biological variability which may comprise unknown subphenotypes among study populations.

- Analytical variables, such as inconsistency in instrument conditions and reagents, resulting in poor reproducibility.

- Measurement imprecision [4,5].

The study conducted by Marshall *et al.* [6] highlights the importance of understanding the effects of preanalytical variation in interpreting data produced through proteomic analysis. By using matrix assisted laser desorption/ionization time-of-flight (MALDI-TOF) MS, these authors analyzed blood samples obtained from patients with myocardial infarction and found that recorded changes in protein profiles were due to protease activities in the serum rather than the result of disease processes. Another source of variation that must also be considered in biomarker discovery is the heterogeneity of the human population (biological variation), referred to as between-subject variation; this is the sum of differences in protein expression between people which results in – but is not limited to – differences in age, gender, or race [7].

The use of control materials (e.g., the addition of standards) is mandatory to control the range of linearity, sensitivity, precision, and accuracy. Once the instrument is calibrated, standardization of the output can be accomplished by running control samples and adjusting parameters such as laser intensity, detector voltage, and detector sensitivity to ensure that the spectra are consistent from run to run. Controls should also be analyzed throughout an experimental run, as either internal or added analytes. The first addition is certainly desirable as these would not interfere at all with the analytical process. A recent study using surface-enhanced laser desorption/ionization time-of-flight (SELDI-TOF) MS on prostate cancer cases and control cases demonstrated that the implementation of such strategies on a SELDI-TOF approach can yield reproducible results [8]. It must be borne in mind that the limitations of this technology include a reduced dynamic range due to the enrichment process, and that SELDI yields protein patterns associated with the presence of the disease rather than potential target proteins. The transition of the MS technology from a research tool to a reliable clinical diagnostic platform certainly requires rigorous chemometric standardization, spectral quality control and assurance, standard operating procedures for robotic and automatic sample application, and standardized controls to insure the generation of highly reproducible spectra. The introduction of peptide standards in defined quantities, as well as the use of quantitative labeling techniques (e.g., stable isotope labeling) for routine applications, may eventually allow the quantitative assessment of selected markers. Currently, laboratories are independently developing their own methods, optimization procedures and in-process controls, although there appears to be a lack of effort to standardize methodology between laboratories at this stage of research and development. At present, the instrumentation is still at the level of "advanced prototypes" which provide research tools to specialized laboratories but do not yet qualify for routine laboratory testing. However, the development of standardized technology MS platforms, reference standards for controls and calibrators will certainly help the field to accelerate the process of evaluating proteomics technologies for clinical applications. However, this transition will require widespread collaborative efforts between government, industry, scientists, healthcare providers, and health insurance companies.

After completing the transition of preclinical proteomics into clinical routine diagnosis, it will be necessary to implement reference values and to identify preanalytical variability. For this, an integrative database such as modern data warehouse systems would ideally function as a strong bioinformatics source, and should be compatible with new developments for many bioinformatic research tools [9]. When data are stored under standardized conditions, subsequent to bioinformatics, meta-analyses using a multi-center approach for validation of new markers are supported [10]. With the successful improvement of analytical instrumentation and high-quality analytical standards, analytical errors are no longer the main factor influencing the reliability and clinical utilization of laboratory diagnostics. In fact, the main sources of variation occur within the preanalytical phase such as a lack of standardized procedures for sample collection, including patient preparation such as fasting, specimen acquisition, handling, and storage. Indeed, these factors account for more than 90% of errors within the entire diagnostic process [11]. An example of a standardized protocol for the sampling and preparation of blood, as established by the National biobank program in Sweden, is available at http://www.biobanks.se (SOP – Collection of blood samples, Gerd Johansson, Göran Hallmans, Medicinska biobanken).

12.1.3
Proteomics in Body Fluids

12.1.3.1 Techniques for Proteomic Analysis
The techniques used for protein analysis can generally be divided into "unbiased" and "biased" methods.

In "unbiased" techniques, the investigator does not specifically preselect the proteins to be examined, but rather searches for potential changes in any protein to be identified. Such methods classically include multiple protein separation techniques such as two-dimensional (2-D) polyacrylamide gel electrophoresis (PAGE), liquid chromatography (LC) methods, where the protein mix flows through a column packed with porous beads of particular binding properties, and capillary electrophoresis (CE), where proteins are separated based on their migration in an electric field. Separation strategies are followed by protein identification, mostly via tandem MS.

The "biased" techniques include antibody-based affinity-binding methods, where the proteins of interest have already been identified and highly specific antibodies generated against them [12]. This allows for a more in-depth profiling on these preselected proteins allowing an approximate four orders of magnitude in protein abundance to be reached.

12.1.3.2 Applications

Proteomics and Peptidomics of Body Fluids for the Clinical Laboratory A fundamental hypothesis of MS-based profiling is that peptidomics, when measured in its complexity, can detect cancer with greater accuracy than any single tumor marker. Certainly, one advantage of a proteomic profile of uncharacterized and unidentified molecules over standard immunoassay measurements is that the fingerprint can be rapidly obtained from as little as 1 µL of raw, unfractionated serum from patients. The small serum sample can be analyzed using MS-type approaches to generate, quite rapidly, a unique proteomic signature of the serum [13]. It is, however, principally limiting, that MS-based methods select for the most abundant peptide ions and peaks, which makes it likely that markers present in minute amounts are left undetected. At present, such immunoassays capable of detecting at a sensitivity of up to 10^{-14} are superior with respect to detecting low-abundant markers.

From an optimistic point of view, peptidomics may mirror disease processes such as tumor development. Communication networks between cells, stroma, and extracellular matrix exist in neoplastic disease. The tumor–host interface system is thought to involve unique enzymatic events, flow of information, and sharing of growth factors and metabolic substrates. The plasma proteome is presumably altered in the diseased population as a consequence of constant perfusion of the diseased organ. It is reasonable to hypothesize that protein fragments associated with the unique and active biological processes occurring at the tumor–stromal interface would be shed into the extracellular interstitium. A unique combination of protein fragments derived from the microenvironment of the tumor–host interaction would be drained from the tumor site via the lymphatics and ultimately shed into the serum. If a sensitive and specific method of detection and characterizing LMW tumor markers were to be developed, it would be possible – in theory – to detect the presence of a tumor while it is still microscopic in size. The early detection of malignancy may be more likely, if a sample is drained close to a lesion. This may be of specific importance for applications which should detect minimal residual disease or tumor dormancy.

The identification of the specific analytes will be very important to enhance an understanding of disease mechanisms, to (potentially) provide new novel effective drug targets, and perhaps to lead to the discovery of analytes that may be measured as a multiplex immunoassay using conventional antibody-based approaches. As it is likely that such biomarkers would be smaller cleaved proteins arising from larger molecules, it will be necessary to develop methods that can effectively distinguish the diagnostic fragments from the larger parental molecules, as epitopes contained in the LMW forms might cross-react with the wild-type normal proteins. Regardless of whether or not these biomarker candidates are measured as unknown entities by MS profiling, or via a multiplexed immunoassay, it could be predicted that these "combinatorial diagnostic" approaches would be inherently superior to single marker antibody-based tests for early disease detection.

Today, several successful approaches of clinical proteomics are already available. For example, MS has been shown capable of discriminating patients having prostate cancer with a specificity and sensitivity greater than the currently used prostate-specific antigen (PSA) test [14,15]. Potential serum biomarkers for early stroke have been identified via SELDI-TOF MS, while the analysis of cerebrospinal fluid (CSF) by MS has led to the identification of potential biomarkers for Alzheimer's disease [16]. Clarke et al. [17] reported the successful analysis of urine by SELDI-TOF MS to distinguish rejection versus no rejection in a renal transplant population. The use of SELDI-TOF MS in these investigations has, however, been questioned as the technique yields a comparably minor sensitivity-specificity profile.

A direct comparison of the SELDI with CE-MS technology by Neuhoff et al. [18], using identical urine samples, resulted in the definition of three potential biomarkers using SELDI and 200 potential biomarkers from the CE-MS analysis. These authors also concluded that it is necessary to characterize any disease by using a panel of well-defined biomarker proteins rather than a few, not too-well-defined peaks. Mischak and coworkers [19–21] established CE coupled to MS together with appropriate software solutions with the goal of analyzing urine (and other body fluids) to diagnose various kidney disorders based on well-defined protein patterns. Using this approach, each protein is defined by its mass and migration time, and the signal intensity serves as measure of its abundance (for reviews, see [22] and [23]). The urine samples were analyzed individually and the data from the individual CE-MS runs combined due to the high reproducibility of the method. This feature allows compilation of groups of datasets and their comparisons (e.g., patients with a specific kidney disease compared to patients with other types of kidney disease or healthy controls). This comparison permits the definition of an array of biomarkers that differentiate healthy subjects from patients and other markers that define the specific (kidney) disease or clinical condition. The latter type of biomarker is thus useful for differential diagnosis. CE-MS permits fast and reproducible analysis and subsequent differentiation of protein patterns based on dozens of protein markers. Panels of 20 to 50 protein markers allowed not only the diagnosis of a specific (primary) kidney disease, but also the discrimination with high sensitivity and specificity between different kidney diseases such as IgA nephropathy, focal-segmental glomerulosclerosis, membranous glomerulonephritis, minimal-change disease, and diabetic nephropathy [24,25].

In this context, urine proteome diagnostics may also represent a diagnostic approach to kidney disease without significant morbidity when compared to standard invasive kidney biopsy. In a recent study, Decramer *et al.* [26] utilized CE-MS-based urinary proteome analysis to define specific biomarker patterns for different grades of ureteropelvic junction obstruction, a frequently encountered pathology in the newborn. (Note: these patients did not have any signs of increased proteinuria.) In a blinded prospective study, these patterns predicted with 95% accuracy the clinical outcome of these newborns some nine months in advance. These data clearly indicate the potential of urinary proteomics not only in enabling diagnosis but also in the prognosis of renal disease. Proteome analysis of the urine has also revealed biomarkers for several non-renal diseases. As in the case of ureteropelvic junction obstruction, these diseases generally do not result in increased proteinuria. Not surprisingly, biomarkers for urothelial cancer have been defined in urine. While the first studies based on the SELDI technology analyzed only few samples and reported different biomarkers for the same disease [27,28], Theodorescu *et al.* [29] recently used CE-MS to assay more than 600 samples, including 180 examined in blinded fashion as a validation set. The discovered biomarkers correctly classified all blinded urothelial cancer samples and normal controls; however, nine of 138 patients with various chronic kidney diseases or nephrolithiasis were incorrectly classified as having urothelial cancer. Kaiser *et al.* [30] found biomarkers for graft-versus-host disease (GvHD) after bone-marrow transplantation using CE-MS-based urine proteomics. GvHD results in, or is the consequence of endothelial dysfunction, which may also alter kidney structure and/or function. This complication will, in turn, influence filtration and generation of the primary urine, and subsequently add disease-specific proteins to the urine.

Protein arrays comprised of immobilized proteins are an emerging biochip format [31,13]. Screening protein arrays with plasma from autoimmune patients would not only allow the identification of potentially new autoantigens, but also enable the diagnosis and subtyping of the autoimmune disease on the basis of the presence of specific autoantibodies. By combining the cDNA expression library approach with protein microarrays, the humoral autoimmune repertoire of dilated cardiomyopathy patients has been profiled and several protein antigens determined which are associated with heart failure [32]. In a different approach, a protein array consisting of 196 structurally diverse biomolecules representing major autoantigens was probed with serum from patients with different autoimmune diseases, including systemic lupus erythematosus, Sjögren syndrome, and rheumatoid arthritis. There were distinct autoantibody patterns for the different autoimmune diseases, indicating their suitability for diagnosis [33].

A new technique, termed "layered peptide array", serves as a screening tool to detect antibodies in a highly multiplexed format. The prototype was capable of producing approximately 5000 measurements per experiment. This platform exhibited both a high sensitivity (100%) and high specificity (94%) for correctly identifying SS-B antigen-positive samples of Sjögren patients [33,34]. Apart from tumor marker or autoantibody identification, cytokine networks in inflammation or transplant rejection may smartly be detected by planar or bead-based protein arrays.

As cardiovascular disease remains the leading cause of death in the industrialized world, protein identification by proteomics may in particular be helpful for the early diagnosis and monitoring of disease. The collaborative effort of the Human Proteome Organization Plasma Proteome Project (HUPO PPP) has analyzed standardized human plasma samples using various proteomic platforms. Through this effort, 3020 proteins were identified, each requiring a minimum of two high-scoring MS/MS spectra. A critical step subsequent to protein identification is functional annotation. A subset of proteins from this project was annotated for relevance to cardiovascular disease. Most of the proteins in the vascular and coagulation system and markers of inflammation had been shown to localize to the plasma, whereas the majority of other groups such as signaling, growth and differentiation, cytoskeleton, transcription factors or channels and receptors hosted a larger number of novel plasma components. Knowledge from the cardiovascular system regarding the actions of these novel plasma constituents may provide insights into their biological roles [35].

12.1.3.3 Preparation of Clinical Samples for Fluidic Proteomics

Besides the previously mentioned challenges associated with the complexity of diseases, the heterogeneity of human clinical samples, together with the expected very low concentrations of potential biomarkers in plasma, places a question mark on whether disease biomarkers can be reliably detected using a proteomics approach in a clinical setting. Therefore, it is necessary to clearly distinguish clinical proteomic applications from more research-oriented proteome studies.

These particular circumstances make it obligatory first to establish SOPs prior to collecting samples. SOPs have been established for analyzing blood samples that contain peripheral blood-isolated cells plasma, liquor [36,37], and urine. The collection of urine appears less challenging, as proteolytic degradation for several hours after collection is of little consideration and no additives are required [29]. Fiedler *et al.* [38] detected 427 different mass signals in the urine of healthy donors with acceptable within- and between-day imprecision using a standardized protocol employing magnetic bead separation and MALDI-TOF MS.

The careful collection of clinical samples, together with disease-related clinical data (case history, diagnostic phenotype, treatment scheme, etc.), provides the basis for the scientific evaluation of these data.

Plasma/Serum In laboratory diagnostics, the most frequently analyzed sample is human blood, consisting of the plasma or serum, respectively, and cellular proteome. Plasma is superior to serum for most applications, as changes occur *in-vitro* in the proteome and peptidome composition in serum during the coagulation processes. Examples are the release of proteins and peptides from thrombocytes and the degradation of coagulation factors. In principle, proteomics offers great promise for the study of proteins in plasma and a number of proteomic databases have already been established. About 1500 different proteins (not including immunoglobulin variants) have recently been identified and several potential new markers of diseases characterized. One challenging problem of plasma proteins is associated with the dynamic range of protein concentrations, which may be as high as 12 orders of magnitude. Albumin, for example, is present in blood at millimolar (10^{-3}) concentration, whereas many cytokines are physiologically active in concentrations between 10^{-12} to 10^{-9} mol, such as tumor necrosis factor (TNF) [39]. Furthermore, 95% of the whole mass of plasma proteins is constituted by only 10 proteins [40]. As biomarker discovery means searching for low-abundance proteins, a high efficiency of the fractionation systems is essential in order to avoid interference from high-abundance proteins, and consequently the development of multidimensional fractionation methods is vital. Currently, two major depletion methods are in use: resin-based and antibody-based depletion (e.g., multiple affinity removal system, MARS). Several different techniques for plasma protein fractionation have been employed, including a 2-D liquid enrichment system (Gradiflow®), plasma fractionation by multichannel electrolyte (MCE) and microscale solution IEF (ZOOM), and by free-flow electrophoresis (FFE) [41,42].

Urine The "urinary peptidome" promises to be a resource at least as dynamic and informative as the "urinary proteome". Urine, as a matrix, must be one of the least desirable biological fluids for both peptidomic and proteomic studies. The composition of urine is, by design, highly variable. Indeed, the variability of its composition is a major way in which the "milieu interieur" is maintained more or less constant in the face of environmental and nutritional changes. Urine volume, salt composition, and pH vary widely in health. In disease, there is frequently proteinuria and the urine is often infected, even in apparently healthy individuals.

The filtered plasma "peptidome" is normally processed by the proximal renal tubule, which removes substantial but ill-defined amounts of peptides and proteins from the filtrate. When this process is ineffective, as in the renal Fanconi syndromes, very large quantities of plasma peptides are found in urine. In order to use urine as a source of reliable peptide biomarkers in disease, it must first be ascertained how certain variables will affect the results. For example, "endogenous" variables within the urine itself, such as salt composition and pH, may influence peptide recoveries, whilst "exogenous" variables relate to sample-processing procedures, such as freeze–thawing. Typically, a single freeze–thaw cycle can produce dramatic changes in the intensities of several urinary peptides.

Three outstanding problems are apparent: (1) as with proteins, the range of peptide concentrations in urine spans several orders of magnitude; (2) most of the peptides still cannot be quantified; and (3) beyond their mass measurement, the structure of the peptides is still unknown.

Another significant potential problem is the presence of large quantities of uromodulin ("Tamm–Horsfall protein") in urine; indeed, uromodulin is the major single protein identified in healthy urine. The problem is that, depending on salts and pH, uromodulin forms fibrils that form a sediment under low-g conditions; this is important because uromodulin is known to bind several LMW proteins and, by implication, plasma peptides that enter the tubular filtrate.

The definition of disease-specific biomarkers in urine is complicated by significant changes in the urinary proteome during the day, most likely due to exercise, variations in the diet, circadian rhythms, etc. [22]. As a consequence, the reproducibility of biological assays is more often reduced due to these physiological changes rather than to the nature of the analytical method itself. In addition, clear differences between first-void and midstream samples can be noted (H. Mischak et al., unpublished results), further highlighting the importance of standardized protocols for the collection. Unfortunately, the choice is not always available of how urine is collected, for example from babies, although non-invasive urine sampling is clearly much more welcome than biopsy.

It appears that CE-MS, due to its ability to detect a large number of peptides, is less sensitive to these variations than other methods. This intrinsic variability in protein profiles increases the problems in establishing a "normal human urine proteome map". By using acetone-precipitated urine samples from healthy volunteers, Thongboonkerd et al. [43] defined the first human proteome map, consisting of 67 proteins and their isoforms, that could be used as a reference. In a subsequent study by Oh et al. [44], pooled urine samples from 20 healthy volunteers were used to annotate 113 proteins on 2-D electrophoresis by peptide mass fingerprinting (PMF). Additional experiments that further expanded the knowledge of the normal urinary proteome have been reported, leading to the identification of approximately 800 proteins in the urinary proteome [45,46].

Cerebrospinal Fluid (CSF) CSF samples must be collected by means of lumbar puncture. To discard the cells, the CSF must be centrifuged (e.g., at $250 \times g$ for 10 min at $4\,^\circ$C) and the supernatant stored at least at $-80\,^\circ$C, with or without the addition of protease inhibitors. Because of the very low protein concentration in CSF, the sample preparation usually includes protein-enrichment techniques such as ultrafiltration. Columns with different molecular weight cut-offs can be used either alone or in combination (e.g., with Centricon YM-50 columns; Millipore, Bedford, CA, USA) with a nominal molecular weight cut-off of 50 kDa and/or Ultrafree columns (Millipore) with a molecular weight cut-off of 5 kDa.

Bronchoalveolar Lavage (BAL) Over the years, the European Respiratory Society (ERS) task force has drafted documents on methods of performing BAL [47]. However, BAL sample preparation and processing needs to be standardized to make the results reproducible and comparable. Operators in this sector should follow international guidelines. BAL, which usually is performed during bronchoscopy, has a variety of indications, such as diagnosis and follow-up [48]. Four to five aliquots (50–60 mL) of phosphate-buffered saline (PBS) are instilled by fiberoptic bronchoscope and the fluid is recovered by gentle aspiration. Recovery varies with the lung site or the disease. The first sample is generally kept separate from the others as it contains more debris and bronchial contamination, and is not used for proteome analysis. The other aliquots are filtered and centrifuged (e.g., at $250 \times g$ for 10 min at $4\,^\circ$C) to separate cells from the fluid component; the supernatant obtained may be frozen at $-80\,^\circ$C until analysis.

Cell differential counts are performed with cytocentrifuge preparations. The phenotype of lymphocytes and macrophages (or other cells) can be analyzed by flow cytometry using monoclonal antibodies [49].

The study of BAL by sophisticated techniques such as proteomics calls for better standardization of the method in order to reduce variability. For example, for 2-D electrophoretic analysis, the BAL samples must be treated to remove salts and concentrated to obtain a suitable protein content. Factors linked to sample variability must be considered: the quantity of instilled fluid, the minimal accepted percentage recovery, the type of aspiration, and the choice of site in the lung. Reference parameters are chosen to express the results (e.g., per mL BAL recovered or per mg total proteins, or albumin or with reference to urea content), though not without problems. The International Scientific Societies of Respiratory Diseases (ATS/ERS) are convening an international group of experts to set up-to-date guidelines and recommendations [50].

12.1.4
Cellular Proteomics (Cytomics)

Blood cells offer unique insights into disease processes and therefore, erythrocytes, granulocytes, monocytes, lymphocytes, and platelets are of special interest for clinical proteomics [51]. Flow cytometry is currently widely used as an analytical tool for clinical cell analysis directly from anticoagulated whole blood and also for cell sorting to generate pure populations of cells from heterogeneous and highly integrated mixtures as are found in a majority of biological environments. Elispot, slide-based cytometry, and tissue arrays, together with high-content screening microscopy, are further upcoming techniques in cytomics. The major challenge for pre-analytical standardization is related to the use of fresh samples, either for direct multiparameter analysis of cellular proteomics in whole blood or body fluids without pre-separation, or for cell sorting and enrichment strategies for subsequent proteomic and functional genomic analysis.

12.1.4.1 Sample Preparation and Standardization for Clinical Cytomics
For clinical analysis, the samples should be rapidly analyzed because long-term transportation and storage could lead to artifacts such as selective damage or aggregation of specific cell subpopulations or the shedding of cell-surface markers. Especially for activation markers or the functionell test, the samples should be analyzed within 8 h, which means that they cannot be dispatched overnight. Most parameters in clinical flow cytometry are measured using EDTA-blood, especially the surface parameters for immunophenotyping. The EDTA-tube can be used in parallel for routine hematology and is available in all clinical settings. Many applications may also be made with heparinized or citrated blood, although for parameters which are dependent upon calcium (e.g., oxidative burst or phagocytosis), heparinized blood must be used.

Currently, for cell preparation, the use of gradient centrifugation has almost been completely substituted by erythrocyte lysis for routine applications before cytometric analysis [52]. Highly automated and standardized preparation stations are available for lysis, and also in part for complete sample handling (www.beckmancoulter.com, www.bdbioscience.com). For the analysis, highly standardized multiparameter flow cytometers with special software application for routine diagnosis (www.beckmancoulter.com, www.bdbioscience.com, www.dakogmbh.com, www.partec.com) are available commercially. Up to eight parameters [forward light scatter (FSC), sideward light scatter (SSC), and six different antibodies] are already implemented in routine diagnosis, and CE-certified kits are available. More complexity with more than 15 parameters is realized for scientific approaches. Clearly, more information is available with high-content analysis of single cells, but more problems occur due to antibody interactions such as unspecific binding or cross-linking of antibodies. With increasing numbers of different fluorescent dyes, the compensation of fluorescence signals in different detector channels is an important aspect. Most up-to-date analyzers use digital compensation methods which are highly standardized due to software-driven compensation algorithms [53].

The strategy to validate antibody-producing cell lines in so-called "Cluster of Differentiation" workshops (CD-nomenclature) has generated more than 300 defined antibody specificities which are of great help when standardizing clinical test development. Standardization is an important issue in clinical cell analysis, and a main topic of the working group "Flow cytometry and quantitative microscopy" of the DGKL is quality control and standardization. This working group is involved in the European Working Group on Clinical Cell Analysis, which already has begun to define regularities for flow cytometric multiparameter analysis. However, and especially with regards to phenotyping and the isolation of cells for genomic or proteomic analysis, the standardization of techniques remains incomplete [54].

12.1.4.2 Tissue Arrays

Tissue arrays are assemblies of multiple patient tissue samples (e.g., tumors) prepared for multiplexed immunohistological serial analysis of tissue specimens. This technology offers the following advantages: amplification of a scarce resource, experimental uniformity (tissues of multiple patients are treated in an identical manner), decreased assay volume; and preservation of the original blocks. Depending on the shape of the tissue sample and the method used to obtain it, multi-tissue array techniques may be classified into two different groups of "rod-shaped" and "core tissue". Certain technical aspects must be considered when deciding which technique to use, including the number, size, and origin of tissue samples, the quality of the paraffin wax, the distance between samples and the depth in the receptor block, antigenicity preservation, and block sectioning. The main applications of this techniques are for screening purposes, quality control, diagnosis, and teaching [55,56]. In clinical proteomics, tissue arrays are already in use in different complexities. In autoimmune diagnosis, for example, commercially distributed tissue arrays with two up to more than 20 different tissue samples for the incubation with the serum of one patient (http://www.euroimmun.de) are available. In cancer studies, multiplexed tissue arrays are in use for the screening for potential therapeutic antibodies, for the prediction of outcome, prognosis, and for biomarker screening and validation [55]. European initiatives are established to design and standardize an infrastructure for a networked tumor tissue bank. The samples are collected according to an SOP used at different institutes, and the data are collected and distributed in a central database [TuBaFrost 4 [57]; http://genome.tugraz.at [58]].

12.1.4.3 Bead-Based Immunoassays for Protein Analysis

Microsphere-based flow cytometric assays have become an attractive alternative to the popular microtiter plate-based enzyme-linked immunosorbent assay (ELISA). Although this methodology is also well suited for single-analyte analysis, it is more desirable for the simultaneous quantitation of multiple analytes from a relatively small sample size in rapid fashion. Sample size becomes a critical factor for the evaluation of multiple analytes, especially when monitoring disease status. The melting of ELISA-based technology with flow cytometry has alleviated most of these issues. The use of fluorescent-labeled microspheres as solid support, coated with antigens or antibodies, has been the key to this success. Beads of different sizes or colors allow multiplexed immunoassays, and with this strategy the available analyzers (www.luminexcorp.com, www.bdfacs.com, www.illumina.com) are now able to distinguish up to 200 different beads in one assay. Furthermore, beads of different sizes can be used as carriers for antigens or antibodies and can then be differentiated by their light-scattering characteristics. By combining both forms of beads, it would be possible to scale up to much more parameters in one test [59]. These bead-based immunoassays are already widely used in the routine laboratory for serum samples of patients to detect cytokines or autoantibodies [60]. Due to their high specificity and high sensitivity, the bead-based microarrays already serve as a primary assay platform in clinical proteomics. Moreover, they are easy to handle, and the analyzers are affordable for most laboratories.

12.1.4.4 Preparative Methods

Fluorescence-Activated Cell Sorting In addition to being an analytical tool, flow cytometry can also be used as an approach to cell isolation using fluorescence-activated cell sorting (FACS). Cell sorting allows the investigator to quantify several fluorescence and light-scattering parameters of individual particles and to purify those events with the desired characteristics for further study. No other technology can separate a heterogeneous cell suspension into purified fractions containing a single cell type with equivalent speed and accuracy.

Different methods of capturing a particle of interest are implemented in different cytometers. The FACSCalibur™ system, for example, uses a mechanical device called a "catcher tube" to sort the cells. This catcher tube, which is located in the upper portion of the flow cell, moves in and out of the sample stream to collect a population of desired cells at a rate of up to 300 cells per second. The particles and cells can be physically separated and deposited onto a defined location for further analysis, or even culture, under sterile conditions. The advantage of this sorting method is a soft handling of cells, although the sorting speed is rather low.

High-speed sorters use techniques such as "stream in the air"; this allows a much higher throughput of up to 100 000 cells per second, but the cells are more stressed and may be less suitable for further analysis.

Applications for high-speed cell sorting are commonly grouped into two categories: bulk sorting and rare-event sorting. In *bulk sorting*, the sort operator is given a large number of cells of which a certain percentage is expressing a given phenotype and is desired for further study. This marker can be a surface antigen bound to a fluorescently labeled antibody, a transient transfection expressing a given protein of interest, activated leukocytes, or any detectable measure of cell state or type. Whatever the measured parameter, the operator must screen through a large number of cells to collect a sufficient amount of material for the next step of the analysis. For example, an investigator may be interested in the top 5% of cells expressing a given intracellular cytokine to perform a proteomic analysis. Typically, a fairly large number of cells is needed for these studies. With readily available MS methods, in the order of 10^7 cells would be needed to detect a protein expressed at roughly 1000 copies per cell, corresponding to a total starting population of approximately 2×10^8 cells. At a sort rate of 40 000 cells per second, this translates to 5000 s or around 1.5 h of sorting. This does not take into account any material losses due to sample handling (e.g., centrifugation) or time losses due to technical issues. Highly standardized flow cytometers with special sorting application for high-throughput sorting are commercially available (www.beckmancoulter.com, www.bdbioscience.com, www.dakogmbh.com, www.partec.com) [61,62].

Magnetic Bead-Based Separation of Cells Isolation techniques must be considered before analysis of the cell proteome. Magnetic bead-based techniques for the isolation of cell subsets according to the expression profile of membrane antigens provides a rapid and convenient approach to cell isolation from blood, bone marrow, and other sources of cells in suspension to select or deplete specific cell populations. These procedures can be performed without major technical instrumentation for low-throughput applications, for example with Dynal magnetic bead systems (http://www.invitrogen.com/dynal). The system has certain disadvantages, however. The first problem is the use of "large" (>5 µm) beads; these grossly activate some cells (e.g., monocytes) and interfere with subsequent analysis of the cell by flow cytometry. A second major problem is that there is no automation for high-throughput sample handling. In clinical proteomics, cell-based analysis requires high-throughput and rapid techniques for cell harvesting and collection, and these are realized by two automated systems for cell isolation from whole-blood without density gradient centrifugation. One system (RoboSep®; https://www.stemcell.com/product_catalog/robosep.asp) uses a lysing protocol prior to magnetic leukocyte preparation. This system has the major disadvantage that some activation pathways are influenced by the lysed erythrocyte membranes. Furthermore, yield and purity is inferior to the manual Dynal or automated Miltenyi System. The Miltenyi System does not require lysing or other manipulation prior to magnetic bead separation (auto-MACS™ Pro Separator®; http://www.miltenyibiotec.com). However, it is mandatory that the separation procedure does not induce major changes in either the expression or the cellular localization of proteins. But, it is also necessary that the separation techniques do not result in unwanted cell loss, as may occur under pathological conditions when cells float outside the density gradient. In this regard, two general approaches can be distinguished:

- *Positive selection* uses one or more cell-specific markers for identification and separation; thus, the purity is usually very high but cell function may be altered.
- *Negative selection*, in contrast, depletes the contaminating cells and leaves the target cell untouched, but may lack accuracy and high purity, as well as a poor recovery.

Based mostly on specific membrane-associated antigens, various cell subsets can be isolated. Unfortunately, one major drawback of most available antibodies in the CD-nomenclature series is their effect on cell activation. Therefore, there is a need for other, non-activating cell-specific targets with stable surface expression for affinity isolation. Sufficient cell counts for genomic, proteomic or lipidomic analysis can be obtained from 10 to 20 mL of peripheral blood, depending on the cell count of the individual subject. Ready-to-use kits for the positive or negative isolation of various cell populations are commercially available. The results must be acceptable in terms of recovery, yield, and purity ($\geq 90\%$) of the final cell population, but this is not yet achieved by all systems [63].

Microdissection of Cells Based on microscopic techniques, the cells of interest can be isolated. For example, in a heterogeneous tumor tissue, the tumor cells can specifically be isolated by microdissection and the dissected cells analyzed using high-content proteomic methods such as MS/MS in parallel with genomic and/or expression analysis. For proteomic analysis, fine needle aspiration is preferable to laser microdissection because fewer artifacts are introduced. Microdissection can be used to design proteome-based studies in combination with serology for the identification of biomarkers and novel targets. Different systems for microscopic microdissection are available (www.leica-microsystems.com, www.arryx.com, www1.qiagen.com/applications/microdissection, www.palm-microlaser.com). Although the standardization and automation of these techniques is still in progress, most preparative steps are still operator-dependent [64].

Preparation of Cellular Samples

- Erythrocytes and Reticulocytes
  Due to their low frequency in peripheral blood, reticulocytes must be isolated for proteomic analysis. In contrast, for the proteomic analysis of mature red blood cells (RBCs), reticulocytes should be allowed to mature, for example by storing peripheral whole blood for 72 to 96 h at 4 °C, without shaking. Although it has been suggested that the maturation of reticulocytes *in vitro* is limited at 4 °C [65], recent data on the analysis of the final material clearly showed that the procedures taken to eliminate reticulocytes were effective [66]. White cells must be eliminated, for example with

filters (Plasmodipur, Euro-diagnostica, Arnhem, The Netherlands). The RBCs may be pelleted by centrifugation (e.g., at $250 \times g$ for 10 min at 4 °C), the supernatant discarded, and the RBCs resuspended. The purity of erythrocytes can be further enhanced by eliminating the top RBC layers after centrifugation. Further density centrifugations, for example on CL5020 (Cedarlane Laboratories, Hornby, ON, Canada), the elimination of granulocytes by passing the RBC fraction through nylon nets, and washing steps to eliminate plasma proteins should be carried out. The RBCs must then be used immediately to prepare membrane and cytoplasmic fractions. The quality of the preparations must be measured by purity assessment of RBC samples. The packed RBC samples should be counted for white blood cells (WBCs), granulocytes, monocytes, platelets, reticulocytes, and RBCs, using a routine hematology analyzer. No reticulocytes or other cell types should be found. Subproteomes of RBCs can be analyzed by membrane preparations and membrane extractions, soluble protein preparations, or cytoskeleton extractions [67]. The most complete study describing the proteome of RBCs has been reported by Kakhniashvili et al. [68]. Erythrocyte membranes as well as cytoplasm were analyzed by ion trap (IT) MS/MS in line with LC. These authors identified a total of 181 unique protein sequences in the membrane fractions and cytoplasmic fractions.

- Mononuclear Cells: Lymphocytes, Monocytes, and Dendritic Cells
  Monocytes are a target of proteomics, especially with regards to clinical diagnosis and the monitoring of atherosclerosis. Seong et al. [69] analyzed the effect of oxidative stress generated at sites of inflammation and injury on the proteomic profiles of monocytes. As a result, 28 identified proteins mainly involved in energy metabolism, translation, and mediation of protein folding were overexpressed. Dupont et al. [70] elaborated 2-D electrophoretic reference maps of human macrophage proteome and secretome in order to elucidate the macrophage dysfunctions involved in inflammatory, immunological, and infectious diseases. These authors showed that macrophages are involved in a wide array of biological functions, including cytoskeletal machinery, carbohydrate metabolism, apoptosis, and protein metabolism. Combined oligonucleotide microarray and proteomic approaches have been used to study genes associated with dendritic cell differentiation. Dendritic cells are antigen-presenting cells essential for the initiation of primary immune responses and are mostly derived from human monocytes. Following a protein analysis of these cells, about 4% of the protein spots separated by 2-D electrophoresis exhibited quantitative changes during differentiation and maturation. The differentially expressed proteins were identified by MS and were seen to represent proteins with calcium-binding, fatty acid-binding, or chaperone activities, as well as cell motility functions [71].

  Several studies have reported on the proteome of lymphocytes. A proteome database of human helper T (Th) cells was established using classical proteomics [72]. The proteins differentially expressed in Th1 and Th2 cells are described by Rautajoki et al. [73]. Detailed proteomic studies have been also reported on lymphoblastoid cells, and strategies for studying signaling pathways in lymphocytes, combining proteomics and genomics, have also been proposed [74]. Signaling via immunoreceptors is orchestrated at specific plasma membrane microdomains, referred to as "lipid rafts"; these are dynamic assemblies which float freely in the surrounding membranes of living cells. The proteins participating in lipid rafts in T lymphocytes have been studied with proteomics, and the subject was recently reviewed by Wollscheid et al. [75]. Mass spectrometry was used by Li et al. [76] specifically to detect proteins depleted from rafts by cholesterol-disrupting drugs. These authors detected a large proportion of signaling molecules in lipid rafts, and provided evidence for a connection between cytoskeletal proteins and lipid rafts.

- Platelets

  Platelets are easily and quickly activated after blood withdrawal; hence, the collected blood should be immediately processed. Especially for platelets, recent studies have shifted away from global profiling to the analysis of subfractions of the proteome and the identification of changes induced upon blood cell activation. For example, studies which focused on platelets have allowed the identification of many more platelet proteins than can be achieved by global profiling, thus providing a more complete view of the platelet proteome. Moreover, the biological information obtained may be of greater relevance. The two important subproteomes of the platelet which have been extensively studied is the phosphoproteome and the secretome [77–79]. Many signaling pathways in platelets are regulated differentially by phosphorylation upon platelet activation. The application of new proteomics approaches to the protein phosphorylation events which occur during the activation of platelets has led to the discovery of new phosphorylated targets proteins, and a new possible phosphorylation site in previously identified phosphorylated proteins [80,81].

  As with many other cells, platelets secrete proteins from preformed storage granules in response to stimuli. The platelets secretome has been analyzed in the supernatants isolated by the differential centrifugation of low-dose thrombin-activated platelets. The supernatant contains secreted proteins but no membrane proteins. By using a complementary multidimensional chromatography approach, the secreted protein fraction may be digested with trypsin and the resulting peptides separated using strong cation-exchange and reverse-phase chromatography before being introduced into an IT-MS. Analysis of the secretome from thrombin-activated platelets has identified over 300 proteins that are secreted upon activation. It was unclear whether several of these proteins were present in, or secreted by, platelets. However, three of the proteins, including secretogranin III, cyclophilin and calumenin, have been identified in atherosclerotic lesions, suggesting a potential role in atherothrombosis [82].

- Circulating Microparticles

  Cell-derived microparticles have been shown to be relevant in a number of diseases, including thrombosis, cardiovascular disease, anti-phospholipid syndrome, and systemic inflammation [systemic inflammatory response syndrome (SIRS), sepsis]. Microparticles have been shown to be a valuable tool in the field of RBC

dysmorphology and dysfunction. Furthermore, it is likely that microparticle release is part of the ageing process of blood cells such as erythrocytes or platelets. This may coincide with other cell alterations such as shape changes or loss of surface protein modifications. In this sense, microparticle release is thought to be an innovative marker for chronic vascular diseases such as atherosclerosis or other ageing disorders.

Microparticles are defined as shed membranous fragments or vesicles with a diameter of less than 1 µm, and bearing markers of the parent cell on their surface. They are surrounded by a phospholipid bilayer that is mainly composed of phosphatidylcholine, sphingomyelin, and phosphatidylethanolamine. Microparticles differ from their parental cells in terms of lipid composition and distribution between the two membrane leaflets. The asymmetrical phospholipid distribution with anionic phospholipids is confined to the inner leaflet, and usually changes during microparticle formation. Activation as well as apoptosis leads to calcium-dependent swelling, budding and microparticle release. Finally, both cell activation and apoptosis result in disruption of the membrane skeleton structure that is necessary for particle release. The principal sites of microparticle release seem to be cell protrusions that, in terms of cell type, resemble microvilli, pseudopodia, filopodia, or proteopodia.

Microparticles are derived from a number of different cell types. In addition to a well-established role for platelet-derived microparticles in hemostasis, an important function in a variety of additional blood cells (e.g., RBCs, monocytes, and lymphocytes) and in other cell types (e.g., endothelial and stromal progenitor cells) has been proposed. RBCs have long been known to shed membrane microparticles under stimulation. It could be shown that, as for other microparticles, this process depends on a rise in intracellular calcium. Additionally, it was found that the microparticle release coincides with the formation of echinocytes, which is characterized by loss of the natural discoid shape to gain a more spherical form with regularly distributed cell membrane protrusions. This shape change, as well as the microparticle release, may resemble the *in-vivo* events that occur during cell ageing. Circulating microparticles of leukocyte origin modulate cellular interactions through the up-regulation of cytokines and cytoadhesins in endothelial cells and monocytes.

Antigenic markers measured by flow cytometry are mainly used to classify and further subdivide membrane microparticles. Alternative approaches include ELISA and solid-phase capture assays. The advantage of flow cytometry lies in its capacity to distinguish different microparticle populations and subpopulations according to the expression of surface and intracellular antigens in a multi-parameter approach. The sensitivity and specificity of flow cytometry may therefore to be used to improve the analysis of blood and other cell types. In this sense, it should be combined with the well- established technique of conventional analyzers, as outlined above.

- Organelles

The introduction of proteomic analysis such as the MS-based identification of proteins has created new opportunities for studies of organelle composition, processing of transport intermediates, and large subcellular structures. Traditional cell biology techniques such as sucrose density gradient centrifugation are used to enrich these structures for proteomic analysis, and such analysis does indeed provide insights into their biology and function. In addition to a good preliminary purification of the sample, the validation of sample purity (as verified by electron microscopy, immunoblotting or other orthogonal methods to exclude significant cross-contamination during organelle separation) is obligatory. The need for a standardization of purification procedures is, therefore, urgent for any successful clinical application [83–86].

One major challenge in the control of purity of the organelle fraction is the lack of specific markers for each organelle [87]. Nonetheless, the study of cell compartments provides a unique way in which to access and identify low-abundance and organelle-specific proteins in a biological sample. Immunoisolation on magnetic beads allows the isolation of highly enriched fractions of certain organelles, whilst a single analysis using MS enables the identification of thousands of peptides, leading to the formal identification of several hundred proteins [88].

The surface plasma membrane of cells shows morphologically distinct regions or domains, such as microvilli, cell–cell junctions, clathrin-coated pits, and lipid raft domains. Each of these domains is specialized for a particular function, such as nutrient absorption, cell–cell communication, and endocytosis. Lipid raft domains include caveolae, characterized by a distinctive membrane coat composed of caveolin-1, and rafts. Both have a high content of cholesterol and sphingolipids, have a light buoyant density, and function in endocytosis and cell signaling.

Classical proteomics does not allow the detection of human endothelial caveolae and raft proteins, as they represent less than 0.5% of the total cellular protein pool and less than 2% of the total plasma membrane proteins. Therefore, fractionation techniques such as cationic silica enrichment of caveolae or detergent solubilization with Triton X-100 at 4 °C are needed to enable the subproteome analysis of these membrane microdomains [89]. The analysis of these distinct regions can be put forward by high-throughput proteomic platforms, together with modern techniques such as stable isotope labeling (SILAC™) [90,91]. The proteomic analysis of purified caveolae has identified a wide assortment of proteins that are localized to these structures. The first detailed proteomic analysis of caveolae was carried out in 1994. Based on their buoyancy and resistance to detergent solubilization, Lisanti et al. [92] used sucrose density ultracentrifugation to purify caveolae-rich membrane domains from murine lung tissue. This procedure allowed for the exclusion of ≥98% of an integral plasma membrane protein marker, while retaining ~85% of total caveolin and ~55% of GPI-linked marker proteins. This initial proteomic analysis of caveolae allowed for the first large-scale characterization of caveolae-enriched protein constituents, and provided the basis for many follow-up investigations into the functional significance that caveolar localization may confer upon these proteins [93]. For clinical proteomics, it is of great interest that a long list of diseases might be associated with lipid rafts and raft-associated proteins [94].

- Phagosomes
  Phagosomes are key organelles for the innate ability of macrophages to participate in tissue remodeling, clear apoptotic cells, and restrict the spread of intracellular pathogens. By using a proteomic approach, more than 140 proteins associated with latex bead-containing phagosomes have been identified. The elaboration of a 2-D gel database of more than 160 identified spots allowed the analysis of how phagosome composition is modulated during phagolysosome biogenesis. The systematic characterization of phagosome proteins provided new insights into phagosome functions, and of the protein or groups of proteins involved in and regulating these functions. Different types of phagocytosis can be distinguished. For example, type I phagocytosis is mediated primarily by Fc-receptors, while type II phagocytosis is mediated by complement receptors, especially CR3. Another type of phagocytosis is associated with deep tubular invaginations. To date, more than 500 proteins have been identified in different phagosome preparations using MALDI-TOF MS and nano-electrospray MS/MS [86,95].

  – Protein Secretion
    The secretion of proteins by exocytosis to the outside of cells is an important subset of protein-trafficking events. Secretion can be either constitutive (occurring continuously) or regulated (occurring on demand) as a result of an extracellular signal. Both genetic and biochemical approaches have combined to produce the current understanding of eukaryotic protein secretion, although there are clearly many questions that remain unanswered [96]. With regards to clinical proteomics, it is of interest, that constituents of the exocytosis pathways are associated with multiple diseases. For example, members of the AP-3 pathway are involved in disorders of lysosome-related organelles such as the Hermansky–Pudlack syndrome complex, Chediak–Higashi syndrome, and the ceroid lipofuscinoses. This provides new opportunities to understand AP-3 pathway-related disorders and their relation to membrane phospholipid processing. Mutations in the *ABCA1* gene, which is one member of the AP-3 pathway, are involved in dysregulated vesicular trafficking from the trans-Golgi compartment to the plasma membrane, and ABCA1 R1925Q was shown to contribute to Scott syndrome, a phospholipid-processing disorder of missing surface exposure of phosphatidylserine [97].

  – Nucleolus/Nucleosome
    The nucleolus/nucleosome is a key organelle that coordinates the synthesis and assembly of ribosomal subunits, and forms in the nucleus around the repeated ribosomal gene clusters. A quantitative analysis of the proteome of human nucleoli was successfully performed using MS-based organellar proteomics and stable isotope labeling. *In-vivo* fluorescent imaging techniques are directly compared to endogenous protein changes. The flux of 489 endogenous nucleolar proteins in response to three different metabolic inhibitors that each affect nucleolar morphology could be registered [98,99]. Both, the nuclear- and mitochondrial-encoded proteins and their genes are summarized in MITOP (http://www.mips.biochem.mpg.de/proj/medgen/mitop/), a comprehensive database for genetic and functional information. The "Human disease catalogue" contains

tables with a total of 110 human diseases related to mitochondrial protein abnormalities, sorted by clinical criteria and age of onset. MITOP should contribute to the systematic genetic characterization of the mitochondrial proteome in relation to human disease [100,98].

12.1.4.5 Clinical Applications in Cytomics

Multiparameter Flow Cytometry   Multiparametric flow cytometry is widely used in the routine laboratory to generate clinical diagnostic information from complex heterogeneous mixtures such as human blood for multiple indications [63]. The main indications for clinical flow cytometry can be classified according to the material used and divided into two major principles, namely immunophenotyping and functional analysis.

Immunophenotyping has become a routine practice in the diagnosis and classification of leukemias. For non-Hodgkin lymphoma, flow cytometry is the method of choice in many laboratories [101]. Further important examples include the analysis of lymphocyte subpopulations, for example in the diagnosis of primary or acquired immunodeficiencies, in the quantitation of hematopoietic stem cells, and of residual leukocytes in erythrocyte preparations [102] and in the analysis of platelets [103]. Diagnostic hematopathology depends on the applications of flow cytometric immunophenotyping (FCI) combined with immunohistochemical immunophenotyping. Select cases may require additional cytogenetic and molecular studies for diagnosis. FCI offers the sensitive detection of antigens for which antibodies may not be available for paraffin immunohistochemical immunophenotyping. However, the latter method offers the preservation of architecture and evaluation of expression of some proteins, which may not be available by FCI. These techniques should be used as complementary tools in diagnostic hematopathology.

The types of specimen suitable for FCI include peripheral blood, bone marrow aspirates, core biopsies, fine-needle aspirates, fresh tissue biopsies, and all types of body fluid. There are many advantages of FCI compared to immunohistochemistry. For example, dead cells may be gated out of the analysis, while weakly expressed surface antigens may be detected. Multicolor (two-, three-, four-) analysis may be performed, allowing for an accurate definition of the surface antigen profile of specific cells. Two simultaneous hematologic malignancies may be detected within the same tissue site. Tissue biopsy may be obviated by the relatively non-invasive diagnostic evaluation of body fluids. A disadvantage of FCI is that both, a sclerotic bone marrow or a markedly hypercellular or "packed" bone marrow, may each yield too few cells for analysis. Also, sclerotic tissue may be difficult to suspend for an individual cellular analysis. There is also a loss of architectural relationships. For example, T-cell lymphomas that do not have an aberrant immunophenotype may not be detected, whereas an aberrant T-cell immunophenotype (i.e., an absence or downregulation of pan-T-cell antigens, particularly CD7) does not necessarily indicate malignancy and may be observed in infectious mononucleosis or inflammatory disorders. Due to these disadvantages, FCI data should always be correlated with light microscopy if no FCI abnormalities are detected.

Immunohistochemistry (IHC) may need to be performed in selected cases. In addition, as mentioned, an aberrant T-cell immunophenotype does not necessarily indicate malignancy and requires correlation with light microscopy, as well as clinical data and additional ancillary studies (i.e., molecular/cytogenetic analysis) in some situations [104]. A "Kompetenznetz Leukämien und Lymphom" is established in Germany to support the standardization of clinical diagnosis (http://www. kompetenznetz-leukaemie.de). The diagnosis of many primary immunodeficiency diseases requires the use of several laboratory tests. Flow cytometry has become an important part of the work-up of individuals suspected of having such a disorder. Flow cytometry is applicable in the initial work-up and in the subsequent management of several primary immunodeficiency diseases [105]. Further indications for immunophenotyping are listed in Table 12.1 [52].

Sample Preparation for Specific Clinical Diagnostic Procedures For certain samples and functional assays, specific sample preparation procedures must be considered, as summarized below:

Parameter	Indication	Sample preparation
Activated, reticulated platelets	Thrombocytopenia, thrombotic events	Freshly drawn citrated blood; CTAD anti-coagulated blood
Bone marrow	Leukemia-, lymphoma-diagnosis	EDTA- or heparin-anticoagulated sample; erythrocyte-lysing methods are preferable to density centrifugation
Bronchoalveolar lavage	Inflammatory, autoimmune lung disease; sarcoidosis, collagenoses, fibrosis	Heparinized sample preferable, stored at 4 °C
Cerebrospinal fluid	Inflammatory, autoimmune diseases of the CNS	Preferably stored at 4 °C
Oxidative burst	Chronic granulomatosis; sepsis, SIRS	Heparinized blood mandatory
Phagocytosis	Hereditary or transient defects in phagocytosis	Heparinized blood mandatory
Basophile degranulation	Allergic or pseudoallergic reactions	EDTA- or heparinized blood
Microparticle analysis	Thrombosis, cardiovascular disease, systemic inflammation (SIRS, sepsis)	Enrichment from anticoagulated blood by two differential centrifugation steps; flow cytometric measurement

CNS, central nervous system; CTAD, citrate-theophylline-adenosine-dipyridamole; SIRS, systemic inflammatory response syndrome.

Table 12.1 Immunophenotyping in peripheral blood.

Parameter	Indication
Lymphocyte subpopulations (T-, B-, NK-cells)	Primary immunodeficiency syndromes, HIV, organ transplantation
Malignant lymphocytes	Non-Hodgkin lymphoma
CD34+ stem cells	Stem cell transplantation
Lymphocyte activation (CD38; HLA-DR)	HIV
T-cell repertoire	Omenn syndrome
CD16, CD66b on granulocytes	Paroxysmal nocturnal hemoglobinuria
CD11a/b on granulocytes	Leukocyte adhesion defects
HLA-DR on monocytes	Sepsis
LDL-receptor	Familial hypercholesterolemia

For the analysis of surface molecules, EDTA-blood is normally used and stored at room temperature (see also "Sample preparation").

Protocol: Evaluation of Monocytes and Platelets by Flow Cytometry Circulating monocytes can be assessed by multiparametric flow cytometry from anticoagulated whole blood, as they are clearly distinguished from other leukocytes upon their high surface expression of CD14 (lipopolysaccharide receptor, LPS-R) and CD33 (siglec-3). In addition, monocytes possess a highly versatile phenotype which reflects their activation/priming; thus, activated monocytes can also be analyzed through expression of a set of activation-dependent antigens, such as CD16 (Fcγ-receptor III), CD142 (tissue factor), CD143 (angiotensin-converting enzyme, ACE), and CD105 (endoglin). Circulating monocytes represent a heterogeneous cell population according to their immunophenotype. In healthy individuals, the majority of circulating monocytes shows "bright" expression of CD14 and CD64 (Fcγ-receptor I) without expression of CD16. Other monocyte subsets are characterized by: (i) the simultaneous expression of both CD14 and CD16; (ii) the isolated expression of CD16 with dim CD14 expression; and (iii) the high expression of CD33 with only dim expression of CD14. The smallest subset of monocytes (2.1 ‹ 0.8%) differs from the predominant population of $CD14^{bright}CD64^+CD16^-$ monocytes by additional expression of CD56 (neural cell adhesion molecule, N-CAM). $CD14^+CD16^+$ monocytes exhibit several features of tissue macrophages, and this population is largely expanded (up to >20%) in acute and chronic infections, systemic inflammatory syndromes, AIDS, or renal failure. In hypercholesterolemic patients, high-density lipoprotein (HDL)-cholesterol levels are negatively correlated to the population size of $CD64^-CD16^+$ monocytes. In both healthy subjects and hypercholesterolemic patients, the apolipoprotein E3/E4 and E4/E4 phenotypes are associated with a tendency toward a larger population of $CD64^-CD16^+$ monocytes. These data suggest that systemic abnormalities in monocyte subpopulations may reflect the role of these cells in the pathogenesis of atherosclerosis.

Like monocytes, circulating platelets can also be analyzed by multiparametric flow cytometry from anticoagulated whole blood, as they are clearly distinguished from other cells by size and their high surface expression of CD61 or CD41. Platelet function is sensitive to alterations in cholesterol metabolism, and hypercholesterolemia is associated with enhanced platelet membrane rigidity. Atherogenic low- density lipoproteins (LDL), in particular oxidized LDL, activate platelets through src-kinase-family-dependent signaling. In contrast, anti-atherogenic HDLs inhibit platelet aggregation and target the phosphatidylinositol phospholipase C (PI-PLC) pathway. Platelet activation can be induced *in vitro* by TRAP-6 and ADP, for example. Activation markers on the surface detected by CD62P, CD63 or PAC-1 can be measured by flow cytometry.

Protocol: EDTA-Blood-Based Cellular Human Monocyte and Platelet Analysis of Parameters Relevant for Atherosclerosis Suggested marker panel for monocyte immunophenotyping (four-color flow cytometry) from whole blood:

- Sample
  - Peripheral blood: 2.5 mL of EDTA-anticoagulated whole blood.
  - Staining and analysis should be performed *within 2 h* of taking the blood sample.

- Materials
  - Antibodies (see Table 12.2).
  - Erythrocyte lysis solution (from Becton Dickinson; 10 × concentrated stock solution, to be diluted with sterile distilled water).
  - Dulbecco's modified phosphate buffer saline (DPBS) without Ca^{2+}/Mg^{2+} (e.g., from Biochrom).
  - Lipoprotein-free bovine serum albumin (BSA) (e.g., BSA fraction-V from Sigma).

- Staining procedure
  - Add 100 µL of EDTA-anticoagulated whole blood to titrated amounts of fluorochrome-labeled antibodies according to Table 12.2.

Table 12.2 Pipetting scheme.

Tube	µL	FITC	µL	R-PE	µL	PerCP	µL	APC
1	20	CD16	5	CD33	20	CD45	5	CD14
2	20	CD11a	20	CD36	20	CD45	5	CD14
3	20	CD18	20	CD163	20	CD45	5	CD14
4	20	CD16	20	CD32	20	CD45	5	CD14
5	20	CD64	20	CD16	20	CD45	5	CD14
6	20	CD62L	10	CD13	20	CD45	5	CD14
7	20	CD26	15	CD11b	20	CD45	5	CD14
8	20	CD284 (TLR4)	20	CD192 (CCR2)	20	CD45	5	CD14
9	20	CD143	20	CD16	20	CD45	5	CD14
10	20	CD142	20	CD169	20	CD45	5	CD14
11	20	CX_3CR1	20	CD56	20	CD45	5	CD14

All fluorochrome-conjugated monoclonal antibodies listed in Table 2 are available commercially.
APC, allophycocyanin; FITC, fluoroscein isothiocyanate;
R-PE, phycoerythrin; PerCP, peridinin–chlorophyll–protein complex.

- Incubate for 15 min at 4 °C in the dark.
- Add 3 mL DPBS containing 0.5% BSA.
- Centrifuge sample at 400 × g for 5 min at 4 °C.
- Discard the supernatant.
- Add 3 mL of erythrocyte lysis solution and vortex gently.
- Incubate for 10 min at room temperature in the dark.
- Centrifuge sample at 400 × g for 5 min at 4 °C.
- Completely remove the supernatant by pipetting.
- Add 3 mL DPBS containing 0.5% BSA.
- Centrifuge the cells (400 × g for 5 min at 4 °C) and completely remove the supernatant by pipetting.
- Add 3 mL DPBS containing 0.5% BSA.
- Centrifuge the cells (400 × g for 5 min at 4 °C) and completely remove the supernatant by pipetting.
- Resuspend the pellet in 400 µL DPBS.
- Acquire cells at the flow cytometer (at least 50 000 leukocytes per tube!) within 1 h.

- Analysis
  - Gating of monocytes: FSC(lin)/SSC(lin) and $CD45^+CD14^{bright}$ and $CD45^+CD14^{dim}$. Cytometer settings *must be kept constant* during the whole project, otherwise data reflecting relative antigen densities are not comparable.
  - Recording: all mean fluorescence intensities (MFI) in all four channels.
  - There are large individual differences in the antigen expression densities, even in healthy probands. Thus, in the final analysis it is not the MFI values but rather the MFI-ratios that must be considered.

Suggested marker panel for platelet immunophenotyping (three-color flow cytometry) from whole blood:

- Sample
  - Peripheral blood: 2.5 mL of citrate-anticoagulated whole blood.
  - Staining and analysis should be performed *immediately* after taking the blood sample.

- Materials
  - Antibodies (see Table 12.3).
  - *In-vitro* stimuli:
    - TRAP-6: Oligopeptide, analogue of thrombin, binds to PAR1, PAR2 and PAR4 receptors (Bachem; final concentration: 1 µM in cell suspension).
    - ADP: Ligand of certain purinoceptors, purchased from DiaMed. Final concentrations: 2 µM (light activation) and 10 µM (massive activation).
  - DPBS without Ca^{2+}/Mg^{2+} (Biochrom).

In the first step, the platelets should be analyzed in a standard hematology analyzer to assess platelet counts, mean platelet volume (MPV) and platelet maturity (residual RNA staining and/or Annexin-V positivity).

Table 12.3 Panel of monoclonal antibody cocktails.

Panel	µL	FITC	µL	R-PE	µL	PerCP
A	10	CD62P	10	CD41	10	CD61
B	10	CX$_3$CR1	10	CD36	10	CD61
C	10	PAC-1	10	CD163	10	CD61
D	10	CD40	10	CD32	10	CD61
E	10	CD107a	10	CD154	10	CD61

All fluorochrome-conjugated monoclonal antibodies listed in Table 3 are available commercially.

In the second step, the platelet counts should be adjusted to 10 000–20 000 µL^{-1} through dilution of the sample with DPBS (i.e., one aliquot of whole blood sample must be diluted, usually 1 : 10).

- Staining procedure
    - Take 10 µL aliquots of diluted citrated blood to polypropylene tubes and add 10 µL of *in-vitro* stimuli. Note: No vortexing here – only *gentle stirring by hand* is needed in order to avoid mechanical platelet activation.
    - Incubate the diluted blood samples with the *in-vitro* stimuli for 15 min at room temperature in the dark.
    - Add 30 µL aliquots of monoclonal antibody cocktails as indicated in Table 12.3, and incubate for a further 15 min at room temperature in the dark.
    - Finally, add 2 mL of DPBS (to stop the activation) and acquire the platelets at the flow cytometer *immediately*.

- Analysis
    - Gating of platelets: FSC(log)/SSC(log) *and* CD61+ together with *exclusion of small aggregates* (higher FSC/SSC).
    - Cytometer settings must be *kept constant* during the whole project, otherwise data reflecting relative antigen densities are not comparable.
    - Recording: all mean fluorescence intensities (MFI) in all three channels.
    - There are large individual differences in the antigen expression densities, even in healthy probands. Thus, in the final analysis it is not the MFI values but rather the MFI ratios that must be considered.

12.1.5
Conclusion

Recent advances in proteomic analysis both, in the field of affinity binder technologies as well as for liquid-phase technologies such as MS/MS, high-resolution liquid-phase separation methods, and advances in chemometry and biometry for large-scale data analysis, form the basis for high-throughput and high-content analyses required in clinical proteomics. They also offer the possibility of introducing these research tools into laboratory diagnostics to screen for risk factors, to identify new disease-specific or stage-specific biomarkers, and to identify novel markers for therapeutic drug monitoring or new therapeutic targets. Therefore, clinical proteomics has the potential to complement genomics, metabolomics, lipidomics, glycomics and transcriptomics including splice variant analysis to gain a better understanding of disease processes.

Current approaches to understand the functional diversity of an organism strive preferentially for a systems approach whereby first, the phenotypic classification of a specific cytome as the cellular systems and subsystems and functional components of the organism is achieved, prior to an attempt to perform proteomic analysis. The fundamental basis for this approach has been well established in studies of cellular systems over many years, and when combined with advanced proteomic approaches, it can achieve a rapid and specific identification of a direct link between biomarkers and their functional roles in complex organisms [4].

The integration of proteomics and cell-based technologies will allow the description of the molecular set-up of normal and abnormal cell systems within a relational knowledge system, thus permitting the standardized discrimination of abnormal cell states in disease. At present, these methods are more exploratory in nature, aiming at advancing scientific knowledge within clinical investigations rather than becoming routine. However, in the future this may gradually change, as some of the methods and their applications become established in clinical routine assays. One consequence of this would be that individualized predictions of further disease courses in patients (i.e., predictive medicine by cytomics) based on characteristic discriminatory data patterns would permit individualized therapies, the identification of new pharmaceutical targets, and the establishment of a standardized framework of relevant molecular alterations in disease [106]. As clinical proteomics becomes routine laboratory diagnosis, the preanalytical aspects will require a high degree of standardization, including SOPs and automated workstations for high-throughput sample preparation.

References

1 Liotta, L.A., Ferrari, M. and Petricoin, E. (2003) Clinical proteomics: written in blood. *Nature*, **425**, 905.

2 Liotta, L.A. and Petricoin, E.F. (2006) Serum peptidome for cancer detection: spinning biologic trash into diagnostic gold. *J. Clin. Invest.*, **116**, 26–30.

3 Conrads, T.P., Hood, B.L. and Veenstra, T.D. (2006) Sampling and analytical strategies for biomarker discovery using mass spectrometry. *Biotechniques*, **40**, 799–805.

4 Diks, S.H. and Peppelenbosch, M.P. (2004) Single cell proteomics for personalised medicine. *Trends Mol. Med.*, **10**, 574–577.

5 Rai, A.J., Gelfand, C.A., Haywood, B.C., Warunek, D.J., Yi, J., Schuchard, M.D., Mehigh, R.J., Cockrill, S.L., Scott, G.B., Tammen, H., Schulz-Knappe, P., Speicher, D.W., Vitzthum, F., Haab, B.B., Siest, G. and Chan, D.W. (2005) HUPO Plasma Proteome Project specimen collection and handling:

towards the standardization of parameters for plasma proteome samples. *Proteomics*, **5**, 3262–3277.

6 Marshall, J., Kupchak, P., Zhu, W., Yantha, J., Vrees, T., Furesz, S., Jacks, K., Smith, C., Kireeva, I., Zhang, R., Takahashi, M., Stanton, E. and Jackowski, G. (2003) Processing of serum proteins underlies the mass spectral fingerprinting of myocardial infarction. *J. Proteome. Res.*, **2**, 361–372.

7 Nedelkov, D. (2005) Population proteomics: addressing protein diversity in humans. *Expert Rev. Proteomics*, **2**, 315–324.

8 Semmes, O.J., Feng, Z., Adam, B.L., Banez, L.L., Bigbee, W.L., Campos, D., Cazares, L.H., Chan, D.W., Grizzle, W.E., Izbicka, E., Kagan, J., Malik, G., McLerran, D., Moul, J.W., Partin, A., Prasanna, P., Rosenzweig, J., Sokoll, L.J., Srivastava, S., Srivastava, S., Thompson, I., Welsh, M.J., White, N., Winget, M., Yasui, Y., Zhang, Z. and Zhu, L. (2005) Evaluation of serum protein profiling by surface-enhanced laser desorption/ionization time-of-flight mass spectrometry for the detection of prostate cancer: I. Assessment of platform reproducibility. *Clin. Chem.*, **51**, 102–112.

9 Moller, S., Mix, E., Bluggel, M., Serrano-Fernandez, P., Koczan, D., Kotsikoris, V., Kunz, M., Watson, M., Pahnke, J., Illges, H., Kreutzer, M., Mikkat, S., Thiesen, H.J., Glocker, M.O., Zettl, U.K. and Ibrahim, S.M. (2005) Collection of soluble variants of membrane proteins for transcriptomics and proteomics. *In Silico. Biol.*, **5**, 295–311.

10 Glocker, M.O., Guthke, R., Kekow, J. and Thiesen, H.J. (2006) Rheumatoid arthritis, a complex multifactorial disease: on the way toward individualized medicine. *Med. Res. Rev.*, **26**, 63–87.

11 Lippi, G., Salvagno, G.L., Montagnana, M. and Guidi, G.C. (2006) Reliability of the thrombin-generation assay in frozen-thawed platelet-rich plasma. *Clin. Chem.*, **52**, 1827–1828.

12 Stahl, S. and Uhlen, M. (1997) Bacterial surface display: trends and progress. *Trends Biotechnol.*, **15**, 185–192.

13 Omenn, G.S., States, D.J., Adamski, M., Blackwell, T.W., Menon, R., Hermjakob, H., Apweiler, R., Haab, B.B., Simpson, R.J., Eddes, J.S., Kapp, E.A., Moritz, R.L., Chan, D.W., Rai, A.J., Admon, A., Aebersold, R., Eng, J., Hancock, W.S., Hefta, S.A., Meyer, H., Paik, Y.K., Yoo, J.S., Ping, P., Pounds, J., Adkins, J., Qian, X., Wang, R., Wasinger, V., Wu, C.Y., Zhao, X., Zeng, R., Archakov, A., Tsugita, A., Beer, I., Pandey, A., Pisano, M., Andrews, P., Tammen, H., Speicher, D.W. and Hanash, S.M. (2005) Overview of the HUPO Plasma Proteome Project: results from the pilot phase with 35 collaborating laboratories and multiple analytical groups generating a core dataset of 3020 proteins and a publicly-available database. *Proteomics*, **5**, 3226–3245.

14 Petricoin, E.F., III, Ornstein, D.K., Paweletz, C.P., Ardekani, A., Hackett, P.S., Hitt, B.A., Velassco, A., Trucco, C., Wiegand, L., Wood, K., Simone, C.B., Levine, P.J., Linehan, W.M., Emmert-Buck, M.R., Steinberg, S.M., Kohn, E.C. and Liotta, L.A. (2002) Serum proteomic patterns for detection of prostate cancer. *J. Natl. Cancer Inst.*, **94**, 1576–1578.

15 Petricoin, E.F., Ornstein, D.K. and Liotta, L.A. (2004) Clinical proteomics: Applications for prostate cancer biomarker discovery and detection. *Urol. Oncol.*, **22**, 322–328.

16 Carrette, O., Demalte, I., Scherl, A., Yalkinoglu, O., Corthals, G., Burkhard, P., Hochstrasser, D.F. and Sanchez, J.C. (2003) A panel of cerebrospinal fluid potential biomarkers for the diagnosis of Alzheimer's disease. *Proteomics*, **3**, 1486–1494.

17 Clarke, W., Silverman, B.C., Zhang, Z., Chan, D.W., Klein, A.S. and Molmenti, E.P. (2003) Characterization of renal allograft rejection by urinary proteomic analysis. *Ann. Surg.*, **237**, 660–664.

18 Neuhoff, N., Kaiser, T., Wittke, S., Krebs, R., Pitt, A., Burchard, A., Sundmacher, A., Schlegelberger, B., Kolch, W. and Mischak, H. (2004) Mass spectrometry for the detection of differentially expressed proteins: a comparison of surface-enhanced laser desorption/ionization and capillary electrophoresis/mass spectrometry. *Rapid Commun. Mass Spectrom.*, **18**, 149–156.

19 Kaiser, T., Hermann, A., Kielstein, J.T., Wittke, S., Bartel, S., Krebs, R., Hausadel, F., Hillmann, M., Golovko, I., Koester, P., Haller, H., Weissinger, E.M., Fliser, D. and Mischak, H. (2003) Capillary electrophoresis coupled to mass spectrometry to establish polypeptide patterns in dialysis fluids. *J. Chromatogr. A*, **1013**, 157–171.

20 Wittke, S., Fliser, D., Haubitz, M., Bartel, S., Krebs, R., Hausadel, F., Hillmann, M., Golovko, I., Koester, P., Haller, H., Kaiser, T., Mischak, H. and Weissinger, E.M. (2003) Determination of peptides and proteins in human urine with capillary electrophoresis-mass spectrometry a suitable tool for the establishment of new diagnostic markers. *J. Chromatogr. A*, **1013**, 173–181.

21 Mischak, H., Kaiser, T., Walden, M., Hillmann, M., Wittke, S., Herrmann, A., Knueppel, S., Haller, H. and Fliser, D. (2004) Proteomic analysis for the assessment of diabetic renal damage in humans. *Clin. Sci. (Lond)*, **107**, 485–495.

22 Fliser, D., Wittke, S. and Mischak, H. (2005) Capillary electrophoresis coupled to mass spectrometry for clinical diagnostic purposes. *Electrophoresis*, **26**, 2708–2716.

23 Kolch, W., Neususs, C., Pelzing, M. and Mischak, H. (2005) Capillary electrophoresis-mass spectrometry as a powerful tool in clinical diagnosis and biomarker discovery. *Mass Spectrom. Rev.*, **24**, 959–977.

24 Weissinger, E.M., Wittke, S., Kaiser, T., Haller, H., Bartel, S., Krebs, R., Golovko, I., Rupprecht, H.D., Haubitz, M., Hecker, H., Mischak, H. and Fliser, D. (2004) Proteomic patterns established with capillary electrophoresis and mass spectrometry for diagnostic purposes. *Kidney Int.*, **65**, 2426–2434.

25 Haubitz, M., Wittke, S., Weissinger, E.M., Walden, M., Rupprecht, H.D., Floege, J., Haller, H. and Mischak, H. (2005) Urine protein patterns can serve as diagnostic tools in patients with IgA nephropathy. *Kidney Int.*, **67**, 2313–2320.

26 Decramer, S., Wittke, S., Mischak, H., Zurbig, P., Walden, M., Bouissou, F., Bascands, J.L. and Schanstra, J.P. (2006) Predicting the clinical outcome of congenital unilateral ureteropelvic junction obstruction in newborn by urinary proteome analysis. *Nat. Med.*, **12**, 398–400.

27 Vlahou, A., Schellhammer, P.F., Mendrinos, S., Patel, K., Kondylis, F.I., Gong, L., Nasim, S. and Wright, J.G., Jr. (2001) Development of a novel proteomic approach for the detection of transitional cell carcinoma of the bladder in urine. *Am. J. Pathol.*, **158**, 1491–1502.

28 Liu, J. and Li, M. (2005) Finding cancer biomarkers from mass spectrometry data by decision lists. *J. Comput. Biol.*, **12**, 971–979.

29 Theodorescu, D., Wittke, S., Ross, M.M., Walden, M., Conaway, M., Just, I., Mischak, H. and Frierson, H.F. (2006) Discovery and validation of new protein biomarkers for urothelial cancer: a prospective analysis. *Lancet Oncol.*, **7**, 230–240.

30 Kaiser, T., Wittke, S., Just, I., Krebs, R., Bartel, S., Fliser, D., Mischak, H. and Weissinger, E.M. (2004) Capillary electrophoresis coupled to mass spectrometer for automated and robust polypeptide determination in body fluids for clinical use. *Electrophoresis*, **25**, 2044–2055.

31 Janzi, M., Odling, J., Pan-Hammarstrom, Q., Sundberg, M., Lundeberg, J., Uhlen, M., Hammarstrom, L. and Nilsson, P. (2005) Serum microarrays for large scale

32 Horn, S., Lueking, A., Murphy, D., Staudt, A., Gutjahr, C., Schulte, K., Konig, A., Landsberger, M., Lehrach, H., Felix, S.B. and Cahill, D.J. (2006) Profiling humoral autoimmune repertoire of dilated cardiomyopathy (DCM) patients and development of a disease-associated protein chip. *Proteomics*, **6**, 605–613.

33 Robinson, W.H., DiGennaro, C., Hueber, W., Haab, B.B., Kamachi, M., Dean, E.J., Fournel, S., Fong, D., Genovese, M.C., de Vegvar, H.E., Skriner, K., Hirschberg, D.L., Morris, R.I., Muller, S., Pruijn, G.J., Van Venrooij, W.J., Smolen, J.S., Brown, P.O., Steinman, L. and Utz, P.J. (2002) Autoantigen microarrays for multiplex characterization of autoantibody responses. *Nat. Med.*, **8**, 295–301.

34 Gannot, G., Tangrea, M.A., Gillespie, J.W., Erickson, H.S., Wallis, B.S., Leakan, R.A., Knezevic, V., Hartmann, D.P., Chuaqui, R.F. and Emmert-Buck, M.R. (2005) Layered peptide arrays: high-throughput antibody screening of clinical samples. *J. Mol. Diagn.*, **7**, 427–436.

35 Berhane, B.T., Zong, C., Liem, D.A., Huang, A., Le, S., Edmondson, R.D., Jones, R.C., Qiao, X., Whitelegge, J.P., Ping, P. and and Vondriska, T.M. (2005) Cardiovascular-related proteins identified in human plasma by the HUPO Plasma Proteome Project pilot phase. *Proteomics*, **5**, 3520–3530.

36 Glocker, M.O., Ringel, B., Gotze, L., Lorenz, S., Wandschneider, V., Fehring, V., Damm, B., Bantscheff, M., Ibrahim, S.M., Lohr, M. and Thiesen, H.J. (2000) Klinische Proteomforschung. in *Laborwelt*, (ed. S. Gabrielczyk), Verlag der BIOCOM AG, Berlin, p. 7–12.

37 Sinz, A., Bantscheff, M., Mikkat, S., Ringel, B., Drynda, S., Kekow, J., Thiesen, H.J. and Glocker, M.O. (2002) Mass spectrometric proteome analyses of synovial fluids and plasmas from patients suffering from rheumatoid arthritis and comparison to reactive arthritis or osteoarthritis. *Electrophoresis*, **23**, 3445–3456.

38 Fiedler, G.M., Baumann, S., Leichtle, A., Oltmann, A., Kase, J., Thiery, J. and Ceglarek, U. (2007) Standardized peptidome profiling of human urine by magnetic bead separation and matrix-assisted laser desorption/ionization time-of-flight mass spectrometry. *Clin. Chem.*, **53**, 421–428.

39 Anderson, N.L. and Anderson, N.G. (2002) The human plasma proteome: history character, and diagnostic prospects. *Mol. Cell Proteomics*, **1**, 845–867.

40 Righetti, P.G., Castagna, A., Antonucci, F., Piubelli, C., Cecconi, D., Campostrini, N., Rustichelli, C., Antonioli, P., Zanusso, G., Monaco, S., Lomas, L. and Boschetti, E. (2005) Proteome analysis in the clinical chemistry laboratory: myth or reality? *Clin. Chim. Acta*, **357**, 123–139.

41 Lee, H.J., Lee, E.Y., Kwon, M.S. and Paik, Y.K. (2006) Biomarker discovery from the plasma proteome using multidimensional fractionation proteomics. *Curr. Opin. Chem. Biol.*, **10**, 42–49.

42 Qian, W.J., Jacobs, J.M., Liu, T., Camp, D.G. and Smith, R.D. (2006) Advances and challenges in liquid chromatography-mass spectrometry-based proteomics profiling for clinical applications. *Mol. Cell Proteomics*, **5**, 1727–1744.

43 Thongboonkerd, V., McLeish, K.R., Arthur, J.M. and Klein, J.B. (2002) Proteomic analysis of normal human urinary proteins isolated by acetone precipitation or ultracentrifugation. *Kidney Int.*, **62**, 1461–1469.

44 Oh, J., Pyo, J.H., Jo, E.H., Hwang, S.I., Kang, S.C., Jung, J.H., Park, E.K., Kim, S.Y., Choi, J.Y. and Lim, J. (2004) Establishment of a near-standard two-dimensional human urine proteomic map. *Proteomics*, **4**, 3485–3497.

45 Pieper, R., Gatlin, C.L., McGrath, A.M., Makusky, A.J., Mondal, M., Seonarain, M., Field, E., Schatz, C.R., Estock, M.A., Ahmed, N., Anderson, N.G. and Steiner, S. (2004) Characterization of the human

urinary proteome: a method for high-resolution display of urinary proteins on two-dimensional electrophoresis gels with a yield of nearly 1400 distinct protein spots. *Proteomics*, **4**, 1159–1174.

46 Sun, T., Ye, F., Ding, H., Chen, K., Jiang, H. and Shen, X. (2006) Protein tyrosine phosphatase 1B regulates TGF beta 1-induced Smad2 activation through PI3 kinase-dependent pathway. *Cytokine*, **35**, 88–94.

47 Haslam, P.L. and Baughman, R.P. (1999) Report of ERS Task Force: guidelines for measurement of acellular components and standardization of BAL. *Eur. Respir. J.*, **14**, 245–248.

48 Magi, B., Bini, L., Perari, M.G., Fossi, A., Sanchez, J.C., Hochstrasser, D., Paesano, S., Raggiaschi, R., Santucci, A., Pallini, V. and Rottoli, P. (2002) Bronchoalveolar lavage fluid protein composition in patients with sarcoidosis and idiopathic pulmonary fibrosis: a two-dimensional electrophoretic study. *Electrophoresis*, **23**, 3434–3444.

49 Wahlstrom, J., Berlin, M., Skold, C.M., Wigzell, H., Eklund, A. and Grunewald, J. (1999) Phenotypic analysis of lymphocytes and monocytes/macrophages in peripheral blood and bronchoalveolar lavage fluid from patients with pulmonary sarcoidosis. *Thorax*, **54**, 339–346.

50 Magi, B., Bargagli, E., Bini, L. and Rottoli, P. (2006) Proteome analysis of bronchoalveolar lavage in lung diseases. *Proteomics*, **6**, 6354–6369.

51 Thadikkaran, L., Siegenthaler, M.A., Crettaz, D., Queloz, P.A., Schneider, P. and Tissot, J.D. (2005) Recent advances in blood-related proteomics. *Proteomics*, **5**, 3019–3034.

52 Sack, U., Rothe, G., Barlage, S. and Gruber, R. (2007) Flow cytometry in clinical diagnostics. *J. Lab. Med.*, **24**, 277.

53 Valet, G. and Tarnok, A. (2004) Potential and challenges of a human cytome project. *J. Biol. Regul. Homeost. Agents*, **18**, 87–91.

54 Schwartz, A., Marti, G.E., Poon, R., Gratama, J.W. and Fernandez-Repollet, E. (1998) Standardizing flow cytometry: a classification system of fluorescence standards used for flow cytometry. *Cytometry*, **33**, 106–114.

55 Lam, J.S., Belldegrun, A.S. and Figlin, R.A. (2004) Tissue array-based predictions of pathobiology prognosis, and response to treatment for renal cell carcinoma therapy. *Clin. Cancer Res.*, **10**, 6304S–6309S.

56 Eguiluz, C., Viguera, E., Millan, L. and Perez, J. (2006) Multitissue array review: a chronological description of tissue array techniques applications and procedures. *Pathol. Res. Pract.*, **202**, 561–568.

57 Lopez-Guerrero, J.A., Riegman, P.H., Oosterhuis, J.W., Lam, K.H., Oomen, M.H., Spatz, A., Ratcliffe, C., Knox, K., Mager, R., Kerr, D., Pezzella, F., van Damme, B., van de Vijver, M., van Boven, H., Morente, M.M., Alonso, S., Kerjaschki, D., Pammer, J., Carbone, A., Gloghini, A., Teodorovic, I., Isabelle, M., Passioukov, A., Lejeune, S., Therasse, P., van Veen, E.B., Dinjens, W.N. and Llombart-Bosch, A. (2006) TuBaFrost 4: access rules and incentives for a European tumour bank. *Eur. J. Cancer*, **42**, 2924–2929.

58 Thallinger, G.G., Baumgartner, K., Pirklbauer, M., Uray, M., Pauritsch, E., Mehes, G., Buck, C.R., Zatloukal, K. and Trajanoski, Z. (2007) TAMEE: data management and analysis for tissue microarrays. *BMC. Bioinformatics*, **8**, 81.

59 Vignali, D.A. (2000) Multiplexed particle-based flow cytometric assays. *J. Immunol. Methods*, **243**, 243–255.

60 Fritzler, M.J. and Fritzler, M.L. (2006) The emergence of multiplexed technologies as diagnostic platforms in systemic autoimmune diseases. *Curr. Med. Chem.*, **13**, 2503–2512.

61 Ashcroft, R.G. and Lopez, P.A. (2000) Commercial high speed machines open

new opportunities in high throughput flow cytometry (HTFC). *J. Immunol. Methods*, **243**, 13–24.
62 Ibrahim, S.F. and van den, E.G. (2003) High-speed cell sorting: fundamentals and recent advances. *Curr. Opin. Biotechnol.*, **14**, 5–12.
63 Bernas, T., Gregori, G., Asem, E.K. and Robinson, J.P. (2006) Integrating cytomics and proteomics. *Mol. Cell Proteomics*, **5**, 2–13.
64 Seliger, B. and Kellner, R. (2002) Design of proteome-based studies in combination with serology for the identification of biomarkers and novel targets. *Proteomics*, **2**, 1641–1651.
65 Neildez-Nguyen, T.M., Wajcman, H., Marden, M.C., Bensidhoum, M., Moncollin, V., Giarratana, M.C., Kobari, L., Thierry, D. and Douay, L. (2002) Human erythroid cells produced ex vivo at large scale differentiate into red blood cells in vivo. *Nat. Biotechnol.*, **20**, 467–472.
66 Pasini, E.M., Kirkegaard, M., Mortensen, P., Lutz, H.U., Thomas, A.W. and Mann, M. (2006a) In-depth analysis of the membrane and cytosolic proteome of red blood cells. *Blood*, **108**, 791–801.
67 Pasini, E.M., Kirkegaard, M., Mortensen, P., Lutz, H.U., Thomas, A.W. and Mann, M. (2006b) In-depth analysis of the membrane and cytosolic proteome of red blood cells. *Blood*, **108**, 791–801.
68 Kakhniashvili, D.G., Bulla, L.A., Jr. and Goodman, S.R. (2004) The human erythrocyte proteome: analysis by ion trap mass spectrometry. *Mol. Cell Proteomics.*, **3**, 501–509.
69 Seong, J.K., Kim, D.K., Choi, K.H., Oh, S. H., Kim, K.S., Lee, S.S. and Um, H.D. (2002) Proteomic analysis of the cellular proteins induced by adaptive concentrations of hydrogen peroxide in human U937 cells. *Exp. Mol. Med.*, **34**, 374–378.
70 Dupont, A., Tokarski, C., Dekeyzer, O., Guihot, A.L., Amouyel, P., Rolando, C. and Pinet, F. (2004) Two-dimensional maps and databases of the human macrophage proteome and secretome. *Proteomics*, **4**, 1761–1778.
71 Richards, J., Le, N.F., Hanash, S. and Beretta, L. (2002) Integrated genomic and proteomic analysis of signaling pathways in dendritic cell differentiation and maturation. *Ann. N.Y. Acad. Sci.*, **975**, 91–100.
72 Rosengren, A.T., Nyman, T.A. and Lahesmaa, R. (2005) Proteome profiling of interleukin-12 treated human T helper cells. *Proteomics*, **5**, 3137–3141.
73 Rautajoki, K., Nyman, T.A. and Lahesmaa, R. (2004) Proteome characterization of human T helper 1 and 2 cells. *Proteomics*, **4**, 84–92.
74 Caron, M., Imam-Sghiouar, N., Poirier, F., Le Caer, J.P., Labas, V. and Joubert-Caron, R. (2002) Proteomic map and database of lymphoblastoid proteins. *J. Chromatogr. B Analyt. Technol. Biomed. Life Sci.*, **771**, 197–209.
75 Wollscheid, B., von Haller, P.D., Yi, E., Donohoe, S., Vaughn, K., Keller, A., Nesvizhskii, A.I., Eng, J., Li, X.J., Goodlett, D.R., Aebersold, R. and Watts, J.D. (2004) Lipid raft proteins and their identification in T lymphocytes. *Subcell. Biochem.*, **37**, 121–152.
76 Li, N., Shaw, A.R., Zhang, N., Mak, A. and Li, L. (2004) Lipid raft proteomics: analysis of in-solution digest of sodium dodecyl sulfate-solubilized lipid raft proteins by liquid chromatography-matrix-assisted laser desorption/ionization tandem mass spectrometry. *Proteomics*, **4**, 3156–3166.
77 Jung, E., Heller, M., Sanchez, J.C. and Hochstrasser, D.F. (2000) Proteomics meets cell biology: the establishment of subcellular proteomes. *Electrophoresis*, **21**, 3369–3377.
78 Marcus, K. and Meyer, H.E. (2004) Two-dimensional polyacrylamide gel electrophoresis for platelet proteomics. *Methods Mol. Biol.*, **273**, 421–434.
79 Dittrich, M., Birschmann, I., Stuhlfelder, C., Sickmann, A., Herterich, S., Nieswandt, B., Walter, U. and Dandekar,

T. (2005) Understanding platelets. Lessons from proteomics genomics and promises from network analysis. *Thromb. Haemost.*, **94**, 916–925.

80 Marcus, K., Moebius, J. and Meyer, H.E. (2003) Differential analysis of phosphorylated proteins in resting and thrombin-stimulated human platelets. *Anal. Bioanal. Chem.*, **376**, 973–993.

81 Garcia, A., Prabhakar, S., Brock, C.J., Pearce, A.C., Dwek, R.A., Watson, S.P., Hebestreit, H.F. and Zitzmann, N. (2004) Extensive analysis of the human platelet proteome by two-dimensional gel electrophoresis and mass spectrometry. *Proteomics*, **4**, 656–668.

82 Coppinger, J.A., Cagney, G., Toomey, S., Kislinger, T., Belton, O., McRedmond, J.P., Cahill, D.J., Emili, A., Fitzgerald, D.J. and Maguire, P.B. (2004) Characterization of the proteins released from activated platelets leads to localization of novel platelet proteins in human atherosclerotic lesions. *Blood*, **103**, 2096–2104.

83 Gruenberg, J. and Howell, K.E. (1989) Membrane traffic in endocytosis: insights from cell-free assays. *Annu. Rev. Cell Biol.*, **5**, 453–481.

84 Lamond, A.I. and Earnshaw, W.C. (1998) Structure and function in the nucleus. *Science*, **280**, 547–553.

85 Verma, R., Chen, S., Feldman, R., Schieltz, D., Yates, J., Dohmen, J. and Deshaies, R.J. (2000) Proteasomal proteomics: identification of nucleotide-sensitive proteasome-interacting proteins by mass spectrometric analysis of affinity-purified proteasomes. *Mol. Biol. Cell*, **11**, 3425–3439.

86 Garin, J., Diez, R., Kieffer, S., Dermine, J.F., Duclos, S., Gagnon, E., Sadoul, R., Rondeau, C. and Desjardins, M. (2001) The phagosome proteome: insight into phagosome functions. *J. Cell Biol.*, **152**, 165–180.

87 Yates, J.R., III, Gilchrist, A., Howell, K.E. and Bergeron, J.J. (2005) Proteomics of organelles and large cellular structures. *Nat. Rev. Mol. Cell Biol.*, **6**, 702–714.

88 Brunet, S., Thibault, P., Gagnon, E., Kearney, P., Bergeron, J.J. and Desjardins, M. (2003) Organelle proteomics: looking at less to see more. *Trends Cell Biol.*, **13**, 629–638.

89 Sprenger, R.R., Speijer, D., Back, J.W., De Koster, C.G., Pannekoek, H. and Horrevoets, A.J. (2004) Comparative proteomics of human endothelial cell caveolae and rafts using two-dimensional gel electrophoresis and mass spectrometry. *Electrophoresis*, **25**, 156–172.

90 Anderson, R.G. and Jacobson, K. (2002) A role for lipid shells in targeting proteins to caveolae rafts, and other lipid domains. *Science*, **296**, 1821–1825.

91 Foster, L.J., de Hoog, C.L. and Mann, M. (2003) Unbiased quantitative proteomics of lipid rafts reveals high specificity for signaling factors. *Proc. Natl. Acad. Sci. USA*, **100**, 5813–5818.

92 Lisanti, M.P., Scherer, P.E., Vidugiriene, J., Tang, Z., Hermanowski-Vosatka, A., Tu, Y.H., Cook, R.F. and Sargiacomo, M. (1994) Characterization of caveolin-rich membrane domains isolated from an endothelial-rich source: implications for human disease. *J. Cell Biol.*, **126**, 111–126.

93 Cohen, A.W., Hnasko, R., Schubert, W. and Lisanti, M.P. (2004) Role of caveolae and caveolins in health and disease. *Physiol Rev.*, **84**, 1341–1379.

94 Simons, K. and Ehehalt, R. (2002) Cholesterol lipid rafts, and disease. *J. Clin. Invest.*, **110**, 597–603.

95 Houde, M., Bertholet, S., Gagnon, E., Brunet, S., Goyette, G., Laplante, A., Princiotta, M.F., Thibault, P., Sacks, D. and Desjardins, M. (2003) Phagosomes are competent organelles for antigen cross-presentation. *Nature*, **425**, 402–406.

96 Alexander, S., Srinivasan, S. and Alexander, H. (2003) Proteomics opens doors to the mechanisms of developmentally regulated secretion. *Mol. Cell Proteomics*, **2**, 1156–1163.

97 Schmitz, G. and Schambeck, C.M. (2006) Molecular defects in the ABCA1 pathway affect platelet function. *Pathophysiol. Haemost. Thromb.*, **35**, 166–174.

98 Andersen, J.S., Lam, Y.W., Leung, A.K., Ong, S.E., Lyon, C.E., Lamond, A.I. and and Mann, M. (2005) Nucleolar proteome dynamics. *Nature*, **433**, 77–83.

99 Andersen, J.S. and Mann, M. (2006) Organellar proteomics: turning inventories into insights. *EMBO Rep.*, **7**, 874–879.

100 Scharfe, C., Zaccaria, P., Hoertnagel, K., Jaksch, M., Klopstock, T., Dembowski, M., Lill, R., Prokisch, H., Gerbitz, K.D., Neupert, W., Mewes, H.W. and Meitinger, T. (2000) MITOP the mitochondrial proteome database: 2000 update. *Nucleic Acids Res.*, **28**, 155–158.

101 Kaleem, Z. (2006) Flow cytometric analysis of lymphomas: current status and usefulness. *Arch. Pathol. Lab Med.*, **130**, 1850–1858.

102 Sack, U., Rothe, G., Barlage, S., Kabelitz, D., Kleines, T.O., Luns, A., Renz, H., Ruf, A. and Schmitz, G. (2000) Flow cytometry in clinical diagnostics. *J. Lab. Med.*, **24**, 277–297.

103 Michelson, A.D. (2006) Evaluation of platelet function by flow cytometry. *Pathophysiol. Haemost. Thromb.*, **35**, 67–82.

104 Dunphy, C.H. (2004) Applications of flow cytometry and immunohistochemistry to diagnostic hematopathology. *Arch. Pathol. Lab Med.*, **128**, 1004–1022.

105 Illoh, O.C. (2004) Current applications of flow cytometry in the diagnosis of primary immunodeficiency diseases. *Arch. Pathol. Lab Med.*, **128**, 23–31.

106 Valet, G., Leary, J.F. and Tarnok, A. (2004) Cytomics – new technologies: towards a human cytome project. *Cytometry A*, **59**, 167–171.

12.2
Stem Cell Proteomics

Regina Ebert, Gabriele Möller, Jerzy Adamski, and Franz Jakob

12.2.1
Introduction

Embryonic stem cells are localized in the inner mass of the blastocyst and have the potential of long-term self renewal. These cells are totipotent, which means that they are capable of developing into cells of all germ layers (endoderm, ectoderm, and mesoderm), and can give rise to all tissues of the body. Embryonic stem cell lines are highly proliferative and usually represent quite homogeneous cell populations. Recently, the first proteomic profile of their basal proteome has been established in some embryonic stem cell lines [107–109].

Adult stem cells or somatic stem cells are localized all over the body, and have been identified in the skin, brain, blood, intestine, gonads, breast, muscle, kidney, and bone. They are responsible for the regeneration of damaged tissues. Adult stem cells are pluripotent, which means that their differentiation capacity is committed. Mesenchymal stem cells (MSC) are a specific subtype of adult stem cell; these can give rise to osteogenic, chondrogenic, myogenic, fibroblastic, and adipogenic cell types, and various subtypes thereof [110–112]. Mesenchymal stem cells are localized in the bone marrow stroma and trabecular bone but, due to the fact that mesenchyme is present in all tissues and organs, MSC can be found in the pancreas, skin, adipose tissue and liver, and many other locations [113–115].

12.2.2
Stem Cell Niches

Stem cells are situated in so-called "niches": these are distinct locations within tissues where the stem cells reside, self-renew, and give rise to their offspring [116]. The present model of stem cell niches is that the niches are provided by specialized somatic cells (cap cells), which inhibit stem cell proliferation and differentiation and keep them in place by expressing adhesion molecules (Figure 12.1) [117]. Stem cells are exposed to factors which inhibit their differentiation and maintain their stemness. These inhibitory molecules are expressed by cap cells [117,118]. Symmetric cell division controls the stem cell pool, whereas asymmetric cell division gives rise to a transient amplifying compartment which is already committed by comparison, and which represents the basis for tissue development and regeneration [119].

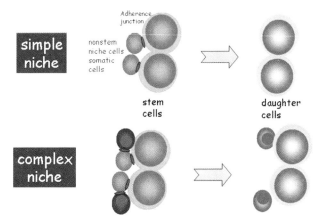

Figure 12.1 Stem cell niches. Non-stem cells (cap cells) keep stem cells in place by expressing adhesion molecules and inhibiting stem cell proliferation and differentiation. A complex population of stem cells resides in the complex stem cell niche. Asymmetric cell division gives rise to a transient amplifying pool of stem cells.

12.2.3
Why Study Proteomes in Stem Cells?

Embryonic stem cells (ESC) and MSC form a fascinating subject for a continuing search for unique biomarkers of their lineage, differentiation processes, and their pluripotency. In recent years, many comparative analyses of transcriptomes, proteomes and even metabolomes have been undertaken [120–125] and, although abundant data have been gathered on the subject, substantial discrepancies persist between the biomarkers found at different levels. The reasons for such discrepancies are manifold, but center on the fact that stem cells have different genetic backgrounds, and have been maintained under a wide variety of different conditions, with biomarkers being identified after distinct challenges. More recently, much progress has been made in unraveling the signaling events by the use of large-scale studies at the transcriptional level. However, the analysis of protein expression, interaction and modification has been more limited, as this requires different strategies to be followed [108].

12.2.4
Technical Challenges and Problems

12.2.4.1 Stem Cell Preparation

Almost all attempts to characterize adult stem cells involve the transient amplifying compartment, since to date it has been very difficult to obtain sufficiently high numbers of non-dividing stem cells on which to perform analyses. Moreover, if this population of cells could be isolated from their niches, the question would immediately arise as to whether the mere separation from their neighbors would alter their proteomic phenotype. Certainly, their transcriptome would change very rapidly. If surface markers (or a combination of such markers) derived by fluorescence-activated cell sorting (FACS) are representative and specific for certain populations apart from their niche (e.g., CD34 expression of hematopoietic stem cells [126]), these could be used for sorting procedures to yield relatively homogeneous cell populations and to identify the intricate differences between populations [127]. In case of MSCs, however, the common preparations used for research – and even for preclinical and clinical therapeutic strategies – are most likely mixed populations, there being no consensus marker proteins which would allow for the isolation of homogeneous populations [113,128].

In order to provide small numbers of cells in basal and comparative analyses, methodological advances have been made using differentially labeled protein samples for analysis on a single gel. Staining with dyes which are more sensitive than Coomassie blue allows for a more efficient detection on two-dimensional (2-D) differential gel electrophoresis [129,130]. Thus, preparations containing only 200 µg of protein can be successfully analyzed [130]. Cysteine labeling of dyes such as Cy3 and Cy5 enhances the sensitivity by one order of magnitude, such that the detection limit is approximately 0.1 ng of protein. Whilst this would certainly be helpful when working with the limited numbers of stem cells available [131], one out of seven proteins lost will not contain cysteines [129,132]. Although this approach seems to be more suitable for cell lines than primary cells, Evans and coworkers reported using only 2.5×10^5 cells to visualize 1000 spots, and subsequently identified differences between two populations [127,130].

12.2.4.2 Cultivation

Although mouse ESC were the first to be established experimentally [133], they proved not to be a substitute for human ESC [134], which require a complex cell culture cultivation arrangement to ensure their pluripotent character and inhibit differentiation. In general, ESC are derived from blastocyst-stage embryos and are cocultured on a monolayer of feeder cells. The feeder cells are in fact mouse fibroblasts arrested at mitosis by irradiation or treatment with mitomycin-C [135,136]. The preparation of cells for further proteomic assays is accomplished by the manual excision of ESC colonies, in order to minimize contamination from the feeder layers.

12.2.4.3 Treatment

It is somewhat impossible to understand interlaced regulatory pathways in ESC or MSC in a steady (undifferentiated) condition. Therefore, specific challenges must be employed to evoke differentiation-specific reactions. Such treatment may include the application of contact-stress (suspension of the culture causes ESC aggregation) [109], exposure to specific drugs such as dexamethasone/β-glycerophosphate and ascorbic acid (for osteogenic MSC differentiation) [120,137], or to all-*trans*-retinoic acid (for human pluripotent embryonal carcinoma stem cells) [138] and further retinoic acid or calcium exposition (differentiation towards neurons of mouse ESC) [139].

12.2.4.4 Whole-Cell Proteome

Many of the investigations performed on ESC proteomes have used whole lysates of cells. This approach was facilitated by standardized methods which were developed in part for transcriptome analyses. The cells are collected under defined conditions (e.g., undifferentiated or differentiated), manually inspected and verified morphologically under the light microscope, washed, sonicated (or ground in a mortar) and lysed in a high-urea/detergent buffer with the addition of proteolysis inhibitors (see below). Whole-proteome analyses were recently reviewed [108,130], and have provided resources which are available on the internet (e.g., http://www.cprmap.com/stem-cell-proteomics/) or profiles for neural stem cells [140], MSC [141] and undifferentiated human ESC seen at different passages [107].

As in all proteomic approaches, a simplification or prefractionation of samples increases the resolution of profiling. Examples using whole-cell lysates include studies on cyclin-kinases, which play a pivotal role in the regulation of the cell cycle in stem cells [142]. In this context, phosphoproteins are important targets in stem cell research, and may be enriched from total cell lysates of differentiated and non-differentiated stem cells by affinity chromatography [143], allowing for analyses of more defined proteomes.

12.2.4.5 Secretory Proteome

The approach using whole-cell lysates provides profiles for all proteins of the cell, among which some are characteristic for the intracellular processes, whilst others are prepared for secretory purposes. The latter are difficult to analyze in the harvested cell culture supernatants, for several reasons:

- Although the cells are well propagated and differentiated in full cell culture media (e.g., with 10% fetal bovine serum, FBS), the secreted proteins may be simultaneously masked by abundant medium additives.
- Serum-free medium is easier to analyze due to its lower protein complexity; however, the cells grow less well and may show symptoms of stress, or even die.
- Some cells will die under any conditions and release intrinsic proteins into the medium.
- The secreted proteins are usually heavily modified post-translationally (e.g., by sugars).

In general, the cells are maintained under full medium conditions to reach 60–80% confluency, and subsequently are cultivated in a serum-free medium. After a limited period of time, the latter medium is collected and undergoes proteome profiling (see Section 12.2.5, Recipes for Beginners). Not unexpectedly, even when experiments were designed to focus on the secretory proteome of human ESC [144] or MSC [137], some of the proteins identified fell into the intracellular class of chaperones, metabolic enzymes, apoptosis, or cytoskeletal elements. However, several extracellular matrix proteins, proteases or growth factors were unequivocally identified using this approach. Ultimately, it remains to be seen whether only the latter class are representative of signal molecules, or both.

The analyses in stem cells were facilitated by technological progress which has been achieved during analyses of the secretory proteome of hepatocytes [145], smooth muscle cells [146], retinal pigment epithelium [147], prostate cancer cell line PC3 [148], or fibroblasts [123,149,150]. *Note*: A web-based database on secreted proteins is maintained at http://spd.cbi.pku.edu.cn/ [151].

One interesting option for the detection of secreted proteins is to perform a metabolic labeling with short-lived proteins using radioisotopically labeled amino acids such as ^{35}S-methionine or cysteine (see Section 12.2.5) [152]. As the secreted proteins have high turnover rates, they will be labeled preferentially *in vivo* and, after media collection, can be visualized using autoradiography on 2-D polyacrylamide gels.

One consequence of a large proportion of the secreted proteins being heavily glycosylated is that glycoproteomics is rapidly becoming a new area in the search for biomarkers in serum [138,153]. This approach includes selective protein isolation and deglycosylation [154–156], in addition to the technological challenges associated with protein identification [157,158].

An example of proteome analysis by 2-D gel electrophoresis of whole-cell lysate and of secreted proteins is shown in Figure 12.2. The point to note here is the distinct levels of complexity and the abundance of proteins.

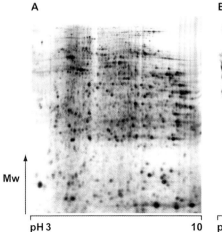 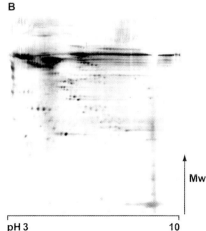

Figure 12.2 Proteome analyses of whole-cell lysate and secreted proteins. (A) Lysate of mouse kidney. The kidney was homogenized in protease inhibitor solution. The homogenate was lyophilized, dissolved in 9 M urea, 2% CHAPS, 1% DTT, 2% IPG-buffer, 10 mM Pefabloc and centrifuged at $150\,000 \times g$ for 30 min at 4 °C. Proteins in the supernatant were separated on an IPG strip pH 3–10 followed by PAGE on a 12% SDS gel. The pH-gradient is indicated below the gels; arrows point in the direction of higher molecular mass. The protein pattern is characteristic for low proteolysis. (B) Proteins secreted from mesenchymal stem cells. MSC cells were cultivated in serum-free medium 2 days before being harvested. Proteins were precipitated by acetone (protease inhibitor added before) resuspended in 8 M urea, 2% CHAPS, 2.5% DTT, 0.5% IPG-buffer, 0.003% bromophenol blue, and separated on an IPG strip pH 3–10 followed by PAGE on a 12% SDS gel. The protein pattern indicates a high content of glycosylated proteins.

12.2.5
Recipes for Beginners

The aim of the "recipes" provided in this section is to provide samples that are compatible with various techniques for protein separation (e.g., gel filtration, reverse-phase chromatography, 2-D gel electrophoresis) and protein identification (e.g., LC-MS/MS, FT-ICR-MS/MS). The procedures do require, however, further sample processing, such as the addition of Immobilines for 2-D gel electrophoresis or of trifluoroacetic acid (TFA) for RP-HPLC. These steps are described in greater details elsewhere in this book.

12.2.5.1 Whole-Cell Lysate

- Collect the cells (ca. 10^6); these may be separated from the feeder cells by microdissection and inspected microscopically, or detached from uncoated dishes by incubation in 0.25% trypsin in phosphate-buffered saline (PBS) for 10 min at 37 °C.

- Wash the cells twice by resuspension in PBS and centrifugation at 500 × g for 2 min at 4 °C to remove the cell culture media.
- Solubilize the cells in 500 µL lysis buffer [9.0 M urea, 4% (w/v) CHAPS, 1% (w/v) dithiothreitol (DTT)] containing protease inhibitor cocktail (1 mM EDTA, 10 mM ε-amino-caproic acid, 50 mM benzamidine, 1 mM PMSF, or ready-to-use pellets from Roche) for 5 min at room temperature.
- Centrifuge the lysate at 100 000 × g for 30 min at room temperature.
- Store the supernatant at −80 °C until used for the analysis.
- Determine the protein concentration using a Bradford assay.

12.2.5.2 Secretory Proteome Procedure

- Cultivate the cells to about 60–80% confluency in full media (e.g., with 10% fetal bovine serum, FBS).
- Wash the cells twice with serum-free media, without removing them from the flask or Petri dish.
- Collect parts of media at different time points from the same incubation, or keep the cells for 24–48 h and collect media from parallel dishes.
- Add protease inhibitors (final concentrations 1 mM EDTA, 10 mM ε-amino-caproic acid, 50 mM benzamidine, 1 mM PMSF, or ready-to-use pellets from Roche) to the collected samples.
- Precipitate the proteins from the medium (see below).

12.2.5.3 Labeling with ^{35}S

- Cultivate the cells to about 60–80% confluency in full media (e.g., with 10% FBS).
- Wash the cells with PBS and twice with methionine, cysteine and serum-free medium (e.g., William's E medium, ICN).
- Incubate the cells with 1–20 µCi [^{35}S]-labeled methionine and cysteine (specific activity >1000 Ci mmol^{-1} (e.g., Trans35-Slabel; Biomedica, MP Biomedicals) for 1–6 h at 37 °C.
- Aspirate the media and filter through low dead-volume 0.2 µm (Millex-GP, Millipore).
- Supplement with protease inhibitors (final concentration 1 mM EDTA, 10 mM ε-amino-caproic acid, 50 mM benzamidine, 1 mM PMSF, or ready-to-use pellets from Roche).
- Precipitate the proteins with ethanol or trichloroacetic acid (TCA). *Note*: different proteins have distinct susceptibilities to these methods.
- Freeze the pellet at −80 °C until used for the analysis.

12.2.5.4 Ethanol Precipitation

- Consider ALL waste to be radioactive if working with ^{35}S.
- Mix 1 mL of media with 8 mL ethanol (99.9%).
- Incubate for 3 h at room temperature.
- Incubate at $-20\,^\circ$C overnight.
- Collect the proteins by centrifugation at $10\,000 \times g$ for 20 min at 4 $^\circ$C and remove the supernatant.
- Wash the pellet once with pure ethanol.
- Air-dry the pellet.
- Use the pellet immediately; alternatively, cover with pure ethanol and store at $-20\,^\circ$C until used for the analysis.

12.2.5.5 TCA Precipitation

- Consider ALL waste to be radioactive if working with ^{35}S.
- Mix 100 µL of medium with 500 µL of ice-cold 6.1 N TCA solution containing 80 mM DTT.
- Incubate for 1 h at 4 $^\circ$C.
- Centrifuge at $10\,000 \times g$ for 10 min at 4 $^\circ$C and discard the supernatant.
- Wash the pellet three times with 1 mL of ice-cold acetone containing 20 mM DTT. Ensure that the pellet (it may be barely visible) remains in the tube by centrifuging at $10\,000 \times g$ for 10 min at 4 $^\circ$C.
- Air-dry the pellet.
- Use the pellet immediately; alternatively, cover with acetone and store at $-20\,^\circ$C until used for the analysis.

References

107 Baharvand, H., Hajheidari, M., Ashtiani, S.K. and Salekdeh, G.H. (2006) Proteomic signature of human embryonic stem cells. *Proteomics*, **6**, 3544–3549.

108 Van Hoof, D., Mummery, C.L., Heck, A.J. and Krijgsveld, J. (2006) Embryonic stem cell proteomics. *Expert Rev. Proteomics*, **3**, 427–437.

109 Van Hoof, D., Passier, R., Ward-Van Oostwaard, D., Pinkse, M.W., Heck, A.J., Mummery, C.L. and Krijgsveld, J. (2006) A quest for human and mouse embryonic stem cell-specific proteins. *Mol. Cell. Proteomics*, **5**, 1261–1273.

110 Bianco, P., Riminucci, M., Gronthos, S. and Robey, P.G. (2001) Bone marrow stromal stem cells: nature biology, and potential applications. *Stem Cells*, **19**, 180–192.

111 Forriol, F. and Shapiro, F. (2005) Bone development: interaction of molecular

components and biophysical forces. *Clin. Orthop. Relat. Res.*, **432**, 14–33.

112 Caplan, A.I. (2005) Review: mesenchymal stem cells: cell-based reconstructive therapy in orthopedics. *Tissue Eng.*, **11**, 1198–1211.

113 Pittenger, M.F., Mackay, A.M., Beck, S.C., Jaiswal, R.K., Douglas, R., Mosca, J.D., Moorman, M.A., Simonetti, D.W., Craig, S. and Marshak, D.R. (1999) Multilineage potential of adult human mesenchymal stem cells. *Science*, **284**, 143–147.

114 Baksh, D., Song, L. and Tuan, R.S. (2004) Adult mesenchymal stem cells: characterization differentiation, and application in cell and gene therapy. *J. Cell Mol. Med.*, **8**, 301–316.

115 Meirelles, L.d.S., Chagastelles, P.C. and Nardi, N.B. (2006) Mesenchymal stem cells reside in virtually all post-natal organs and tissues. *J. Cell Sci.*, **119**, 2204–2213.

116 Ohlstein, B., Kai, T., Decotto, E. and Spradling, A. (2004) The stem cell niche: theme and variations. *Curr. Opin. Cell Biol.*, **16**, 693–699.

117 Fuchs, E., Tumbar, T. and Guasch, G. (2004) Socializing with the neighbors: stem cells and their niche. *Cell*, **116**, 769–778.

118 Nystul, T.G. and Spradling, A.C. (2006) Breaking out of the mold: diversity within adult stem cells and their niches. *Curr. Opin. Genet. Dev.*, **16**, 463–468.

119 Morrison, S.J. and Kimble, J. (2006) Asymmetric and symmetric stem-cell divisions in development and cancer. *Nature*, **441**, 1068–1074.

120 Wagner, W., Feldmann, R.E., Jr., Seckinger, A., Maurer, M.H., Wein, F., Blake, J., Krause, U., Kalenka, A., Burgers, H.F., Saffrich, R., Wuchter, P., Kuschinsky, W. and Ho, A.D. (2006) The heterogeneity of human mesenchymal stem cell preparations – evidence from simultaneous analysis of proteomes and transcriptomes. *Exp. Hematol.*, **34**, 536–548.

121 Jeong, J.A., Lee, Y., Lee, W., Jung, S., Lee, D.S., Jeong, N., Lee, H.S., Bae, Y., Jeon, C.J. and Kim, H. (2006) Proteomic analysis of the hydrophobic fraction of mesenchymal stem cells derived from human umbilical cord blood. *Mol. Cell*, **22**, 36–43.

122 Jeong, J.A., Hong, S.H., Gang, E.J., Ahn, C., Hwang, S.H., Yang, I.H., Han, H. and Kim, H. (2005) Differential gene expression profiling of human umbilical cord blood-derived mesenchymal stem cells by DNA microarray. *Stem Cells*, **23**, 584–593.

123 Kim, K.H., Park, G.T., Lim, Y.B., Rue, S.W., Jung, J.C., Sonn, J.K., Bae, Y.S., Park, J.W. and Lee, Y.S. (2004) Expression of connective tissue growth factor a biomarker in senescence of human diploid fibroblasts, is up-regulated by a transforming growth factor-beta-mediated signaling pathway. *Biochem. Biophys. Res. Commun.*, **318**, 819–825.

124 Becker, R.C. and Andreotti, F. (2006) Preoteomics metabolomics and progenitor cells in acute coronary syndromes. *J. Thromb. Thrombolysis*, **22**, 85–88.

125 Unwin, R.D. and Whetton, A.D. (2006) Systematic proteome and transcriptome analysis of stem cell populations. *Cell Cycle*, **5**, 1587–1591.

126 Gangenahalli, G.U., Singh, V.K., Verma, Y.K., Gupta, P., Sharma, R.K., Chandra, R. and Luthra, P.M. (2006) Hematopoietic stem cell antigen C D34:Role in adhesion or homing. *Stem Cells Dev.*, **15**, 305–313.

127 Evans, C.A., Tonge, R., Blinco, D., Pierce, A., Shaw, J., Lu, Y., Hamzah, H.G., Gray, A., Downes, C.P., Gaskell, S.J., Spooncer, E. and Whetton, A.D. (2004) Comparative proteomics of primitive hematopoietic cell populations reveals differences in expression of proteins regulating motility. *Blood*, **103**, 3751–3759.

128 Song, L., Young, N.J., Webb, N.E. and Tuan, R.S. (2005) Origin and characterization of multipotential mesenchymal stem cells derived from

adult human trabecular bone. *Stem Cells Dev.*, **14**, 712–721.

129 Tonge, R., Shaw, J., Middleton, B., Rowlinson, R., Rayner, S., Young, J., Pognan, F., Hawkins, E., Currie, I. and Davison, M. (2001) Validation and development of fluorescence two-dimensional differential gel electrophoresis proteomics technology. *Proteomics*, **1**, 377–396.

130 Unwin, R.D., Gaskell, S.J., Evans, C.A. and Whetton, A.D. (2003) The potential for proteomic definition of stem cell populations. *Exp. Hematol.*, **31**, 1147–1159.

131 Shaw, J., Rowlinson, R., Nickson, J., Stone, T., Sweet, A., Williams, K. and Tonge, R. (2003) Evaluation of saturation labelling two-dimensional difference gel electrophoresis fluorescent dyes. *Proteomics*, **3**, 1181–1195.

132 Thierry, R. (2002) Two-dimensional gel electrophoresis in proteomics: Old, old fashioned, but it still climbs up the mountains. *Proteomics*, **2**, 3–10.

133 Evans, M.J. and Kaufman, M.H. (1981) Establishment in culture of pluripotential cells from mouse embryos. *Nature*, **292**, 154–156.

134 Thomson, J.A., Itskovitz-Eldor, J., Shapiro, S.S., Waknitz, M.A., Swiergiel, J.J., Marshall, V.S. and Jones, J.M. (1998) Embryonic stem cell lines derived from human blastocysts. *Science*, **282**, 1145–1147.

135 Reubinoff, B.E., Pera, M.F., Fong, C.Y., Trounson, A. and Bongso, A. (2000) Embryonic stem cell lines from human blastocysts: somatic differentiation in vitro. *Nat. Biotechnol.*, **18**, 399–404.

136 Mummery, C. Ward-van Oostwaard, D., Doevendans, P., Spijker, R., van den Brink, S., Hassink, R., van der Heyden, M., Opthof, T., Pera, M., de la Riviere, A.B., Passier, R. and Tertoolen, L. (2003) Differentiation of human embryonic stem cells to cardiomyocytes: role of coculture with visceral endoderm-like cells. *Circulation*, **107**, 2733–2740.

137 Sun, H.J., Bahk, Y.Y., Choi, Y.R., Shim, J.H., Han, S.H. and Lee, J.W. (2006) A proteomic analysis during serial subculture and osteogenic differentiation of human mesenchymal stem cell. *J. Orthop. Res.*, **24**, 2059–2071.

138 Hayman, M.W., Christie, V.B., Keating, T.S. and Przyborski, S.A. (2006) Following the differentiation of human pluripotent stem cells by proteomic identification of biomarkers. *Stem Cells Dev.*, **15**, 221–231.

139 Wang, D. and Gao, L. (2005) Proteomic analysis of neural differentiation of mouse embryonic stem cells. *Proteomics*, **5**, 4414–4426.

140 Maurer, M.H., Feldmann, R.E., Jr., Futterer, C.D. and Kuschinsky, W. (2003) The proteome of neural stem cells from adult rat hippocampus. *Proteome Sci.*, **1**, 4.

141 Feldmann, R.E., Jr., Bieback, K., Maurer, M.H., Kalenka, A., Burgers, H.F., Gross, B., Hunzinger, C., Kluter, H., Kuschinsky, W. and Eichler, H. (2005) Stem cell proteomes: a profile of human mesenchymal stem cells derived from umbilical cord blood. *Electrophoresis*, **26**, 2749–2758.

142 Becker, K.A., Ghule, P.N., Therrien, J.A., Lian, J.B., Stein, J.L., van Wijnen, A.J. and Stein, G.S. (2006) Self-renewal of human embryonic stem cells is supported by a shortened G1 cell cycle phase. *J. Cell Physiol.*, **209**, 883–893.

143 Puente, L.G., Borris, D.J., Carriere, J.F., Kelly, J.F. and Megeney, L.A. (2006) Identification of candidate regulators of embryonic stem cell differentiation by comparative phosphoprotein affinity profiling. *Mol. Cell. Proteomics*, **5**, 57–67.

144 Prowse, A.B., McQuade, L.R., Bryant, K.J., Van Dyk, D.D., Tuch, B.E. and Gray, P.P. (2005) A proteome analysis of conditioned media from human neonatal fibroblasts used in the maintenance of human embryonic stem cells. *Proteomics*, **5**, 978–989.

145 Farkas, D., Bhat, V.B., Mandapati, S., Wishnok, J.S. and Tannenbaum, S.R. (2005) Characterization of the secreted

proteome of rat hepatocytes cultured in collagen sandwiches. *Chem. Res. Toxicol.*, **18**, 1132–1139.

146 Dupont, A., Corseaux, D., Dekeyzer, O., Drobecq, H., Guihot, A.L., Susen, S., Vincentelli, A., Amouyel, P., Jude, B. and Pinet, F. (2005) The proteome and secretome of human arterial smooth muscle cells. *Proteomics*, **5**, 585–596.

147 An, E., Lu, X., Flippin, J., Devaney, J.M., Halligan, B., Hoffman, E., Csaky, K. and Hathout, Y. (2006) Secreted proteome profiling in human RPE cell cultures derived from donors with age related macular degeneration and age matched healthy donors. *J. Proteome Res.*, **5**, 2599–2610.

148 Andersen, H., Jensen, O.N. and Eriksen, E.F. (2003) A proteome study of secreted prostatic factors affecting osteoblastic activity: identification and characterisation of cyclophilin A. *Eur. J. Cancer*, **39**, 989–995.

149 Puricelli, L., Iori, E., Millioni, R., Arrigoni, G., James, P., Vedovato, M. and Tessari, P. (2006) Proteome analysis of cultured fibroblasts from type 1 diabetic patients and normal subjects. *J. Clin. Endocrinol. Metab.*, **91**, 3507–3514.

150 Shimmura, S., Miyashita, H., Higa, K., Yoshida, S., Shimazaki, J. and Tsubota, K. (2006) Proteomic analysis of soluble factors secreted by limbal fibroblasts. *Mol. Vis.*, **12**, 478–484.

151 Chen, Y., Zhang, Y., Yin, Y., Gao, G., Li, S., Jiang, Y., Gu, X. and Luo, J. (2005) SPD – a web-based secreted protein database. *Nucleic Acids Res.*, **33**, D169–D173.

152 Zwickl, H., Traxler, E., Staettner, S., Parzefall, W., Grasl-Kraupp, B., Karner, J., Schulte-Hermann, R. and Gerner, C. (2005) A novel technique to specifically analyze the secretome of cells and tissues. *Electrophoresis*, **26**, 2779–2785.

153 Hayman, M.W. and Przyborski, S.A. (2004) Proteomic identification of biomarkers expressed by human pluripotent stem cells. *Biochem. Biophys. Res. Commun.*, **316**, 918–923.

154 Petra, P.H., Griffin, P.R., Yates, J.R., III, Moore, K. and Zhang, W. (1992) Complete enzymatic deglycosylation of native sex steroid-binding protein (SBP or SHBG) of human and rabbit plasma: effect on the steroid-binding activity. *Protein Sci.*, **1**, 902–909.

155 Zhang, H., Li, X.J., Martin, D.B. and Aebersold, R. (2003) Identification and quantification of N-linked glycoproteins using hydrazide chemistry stable isotope labeling and mass spectrometry. *Nat. Biotechnol.*, **21**, 660–666.

156 Zhao, J., Simeone, D.M., Heidt, D., Anderson, M.A. and Lubman, D.M. (2006) Comparative serum glycoproteomics using lectin selected sialic acid glycoproteins with mass spectrometric analysis: application to pancreatic cancer serum. *J. Proteome Res.*, **5**, 1792–1802.

157 Frolov, A., Hoffmann, P. and Hoffmann, R. (2006) Fragmentation behavior of glycated peptides derived from D-glucose D-fructose and D-ribose in tandem mass spectrometry. *J. Mass Spectrom.*, **41**, 1459–1469.

158 Brancia, F.L., Bereszczak, J.Z., Lapolla, A., Fedele, D., Baccarin, L., Seraglia, R. and Traldi, P. (2006) Comprehensive analysis of glycated human serum albumin tryptic peptides by off-line liquid chromatography followed by MALDI analysis on a time-of-flight/curved field reflection tandem mass spectrometer. *J. Mass Spectrom.*, **41**, 1179–1185.

13
Bioinformatics

13.1
Bioinformatics Support for Mass Spectrometric Quality Control

Knut Reinert, Tim Conrad, and Oliver Kohlbacher

13.1.1
Introduction

In this chapter, the computational aspects of the sample preparation process are examined. Data preparation and evaluation must be considered an integral part of the whole experimental set-up, and as such must be included into quality assessment and control. Data processing is often left to the standard (default) settings of the software provided by the instrument manufacturer. However, data processing and analysis algorithms are complex and their applicability, as well as the proper choice of parameters, must be validated. This validation can be made on different levels, corresponding to the potential error sources. Problematic data can arise either from problematic measurements (i.e., problems within the sample or within the instrument) or from incorrect processing of the data. Due to the immense volume of the data produced by modern mass spectrometers (in the range of several Gigabytes), it is not viable to do this validation by hand. Rather, automatic methods must be used.

In the following, some of the problems that occur during data processing will be highlighted, and several techniques that are able to detect problematic data automatically described.

13.1.2
Problem Description

Mass spectrometric measurements in proteomics are typically made with one of two different goals in mind: the *identification* of peptides using peptide mass fingerprints (PMF) or tandem mass spectrometric measurements (MS/MS); or the relative *quantification* of peptides using single MS measurements.

Identification alone can be sufficient for many problems; however, biomarker detection or systems biology applications also often require absolute or relative

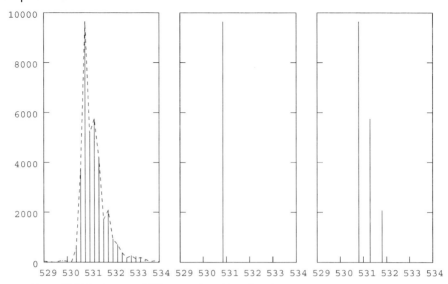

Figure 13.1 The impact of different peak- picking algorithms (OpenMS and Bruker Apex) on raw data is from an ion trap mass spectrometer. The raw data on the left is converted to peak lists in different ways by two peak picking algorithms (middle and right).

quantification of peptides. While different experiments are conducted to address identification and quantification, the underlying data are virtually identical in their structure. In both cases, it is a series of acquired *raw mass spectra*, which are then processed by (often machine-dependent) signal processing software to produce *peak lists* (e.g., see Figure 13.1, right). Those peak lists are then fed to the identification algorithms.

For quantification purposes a sequence of MS spectra is acquired over time. Each spectrum provides a snapshot of the peptide content of the sample during a fixed time interval. It consists of ion counts or *intensities* measured by the mass spectrometer within a certain interval of mass/charge ratios. The scans are acquired at (more or less equally spaced) periodic time intervals, and the collection of scans represents what we will call an *MS map*. The goal is to integrate all signal intensities caused by isotopic variants of the same peptide eluting over a certain time (e.g., see Figure 13.2).

In the following, three problems that occur during data acquisition and processing will be described.

13.1.2.1 Signal Processing Pitfalls

Raw data usually consists of a series of ion counts for different mass/charge ratios (c.f. Figure 13.1, left). The standard preprocessing steps include *noise filtering, baseline reduction,* and *finally peak picking*. The type of method employed for each of these steps – and their parameters – can have a significant impact on the final result, as can be seen in Figure 13.1. This illustrates how the application of two standard algorithms to the same mass spectrum can yield quite different peak lists. Similar effects can also occur during other preprocessing steps, for example the aggressive removal of a measurement's baseline can result in loss of weak signals.

13.1.2.2 Map Quality Control

Quantitative proteomics is increasingly developing into one of the cornerstones of fundamental research in the life sciences and of clinical studies [1–3]. In a typical experimental setting, the protein sample is subjected to a proteolytic digestion yielding a mixture of peptides which is separated, for example, via a chromatographic column or by capillary electrophoresis. The peptides elute at different times and are thus separated according to a physical property, such as the peptide's hydrophobicity in the case of reversed-phase (RP) liquid chromatography. The raw data collected in such an experiment is depicted in Figure 13.2. Since the size of the raw data is usually too large, data-reduction techniques must be applied which convert the raw data into a collection of peak lists (see Figure 13.2, middle) or into a single representative *feature* (see Figure 13.2, right, where all signals belonging to a charge variant of a peptide are integrated). Depending on the parameters of the peak picking and quantification algorithms applied to these data, the results might be as different as illustrated in the peak-picking example. The goal is hence to identify wrong settings or bad data via inspection of a single map, or via comparison to known good maps with similar content.

13.1.2.3 Statistical Validation Results

In a clinical setting it is more important to derive quality values for these measurements. If the aim is to identify a biomarker this cannot be done by inspecting data that is of dubious quality. Hence, it is very important to have an assessment of how much trust can be placed in the results of previous computational stages.

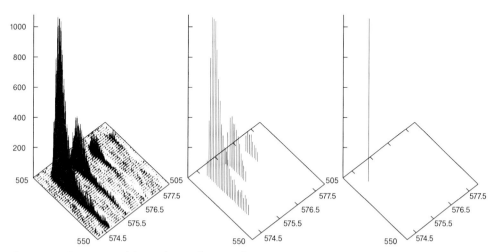

Figure 13.2 The amount of data reduction from raw data over picked peaks to a single representative of a charge variant.

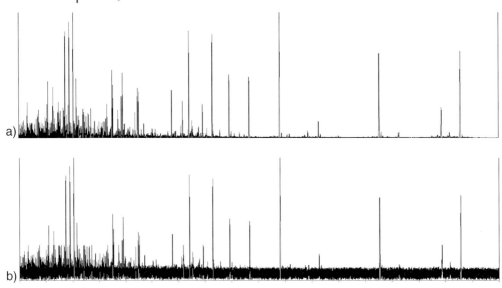

Figure 13.3 Example for "good" (a) and "bad" (b) spectra. In (b), the signal-to-noise ratio is worse than in (a), as the data are too noisy. Therefore, small peaks are no longer detectable.

In the following, focus is centered on describing methods and measures to determine the quality of individual mass spectra (Section 13.1.3) as the quality control for maps is still in its infancy. However, the concepts used in the quality assessment for single spectra can often be applied to two-dimensional data. In Section 13.1.4, a real-world example is provided illustrating how these techniques can be employed in larger projects. We summarize and give some perspectives in Sections 13.1.5 and 13.1.6. Finally, Section 13.1.7 provides a recipe that might guide the reader in employing quality control techniques in the processing of MS data.

13.1.3
Quality Assessment for One-Dimensional (1-D) MS Data

The purpose of quality measures is to determine the reliability of a spectrum's information content (and therefore the subsequent analyses thereof). These measures reflect the quality of a spectrum (see, for example, Figure 13.3) regarding particular aspects (such as "number of peaks" or "average peak size") and are usually combined into a total (subjective) quality value. Although a variety of attributes has been proposed for this task, only few have proven useful [4–6]. Some of these are as follows:

- *Simple Peak-Based Statistics:* While being simple and quickly calculated, these measures give a first hint of the spectra quality. Note that they heavily depend on the parameters used while applying the peak picking algorithm (e.g., minimum intensity threshold).

- Total Number of Peaks (TNP): The number of peaks in this spectrum.
- Total Ion Count (TIC)/Total Intensity: The sum of the intensities of all peaks in this spectrum. This reflects the concentration and degree of ionization of the analytes in the sample. It is quite an important measure for the normalization of spectra.
- Average Peak Intensity (API): The average height of all peaks is calculated by $API = \frac{TNP}{TIC}$.

- *Signal-to-Noise Ratio (SNR):* Ratio of API and the estimated noise level of this spectrum (see e.g., Ref. [7] for calculation of noise level). Together with TNP, this provides a rough idea of how much information this spectrum carries. This criteria is mostly applied unconsciously by human experts when manually determining spectrum quality.

- *Sum of "Satellite Peaks":* Many peaks in a spectrum can be related to each other because they stem from modified versions of the same peptide. Examples for this are losses of small groups or atoms (e.g., neutral side chains, NH_3 or H_3PO_4, see Refs. [8,9] or [10] for occurrence probabilities). Since the mass differences of these "satellite" peaks are known they can easily be identified. The total sum of all of these peaks gives another quality criteria.

- *Sum of Isotope Peaks:* The sum of intensities of peaks which are part of an isotopic envelope (see e.g., Figure 13.1). This measure is a special case of the preceding measure accounting for isotope clusters only.

13.1.3.1 Filter

This section describes two common classes of techniques for filtering spectra prior to actual content analysis using the measures introduced in the previous section. Filtering spectra in this context means to filter out spectra of low quality.

There exist two predominant approaches: *thresholding* and *clustering*, which are explained in more detail in the remainder of this section.

- *Thresholding:* In a simple version of the threshold-driven approach a linear combination of n feature values (see Section 13.1.3) are used to calculate the quality score Q for spectrum s: $s: Q(s) = \sum_i f_i(s) \cdot w_i$.[1] The w_is are weights which are usually learned automatically from training data or manually determined by experts. If no further knowledge is available they are simply set to "1". The function $f_i(s)$ specifies the quality value for a given feature i of spectrum s. In a more complex setting, another function $k_j(s)$ might be introduced, which can – for example – become zero if some criteria are not met (knock-out criterion) and therefore sets the total quality to zero (for an application, see e.g., Ref. [11].[2] This results in $Q(s) = \prod_{k_j(s)} \cdot \sum_i f_i(s) \cdot w_i$.

---

[1] In statistics there exists a very similar concept called "Linear Discriminant Analysis", see e.g., Refs. [16,17].

[2] Of course, introducing non-linearity and/or adding other concepts lead to many well-known machine learning algorithms not explicitly named here.

There exist two implicit problems in this approach, namely:
(1) the determination of a threshold below which a spectrum is classified as "bad"; and (2) the actual choice of features to use ("Feature Selection"). While (1) can be learned from training data relatively easily, (2) is a branch of research by its own and still not finally solved (see e.g., Refs. [12–15] and references therein).

In Ref. [17] this approach was used to filter MS^2 spectra. A combination of "charge state differentiation score", "sum of total signal intensity" and a "signal-to-noise score" was used. The authors claim to be able to correctly identify the spectra as "good" or "bad" with a statistical sensitivity of 96.8% and a specificity of 98%.

- *Clustering:* The base assumption for this approach is that spectra from the same class (e.g., "good" or "bad") will have similar quality values of their respective attributes (see Section 13.1.3) and therefore build clusters in an n-dimensional space (n being the number of features). In this setting, no weights or thresholds have to be determined but, as described in the previous paragraph, the "Feature Selection" task must be performed prior to analysis. Furthermore, clustering fully relies on the definition or choice of some metric (such as Manhatten or Euclidian) determining the distance between any two elements. After choosing the metric and the set of features, the spectra are projected in the so-called "feature space". Here, the detection and interpretation of emerging clusters is sought for, which can become quite challenging. To achieve this, either specialized classes of high-dimensional clustering algorithms are applied (e.g., subspace clustering, subspace ranking, projected clustering, or correlation clustering – see Ref. [18]) or the data are first projected down to some lower-dimensional space (PCA, MDS, ISOMAP) where clustering becomes easier (by e.g., k-means clustering, Deterministic Annealing-based clustering, hierarchical clustering).

13.1.4
Application Example: Absolute Quantification of an Unknown Peptide Content

The use of some of the above concepts will now be illustrated in a real-life example, a proteomics pipeline for absolute quantification of myoglobin in complex serum. In this example, the consequent use of automated data processing and quality control during this process yields very low error bounds (lower than those achieved in a manual evaluation). The details of the study are described elsewhere [19]. Here, focus will be centered on the data processing aspects and how appropriate methods for data handling can enable more complex experiments.

The efficient analysis of data relies on modular software for the individual analysis steps. There exist several packages for constructing modular proteomics workflows, of which only the Trans Proteomics Pipeline [20] and TOPP, the OpenMS Proteomics Pipeline, will be mentioned [21]. TOPP provides algorithms for processing data as individual, independent modules that can be stringed into complex analysis pipelines. Communication between these modules is enabled by the use of (mostly standardized) XML formats. The individual modules can be easily combined into a coherent analysis pipeline fitted towards a specific analysis task without any skill in programming. In the case described here, it was necessary to construct a pipeline for absolute quantification of myoglobin in human blood serum with very high accuracy.

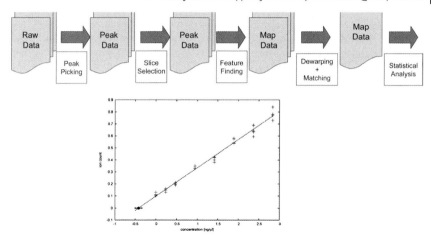

Figure 13.4 A simple analysis pipeline constructed from TOPP modules managing all major steps from raw data conversion to the statistical evalutaion of the standard addition protocol.

A complex protocol consisting of two-dimensional (2-D) chromatographic separation and subsequent HPLC-MS analysis was established. Low error bounds and absolute quantitation was achieved through a standard addition protocol, quadruplicate measurements and the use of internal standards. This set-up results in 36 HPLC-MS runs per sample, an amount of data that is difficult to analyze manually. An automated analysis pipeline can solve this, and yield results in a short period of time. In this case study, the pipeline (shown in Figure 13.4) accepts 36 HPLC-MS raw data files as its input. Each of these files in the run is then reduced to peak data, and these peak data are reduced to include only those regions in m/z and retention time dimension containing the peptides required for quantitation. Running these data through a feature-finding algorithm will yield a set of concentrations for both, myoglobin and reference peptides. After automatically correcting for retention time shifts using a dewarping algorithm, matching features are identified. These data are then automatically analyzed by a further module using standard addition statistics, which results in an individual value for the analyte concentration. Running the data through this pipeline requires less time than performing the actual experiments, and is significantly faster (and also more accurate) than analyzing these data manually.

Automated data processing can thus enable more complex experiments, increase accuracy, and significantly reduce analysis time.

13.1.5
Summary

This exhibition should have made it clear that quality control and assessment of possible data corruption through pre-processing the data are essential steps to guarantee the validity of all subsequent processing steps.

Problematic data can arise either from problematic measurements (i.e., problems within the sample or within the instrument) or from – possibly incorrect – processing of the data. Different attributes we described of the measurements that can be used in conjunction with machine learning approaches to detect bad quality data.

13.1.6
Perspective

Filtering data may become increasingly important as new MS instruments produce many Gigabytes of data per measurement. Hence, there is a need to assess the errors made by such filters. Also, as proteomics is pushed more and more towards clinical applications, quality assessment and reproducibility of results must be taken to new levels.

In the case of protein identification it might become possible to directly determine the sequence from the measurement if the accuracy of the instruments continues to improve. In the realm of quantification, multi-dimensional nano HPLC opens new avenues to analyze large data sets while placing heavy demands on computational methods for data evaluation and quality control.

13.1.7
Recipe for Beginners

Here, some practical recommendations will be provided regarding the quality control of MS data. For more information, the interested reader is referred to [22].

13.1.7.1 Acquiring the Raw Data

- *Try to work with data as raw as possible.* If adequate and possible, try to disable the machine's automatic preprocessing features such as baseline reduction, calibration, etc. This allows the steps to be controlled. After that, any possible errors introduced during preprocessing can be assessed more easily. Often, the algorithms of the vendor software are not described in sufficient detail to do this.

- *Obtain as much information about the data as possible.* This includes:
  - *Clinical meta-information* of the patients (age, gender, anamnesis, blood parameters, etc.).
  - *Technical information* of the machines used and method/parameters of acquisition (laser rate, type of matrix, duration of acquisition, etc.). This information is essential if comparison is to be made between studies or between findings of different labs or data acquired by different machines.
  - *Order of acquisition.* This allows for simple checks whether differences in the data really reflect biological changes or are simply an artifact occurring over time.

- *Use standard formats.* Formats such as mzData or mzXML allow for easy exchange of the data and permit the use of different software packages for preprocessing and analyzing.

- *Verify the RAW data.* Insist on initial assessment of accuracy using replicates and simplified (blind) samples before running the actual study.

- *Establish a data management system.* This comprises storage, archival and back-up mechanisms as well as databases and user-friendly query applications for the meta-information.

13.1.7.2 Preprocessing the Data

- Check that your software package implements verified methods and algorithms that fit the data properties and machine resolution.

- If possible, verify the results with a second analysis software.

- Verify the standard parameters of the analysis tools. They are often tuned to particular needs that the designers had in mind when setting them.

13.1.7.3 Analyzing the Preprocessed Data

- Be aware of statistical pitfalls. Simple things such as different sizes of patient groups can render the results useless if they are not handled correctly by the algorithms.

- Work with simple models. By creating patient groups with very heterogeneous members, such as different gender, age or diseases, the model complexity grows exponentially. Try to avoid this by splitting the groups into smaller subgroups with well-defined patient properties. For example, from a group "ovarian cancer" create subgroups "cancer::female::young", "cancer::female:: middle-aged", "cancer::female::old::diabetes", "cancer::female::old::otherwise_ healthy".

- Be aware of statistical significance. If small groups are used, take this into account in the analysis. If the sample number is too small, the results are not significant.

References

1 MacCoss, M. and Matthews, D.E. (2005) Quantitative MS for proteomics: Teaching a new dog old tricks. *Anal. Chem.*, **77**, 294A–302A.

2 Mann, M. and Aebersolf, R. (2003) Mass spectrometry -based proteomics. *Nature*, **422**, 198–207.

3 Ong, S.-E. and Mann, M. (2005) Mass spectrometry-based proteomic turns quantitative. *Nature Chem. Biol.*, **1**, 252–262.

4 Tabb, D.L., Eng, J.K. and Yates, J.R. III (2001) *Proteome Research: Mass Spectrometry*, Springer, Berlin.

5 Flikka, K., Martens, L., Vanderkerckhove, J., Gevaert, K. and Eidhammer, I. (2006) Improving the reliability and throughput of mass spectrometry-based proteomics by spectrum quality filtering. *Proteomics*, **6**, 2086–2094.

6 Bern, M., Goldberg, D., McDonald, W.H. and Yates, J.R. (2004) Automatic quality

assessment of Peptide tandem mass spectra. *Bioinformatics*, **20**, Suppl 1 I49–I54.

7 McDonough, R.N. and Whale, A.D. (1995) *Detection of Signals in Noise*, Academic Press, San Diego.

8 Salek, M. and Lehmann, W.D. (2003) Neutral loss of amino acid residues from protonated peptides in collision-induced dissociation generates N- or C-terminal sequence ladders. *J. Mass Spectrom.*, **38**: 1143–1149.

9 Martin, D.B., Eng, J K., Nesvizhskii, A.I., Gemmill, A. and Aebersold, R. (2005) Investigation of neutral loss during collision-induced dissociation of peptide ions. *Anal. Chem.*, **77**, 4870–4882.

10 Fridman, T., Day, R., Razumovsbya, J., Xu, D. and Gorin, A. (2003) Probability Profiles – Novel Approach in Tandem Mass Spectrometry De Novo Sequencing.CSB '03: Proceedings of the IEEE Computer Society Conference on Bioinformatics IEEE Computer Society P. 415.

11 Purvine, S., Kolker, N. and Kolker, E. (2004) Spectral quality assessment for high-throughput tandem mass spectrometry proteomics. *OMICS.*, **8**, 255–265.

12 Shin, H. and Markey, M.K. (2006) A machine learning perspective on the development of clinical decision support systems utilizing mass spectra of blood samples. *J. Biomed. Inform.*, **39**, 227–248.

13 Zhang, X., Lu, X., Shi, Q., Xu, X..-Q., Leung, H..-C.E., Harris, L.N., Iglehart, J. D., Miron, A., Liu, J.S. and Wong, W.H. (2006) Recursive SVM feature selection and sample classification for mass-spectrometry and microarray data. *BMC Bioinformatics*, **7**, 197.

14 Li, L., Tang, H., Wu, Z., Gong, J., Gruidl, M., Tockman, M. and Clark, R.A. (2004) Data mining techniques for cancer detection using serum proteomic profiling. *Artif. Intell. Med.*, **32**, 71–83.

15 Walters, W.P. and Goldman, B.B. (2005) Feature selection in quantitative structure-activity relationships. *Curr. Opin. Drug Discov. Devel.*, **8**, 329–333.

16 Fisher, R. (1936) The use of multiple measurements in Taxonomic Problems. *Ann. Eugenics*, **7**, 179–188.

17 Mika, S., Ratsch, E., Weston, J., Scholkopf, B. and Muller, K. (1999) Fisher Discriminant Analysis with Kernels. Proceedings of IEEE Conference on Neural Networks for Signal Processing.

18 Böhm, C., Kailing, K., Kriegel, H.P. and Kröger, P. (2004) Density Connected Clustering with Local Subspace Preferences. Proceedings 4th IEEE International Conference on Data Mining (ICDM'04), 27–34.

19 Mayr, B., Kohlbacher, O., Reinert, K., Sturm, M., Grpl, C., Lange, E., Klein, C. and Huber, C.G. (2006) Absolute myoglobin quantitation in serum by combining two-dimensional liquid chromatography-electrospray ionization mass spectrometry and novel data analysis algorithms. *J. Proteome Res.*, **5**, 414–421.

20 Keller, A., Eng, J., Zhang, N., Li, X. and Aebersold, R. (2005) A uniform proteomics MS/MS analysis pipeline utilizing open XML file formats. *Mol. Syst. Biol.*, **1**, 2005–2017.

21 Kohlbacher, O., Reinert, K., Grpl, C., Lange, E., Pfeifer, N., Schulz-Trieglaff, O. and Sturm, M. (2007) TOPP – The OpenMS Proteomics Pipeline. *Bioinformatics*, **23**, e191–e197.

22 Mischak, H., Apweiler, R., Banks, R., Conaway, M., Coon, J., Dominiczak, A., Ehrich, J., Fliser, D., Girolami, M., Goodsaid, F., Hermjakob, H., Hochstrasser, D., Jankowski, J., Julian, B.A., Kolch, W., Massy, Z., Neusuess, C., Novak, J., Peter, K., Rossing, K., Schanstra, J., Semmes, J., Theodorescu, D., Thongboonkerd, V., Weissinger, E., Van Eyk, J.E. and Yamamoto, T. (2007) Clinical Protoemics: A need to define the field and to begin to set adequate standards. *Proteomics*. Published online 22 Jan 2007. DOI: 10.1002/prca. 200600771.

13.2
Use of Physico-Chemical Properties in Peptide and Protein Identification

Anastasia K. Yocum, Peter J. Ulintz, and Philip C. Andrews

13.2.1
Introduction

The most widely used methods for the identification of peptides in proteomics experiments rely on automated methods that match theoretical to observed spectra, with subsequent independent probability calculations. Increasing scores from database search algorithms typically indicate an increasing confidence in the identification; however, the point of transition between a good score and a poor score is often difficult to ascertain. Unfortunately, most probability calculations do not take into consideration all possible biological factors, which detract from their usefulness as a real measurement of confidence in a peptide identification. It has been proposed by a number of investigators that one can use additional properties – the physico-chemical properties of the peptides of interest – to independently validate and verify the peptide identification [23–36]. Such properties are calculable from the amino acid sequence of the peptide, assuming that the identity of the peptide is known, and include the relative isoelectric point, hydrophobicity, amino acid composition, and accurate mass. What is commonly obtained during proteomics experiments are multidimensional signatures in the data that correspond to m/z, intensity, pI, and chromatographic elution time. By comparing these measured values to calculated theoretical values for identified peptides, it is possible to gain additional information that is useful for judging the accuracy of an identification and increasing the discrimination between true and false positives.

Chromatographic or gel electrophoretic separation serves to reduce the complexity of samples prior to mass spectrometry (MS). Most often, due to the perceived low resolving power of high-performance liquid chromatography (HPLC), this technique is thought of as simply a necessary preparation step, and the valuable physical information revealed by separation techniques is often not utilized. Isoelectric points and peptide retention times (measured during isoelectric focusing, ion exchange, or reverse-phase chromatography) can also aid in the identification of peptides during shotgun proteomic experiments. As these properties are intrinsic properties of peptides and can be calculated, they may be utilized to validate database searches and can be useful for optimization of separation methods. A number of studies have developed algorithms that model these attributes for a variety of separations [37,38]. In addition, mass resolution and accuracy can directly contribute to peptide identification and validation during database searches. Most search algorithms, for example, allow the user to specify which isotopic peak to utilize for analysis, as well as an acceptable delta mass range, where delta mass is defined as the difference in mass between the observed and theoretical masses of the identified peptide. Generally, the smaller this delta mass and the more precise the measurement of mass, the more confidence one has in the identified peptide, and as such this parameter is utilized in the scoring model of most algorithms. However, a thorough assessment of mass resolution during the design of the instrument methodology and acquisition of mass spectra, and conscientious scrutiny of mass accuracy during data analysis is often not undertaken. Variations in mass accuracy and resolution can arise for instrument, environmental, or operator reasons, and appropriate controls are necessary to minimize the effects of this variation through good laboratory practice. These variations, when they occur, can often be identified through examination of the details of search results.

Outlined below is a summary of some selected recent studies for utilizing the physico-chemical properties of isoelectric point, chromatographic retention time, amino acid composition, and accurate mass measurement of peptides, and their use in the identification and validation of peptide sequences in proteomics experiments. It is hoped that this chapter will provide the reader with some of the central and fundamental theory, a relevant description of its application, and finally provide literature examples of the more prominent practical uses.

13.2.2
Isoelectric Point

Various acidic and basic functional groups on amino acid side chains and termini of proteins and peptides may be charged, and the presence of a charge is dependent on the pK_a values of the groups and the pH of the solution. The isoelectric point (pI) of a peptide is defined as the pH at which the peptide has a net charge of zero. At pH values below the pI, peptides carry a net positive charge; above the pI they carry a net negative charge. The pI-values of peptides can be utilized as a basis for separation in solution. If a voltage is applied to a complex peptide mixture in a pH gradient, the peptides will migrate to the pH at which they are neutrally charged. The pI of a peptide may be accurately calculated from its sequence, and provides information that complements mass spectrometric methods for protein and peptide identifications. If the calculated pI of the peptide correlates well with the experimental value, this information can provide an extra filter for false-positive rate calculations, in that truly random false-positive matches would be expected to produce correspondingly random pI measurements. This information might then be used to extend peptide identifications below normal score thresholds from MS database search engines such as Mascot and SEQUEST, resulting in more peptide identifications, or increased confidence in "borderline" matches. The same approach could be applied to peptides above the threshold, to reduce the false-positive rate. Current popular database search platforms typically provide filters which restrict matching protein sequences to those within a defined pI range, but typically do not incorporate this information into the actual scoring algorithm. The application of a pI filter to intact proteins can also be problematic, as many post-translational modifications result in changes in pI-values (e.g., phosphorylation, deamidation, acetylation). The development of the next-generation database search algorithms to incorporate this information is desirable. The PeptideProphet tool is an example of current software which can accommodate pI information into a probability score for peptide identification [29]. When the p*I* feature is turned on in this software, the tool assumes that the input sample consists of an isoelectrically focused set of peptides. The software calculates the average p*I* of all peptides, and refines the probability score for each peptide based on how close or far the calculated p*I* of the peptide is from the average [37].

It has been established that the p*I* of a peptide can be determined experimentally with high reproducibility; the resolving power of isoelectric focusing often results in experimental values well within predicted calculated values – that is, within 0.1 standard deviations [24]. This high reproducibility can be attributed to the commercial availability of high-quality pH gradient stationary phases [immobilized pH gradient (IPG) strips] which provide a means of determining the p*I* of separated peptides without using internal standards. In fact, when used in conjunction with tandem MS, these gradients can be calibrated from the peptide identifications. Newer instrumentation such as the BD Free Flow electrophoresis instrument [38] and the Agilent 3100 OFFGEL device [39] also show promise in this domain, providing a simpler means for the isoelectric focusing of peptide mixtures.

A number of algorithms exist for p*I* calculation, including online calculators (http://expasy.org/tools/pi_tool.html, http://www.embl-heidelberg.de/cgi/pi-wrapper.pl, http://www.bioinformatics.vg/sms/protein_iep.html, http://www.nihilnovus.com/Palabra.html) [40–43]. The most typical method for calculating p*I* is in the form of a theoretical titration, in which the net charge on the peptide sequence is calculated for a number of pH values; the pH value that generates a net charge closest to zero is returned as the p*I*. A "divide and conquer" approach works well, in which an algorithm successively divides the pH range in half, iteratively narrowing in on the neutral pH. For example, if the p*I* of a peptide is 4.2, the algorithm would begin by calculating the charge for the peptide at the neutral pH, 7.0. The algorithm would find that the charge is negative, indicating that pH is lower than 7.0, and narrow the initial pH range of 0.0 to 14.0 to 0.0 to 7.0. Splitting the difference, the algorithm would next test pH 3.5, find the net charge to be positive (thus narrowing the pH range to 3.5–7.0), and then test 5.25, etc. Complete descriptions for these types of calculation are described in the works of Bjellqvist *et al.* [42] and Sillero *et al.* [43].

Two different p*I* prediction algorithms [40,41] were employed in the investigations of Benjamin Cargile and colleagues. Both prediction algorithms show similar trends when compared with experimental [44]. However, there was a closer correlation between predicted and experimental for peptide p*I* values in the acidic to neutral regions, but the calculated values diverged from the experimental values in the basic region. This is a known issue, in that common pK_a values used in p*I* prediction algorithms are often extrapolated using data generated from proteins focused only in the acidic pH range [42]. The following are some theoretically derived trends that have been confirmed experimentally correlating the amino acid composition of a peptide and its behavior during isoelectric focusing [23,27,44–46]:

1. Only a few fully tryptic peptides are predicted to have a p*I* higher than 7, and this is corroborated by the finding in number 2 below.

2. Diffusion of tryptic peptides within IPG strips occurs mostly at the basic end.
3. The bulk of fully tryptic peptides should have a pI within 3.5 to 4.5.
4. As the number of positively charged amino acids increases, the pI increases.
5. As fully tryptic peptides increase in mass, the average pI generally decreases. This is believed to be caused by the additive effects of an increasing number of negative charged amino acids against a constant number of positive charges (N-terminus and C-terminal residue).
6. The corollary to number 4 is for any mass range, the pI increases as the number of tryptic missed cleavages increases.
7. With two missed tryptic cleavages, most peptides have pI values between 3 and 5.
8. With three missed cleavages, most peptides measured pI is between 8 and 9.

As mentioned above, the use of pI as a discriminator can lower thresholds for identification, decreasing false-positive hits, and thus increasing sensitivity. In the study conducted by Cargile et al., the correlation of calculated pI values to experimental observations enabled approximately 23% more peptides to be identified than using thresholds alone [24]. Similarly, by utilizing *Drosophila* Kc167 cells, Heller *et al.* employed isoelectric focusing information for the validation of peptide identifications with two different database search algorithms, PHENYX and SEQUEST [25–27,47,48]. In this study, it was found that identifications from the PHENYX search algorithm corroborated and aligned nicely with reproducible pI clustering. The authors then used these pI data to determine a SEQUEST DeltaCn threshold and compared the PHENYX and SEQUEST results to PeptideProphet confidence calculations. Overall, they were able to increase the number of validated peptide identifications from 84.4% to 92.2% of identified spectra by utilizing pI as an additional parameter. An interesting finding in this study was that peptides containing the amino acid proline focused poorly. It was hypothesized that cis and trans proline bonds exist in equilibrium under the isoelectric focusing conditions used, and that the resulting steric effects alter the pI-values of the two forms. This would be expected to be both sequence- and temperature-dependent.

13.2.3
Ion-Exchange Chromatography

Ion-exchange chromatography is another fractionation technique which exploits the charge state of a peptide. In this technique, a charged stationary phase is used to retain analyte peptides of opposite charge. Peptides of varying charge will have greater or lesser affinity to the stationary phase, and bound peptides will elute from the stationary phase at different rates depending on the degree of their affinity. Analyte molecules that are negatively charged are retained on anion exchangers, while positively charged analytes are retained on cation exchangers, respectively. While both types of chromatography are used in proteomics studies, strong cation-exchange (SCX) chromatography is most often used because, at low pH values, the vast majority of tryptic peptides bind to SCX and elution with a salt or pH gradient allows peptides to be resolved by their relative charges.

Computationally, the chromatographic properties of a peptide, which is based on the pI of the peptide in a particular solution, can be predicted and utilized in a manner similar to the isoelectric point. One important practical difference is that the inherent resolution of ion-exchange chromatography is much lower than isoelectric focusing, and thus is a weaker discriminator. While a calculated pI can be utilized for both isoelectric focusing and SCX chromatography, it is not the calculated pI of any peptide but rather the number of theoretical charges that the peptide can take on in solution that influences its behavior in ion-exchange chromatography. Thus, pI and charge are related; but charge varies dependent on the number of ionizable groups. There is considerable correlation between solution-phase charge and the relative elution time, but the SCX retention time has considerable variance between peptides because there are often additional interactions (hydrophobic, H-bonding, etc.) with the stationary media. This is compounded by the fact that the apparent number of positive charges of a particular peptide is dependent on the pH of the buffering system in the mobile phase, the presence of organic solvents, and the concentrations of counterions. Even though the general trend is that the average pI of eluting peptides increases as the salt concentration of the mobile phase increases, there is great variability of pI within each salt fraction. Additionally, the pI is difficult to predict in these experiments due to the different influences of the surrounding matrix, including the ionic strength of solution, the type and concentration of denaturing agents, and the type of stationary phase. Further, intrinsic to the peptide are the direct influences of the chemical and biological properties of the peptides, including PTMs and experimental artifacts that also make predicting the pI a challenge.

Peng *et al.* were among the first to describe the use of SCX fractionation information to assist in the identification of peptides [33]. These authors showed that peptides can be separated based on their ionic charge. At pH 3.0, the amino acids lysine, arginine and histidine, as well as the N-terminus, contribute most to the solution charge state of a peptide. However, peptides are not separated by SCX alone; rather, there was an additional hydrophobic contribution that affected retention and elution times [49]. By increasing the salt concentration, the retention time decreased due to electrostatic interactions and the increasing ionic strength. If the ionic strength was increased beyond a threshold, the hydrophobic interaction with the resin increased, causing a shift in the retention time and thus affecting the peptide elution order beyond simple net charge. This mixed-mode chromatography contributes to an uncertainty in predicting retention times, but also reduces chromatographic resolution. Peng *et al.* concluded that, by analyzing every other SCX fraction collected, it was possible to decrease the analysis time by 50% while decreasing peptide identifications by only 16%. Likewise, analyzing every third SCX fraction would decrease the number of identifications by 26%. The low resolution implied by this study can be attributed to not compensating for mixed-mode effects, inefficiency of the packed cartridges, the use of non-optimized step gradients, and possibly also the need to load sufficient peptide on the SCX columns in order to detect low-abundance proteins [33].

In another study by the same group, SCX is utilized to enrich for phosphopeptides. The negatively charged phosphate moiety functions to weaken the binding of phosphopeptides to an SCX column. Ballif *et al.* exploited this fact by selecting early-eluting salt fractions from the SCX column which were enriched for phosphopeptides [50].

In general, p*I* is considered a more discriminative and reproducible property than SCX retention time for peptide separations. Cargile *et al.* [44] performed a comprehensive comparison of isoelectric focusing (IEF) methodology to SCX, and found that IEF methodology identifies more peptides when compared to SCX, but there was no considerable overlap, implying that neither are comprehensive. The study by Essader *et al.* also compared the use of immobilized pH gradient IEF with SCX chromatography for the first dimension of a two-dimensional (2-D) peptide separation [46]. These authors concluded that the retention time for eluting peptides from SCX cannot be accurately predicted because the elution time depends on interactions between the charged packing material and peptides, in addition to the non-specific peptide adsorption to support resin. Data from these studies indicate that a majority of peptides with two tryptic termini – that is, the last amino acid is lysine or arginine and the first amino acid is immediately adjacent to a lysine or arginine – have a p*I* range of 4 to 5. It was concluded that the most effective way of using this additional information is as a p*I* filter. Peng *et al.* [33] also demonstrated the utilization of similar information during SCX fractionation, showing in their study that peptides from non-specific tryptic cleavages resulted in 100% false-positive hits, no matter which thresholds were chosen. However, partial tryptic peptides gave 0.4% false-positive when filtered based on p*I*, compared to 9.4% false-positive, indicating that p*I* can be utilized to filter spurious results, thus increasing confidence and decreasing threshold. These results demonstrate that an additional property of the peptide identification process – namely the number of tryptic termini of a potential peptide sequence – can be an effective discriminating measure of the correctness of an identification.

13.2.4
Reversed-Phase Chromatography

Efforts to accurately predict peptide elution times on reversed-phase (RP) chromatography are not new; the calculation and prediction of peptide retention time in RP chromatography was first reported by Martin *et al.* in 1948 [51]. The simplest form of retention time prediction relies on the assumption that the chromatographic behavior of a peptide is mainly dependent on its amino acid composition [30], and thus the retention time of peptides on a RP column can be predicted by summing the relative hydrophobic contributions of each constitutive amino acid residue. Most recent studies have relied on the calculation of a "retention coefficient" which is based on the summation of empirically determined amino acid residue retention coefficients, which in turn is based on the assumption that the chromatographic behavior of peptides is based solely on its amino acid composition and the location of amino acids in the peptide [52]. While this approach seems to work for most 10- to 20-mer peptides, it does not work for isomeric peptides in this size range, thereby indicating that the sequences of the peptides play a role [53,64]. It is clear that the factors affecting the elution of peptides include amino acid composition, peptide length [30,55], and sequence-dependent effects [56] – that is, the nearest-neighbor and conformation effects. A recent methodology and its application was completed by Guo and Jaitly, and is discussed in detail below. In a more extensive analysis, Baczek, Kaliszan, and coworkers used quantitative structure–retention relationships (QSRRs) including the octanol–water partition coefficient, van der Waals volume, and the amino acid composition, to predict the retention times of peptides [57]. A yet more technical analysis involving protein structure calculations was completed by Makrodimitris *et al.*, by utilizing Langevin dipoles on the solvent lattice and partial charges for the solute to estimate free energies of adsorption of peptides on the stationary phase [58,59]. Most recently, initially completed for amino acid composition and then later extended to include partial peptide sequence information, Petritis *et al.* have developed an artificial neural network method for predicting the elution time of peptides [52,60].

Most of the methods of retention time prediction listed above contain fundamentally the basic summation of individual amino acid hydrophobicities, as initially reported by Guo *et al.* [61]. In this model [61], the prediction of retention time at a gradient of 1% acetonitrile per minute is equal to the sum of all the retention coefficients for the amino acids residues and end groups, as well as the time for elution of non-retained compounds and the time correction for peptide standards. The percent of acetonitrile at the time of elution is related to the ratio of non-polar amino acids [62]. Unfortunately, these retention coefficients were developed with a limited training set with modified peptide N- and C-terminals, and displayed non-linearity for long peptides.

Deviations from this model were demonstrated by Jaitly *et al.* [28]. These deviations included positive deviations (i.e., the peptide eluted later than predicted when the peptides contained a hydrophilic amino acid at the N-terminus) and negative deviations in which the peptides eluted earlier than predicted when there were hydrophobic amino acids at the N-terminus. Guo *et al.* also demonstrated evidence of deviations from predicted retention times, with the largest positive shifts seen in their data for peptides having moderate to highly hydrophobic and acidic acid amino acids such as aspartic and glutamic acid near the N-terminus. The observed shift was explained by suggesting that peptides carry a charged N-terminus in low-pH mobile phases containing trifluoroacetic acid or formic acid. The charge causes the amino acid at the N-terminus to be partially shielded from the stationary phase interactions, causing a negative shift for a hydrophobic N-terminus and a positive shift for hydrophilic N-terminus amino acids. Additionally, ion pair formation between the amino acid at the N-terminus and the buffers in the mobile phase affect the retention time: a smaller ion pair permits exposure of hydrophobic residues for interaction, while weaker ion pairs allow more of the charge to be exposed. This effect is analogous to N-terminal acetylation. The size of the amino acid side chain has an effect on retention time: smaller amino acids such as leucine and isoleucine have low retention coefficients, while tryptophan and phenylalanine have high retention coefficients. Amino acids with aromatic or aliphatic side chains have a positive effect on retention time, while neutral or polar side chains make weaker or smaller negative contributions to retention time [36].

Differences in analyte retention time observed during experimental replicate analyses in the study conducted by Jaitly et al. were attributed to differences in dead volume and in relative separation speeds – that is, the time difference between the start of the analysis and the first analyte elution which is dependent on the speed of the linear gradient [28]. For constant-pressure liquid chromatography (LC) systems, there are expected variations of flow due to changes of porosity in the stationary phase, viscosity of the solvent, and temperature [28]. These systematic differences can be corrected for, and thus should not significantly affect any comparison of predicted to experimental retention times.

The sequence-specific retention calculator (SSRCalc) [47] differs from other retention prediction models because it not only takes into account the hydrophobicity summation of amino acids in the peptide, the specificity of the N- and C-terminal residues, and peptide length, but also includes the relative position of amino acids, pI and secondary structures [63]. The authors used approximately 2000 peptides to refine current prediction models. An initial prediction utilizing a summation of the amino acids, the specificity of the N- and C-terminal residues, and peptide length resulted in a R^2-value of 0.9585. However, when these authors utilized the SSRCalc including the additional information, the relative position of each amino acid, the calculated pI, length of peptide and secondary structure, the R^2-value increased from 0.9585 to 0.9800.

The performance of retention time prediction algorithms has not been as effective with unknown samples as with training sets of known peptides, however. Krokin et al. performed an analysis with the SSRCalc algorithm on eukaryotic K562 whole-cell protein lysate that was affinity-purified with lectin and digested with trypsin, a common analytical procedure. Although a lower correlation was found with this dataset compared to the training set, the improved prediction capabilities of the newer model was reflected in the unknown sample: an R^2-value of 0.92 was achieved with the simple model, and 0.95 for the more complex model [63]. The retention times for 97, 78, and 49% of the peptides were predicted with an accuracy of ±4, ±2, and ±1 min, respectively.

One goal of retention time prediction is to minimize the instrument time required for MS/MS analysis. The approach is effective for identifications of proteins with a minimal number of PTMs and non-specific cleavage, and is most applicable to HPLC-MALDI analysis, since the off-line nature of this instrument combination allows for enough time to perform the necessary calculations. However, any additional information that can be utilized to solve the increasing demands for more rigorous approaches to protein identification is worthy of effort. In the case of retention time prediction, the necessary empirical information is retained by default during the course of an MS/MS analysis, and can be easily extracted and compared to predicted values.

The pertinent goal of these empirical approaches is to predict elution times from reversed-phase HPLC columns and to use these experimental versus prediction values to aid in the identification of peptide sequences. Often, during proteomics experiments, the identification of missing peptides, or distinguishing between different isobaric peptides, those peptides with similar mass during single-stage MS [i.e., peptide mass fingerprinting (PMF)] can be a challenge. Even the highest mass-resolving instruments may be unable to fully distinguish these peptides. The use of HPLC retention time predictions has been shown as valuable in this context; the resolving capabilities of HPLC can be used to differentiate between isobaric peptides that could not otherwise be distinguished in high mass accuracy measurements. In addition, HPLC retention times can increase the confidence of protein identifications made by PMF alone, the latter being defined as the identification of a protein based on a combination of peptide masses. As a technique, PMF is sensitive to the complexity of a genome: the more complex the genome, the greater the probability that any identified peptide mass may map to more than one protein. Hence, the information content of a PMF spectrum is usually not high enough for the unique identification of a protein in a complex genome [64]. For example, Cavalcoli *et al.* have shown that even in the less complex *Escherichia coli* proteome, when mass accuracy is increased to 0.1%, clusters of similar nominal mass proteins may still be as large as 10 individual proteins [65]. Whilst this study was completed at the protein level, it is expected that an even greater level of mass accuracy must be obtained in order to differentiate successfully between peptides (see below) [65]. Even more importantly, the complete information regarding the translated protein product, including PTMs and splice variants, must be taken into account for theoretical calculation of mass [66]. Therefore, confidence in identifications based on PMF, in particular, benefits from additional supporting evidence from physico-chemical data. False-positive identifications can be reduced if the retention times of the peptides are taken into account.

An additional application for retention times is to distinguish between degradation products with similar mass shifts. For example, the N-terminal glutamine and carboxamidomethylated cysteine residues undergo a cyclization reaction yielding pyroglutamic acid and 5-oxo-thiomorpholine-3-carboxylic acid, respectively. Both of these degradation products yield a mass shift of -17.026 Da. However, the retention time of these degradation products is much later than their unmodified counterparts, allowing further confidence in identification.

Strittmatter *et al.* showed that confident identifications based on mass spectra were increased by up to 30% when the retention time of the peptide was incorporated as a discriminator [35]. Because the information from retention time is independent of MS, the information can augment multivariate discrimination methods. This combination shows promise for both eliminating errant matches and providing a basis for improved identifications from MS/MS spectra that contain poor fragmentation efficiencies, or patterns that differ from the models used by the current major database search algorithms. In this way, false-positive rates are reduced by approximately 50%.

Recent improvements in instrumentation have allowed even more detailed mass accuracy measurement. Thus, the idea of utilizing a peptide's mass as an additional parameter for peptide identification has garnered recent attention. Most notably, the studies of Richard D. Smith have revived the use of this analytical parameter in peptide identifications, and will be discussed further on its own merit in the following section. However, when used in conjunction with a peptide retention time, this additional information allows more confidence in the identification of peptides [32]. Kawakami *et al.* developed a program in which peptide assignments are validated solely on the correlation between measured and predicted LC elution time [67]. In the model developed by Petritis *et al.*, more complex artificial neural network architectures were used to examine peptide length, hydrophobicity, peptide sequence, amphipathicity, and nearest-neighbor effects on peptide retention time. Indeed, when compared with simpler predictive models their models showed an improvement [52]. Unlike previously developed models of retention time predictions, Petritis *et al.* [52] were able to predict with high accuracy the retention times of both isobar and isomer peptides. Such capacity allows for a higher confidence in the identification of peptides that otherwise are indistinguishable by accurate mass measurement alone.

13.2.5
Mass Accuracy

The utility of MS for protein identification derives from the specificity of mass measurements for peptides and their fragments from MS/MS measurements. The latter can provide partial or sometimes complete sequence determination; an MS/MS spectrum contains more information than an MS spectrum of a peptide, but at the expense of more complex and time-consuming analysis. The distinctiveness of a peptide increases with increasing size and mass [65,66]. The most straightforward approach to validate the correctness of peptide identification from mass alone is to use accurate mass; the false-positive rate decreases if the search is constrained to a few parts per million. However, this capability is instrument-specific, requiring instruments designed to provide high resolution and mass accuracy (e.g., Fourier transform mass spectrometers (FTMS), certain electrospray time of flight mass spectrometers (ESTOFMSs), and the Orbitrap™).

In a report published in 2002 in *Proteomics* [34], Richard D. Smith describes and demonstrates a strategy that improves the sensitivity, dynamic range, comprehensiveness and throughput based upon the use of peptide "accurate mass and time" (AMT) tags. The first step in AMT tagging is to utilize spectral information from multiple tandem MS analyses of related samples to create a reference database of masses and observed retention times for each confidently identified peptide. When peptides are identified in multiple analyses, their mass and retention time values are averaged, which can then provide distribution statistics. These AMT tags are then used to identify peptides in subsequent samples in a high-throughput, single mass spectrum analysis. The two-stage strategy uses high-resolution capillary LC separation combined with tandem MS to create "potential mass tags" that are used along with high mass resolution and accuracy MS to facilitate subsequent analysis. This strategy provides high confidence for protein identifications and can greatly increase the throughput compared to conventional technologies.

The AMT approach assumes that, given sufficient mass measurement accuracy (MMA), a useful number of unique peptides can be attributed to a protein. Time of flight (TOF) instruments have accuracies of <10 ppm, and the resolution of $m/\Delta m$ on current instruments now exceeds 10 000. The Fourier transform ion cyclotron resonance (FT-ICR) instruments can achieve 100 ppb to 1 ppm accuracy for peptides, although this can fluctuate with ion charge densities under the effect of a large ion population. It is expected that, when analyzing samples from eukaryotic organisms with complex proteomes, an instrument must have a mass accuracy of less than 1 ppm in order uniquely to identify peptides [68]. The use of retention time information can alleviate such constraint, however, thus facilitating the analysis of these samples. The strategy diminishes the need for MS/MS analysis, thereby increasing throughput and reducing sample requirements.

The diminished need for MS/MS then affords a greater coverage of protein sequence because additional time spent accumulating and averaging MS/MS spectra for low-abundant peptides is not consumed. However, the analysis of complex samples by only single-stage MS requires a fairly large number of detectable peptides per spectrum, and a very large dynamic range of peptide abundance. In practice, this is dependent on the capacity of the mass spectrometer. The limiting factor of peak capacity in high-resolution instruments (e.g., a FT-ICR) will be limited to the trap charge capacity, which is in turn based on the size of the magnet. The higher the magnet power, the higher the charge capacity, with the additional consequence of a larger dynamic range and improved sensitivity. Smith et al. utilized an 11.5 Tesla magnet in their studies; this has the capacity to provide a dynamic range of 10^7, compared with 10^3 in a linear ion trap. Further, it is shown that a full-scan MS in the FT-ICR had an order of magnitude higher sensitivity for detecting peptides when compared to analyses that utilize both MS and MS/MS [34]. It is therefore easier to validate a peptide as an AMT if its observed mass is within 1 ppm of a theoretically calculated mass, with a corresponding correlation in LC elution times. Perhaps the most notable example of AMT methodology is that reported by Lipton and coworkers [31], whereby approximately 60% of the predicted proteome of bacterium *Deinococcus radiodurans* was identified. This analysis represented the most extensive proteome coverage for any organism of that time, and thus indicated the methodology's capability of identifying numerous proteins with high confidence and high throughput [31].

13.2.6
Summary

Depending upon sample complexity, an accurate mass and a reversed-phase retention time can be used together as a rapid means of identifying peptides and proteins in shotgun proteome experiments, and consequently also reduce the time spent on tandem MS experiments. The use of p*I* as an additional constraint allows the stringency of mass accuracy to be relaxed somewhat while maintaining an acceptable false-positive rate. The use of additional intrinsic peptide information can provide significantly improved discrimination that may be used to raise confidence in peptide identification. As described above, this advance towards increased discrimination can be used to generate a greater sensitivity, or improved speed of analysis.

References

23 Cargile, B.J., Bundy, J.L. and Stephenson, J.L. Jr. (2004) Potential for false positive identifications from large databases through tandem mass spectrometry. *J. Proteome Res.*, **3** (5), 1082–1085.

24 Cargile, B.J., Bundy, J.L., Freeman, T.W. and Stephenson, J.L. Jr. (2004) Gel based isoelectric focusing of peptides and the utility of isoelectric point in protein identification. *J. Proteome Res.*, **3** (1), 112–119.

25 Colinge, J., Masselot, A., Cusin, I., Mahe, E., Niknejad, A., Argoud-Puy, G., Reffas, S., Bederr, N., Gleizes, A., Rey, P.A. and Bougueleret, L. (2004) High-performance peptide identification by tandem mass spectrometry allows reliable automatic data processing in proteomics. *Proteomics*, **4** (7), 1977–1984.

26 Eng, J.K. McCormack, Ashley L. and Yates, J.R. III (1994) An approach to correlate tandem mass spectral data of peptides with amino acid sequences in a protein database. *J. Am. Soc. Mass Spectrom.*, **5** (11) 976–989.

27 Heller, M., Ye, M., Michel, P.E., Morier, P., Stalder, D., Junger, M.A., Aebersold, R., Reymond, F. and Rossier, J.S. (2005) Added value for tandem mass spectrometry shotgun proteomics data validation through isoelectric focusing of peptides. *J. Proteome Res.*, **4** (6), 2273–2282.

28 Jaitly, N., Monroe, M.E., Petyuk, V.A., Clauss, T.R., Adkins, J.N. and Smith, R.D. (2006) Robust algorithm for alignment of liquid chromatography-mass spectrometry analyses in an accurate mass and time tag data analysis pipeline. *Anal. Chem.*, **78** (21), 7397–7409.

29 Keller, A., Nesvizhskii, A.I., Kolker, E. and Aebersold, R. (2002) Empirical statistical model to estimate the accuracy of peptide identifications made by MS/MS and database search. *Anal. Chem.*, **74** (20), 5383–5392.

30 Krokhin, O.V., Craig, R., Spicer, V., Ens, W., Standing, K.G., Beavis, R.C. and Wilkins, J.A. (2004) An improved model for prediction of retention times of tryptic peptides in ion pair reversed-phase HPLC: its application to protein peptide mapping by off-line HPLC-MALDI MS. *Mol. Cell. Proteomics*, **3** (9), 908–919.

31 Lipton, M.S., Pasa-Tolic, L., Anderson, G.A., Anderson, D.J., Auberry, D.L., Battista, J.R., Daly, M.J., Fredrickson, J., Hixson, K.K., Kostandarithes, H., Masselon, C., Markillie, L.M., Moore, R.J., Romine, M.F., Shen, Y., Stritmatter, E., Tolic, N., Udseth, H.R., Venkateswaran, A., Wong, K.K., Zhao, R. and Smith, R.D. (2002) Global analysis of the *Deinococcus radiodurans* proteome by using accurate mass tags. *Proc. Natl. Acad. Sci. USA*, **99** (17), 11049–11054.

32 Palmblad, M. (2006) Retention time prediction and protein identification. *Methods Mol. Biol.*, **367**, 195–208.

33 Peng, J., Elias, J.E., Thoreen, C.C., Licklider, L.J. and Gygi, S.P. (2003) Evaluation of multidimensional chromatography coupled with tandem mass spectrometry (LC/LC-MS/MS) for large-scale protein analysis: the yeast proteome. *J. Proteome Res.*, **2** (1), 43–50.

34 Smith, R.D., Anderson, G.A., Lipton, M.S., Pasa-Tolic, L., Shen, Y., Conrads, T.P., Veenstra, T.D. and Udseth, H.R. (2002) An accurate mass tag strategy for quantitative and high-throughput proteome measurements. *Proteomics*, **2** (5), 513–523.

35 Strittmatter, E.F., Kangas, L.J., Petritis, K., Mottaz, H.M., Anderson, G.A., Shen, Y., Jacobs, J.M., Camp, D.G. II and Smith, R.D. (2004) Application of peptide LC retention time information in a discriminant function for peptide identification by tandem mass spectrometry. *J. Proteome Res.*, **3** (4), 760–769.

36 Wang, Y., Zhang, J., Gu, X. and Zhang, X.M. (2005) Protein identification assisted by the prediction of retention time in liquid chromatography/tandem mass spectrometry. *J. Chromatogr. B Analyt. Technol. Biomed. Life Sci.*, **826** (1–2), 122–128.

37 Put, R., Daszykowski, M., Baczek, T. and Vander Heyden, Y. (2006) Retention prediction of peptides based on uninformative variable elimination by partial least squares. *J. Proteome Res.*, **5** (7) 1618–1625.

38 Malmstrom, J., Lee, H., Nesvizhskii, A.I., Shteynberg, D., Mohanty, S., Brunner, E., Ye, M., Weber, G., Eckerskorn, C. and Aebersold, R. (2006) Optimized peptide separation and identification for mass spectrometry based proteomics via free-flow electrophoresis. *J. Proteome Res.*, **5** (9), 2241–2249.

39 Horth, P., Miller, C.A., Preckel, T. and Wenz, C. (2006) Efficient fractionation and improved protein identification by peptide OFFGEL electrophoresis. *Mol. Cell. Proteomics*, **5** (10), 1968–1974.

40 Bjellqvist, B., Pasquali, C., Ravier, F., Sanchez, J.C. and Hochstrasser, D. (1993) A nonlinear wide-range immobilized pH gradient for two-dimensional electrophoresis and its definition in a relevant pH scale. *Electrophoresis*, **14** (12), 1357–1365.

41 Shimura, K., Zhi, W., Matsumoto, H. and Kasai, K. (2000) Accuracy in the determination of isoelectric points of some proteins and a peptide by capillary isoelectric focusing: utility of synthetic peptides as isoelectric point markers. *Anal. Chem.*, **72** (19), 4747–4757.

42 Bjellqvist, B., Hughes, G.J., Pasquali, C., Paquet, N., Ravier, F., Sanchez, J.C., Frutiger, S. and Hochstrasser, D. (1993) The focusing positions of polypeptides in immobilized pH gradients can be predicted from their amino acid sequences. *Electrophoresis*, **14** (10), 1023–1031.

43 Sillero, A. and Ribeiro, J.M. (1989) Isoelectric points of proteins: theoretical determination. *Anal. Biochem.*, **179** (2), 319–325.

44 Cargile, B.J., Talley, D.L. and Stephenson, J.L. Jr. (2004) Immobilized pH gradients as a first dimension in shotgun proteomics and analysis of the accuracy of pI predictability of peptides. *Electrophoresis*, **25** (6), 936–945.

45 Cargile, B.J., Sevinsky, J.R., Essader, A.S., Stephenson, J.L. Jr. and Bundy, J.L. (2005) Immobilized pH gradient isoelectric focusing as a first-dimension separation in shotgun proteomics. *J. Biomol. Tech.*, **16** (3), 181–189.

46 Essader, A.S., Cargile, B.J., Bundy, J.L. and Stephenson, J.L. Jr. (2005) A comparison of immobilized pH gradient isoelectric focusing and strong-cation-exchange chromatography as a first dimension in shotgun proteomics. *Proteomics*, **5** (1), 24–34.

47 Yates, J.R., III, Eng, J.K., McCormack, A.L. and Schieltz, D. (1995) Method to correlate tandem mass spectra of modified peptides to amino acid sequences in the protein database. *Anal. Chem.*, **67** (8), 1426–1436.

48 Colinge, J., Masselot, A., Giron, M., Dessingy, T. and Magnin, J. (2003) OLAV: towards high-throughput tandem mass spectrometry data identification. *Proteomics*, **3** (8), 1454–1463.

49 Alpert, A.J. and Andrews, P.C. (1988) Cation-exchange chromatography of peptides on poly(2-sulfoethyl aspartamide)-silica. *J. Chromatogr.*, **443**, 85–96.

50 Ballif, B.A., Villen, J., Beausoleil, S.A., Schwartz, D. and Gygi, S.P. (2004) Phosphoproteomic analysis of the

developing mouse brain. *Mol. Cell. Proteomics*, **3** (11), 1093–1101.

51 Martin, A.J. and Mittelmann, R. (1948) Quantitative micro-analysis of amino-acid mixtures on paper partition chromatograms. *Biochem. J.*, **43** (3), 353–358.

52 Petritis, K., Kangas, L.J., Yan, B., Monroe, M.E., Strittmatter, E.F., Qian, W.J., Adkins, J.N., Moore, R.J., Xu, Y., Lipton, M.S., Camp, D.G. II and Smith, R.D. (2006) Improved peptide elution time prediction for reversed-phase liquid chromatography-MS by incorporating peptide sequence information. *Anal. Chem.*, **78** (14), 5026–5039.

53 Hearn, M.T. and Aguilar, M.I. (1987) High-performance liquid chromatography of amino acids peptides and proteins. LXXIII. Investigations on the relationships between molecular structure, retention and band-broadening properties of polypeptides separated by reversed-phase high-performance liquid chromatography. *J. Chromatogr.*, **397**, 47–70.

54 Terabe, S., Konaka, R. and Inouye, K. (1979) Separation of some polypeptide hormones by high-performance liquid chromatography. *J. Chromatogr.*, **172**, 163–177.

55 Mant, C.T. and Hodges, R.S. (2006) Context-dependent effects on the hydrophilicity/hydrophobicity of side-chains during reversed-phase high-performance liquid chromatography: Implications for prediction of peptide retention behaviour. *J. Chromatogr. A*, **1125** (2), 211–219.

56 Liu, H.X., Xue, C.X., Zhang, R.S., Yao, X.J., Liu, M.C., Hu, Z.D. and Fan, B.T. (2004) Quantitative prediction of logk of peptides in high-performance liquid chromatography based on molecular descriptors by using the heuristic method and support vector machine. *J. Chem. Inf. Comput. Sci.*, **44** (6), 1979–1986.

57 Kaliszan, R., Baczek, T., Cimochowska, A., Juszczyk, P., Wisniewska, K. and Grzonka, Z. (2005) Prediction of high-performance liquid chromatography retention of peptides with the use of quantitative structure-retention relationships. *Proteomics*, **5** (2), 409–415.

58 Makrodimitris, K., Fernandez, E.J., Woolf, T.B. and O'Connell, J.P. (2005) Simulation and experiment of temperature and cosolvent effects in reversed phase chromatography of peptides. *Biotechnol. Prog.*, **21** (3), 893–896.

59 Makrodimitris, K., Fernandez, E.J., Woolf, T.B. and O'Connell, J.P. (2005) Mesoscopic simulation of adsorption of peptides in a hydrophobic chromatography system. *Anal. Chem.*, **77** (5), 1243–1252.

60 Petritis, K., Kangas, L.J., Ferguson, P.L., Anderson, G.A., Pasa-Tolic, L., Lipton, M.S., Auberry, K.J., Strittmatter, E.F., Shen, Y., Zhao, R. and Smith, R.D. (2003) Use of artificial neural networks for the accurate prediction of peptide liquid chromatography elution times in proteome analyses. *Anal. Chem.*, **75** (5), 1039–1048.

61 Guo, D., Mant, C.T., Taneja, A.K., Parker, J.M.R. and Hodges, R.S. (1986) Prediction of peptide retention times in reversed-phase high-performance liquid chromatography I. Determination of retention coefficients of amino acid residues of model synthetic peptides. *J. Chromatogr.*, **359**, 499–517.

62 Wall, D.B., Parus, S.J. and Lubman, D.M. (2002) Three-dimensional protein map according to pI hydrophobicity and molecular mass. *J. Chromatogr. B Analyt. Technol. Biomed. Life Sci.*, **774** (1), 53–58.

63 Krokhin, O.V., Ying, S., Cortens, J.P., Ghosh, D., Spicer, V., Ens, W., Standing, K.G., Beavis, R.C. and Wilkins, J.A. (2006) Use of peptide retention time prediction for protein identification by off-line reversed-phase HPLC-MALDI MS/MS. *Anal. Chem.*, **78** (17), 6265–6269.

64 Clauser, K.R., Baker, P. and Burlingame, A.L. (1999) Role of accurate mass measurement (± 10 ppm) in protein

identification strategies employing MS or MS/MS and database searching. *Anal. Chem.*, **71** (14), 2871–2882.

65 Cavalcoli, J.D., Van Bogelen, R.A., Andrews, P.C. and Moldover, B. (1997) Unique identification of proteins from small genome organisms: theoretical feasibility of high throughput proteome analysis. *Electrophoresis*, **18** (15), 2703–2708.

66 Wang, R. and Chait, B.T. (1994) High-accuracy mass measurement as a tool for studying proteins. *Curr. Opin. Biotechnol.*, **5** (1), 77–84.

67 Kawakami, T., Tateishi, K., Yamano, Y., Ishikawa, T., Kuroki, K. and Nishimura, T. (2005) Protein identification from product ion spectra of peptides validated by correlation between measured and predicted elution times in liquid chromatography/mass spectrometry. *Proteomics*, **5** (4), 856–864.

68 Strittmatter, E.F., Ferguson, P.L., Tang, K. and Smith, R.D. (2003) Proteome analyses using accurate mass and elution time peptide tags with capillary LC time-of-flight mass spectrometry. *J. Am. Soc. Mass Spectrom.*, **14** (9), 980–991.

Index

a

absolute quantification (AQUA)   13, 111
abundance   12
abundant   6
accurate mass and time (AMT)   442
affinity chromatography   251
affinity purification   364
albumin   6
algorithms   435
alkylation   136
amino acid composition   438, 439
amount   14
ampholytes   135
antibodies   6
ASB14   135
average peak intensity   427

b

background electrolyte (BGE)   171
bacteria   358
bacterial   148
baseline reduction   424
bead-based immunoassays   391
biased   382
biodiscovery   9
biological variance   11
biomarker   9
bioorthogonal non-canonical amino acid tagging (BONCAT)   107
biotin   372
blue native (BN)-PAGE   304
blue-native electrophoresis (BNE)   144
body fluids   382
boronic acid   333
bottom-up approach   148
bronchoalveolar lavage (BL)   388

c

^{13}C   108
capillary electrophoresis (CE)   171
capillary gel electrophoresis (CGE)   172
capillary isoelectric focusing (CIFF)   172
capillary isotachophoresis (CITP)   172
capillary zone electrophoresis (CZE)   172
carbonate washing   123
carrier ampholytes   137
cation-exchange chromatography   321
cell disruption   130
cell walls   344
cellulases   132
cerebrospinal fluid (CSF)   388
chaotropes   135
chaotropic agents   135
CHAPS   135
chemical isotopic labeling   108
chemical shift perturbation (CSP)   274
chemical tags   106
chromolith CapRod   259
chromolith guard   259
circulating microparticles   394
citrated blood   389
clear-native electrophoresis (CNE)   144
clustering   428
collision-induced dissociation (CID)   46, 111
column   253
combined fractional diagonal chromatography (COFRADIC)   323
comparative analysis   16
confidence level   60
coomassie blue G-250   144
crushed crystal method   81
crystallography   281
culture-derived isotope tag (CDIT)   108

α-cyano-4-hydroxy-cinnamic acid   74
cyanogen bromide   362
cytomics   389

d
DAPI   239
DDM   146
deconvolution   53
decoy-database   59
deglycosylation   335
dendritic cells   394
depletion   11
detection limits   13
detector   172
detergent lysis   132
detergents   135
dialysis   288
difference gel electrophoresis (DIGE)   105, 140
differential in gel electrophoresis   13
differential display   140
digitonin   146
dihydroxybenzoic acid (DHB)   74, 320
dissociation constants (K_D)   273
disulfide bonds   136
dithiothreitol (DTT)   136
dodecyl maltoside   135
dodecyl-β-D-maltoside (DDM)   146
dried droplet method   80
dried droplet preparation   90
drop dialysis   83, 90
dye-ligand chromatography   251
dynamic range   6, 346

e
EDTA-blood   389
effective length (L)   178
electro-endosmosis   158
electron transfer dissociation   46
electrospray ionization   43
β-elimination   322
enrichment   11, 265
equilibrium density gradient centrifugation   119, 122
erythrocytes and reticulocytes   393
experiment   23

f
false-discovery rates   59
field-amplified injection (FAI)   175
filter   427
flow cytometric immunophenotyping (FC)   399
flow cytometry   234

flow sorting   235
fluorescence lifetime imaging microscopy (FLIM)   209
fluorescence resonance energy transfer (FRET)   208
fluorescence-activated cell sorting (FACS)   140, 391
fluorescent difference gel electrophoresis (DIGE)   137
free-flow electrophoresis (FFE)   155
freeze/thawing   132
Fungi   371

g
gel-based approach   60
global internal standard technology (GIST)   110
1,3-glucanases   132
glycosylation   12, 328
glycosylphosphatidylinositol-(GPI-) anchor   371
GRAVY (grand average of hydrophobicity)   304, 363
GST pull-down method   296

h
heparinized blood   389
heteronuclear single quantum correlation (HSQC)   274
high resolution two-dimensional electrophoresis   171
highly abundant proteins   134
high-performance liquid chromatography (HPLC)   171, 245
high-resolution, clear native electrophoresis (hrCNE)   152
Human Proteome Organisation (HUPO)   174, 386
hydrolytic enzymes   132
hydrophilic interaction liquid chromatography (HILIC)   333
hydrophobic protein domains   135
hydroxyethyldisulfide   136
hydroxypropyl methyl cellulose (HPC)   158

i
immobilized metal-affinity chromatography (IMAC)   12, 251, 319
immobilized pH gradients   129
immunophenotyping   399
immunoprecipitation   266, 319
in-gel digestion   51
in-silico   11
insoluble material   134

in-solution digestion   51
interferometric quadrature   189
interferometry   189
iodixanol (Optiprep)   120
iodixanol cushion   122
iodoacetamide   136
ion suppression   48
ion trapping   45
ion-exchange chromatography   436
isoelectric focusing (IEF)   130, 158, 438
isoelectric point (pI)   156, 434
isotachophoresis (ITP)   175
isotope tags for relative and absolute quantitation (iTRAQ)   110, 119, 364
isotope-coded affinity tags (ICAT)   108
isotope-coded protein label (ICPL)   111

l

laser capture microdissection   140
laser microdissection and pressure catapulting (LMPC)   219
leading electrolyte (LE)   176
lectin affinity chromatography   332
lectins   12
lignins   345
limit of detection (LOD)   175
lipid rafts   397
lipidomic   380
lipids   134
liquid chromatography   46
liquid chromatography Fourier transform ion cyclotron resonance (LC-FT-ICR)   95
liquid chromatography/mass spectrometry (LC/MS)   245
liquid-based homogenization   131
liquid-liquid extraction   249
liquid-phase transfection (LPT)   209
localization   118
localization of organelle proteins by isotope tagging (LOPIT)   118f.
lower limit of detection   14
low-molecular-weight (LMW)   379
lymphocytes   394
LysC   362
lysozyme   132

m

magnetic bead-based separation   392
magnetic beads   234
MALDI matrix   75
mammalian   148
mannitol   158
mass accuracy   46, 442
mass calibration ladder   151

mass measurement accuracy (MMA)   443
mass spectrometry   43
matrix-assisted laser desorption/ionization mass spectrometry (MALDI MS)   73
mechanical cell disruption   130
membrane protein   361
membrane proteomes   303
membrane-protein complexes   303
metabolomic   380
micellar electrokinetic chromatography (MEKC)   172
micro-batch   288
microdissection   393
6-1minohexanoic acid   148
molecular imprinted polymer (MIP)   176
monocytes   394
mononuclear cells   394
multidimensional protein identification technology (MudPIT)   50, 56, 245, 353, 358
multidimensional-nano-LC/MS   56
multiple-reaction monitoring (MRM)   111

n

^{15}N   108
nano-ESI   45
nano-HPLC   46
native electrophoresis   144
native isoelectric focusing (native IEF)   144
n-butyl (C4)   56
N-hydroxysuccinimide (NHS)-ester   372
N-linked glycopeptide   330
n-octadecyl (C18)   56
n-octyl (C8)   56
noise filter   424
NP-40   135
nuclear magnetic resonance (NMR)   273
nucleic acids   134
nucleolus/nucleosome   398

o

^{16}O/^{18}O Exchange   110
octyl-β-glycoside   135
O-linked glycopeptide   330
open reading frames (ORFs)   295
organelles   396
orthogonal separations   259
osmolysis   131

p

peak lists   424
peak-based statistics   426
pectins   345
peptide fragment analysis   75

peptide mass fingerprint (PMF) 75, 111
peptidome 386
phagosomes 398
phenol extraction 351
phenols 134
phosphatase inhibitors 137
phosphoamidates 322
phosphorylation 12, 317
physico-chemical properties 433
plant proteomics 343
plant tissue 345
plasma proteome 379
plasma proteome project (PPP) 386
plasma/serum 386
platelets 395
PNGase 335
polyaccharides 134
polyacrylamide gel electrophoresis 46
polyethylene glycol (PEG) 353
polyphenolics 345
Ponceau S 149
post-translation modifications (PTMs)
    11f., 317, 328
pre-column technique 253
primary amine 124
prolytes 159
propidium iodide 239
proteases 133, 137
protein database (PDB) 282, 309
protein digestion 51
protein equalizer technology 252
protein localization 118
protein secretion 398
protein solubilization 135
proteinase K 333, 362
protein-ligand interaction 273
proteolytic digestion 333
proteome 21
purification (TAP) 295

q

quadrupole time-of-flight 45, 46
quality control 423
quantification 13
quantitative structure-retention
    relationships (QSRRs) 439
Quantitave MudPIT 363

r

raft-associated proteins 397
RAM-SCX particulate column 253
raw mass spectra 424
recalcitrant 345
recovery 12

reducing agents 136
reproducibility 23
restricted access material (RAM) 255
retention coefficient 438
reverse transfection (RT) 209
reversed-phase (RP) 319, 359, 438
reversed-phase liquid chromatography
    (RPLC) 157, 438
reverse-phase microcolumns 333
RPLC/ESI-MS/MS 50
rubisco 346

s

salt 133
sample clean-up 133
sample recovery 12
sample size 14
saturation transfer difference (STD) 276
scaling up 14
SCX-RPLC 365
SEC-RPLC 365
SELDI-TOF MS 224
selective multiple-reaction monitoring
    (SRM) 111
separation 157
sequence coverage 110
sequence-specific retention calculator
    (SSRCalc) 440
sequential extraction 140
serial extraction 351
shaving 307
shotgun approach 307
signal-to-noise 53
signal-to-noise ratio (SNR) 427
silicamonolithic 253
sinapinic acid 74
small interfering (si) RNA 211
solid-phase extraction (SPE) 163, 245
solid-phase micro-extraction (SPME) 175
sonication 131
SPADNS (2-(4-sulfophenylazo) 1,8-dihydroxy-
    3,6-naphthalenedisulfonic acid) 159
spiking of labeled peptides 111
spinning disc interferometry (SDI) 187
spotter 213
stable isotope 106
stable isotope labeling (SILAC) 13, 107,
    364
stable isotope tagging 118
standard operating procedure (SOP) 380
statistical validation 425
storage 12
strong cation-exchange (SCX) 359, 436
subcellular compartments 118

subcellular fractionation   8
subcellular localization   119
subcellular protein extraction   17
sucrose density gradient centrifugation   397
sum of isotope peaks   427
sum of satellite peaks   427
surfactants   135

t

tandem affinity   295
tandem affinity purification (TAP)   308
tandem mass spectrometry (MS/MS)   46, 157
technical variance   11
terminating buffer (TE)   176
terpenes   345
thin-layer chromatography   80, 265
thin-layer peptide mapping electrophoresis   271
thin-layer preparation   90
three-dimensional (3-D) structure   273, 281
thresholding   427
time-of-flight   74
tissue arrays   390
titanium dioxide   12, 320, 333
titanium oxide   251
top-down approach   248
total ion count (TIC)   427
total length (TL)   178
total number of peaks (TNP)   427

transmembrane   362
tributylphosphine (TBP)   137
Tris-(2-carboxyethyl)phosphine (TCEP)   137
Tris(carboxyethyl) phosphine   124
Triton X-100   135, 146
trypsin   51, 333
two-dimensional (2-D) LC   47
two-dimensional (2-D) phosphopeptide mapping   265
two-dimensional gel electrophoresis (2-DGE)   129

u

ubiquitination   12
unbiased   382
urine   387

v

vapour diffusion, batch   288

w

whole proteome   14

y

yeast   148, 371
yeast two-hybrid (Y-2-H) method   308

z

ZIC-HILIC   333
zirconium dioxide   320